Spectral Theory of Differential Operators

M. Sh. Birman 80th Anniversary Collection

American Mathematical Society

TRANSLATIONS

Series 2 • Volume 225

Advances in the Mathematical Sciences—62

(*Formerly Advances in Soviet Mathematics*)

Spectral Theory of Differential Operators

M. Sh. Birman 80th Anniversary Collection

T. Suslina
D. Yafaev
Editors

American Mathematical Society
Providence, Rhode Island

2000 *Mathematics Subject Classification.* Primary 47-06, 47F05; Secondary 81-06.

Library of Congress Card Number 91-640741
ISBN-13: 978-0-8218-4738-1
ISBN-10: 0-8218-4738-4
ISSN 0065-9290

♾ The paper used in this book is acid-free and falls within the guidelines
established to ensure permanence and durability.
Visit the AMS home page at `http://www.ams.org/`

10 9 8 7 6 5 4 3 2 1 13 12 11 10 09 08

Mikhail Shlemovich Birman

Contents

Amer. Math. Soc. Transl.
(2) Vol. **225**, 2008

On the Scientific Work of M. Sh. Birman in 1998–2007

M. Solomyak and T. Suslina

January 17, 2008 is the 80th birthday of the outstanding mathematician Mikhail Shlëmovich Birman. He is the author of many of the most important results in the spectral theory of abstract and differential operators and in scattering theory. He contributed much to other areas of Real Analysis: to approximation theory, function theory, the theory of integral operators, etc.

Papers [**1***, **3***, **4***] are devoted to the scientific achievements of M. Birman. In the present paper, we give a survey of his results obtained during the last ten years (1998–2007). The list of publications of M. Birman for the period before 1998 can be found in [**2***]. The next item of this collection is the continuation of this list. Herewith, the numbering of publications continues numbering from [**2***]. References [**130, 131**], which will be discussed below in detail, are repeated. We also repeat references [**138, 139, 140, 141**] from [**2***], since they contained noncomplete data.

The main subject which M. Birman was working on during the last ten years (and is currently working on) is the spectral theory of periodic differential operators (DO's). Precisely, the following three themes can be named: 1) the problem of absolute continuity of the spectrum of the periodic operators of mathematical physics; 2) threshold properties and homogenization of periodic DO's; 3) discrete spectrum in the gaps of a periodic elliptic operator perturbed by potential decaying at infinity.

Then, M. Birman returned to his "old" subject, namely, to the study of asymptotics of the discrete spectrum for the Maxwell operator in a resonator in a nonsmooth situation.

§1. Absolute continuity of the spectrum of the periodic operators of mathematical physics

We start with the series of papers [**131, 138, 140, 141, 142, 143, 148**] devoted to the problem of absolute continuity of the spectrum of periodic differential operators. All papers of this series are joint with T. Suslina. Below in Subsections 1.1–1.3 we give necessary preliminary information.

1.1. Direct integral. Let P be a selfadjoint elliptic (in an appropriate sense) lower semi-bounded operator in $L_2(\mathbb{R}^d)$ generated by a second-order linear differential expression $\mathcal{P}(\mathbf{x}, \mathbf{D})$ ($\mathbf{x} \in \mathbb{R}^d$, $\mathbf{D} = -i\nabla$), which is periodic in $\mathbf{x}$ with respect

2000 *Mathematics Subject Classification.* Primary 01A70.

to some lattice $\Gamma \subset \mathbb{R}^d$. By Ω we denote the elementary cell of the lattice Γ. Let $\widetilde{\Gamma}$ be the dual lattice, and let $\widetilde{\Omega}$ be the central Brillouin zone of the lattice $\widetilde{\Gamma}$.

The operator P is decomposed in the direct integral of the operators $P(\mathbf{k})$ acting in $L_2(\Omega)$ and depending on the parameter $\mathbf{k} \in \mathbb{R}^d$ (the quasi-momentum). The operator $P(\mathbf{k})$ is given by the differential expression $\mathcal{P}(\mathbf{x}, \mathbf{D}+\mathbf{k})$ with periodic boundary conditions. The operator P is unitarily equivalent to the direct integral of operators $P(\mathbf{k})$:

$$\mathcal{U}P\mathcal{U}^{-1} = \int_{\widetilde{\Omega}} \oplus P(\mathbf{k})\, d\mathbf{k}. \tag{1.1}$$

Here $\mathcal{U}$ is the Gelfand transformation.

The spectrum of each operator $P(\mathbf{k})$ is discrete. Let $E_j(\mathbf{k})$, $j \in \mathbb{N}$, be the eigenvalues of the operator $P(\mathbf{k})$ numbered in nondecreasing order counting multiplicities. The *band functions* $E_j(\cdot)$ are continuous and $\widetilde{\Gamma}$-periodic. The spectrum of the operator P has a *band structure*: it consists of the closed intervals $\operatorname{Ran} E_j$, $j \in \mathbb{N}$. Spectral bands can overlap. The intervals free of the spectrum are called *gaps*.

1.2. The problem of the absence of degenerate spectral bands. If some band function is constant: $E_j(\mathbf{k}) = \lambda = \text{Const}$, then the corresponding band degenerates into the point λ, which is an eigenvalue of infinite multiplicity for the operator P. For instance, such a case can be realized for an elliptic fourth-order operator (an example of such an operator is given in the book by P. Kuchment), and also for the operator $-\operatorname{div} g(\mathbf{x})\nabla$ with $d \geq 3$ and a non-Lipschitz matrix g (such an example was found in 2001 by N. Filonov).

If a periodic elliptic operator has no degenerate bands, then its spectrum *is absolutely continuous*.

1.3. The Thomas approach. For the first time, the absence of degenerate spectral bands for the Schrödinger operator $H = -\Delta + V(\mathbf{x})$ in $L_2(\mathbb{R}^3)$ with a real-valued periodic potential V was proved in the famous paper by L. Thomas in 1973. (For arbitrary dimension $d \geq 2$, the Thomas proof is presented in the book by M. Reed and B. Simon.) Later the general Thomas approach was used in almost all papers on this subject. The operator family $H(\mathbf{k}) = (\mathbf{D}+\mathbf{k})^2 + V(\mathbf{x})$ (with periodic boundary conditions) acting in $L_2(\Omega)$ is analytically extended to complex values of the parameter $\mathbf{k} \in \mathbb{C}^d$. We fix a unit vector $\mathbf{e} \in \mathbb{R}^d$ and a vector $\mathbf{k}' \in \mathbb{R}^d$, $\mathbf{k}' \perp \mathbf{e}$, and put $\mathbf{k} = k_1\mathbf{e} + \mathbf{k}'$, where $k_1 \in \mathbb{C}$. With respect to the parameter $k_1 \in \mathbb{C}$, the family $H(\mathbf{k}) = H(k_1\mathbf{e} + \mathbf{k}')$ is an analytic operator family with compact resolvent. Let $k_1 = \mu + iy$, where $\mu \in \mathbb{R}$ is fixed, and $y \in \mathbb{R}$ is the main parameter. We denote $H(\mathbf{k}) = H(y)$. Thomas showed that, for appropriate $\mathbf{e}$ and μ, the operator $H(y)$ is invertible for large values of $|y|$, and $\|H(y)^{-1}\| \to 0$ as $|y| \to \infty$. Precisely, the following *Thomas estimate* is true:

$$\|H(y)^{-1}\| \leq C|y|^{-1}, \quad |y| \geq y_0 > 0. \tag{1.2}$$

By the analytic Fredholm alternative, this easily implies the absence of degenerate bands.

First, estimate (1.2) is proved (by the Fourier series) for the "free" operator $H_0(y)$ corresponding to the case $V = 0$. Next, the potential V can be taken into account as an additive perturbation, and estimate (1.2) is carried over to the case of the "perturbed" operator $H(y) = H_0(y)+V$. Herewith, the smaller the assumptions on V are, the greater are the efforts needed for adding V. In the initial paper by Thomas, it was assumed that $V \in L_2(\Omega)$ (for $d = 3$).

1.4. Two-dimensional magnetic Schrödinger operator. The periodic *magnetic Schrödinger operator* $M = (\mathbf{D} - \mathbf{A}(\mathbf{x}))^2 + V(\mathbf{x})$ (the magnetic Hamiltonian) is a much more difficult case. Here $\mathbf{A}(\mathbf{x})$ is the vector-valued (magnetic) potential. Due to the gauge transformation, the potential $\mathbf{A}$ can be subject to the gauge conditions $\operatorname{div}\mathbf{A} = 0$, $\int_\Omega \mathbf{A}(\mathbf{x})\,d\mathbf{x} = 0$. If M is considered as a perturbed operator with respect to H_0, then the perturbation $M - H_0$ is a first-order differential operator. For the corresponding operators in $L_2(\Omega)$ depending on the parameter y, the perturbation $M(y) - H_0(y)$ contains terms of order $|y|$. That is why it is impossible to prove an estimate of the form (1.2) for the operator $M(y)$ considering $M(y)$ as an additive perturbation of the "free" operator $H_0(y)$. This difficulty was not overcome for almost 25 years.

In 1997 in the breakthrough paper [**131**], absolute continuity of the spectrum of the periodic magnetic Hamiltonian was proved in dimension $d = 2$. The magnetic operator M was interpreted as a "multiplicative perturbation" of the operator H_0. Let us explain this in detail. For simplicity, suppose that $\mathbf{A} \in C^1$. The approach of [**131**] is based on the study of the operator $P = (\mathbf{D} - \mathbf{A}(\mathbf{x}))^2 + \partial_1 A_2(\mathbf{x}) - \partial_2 A_1(\mathbf{x})$ (one of two blocks of the corresponding Pauli operator). The operator P admits a convenient factorization. Indeed, there exists a real-valued periodic function $\varphi(\mathbf{x})$ such that $\nabla\varphi = \{A_2, -A_1\}$. In terms of φ, the operator P is factorized into five factors: $P = e^{-\varphi}(D_1 + iD_2)e^{2\varphi}(D_1 - iD_2)e^{-\varphi}$. Each of them is either a differential operator with constant coefficients, or multiplication by a positive function. This allows one to consider the operator $P(y)$ (acting in $L_2(\Omega)$) as a multiplicative perturbation of the operator $H_0(y)$ and to prove an analogue of estimate (1.2) for the operator $P(y)$. After that, it is easy to take the scalar potential $W = V - \partial_1 A_2 + \partial_2 A_1$ into account as an additive perturbation and to prove the required estimate for the operator $M(y) = P(y) + W$.

In [**131**], it was assumed that the vector-valued potential $\mathbf{A}$ is continuous, and that $V \in L_2(\Omega)$. Later in [**138**] the scheme was modified and adapted to the operators defined in terms of quadratic forms. This allowed the authors to relax conditions on coefficients: in [**138**] it was assumed that $\mathbf{A} \in L_\rho(\Omega)$ with $\rho > 2$ and $V \in L_r(\Omega)$ with $r > 1$. These conditions are optimal in the L_p-scale. Thus, the result on absolute continuity of the spectrum of the two-dimensional periodic magnetic Hamiltonian was extended to the case of discontinuous vector-valued potentials.

1.5. Further development of the subject. After almost 25 years of stagnation, the paper [**131**] gave rise to a series of papers by many authors on absolute continuity of the spectrum of periodic operators.

For $d \geq 3$, the problem of absolute continuity of the spectrum of a periodic magnetic Hamiltonian turned out to be much more difficult than for $d = 2$. (If $d \geq 3$, the Pauli operator does not admit a convenient factorization.) This problem was solved in 1997 in the remarkable paper by A. Sobolev. He applied the technique of pseudodifferential operators on a torus and showed that, for an appropriate choice of direction of the complex quasi-momentum (i.e., the direction of the vector $\mathbf{e}$) depending on the potential $\mathbf{A}$, an analogue of estimate (1.2) for the operator $M(y)$

is still true for $d \geq 3$. It turned out that a certain analogue of factorization (up to lower terms) still takes place, but on the "pseudodifferential level".

In [**141**], a survey of the results known at that time on absolute continuity of the spectrum of periodic operators was given. Besides, the result on absolute continuity of the spectrum of the magnetic Schrödinger operator for $d \geq 3$ was extended to a wider class of potentials. In particular, for $d = 3, 4$ the condition $V \in L^0_{d/2,\infty}(\Omega)$ from [**141**] is optimal in the Lorentz scale.

Later, conditions on the coefficients were relaxed in the papers by I. Lapin, R. Shterenberg and in a series of papers by Z. Shen.

The most difficult case is an operator with variable leading coefficients. However, the case of a scalar metric analyzed in [**141**] is relatively simple. Then the operator has the form

$$H(g, \mathbf{A}, V) = (\mathbf{D} - \mathbf{A}(\mathbf{x}))^* g(\mathbf{x})(\mathbf{D} - \mathbf{A}(\mathbf{x})) + V(\mathbf{x}), \tag{1.3}$$

where $g(\mathbf{x}) = \omega^2(\mathbf{x})a$, a is a constant positive matrix, and $\omega(\mathbf{x})$ is a bounded and positive definite (scalar) function. In [**141**], it was shown that the result on absolute continuity can be easily carried over from the case of the magnetic Schrödinger operator to this case (under some smoothness assumptions on ω), by using the identity $H(\omega^2 a, \mathbf{A}, V) = \omega^{-1} H(a, \mathbf{A}, \omega^{-2}V + V_\omega)\omega^{-1}$, where $V_\omega = \omega^{-1} \operatorname{div} a\nabla\omega$.

The case of a metric $g(\mathbf{x})$ of general type turned out to be much more difficult. Up to now, the problem of absolute continuity for the operator with variable metric has been solved only in the two-dimensional case (here the first result belongs to A. Morame, 1998), and for $d \geq 3$, it has been solved in the case where the operator has a special symmetry: it must be invariant under reflection with respect to some axis (the result of L. Friedlander, 2001). For $d \geq 3$, the problem of absolute continuity of the spectrum of the operator (1.3) with metric g of general type still remains open.

1.6. Singular potentials. In [**142**], the absolute continuity of the spectrum was established for the two-dimensional Schrödinger operator of the form

$$H_\Sigma = (\mathbf{D} - \mathbf{A})^2 + V(\mathbf{x}) + \sigma(\mathbf{x})\delta_\Sigma(\mathbf{x}). \tag{1.4}$$

The potential in (1.4) includes the delta-like term $\sigma(\mathbf{x})\delta_\Sigma(\mathbf{x})$ supported on a periodic system Σ of piecewise-smooth curves. Such operators are of interest in the theory of photonic crystals. Using the version of the Thomas approach adapted for the operators defined in terms of quadratic forms (this version was suggested in [**138**]), it is possible to take a delta-like potential into account as an additive perturbation.

Later, absolute continuity of the spectrum for the operator (1.4) was obtained in dimension $d \geq 3$ by T. Suslina and R. Shterenberg in 2001 (under some restrictions on a periodic system Σ of $(d-1)$-dimensional surfaces). For $d = 2$, the case of more general singular potentials (given in terms of measures) was studied by R. Shterenberg (2000, 2001).

1.7. Vector periodic problems. Progress for the magnetic Schrödinger operator allowed one to prove absolute continuity of the spectrum also for the periodic Dirac operator containing both electric and magnetic potentials. This was done in [**140**] on the basis of a relation between the square of the Dirac operator and the magnetic Schrödinger operator. (For the Dirac operator with periodic electric potential only, absolute continuity of the spectrum was obtained in 1990 by L. Danilov.)

In [**148**], absolute continuity of the spectrum was proved for the periodic isotropic operator of elasticity theory in the case where the shear modulus is constant (the so-called "Hill body"). In this case, it is possible to reduce the problem to the known results for a scalar elliptic operator. If the shear modulus is variable, the problem of absolute continuity of the spectrum for the isotropic periodic operator of elasticity theory remains open.

Note that, in spite of the considerable progress, there remain many unsolved problems on absolute continuity of the spectrum of periodic operators.

§2. Threshold properties and homogenization for periodic differential operators

An important series of papers [**139, 145, 150, 151, 153, 155, 156, 158, 159, 161, 162, 163**] by M. Birman and T. Suslina is devoted to the study of threshold properties and homogenization problems for periodic differential operators. The starting point was the paper [**139**], where the spectral characteristics of the two-dimensional periodic Pauli operator at the bottom of the spectrum (the *threshold characteristics*) were studied. It became clear that "good" properties of the threshold characteristics are related not only to specific features of the Pauli operator, but also to existence of factorization of the form $\mathcal{X}^*\mathcal{X}$, where $\mathcal{X}$ is a homogeneous first-order DO. This led to the singling out of a wide class of matrix periodic DO's admitting a factorization of the form $\mathcal{X}^*\mathcal{X}$ and to the study of their threshold characteristics (see Subsection 2.3). During this work the idea was born that threshold properties of periodic DO's must be related to homogenization theory in the small period limit. It was realized that the homogenization procedure for a periodic operator can be studied as a threshold effect at the bottom of the spectrum. In this way, results of a new type in homogenization theory were obtained.

2.1. Differential operators in $L_2(\mathbb{R}^d;\mathbb{C}^n)$. In the papers [**145, 151, 156, 158**], the class of matrix periodic differential operators $\mathcal{A}$ acting in $L_2(\mathbb{R}^d;\mathbb{C}^n)$ and admitting a factorization of the form $\mathcal{A} = \mathcal{X}^*\mathcal{X}$ was introduced and studied in detail. Here $\mathcal{X}$ is a homogeneous first-order DO. Suppose that $\mathcal{X} = h(\mathbf{x})b(\mathbf{D})f(\mathbf{x})$, where an $(n\times n)$-matrix-valued function $f(\mathbf{x})$ and an $(m\times m)$-matrix-valued function $h(\mathbf{x})$ are periodic with respect to some lattice Γ and bounded together with their inverses. It is assumed that $m \geq n$. The operator $b(\mathbf{D})$ is a homogeneous first-order DO; its symbol $b(\boldsymbol{\xi})$ is a linear homogeneous $(m\times n)$-matrix-valued function of $\boldsymbol{\xi}\in\mathbb{R}^d$ such that $\operatorname{rank} b(\boldsymbol{\xi}) = n$ for $\boldsymbol{\xi}\neq 0$. Formally, the operator $\mathcal{A}$ is given by the differential expression

$$\mathcal{A} = \mathcal{A}(g,f) = f(\mathbf{x})^* b(\mathbf{D})^* g(\mathbf{x}) b(\mathbf{D}) f(\mathbf{x}), \quad g(\mathbf{x}) = h(\mathbf{x})^* h(\mathbf{x}). \tag{2.1}$$

The precise definition of $\mathcal{A}$ is given in terms of the corresponding quadratic form. Below, in the case where $f = \mathbf{1}_n$, we use the notation

$$\widehat{\mathcal{A}} = \widehat{\mathcal{A}}(g) = b(\mathbf{D})^* g(\mathbf{x}) b(\mathbf{D}). \tag{2.2}$$

Many operators of mathematical physics admit such a factorization. The acoustics operator and the operator of elasticity theory have the form (2.2), while the periodic Schrödinger operator and the two-dimensional Pauli operator can be written in the form (2.1) (with nontrivial f).

2.2. Main results on homogenization. We use the notation $\phi^\varepsilon(\mathbf{x}) = \phi(\varepsilon^{-1}\mathbf{x})$ for any Γ-periodic function ϕ and put $\mathcal{A}_\varepsilon = \mathcal{A}(g^\varepsilon, f^\varepsilon)$, $\widehat{\mathcal{A}}_\varepsilon = \widehat{\mathcal{A}}(g^\varepsilon)$. The coefficients of these operators are rapidly oscillating as $\varepsilon \to 0$. The *homogenization problem* for the operator $\mathcal{A}_\varepsilon$ is to study the behavior of the resolvent $(\mathcal{A}_\varepsilon + I)^{-1}$ as $\varepsilon \to 0$.

We start with the results for the simpler operator $\widehat{\mathcal{A}}_\varepsilon$. *There exists an operator* $\widehat{\mathcal{A}}^0 = b(\mathbf{D})^* g^0 b(\mathbf{D})$ *with constant coefficients such that*

$$\|(\widehat{\mathcal{A}}_\varepsilon + I)^{-1} - (\widehat{\mathcal{A}}^0 + I)^{-1}\|_{L_2 \to L_2} \leq C\varepsilon. \tag{2.3}$$

The constant positive matrix g^0 is called the *effective matrix*, and the operator $\widehat{\mathcal{A}}^0$ is called the *effective operator*. The existence of the effective operator and the *strong* resolvent convergence in L_2 have been known in traditional homogenization theory. However, the resolvent convergence *in the operator norm* was not known before. The estimate (2.3) established for the first time in [**145, 151**] is order-sharp; the constant C is controlled explicitly in terms of the problem data.

The effective matrix g^0 is defined by the following rule. Let $\Lambda(\mathbf{x})$ be the periodic $(n \times m)$-matrix-valued function satisfying the equation

$$b(\mathbf{D})^* g(\mathbf{x})(b(\mathbf{D})\Lambda(\mathbf{x}) + \mathbf{1}_m) = 0, \quad \int_\Omega \Lambda(\mathbf{x})\, d\mathbf{x} = 0. \tag{2.4}$$

We put $\widetilde{g}(\mathbf{x}) = g(\mathbf{x})(b(\mathbf{D})\Lambda(\mathbf{x}) + \mathbf{1}_m)$. Then $g^0 = |\Omega|^{-1} \int_\Omega \widetilde{g}(\mathbf{x})\, d\mathbf{x}$.

For the operator $\mathcal{A}_\varepsilon$ (with $f \neq \mathbf{1}_n$), it is impossible to find a DO with constant coefficients such that the resolvent $(\mathcal{A}_\varepsilon + I)^{-1}$ converges to the resolvent of this operator. However, it is possible to approximate the resolvent $(\mathcal{A}_\varepsilon + I)^{-1}$ by the generalized resolvent of the effective operator $\widehat{\mathcal{A}}^0$ sandwiched between rapidly oscillating factors. The following approximation was found in [**151**]:

$$\|(\mathcal{A}_\varepsilon + I)^{-1} - (f^\varepsilon)^{-1}(\widehat{\mathcal{A}}^0 + \overline{Q})^{-1}((f^\varepsilon)^*)^{-1}\|_{L_2 \to L_2} \leq C\varepsilon. \tag{2.5}$$

Here $Q(\mathbf{x}) = (f(\mathbf{x})f(\mathbf{x})^*)^{-1}$ and $\overline{Q} = |\Omega|^{-1} \int_\Omega Q(\mathbf{x})\, d\mathbf{x}$. It is important that, in the approximation (2.5), the inverse is taken for the DO with constant coefficients, though this approximation contains rapidly oscillating factors from both sides. One can get rid of these factors only by passing to the weak operator limit.

In the paper [**156**], a qualitatively new result in homogenization theory has been obtained. Namely, a more precise approximation for the resolvent $(\mathcal{A}_\varepsilon + I)^{-1}$ in the operator norm in $L_2(\mathbb{R}^d; \mathbb{C}^n)$ with error estimate of order ε^2 was found. Here we describe this result only for the simpler operator $\widehat{\mathcal{A}}_\varepsilon$. The term $\varepsilon K(\varepsilon)$ should be added to the resolvent of the effective operator, where $K(\varepsilon)$ is the so-called *corrector*. *The following estimate holds:*

$$\|(\widehat{\mathcal{A}}_\varepsilon + I)^{-1} - (\widehat{\mathcal{A}}^0 + I)^{-1} - \varepsilon K(\varepsilon)\|_{L_2 \to L_2} \leq C\varepsilon^2. \tag{2.6}$$

The operator $K(\varepsilon)$ is the sum of *three* terms: $K(\varepsilon) = K_1(\varepsilon) + K_1(\varepsilon)^* + K_3$. The first two terms of the corrector are mutually adjoint and contain a rapidly oscillating factor Λ^ε, where Λ is the periodic solution of the equation (2.4). The operator $K_1(\varepsilon)$ is given by $K_1(\varepsilon) = \Lambda^\varepsilon \Pi_\varepsilon b(\mathbf{D})(\widehat{\mathcal{A}}^0 + I)^{-1}$. Here Π_ε is an auxiliary smoothing operator. (In some cases it is possible to get rid of the smoothing operator and to put $K_1(\varepsilon) = \Lambda^\varepsilon b(\mathbf{D})(\widehat{\mathcal{A}}^0 + I)^{-1}$.) The third term of the corrector does not depend on ε and is given by $K_3 = -(\widehat{\mathcal{A}}^0 + I)^{-1} b(\mathbf{D})^* L(\mathbf{D}) b(\mathbf{D})(\widehat{\mathcal{A}}^0 + I)^{-1}$, where

$L(\mathbf{D})$ is the first-order DO with the symbol $L(\boldsymbol{\xi}) = |\Omega|^{-1} \int_\Omega (\Lambda(\mathbf{x})^* b(\boldsymbol{\xi})^* \widetilde{g}(\mathbf{x}) + \widetilde{g}(\mathbf{x})^* b(\boldsymbol{\xi}) \Lambda(\mathbf{x}))\, d\mathbf{x}$.

The estimate (2.6) is order-sharp. As $\varepsilon \to 0$, the weak limit of the corrector $K(\varepsilon)$ is equal to K_3. In this respect, the third term of the corrector is the most important. In some cases, it may happen that $K_3 = 0$. In particular, this is the case for the scalar operator $\widehat{\mathcal{A}} = -\operatorname{div} g(\mathbf{x})\nabla$, where the matrix g has real entries. However, in the general case the third term of the corrector is nontrivial. This is typical for matrix operators, and also for scalar operators with complex-valued coefficients. Note that the traditional corrector used in homogenization theory corresponds to the term $K_1(\varepsilon)$. The estimate (2.6) shows that, in order to approximate the resolvent in the L_2-operator norm with an error term of order $O(\varepsilon^2)$, it is not sufficient to take into account only the traditional corrector $K_1(\varepsilon)$, but it is necessary to include all three terms in the corrector. This result was quite unexpected.

In [**156**], an analogue of estimate (2.6) for the more general operator $\mathcal{A}_\varepsilon$ was also obtained. In this case, the approximation contains rapidly oscillating factors from both sides.

In [**158**], the resolvent $(\widehat{\mathcal{A}}_\varepsilon + I)^{-1}$ was approximated in the norm of operators acting from L_2 to the Sobolev space H^1, with the error term of order ε. It turned out that, to this end, it suffices to take into account only the first term of the corrector. *The following estimate is true*:

$$\|(\widehat{\mathcal{A}}_\varepsilon + I)^{-1} - (\widehat{\mathcal{A}}^0 + I)^{-1} - \varepsilon K_1(\varepsilon)\|_{L_2 \to H^1} \leq C\varepsilon. \tag{2.7}$$

Comparing estimates (2.6) and (2.7), we see that the form of the corrector depends on the question considered. All the estimates under consideration are order-sharp, and the constants in the estimates are controlled explicitly in terms of L_∞-norms of the coefficients g, g^{-1}, f, f^{-1}, the lower and upper bounds for the symbol $b(\boldsymbol{\xi})^* b(\boldsymbol{\xi})$ on the sphere $|\boldsymbol{\xi}| = 1$, and the parameters of the lattice Γ.

2.3. Method of investigation. Let T_ε be the *unitary* scaling transformation in $L_2(\mathbb{R}^d; \mathbb{C}^n)$ given by $(T_\varepsilon \mathbf{u})(\mathbf{x}) = \varepsilon^{d/2}\mathbf{u}(\varepsilon\mathbf{x})$. The identity

$$(\mathcal{A}_\varepsilon + I)^{-1} = \varepsilon^2 T_\varepsilon^* (\mathcal{A} + \varepsilon^2 I)^{-1} T_\varepsilon$$

shows that, in order to approximate the resolvent $(\mathcal{A}_\varepsilon + I)^{-1}$ in the L_2-operator norm with an error of order $O(\varepsilon)$, it suffices to approximate the operator $(\mathcal{A}+\varepsilon^2 I)^{-1}$ with an error term of order $O(\varepsilon^{-1})$. The presence of the unitary factors T_ε is not an obstacle for proving estimates *in the operator norm*. At the same time, in the study of weak or strong convergence, the influence of these operators cannot be controlled.

The bottom of the spectrum of the operator $\mathcal{A}$ is the point $\lambda = 0$. It is natural that the behavior of the resolvent of $\mathcal{A}$ in the point $\lambda = -\varepsilon^2$ (close to the bottom of the spectrum) can be described in terms of the threshold characteristics of $\mathcal{A}$. Thus, the *homogenization procedure can be studied as a threshold effect at the bottom of the spectrum.*

The operator $\mathcal{A}$ expands in the direct integral of operators $\mathcal{A}(\mathbf{k})$ acting in $L_2(\Omega; \mathbb{C}^n)$ (cf. Subsection 1.1). Let $E_j(\mathbf{k})$, $j \in \mathbb{N}$, be the consecutive eigenvalues of the operator $\mathcal{A}(\mathbf{k})$ (band functions). It turns out that for the operator (2.1), the first n spectral bands overlap and have the common bottom $\lambda = 0$. Herewith, the minimum of band functions $E_j(\mathbf{k})$ is reached only at the point $\mathbf{k} = 0$:

$$\min_{\mathbf{k}\in\widetilde{\Omega}} E_j(\mathbf{k}) = E_j(0) = 0, \quad j = 1, \dots, n.$$

The edge of the $(n+1)$-th band is separated from zero: $\min E_{n+1}(\mathbf{k}) > 0$. The minimum point $\mathbf{k} = 0$ for each function E_j is nondegenerate: $E_j(\mathbf{k}) \geq c_*|\mathbf{k}|^2$, $j = 1, \dots, n$, $c_* > 0$. By the *threshold characteristics* of the operator $\mathcal{A}$ we mean the asymptotic behavior of the eigenvalues $E_j(\mathbf{k})$, $j = 1, \dots, n$, and the corresponding eigenfunctions of the operator $\mathcal{A}(\mathbf{k})$ near $\mathbf{k} = 0$.

The study of the resolvent $(\mathcal{A} + \varepsilon^2 I)^{-1}$ is reduced to the study of the resolvent $(\mathcal{A}(\mathbf{k}) + \varepsilon^2 I)^{-1}$ of the operator family $\mathcal{A}(\mathbf{k})$. This family depends on the parameter $\mathbf{k} \in \mathbb{R}^d$ analytically. For $\mathbf{k} = 0$, the kernel $\mathfrak{N}$ of the operator $\mathcal{A}(0)$ is n-dimensional. The natural desire is to use methods of analytic perturbation theory. However, for $d > 1$ and $n > 1$ (multidimensional parameter and multiple eigenvalue), analytic perturbation theory is not applicable. The solution suggested in [**145, 151**] is to introduce the one-dimensional parameter $t = |\mathbf{k}|$, putting $\mathbf{k} = t\boldsymbol{\theta}$, $\boldsymbol{\theta} \in \mathbb{S}^{d-1}$. The operator family $\mathcal{A}(\mathbf{k}) = \mathcal{A}(t\boldsymbol{\theta}) = A(t, \boldsymbol{\theta})$ is studied by methods of analytic perturbation theory with respect to the one-dimensional parameter t.

Herewith, many constructions can be carried over in *abstract* operator-theoretic terms. A crucial notion of the abstract scheme is the notion of the *spectral germ* $S(\boldsymbol{\theta})$ of the operator family $A(t, \boldsymbol{\theta})$. The germ is a selfadjoint operator acting in the n-dimensional space $\mathfrak{N} = \operatorname{Ker} \mathcal{A}(0)$. Let us give the spectral definition of the germ. By analytic perturbation theory, for $t \leq t_0$, there exist real-analytic branches of eigenvalues $\lambda_l(t, \boldsymbol{\theta})$ and real-analytic branches of eigenvectors $\varphi_l(t, \boldsymbol{\theta})$ (orthonormal in $L_2(\Omega; \mathbb{C}^n)$), $l = 1, \dots, n$, of the operator $A(t, \boldsymbol{\theta})$. For sufficiently small t_*, the following convergent power series expansions hold:

$$(2.8)\qquad \begin{aligned} \lambda_l(t, \boldsymbol{\theta}) &= \gamma_l(\boldsymbol{\theta}) t^2 + \mu_l(\boldsymbol{\theta}) t^3 + \dots, \\ \varphi_l(t, \boldsymbol{\theta}) &= \omega_l(\boldsymbol{\theta}) + \varphi_l^{(1)}(\boldsymbol{\theta})\, t + \dots, \quad l = 1, \dots, n, \quad t \leq t_*. \end{aligned}$$

The coefficients $\gamma_l(\boldsymbol{\theta})$ are positive: $\gamma_l(\boldsymbol{\theta}) \geq c_* > 0$. The vectors $\{\omega_1(\boldsymbol{\theta}), \dots, \omega_n(\boldsymbol{\theta})\}$ form an orthonormal basis in $\mathfrak{N}$. The coefficients $\gamma_l(\boldsymbol{\theta})$ and the vectors $\omega_l(\boldsymbol{\theta})$ are *threshold characteristics* of the operator $\mathcal{A}$ near the bottom of the spectrum.

According to [**145, 151**], the selfadjoint operator $S(\boldsymbol{\theta}) : \mathfrak{N} \to \mathfrak{N}$ such that

$$S(\boldsymbol{\theta})\omega_l(\boldsymbol{\theta}) = \gamma_l(\boldsymbol{\theta})\omega_l(\boldsymbol{\theta}), \quad l = 1, \dots, n,$$

is called the *spectral germ* of the operator family $A(t, \boldsymbol{\theta})$. Thus, the spectral germ contains information about threshold characteristics of the operator $\mathcal{A}$. Therefore, the germ is responsible for threshold effects.

The key result from [**145, 151**] is the following approximation for the resolvent of the operator $A(t, \boldsymbol{\theta})$ by finite rank operators given in terms of the spectral germ:

$$(2.9)\qquad \|(A(t, \boldsymbol{\theta}) + \varepsilon^2 I)^{-1} - (t^2 S(\boldsymbol{\theta}) + \varepsilon^2 I_{\mathfrak{N}})^{-1} P\|_{L_2(\Omega) \to L_2(\Omega)} \leq C\varepsilon^{-1}, \quad t \leq t_0.$$

Here P is the orthogonal projection of the space $L_2(\Omega; \mathbb{C}^n)$ onto $\mathfrak{N}$. The constants C and t_0 are controlled explicitly.

It is possible to calculate the spectral germ. Here we formulate the result for the simpler operator $\widehat{\mathcal{A}}$. In this case, the germ is represented as $S(\boldsymbol{\theta}) = b(\boldsymbol{\theta})^* g^0 b(\boldsymbol{\theta})$, $\boldsymbol{\theta} \in \mathbb{S}^{d-1}$, where g^0 is the effective matrix. It turns out that the effective operator $\widehat{\mathcal{A}}^0$ has the same spectral germ as $\widehat{\mathcal{A}}$. Then, using estimate (2.9), it is possible to approximate the resolvent of the operator $\widehat{\mathcal{A}}$ by the resolvent of $\widehat{\mathcal{A}}^0$:

$$\|(\widehat{\mathcal{A}} + \varepsilon^2 I)^{-1} - (\widehat{\mathcal{A}}^0 + \varepsilon^2 I)^{-1}\|_{L_2(\mathbb{R}^d)\to L_2(\mathbb{R}^d)} \leq C\varepsilon^{-1}.$$

This implies estimate (2.3) by a simple scale transformation.

In order to obtain a more precise approximation (2.6), one has to refine (2.9) and find an approximation of the resolvent $(A(t,\boldsymbol{\theta}) + \varepsilon^2 I)^{-1}$ by some finite rank operators with an error term of order $O(1)$. In abstract terms, such a result was obtained in [**155**]. For this, it is necessary to take into account terms up to order t^3 in the expansions for eigenvalues and terms up to order t in the expansions for eigenvectors in (2.8). In [**156**], these abstract results were applied for the proof of estimate (2.6).

In order to approximate the resolvent in the norm of operators acting from L_2 to H^1, it is necessary to approximate the operator $A(t,\boldsymbol{\theta})^{1/2}(A(t,\boldsymbol{\theta}) + \varepsilon^2 I)^{-1}$ in the operator norm in $L_2(\Omega;\mathbb{C}^n)$. In this way, estimate (2.7) was obtained in [**158**] .

Finally, in order to study the operator $\mathcal{A}_\varepsilon$ (with $f \neq \mathbf{1}$), the identity

$$(\mathcal{A}_\varepsilon + I)^{-1} = (f^\varepsilon)^{-1}(\widehat{\mathcal{A}}_\varepsilon + Q^\varepsilon)^{-1}((f^\varepsilon)^*)^{-1}$$

was used, and the generalized resolvent $(\widehat{\mathcal{A}}_\varepsilon + Q^\varepsilon)^{-1}$ of $\widehat{\mathcal{A}}_\varepsilon$ was studied. For the generalized resolvent the analogues of estimates (2.3), (2.6), (2.7) were obtained.

2.4. In [**145, 151, 156, 158**], general results were applied to specific periodic operators of mathematical physics: to the acoustics operator, the operator of elasticity theory, the Schrödinger operator, the two-dimensional Pauli operator, and also to the stationary Maxwell system.

Homogenization for the stationary periodic Maxwell system is the most difficult problem. In the case where one of the coefficients (dielectric permittivity or magnetic permeability) is constant, the problem is (partially) reduced to the study of the second-order operator $\widehat{\mathcal{A}} = \operatorname{curl} \eta(\mathbf{x})^{-1} \operatorname{curl} - \nabla \operatorname{div}$, which admits factorization of the form (2.2). Then general results for this class of operators are applicable. Such a case was studied in [**151**, Chapter 7], but only in [**159**] were approximations in the $L_2(\mathbb{R}^3)$-norm for all physical fields obtained.

The problem is even more difficult if both coefficients are not constant. The corresponding "model" second-order operator does not belong to the class of operators of the form (2.2) or (2.1). Nevertheless, it is possible to apply the results of the abstract scheme from [**151, 158**]. The homogenization problem for the Maxwell system in the case where both coefficients are variable periodic matrix-valued functions was studied in the papers by T. Suslina (2004, 2007), where approximations in the $L_2(\mathbb{R}^3)$-norm for all fields were obtained.

We also mention the papers [**150, 153**], where an analogue of a homogenization procedure for the scalar elliptic operator $\mathcal{A} = -\operatorname{div} g(\mathbf{x})\nabla + p(\mathbf{x})$ (with real-valued periodic coefficients) near the edge of an *internal gap* was considered. Let $\mathcal{A}_\varepsilon = -\operatorname{div} g^\varepsilon \nabla + \varepsilon^{-2} p^\varepsilon$. Let λ be the right edge of a gap in the spectrum of the operator $\mathcal{A}$. For the operator $\mathcal{A}_\varepsilon$, the edge of the corresponding gap turns into the point $\varepsilon^{-2}\lambda$ (lying in the high-energy area). The problem is to approximate the operator $(\mathcal{A}_\varepsilon - (\varepsilon^{-2}\lambda - \varkappa^2)I)^{-1}$ for small ε in the operator norm in $L_2(\mathbb{R}^d)$. (Here the number $\varkappa$ is such that the point $\lambda - \varkappa^2$ lies in the initial gap in the spectrum of $\mathcal{A}$.) Such an approximation with a sharp-order error estimate was obtained in [**153**] under some assumptions about the spectral characteristics of $\mathcal{A}$ near the edge of the gap.

(The one-dimensional case was studied before in [**150**].) In problems of this type, an interaction between the threshold effects and the high-energy effects is observed.

In the very recent paper [**162**] by M. Birman and T. Suslina a spectral approach suggested by the authors was developed further and applied to homogenization of the Cauchy problem for a periodic non-stationary Schrödinger type equation and the Cauchy problem for a periodic hyperbolic equation.

2.5. The papers [**145, 151**] stimulated interest in homogenization theory to approximations of the resolvent with error estimates in the operator norm. V. Zhikov suggested a different method for obtaining operator estimates in homogenization theory. By this method estimates of the form (2.3) and (2.7) were obtained for the scalar operator $\widehat{\mathcal{A}} = -\operatorname{div} g(\mathbf{x})\nabla$ (V. Zhikov, 2005) and for the operator of elasticity theory (V. Zhikov and S. Pastukhova, 2005). In the recent paper by D. Borisov (2008), estimates of the form (2.3) and (2.7) were obtained for an operator with coefficients depending both on rapid and slow variables.

§3. Discrete spectrum of a perturbed periodic elliptic operator

After 1998, M. Birman continued to study the discrete spectrum in the gaps of a periodic elliptic operator perturbed by a potential decaying at infinity. Two papers by M. Birman were devoted to this subject. One of them is the paper [**144**] joint with A. Laptev and T. Suslina, and the second one is the paper [**146**] joint with M. Solomyak.

3.1. Setting of the problem. Consider the operator $A(\alpha) = A - \alpha V$ acting in $L_2(\mathbb{R}^d)$, where A is an "unperturbed" operator, V is a "perturbation", and $\alpha \geq 0$ is a *coupling constant.*

The unperturbed operator A is an elliptic second-order operator of the form $A = -\operatorname{div} g(\mathbf{x})\nabla + p(\mathbf{x})$ with real-valued coefficients that are periodic with respect to some lattice $\Gamma \subset \mathbb{R}^d$. The matrix-valued function $g(\mathbf{x})$ is bounded and uniformly positive definite, and the potential $p(\mathbf{x})$ is bounded. The precise definition of the operator is given in terms of the quadratic form. One may assume that the bottom of the spectrum of A is the point $\lambda = 0$. The spectrum of the operator A has a band structure (see Subsection 1.1). There exists a semi-bounded gap $(-\infty, 0)$; there may also be internal gaps in the spectrum of A.

The perturbation V is the operator of multiplication by a real-valued function $V(\mathbf{x})$. It is assumed that $V(\mathbf{x}) \to 0$ as $|\mathbf{x}| \to \infty$, in an appropriate sense. In general, for the perturbed operator $A(\alpha)$, eigenvalues appear in the semi-bounded gap $(-\infty, 0)$ and also in the internal gaps. Let us discuss the negative discrete spectrum. If the potential V_+ decays sufficiently fast, then the number of negative eigenvalues is finite. Here $2V_+(\mathbf{x}) = |V(\mathbf{x})| + V(\mathbf{x})$. Let $N(\alpha, \lambda)$ be the number of eigenvalues (counting multiplicities) of the operator $A(\alpha)$ lying below a point $\lambda \leq 0$ called the observation point. The full number of negative eigenvalues $N(\alpha) = N(\alpha, 0)$ is of special interest. The asymptotics of $N(\alpha, \lambda)$ as $\alpha \to +\infty$ is studied.

The corresponding asymptotic formulas are rather diverse, depending on the dimension d, on the rate of decay of V and on the nature of the operator A. They were studied in a series of papers of M. Birman in 1991–1997. See, e.g., [**110, 127, 130, 134**] and, especially, the survey [**128**] and the references therein. (Here and in what follows, references correspond to the list [**2***].) The cases $d \geq 3$ and $d = 2$ turn out to be essentially different.

3.2. The case $d \geq 3$. For $d \geq 3$, a "regular" case is distinguished by the condition $V \in L_{d/2}(\mathbb{R}^d)$. If this condition is satisfied, then the following Rozenblum-Lieb-Cwikel estimate is true:

$$N(\alpha, \lambda) \leq c(d)\alpha^{d/2} \int_{\mathbb{R}^d} V_+(\mathbf{x})^{d/2}\, d\mathbf{x}, \quad \lambda \leq 0, \tag{3.1}$$

and the Weyl asymptotics holds:

$$\begin{aligned} &N(\alpha, \lambda) \sim \alpha^{d/2} J_d(V, g), \quad \alpha \to +\infty, \quad \lambda \leq 0, \\ &J_d(V, g) = \frac{\omega_d}{(2\pi)^d}\alpha^{d/2} \int_{\mathbb{R}^d} V_+(\mathbf{x})^{d/2}(\det g(\mathbf{x}))^{-1/2}\, d\mathbf{x}. \end{aligned} \tag{3.2}$$

Here ω_d is the volume of the unit ball in $\mathbb{R}^d$.

If $V(\mathbf{x}) \geq 0$, then the following properties are true: 1) the finiteness of the Weyl coefficient J_d (which is equivalent to the condition $V \in L_{d/2}$) implies the Weyl asymptotics; 2) the asymptotics (3.2) is uniform in λ up to the edge $\lambda = 0$; 3) if $N(\alpha, \lambda) \leq C\alpha^{d/2}$ for some λ, then $V \in L_{d/2}$, and hence, the asymptotics (3.2) holds (i.e., the Weyl order $\alpha^{d/2}$ ensures the Weyl asymptotics).

In the "non-regular" case $V \notin L_{d/2}(\mathbb{R}^d)$, the Weyl asymptotics is violated. Moreover, the Weyl asymptotic order $\alpha^{d/2}$ is impossible. However, for any $q > d/2$, there are examples of potentials V such that $N(\alpha) \sim C\alpha^q$.

In the regular case, asymptotics has a "high-energy" origin, and in the non-regular case it has a "threshold" origin. The regular asymptotics was studied in [**127**], and the nonregular asymptotics was studied in [**134**].

3.3. Two-dimensional case. First, let $V(\mathbf{x}) \geq 0$. If $d = 2$, the three properties mentioned above are violated: 1) the condition $V \in L_1(\mathbb{R}^2)$ (which is equivalent to the finiteness of the Weyl coefficient J_2) is not sufficient for the validity of the Weyl asymptotics; 2) the asymptotics is not uniform in $\lambda \leq 0$ (for any $q > 1$ there are examples of potentials such that $N(\alpha, \lambda)$ has the Weyl asymptotics for $\lambda < 0$, but $N(\alpha, \lambda)$ is of order α^q for $\lambda = 0$); 3) the estimate $N(\alpha) \leq C\alpha$ does not ensure the Weyl asymptotics (there are examples of potentials such that $N(\alpha) \sim c\alpha$, but $c \neq J_2$).

All these effects were discovered in the paper [**130**] by M. Birman and A. Laptev in the case where $A = -\Delta$. It is assumed that the potential $V(\mathbf{x})$ satisfies the condition

$$\Big(\int_{|\mathbf{x}|\leq 1} |V|^\sigma\, d\mathbf{x}\Big)^{1/\sigma} + \sum_{k=1}^{\infty}\Big(\int_{e^{k-1}\leq|\mathbf{x}|\leq e^k} |V|^\sigma |\mathbf{x}|^{2(\sigma-1)}\, d\mathbf{x}\Big)^{1/\sigma} < \infty, \quad \sigma > 1, \tag{3.3}$$

with some $\sigma > 1$. Relation (3.3) implies that $V \in L_1(\mathbb{R}^2)$. Condition (3.3) ensures the Weyl asymptotics for the function $N(\alpha, \lambda)$ if $\lambda < 0$, but not for $N(\alpha)$. In [**130**], for each $q > 1$, the class of potentials V such that the asymptotics of the function $N(\alpha)$ is of order α^q was distinguished. Then the asymptotics has a threshold origin, and the answer is formulated in terms of an auxiliary problem on the semi-axis, which is obtained by restricting the operator $-\Delta - \alpha V$ on the subspace of functions depending only on $r = |\mathbf{x}|$. Simultaneously, the potential V is averaged over the polar angle. In [**130**], the class of potentials such that the asymptotics of $N(\alpha)$

is of order α, and the asymptotic coefficient is the sum of the Weyl term and the threshold term, was also distinguished.

For the case where $d = 2$ and A is a *periodic elliptic operator*, the nonregular asymptotics was studied in [**144**]. Here two different cases are observed: the case of the purely threshold asymptotics of order $q > 1$, and the case of a competition between the Weyl asymptotics and the threshold asymptotics. The threshold term is described in terms of the auxiliary problem on the semi-axis, which is now determined not only by the potential V, but also by the threshold characteristics of the operator A at the bottom of the spectrum. By the threshold characteristics, we mean the positive matrix g^0 and the periodic function $\varphi(\mathbf{x})$ defined as follows. The matrix g^0 determines the asymptotics of the first band function $E_1(\mathbf{k})$ as $|\mathbf{k}| \to 0$ (the asymptotics has the form $E_1(\mathbf{k}) \sim \langle g^0\mathbf{k}, \mathbf{k}\rangle$), and $\varphi(\mathbf{x})$ is a positive periodic solution of the equation $A\varphi = 0$.

For $d = 2$, the nonregular asymptotics for the spectrum in internal gaps was studied by T. Suslina (2003).

3.4. Periodic waveguide. The paper [**146**] is close to [**144**] both by the nature of the problem considered and by the character of the results. The main difference is in the geometry: in [**146**] the operators act in the space $L_2(X)$, where the domain $X \subset \mathbb{R}^d$ is periodic in only one direction. The unperturbed operator is again $A = -\mathrm{div}\, g(\mathbf{x})\nabla + p(\mathbf{x})$, with the Dirichlet or the Neumann boundary condition on ∂X. The functions $g(\mathbf{x})$ and $p(\mathbf{x})$ are also assumed to be periodic, which gives one the possibility of applying the Floquet-Bloch theory. By adding a constant to $p(\mathbf{x})$, one can always assume that $\lambda = 0$ is the lower point of the spectrum of A.

The perturbation is introduced by the potential $V(\mathbf{x})$ decaying, in an appropriate sense, along the same direction. The assumptions about $V(\mathbf{x})$ guarantee that for the operator $A - \alpha V$ the function $N(\alpha, \lambda)$ (see its definition in Subsection 3.1) has the Weyl-type asymptotic behavior (as $\alpha \to +\infty$) for any $\lambda < 0$. The main problem consists in studying $N(\alpha, 0)$.

Unlike the case $X = \mathbb{R}^d$, here the threshold effect arises in any dimension $d > 1$. To describe its influence on the spectrum, one introduces an auxiliary Schrödinger operator $A'_{\alpha Q} = -\partial_t^2 - \alpha Q$ acting in $L_2(\mathbb{R})$. The "effective potential" $Q(t)$ is expressed in terms of the original potential $V(\mathbf{x})$ and the Floquet data for the unperturbed operator A.

The main result can be written in the form

$$N(\alpha, 0) \sim [\text{Weyl's term}] + N'(\alpha, 0), \qquad \alpha \to +\infty,$$

where the second term corresponds to the operator $A'_{\alpha Q}$. All the effects mentioned in Subsection 3.3 are exhibited here. So, as in [**144**] (and earlier, in [**130**]), one encounters a sort of competition between the Weyl term in asymptotics and another term, of the threshold origin.

§4. Asymptotic behavior of the spectrum of the nonsmooth Maxwell operator

For the empty electromagnetic resonator in a bounded domain Ω with smooth, ideally conducting boundary, the behavior of the eigenfrequencies was found by H. Weyl as far back as in 1912. Results for the nonsmooth situation (filled resonator with nonsmooth dielectric permittivity $\varepsilon(\mathbf{x})$ and magnetic permeability $\mu(\mathbf{x})$, in the case of a nonsmooth boundary) were obtained only recently by M. Birman and his colleagues and students. The reason for such a long delay is not purely technical; its roots are of a rather fundamental nature.

Formally, the stationary Maxwell operator $M_{\varepsilon,\mu}$ acts as

$$M_{\varepsilon,\mu}\{\mathbf{u},\mathbf{v}\} = \{i\varepsilon^{-1}\operatorname{curl}\mathbf{v}, -i\mu^{-1}\operatorname{curl}\mathbf{u}\},$$

under the divergence-free conditions

$$\operatorname{div}(\varepsilon\mathbf{u}) = 0, \qquad \operatorname{div}(\mu\mathbf{v}) = 0,$$

and the boundary conditions

$$\mathbf{u}_\tau|_{\partial\Omega} = 0, \qquad (\mu\mathbf{v})_\nu|_{\partial\Omega} = 0.$$

Here $\mathbf{u}$ and $\mathbf{v}$ stand for the electric and the magnetic components of an electromagnetic field, and τ and ν indicate the tangent and the normal components of a vector field on $\partial\Omega$. However, all the operations involved have to be defined in an appropriate way, and this turns out to be nontrivial if the boundary $\partial\Omega$ or the positive definite matrix-valued functions $\varepsilon(\mathbf{x})$, $\mu(\mathbf{x})$ are nonsmooth. Such nonsmooth situations are important for many applications (stratified media, domains with edges, conical points, or screens, etc.).

Prior to calculating the spectrum, one has to define $M_{\varepsilon,\mu}$ as a selfadjoint operator acting in an appropriate Hilbert space. In the case of a filled resonator, the latter is an L_2-space of vector-valued functions, with the matrix weights generated by $\varepsilon(\mathbf{x})$, $\mu(\mathbf{x})$. The difficulties arise when one attempts to describe the operator domain of $M_{\varepsilon,\mu}$. The standard approaches, which include using the Sobolev spaces and quadratic forms, do not work here, since the operator is neither elliptic, nor semi-bounded. The choice of the "correct" selfadjoint realization is determined by the condition of finiteness of the electromagnetic energy. In the nonsmooth cases, it cannot be described in terms of Sobolev spaces. Ignoring these important facts led to errors in several papers, whose authors claimed that they found the spectral asymptotics of the Maxwell operator in the nonsmooth domains.

Due to the block structure of the Maxwell operator, its spectrum is symmetric with respect to zero. For its calculation it is sufficient to work with the electric component of the electromagnetic field. The above-mentioned condition of finiteness of the energy dictates that this component must belong to the space

$$\Phi(\tau,\varepsilon) = \{\mathbf{u}\in L_2(\Omega;\mathbb{C}^3) : \operatorname{curl}\mathbf{u}\in L_2(\Omega;\mathbb{C}^3),\ \operatorname{div}(\varepsilon\mathbf{u}) = 0,\ \mathbf{u}_\tau|_{\partial\Omega} = 0\}.$$

Here the operators curl and div are defined in the sense of distributions, and the equality $\mathbf{u}_\tau|_{\partial\Omega} = 0$ is understood in an appropriate generalized sense.

If the boundary $\partial\Omega$ and the matrix-valued function $\varepsilon(\mathbf{x})$ are smooth, the space $\Phi(\tau,\varepsilon)$ can be easily described in classical terms. Then the asymptotic behavior of the eigenfrequencies can be studied by various standard tools. The resulting asymptotic formula looks as follows. Let $r(\boldsymbol{\xi})$ be the symbol of curl, i.e.

$$r(\boldsymbol{\xi}) = i\begin{pmatrix} 0 & \xi_3 & -\xi_2 \\ -\xi_3 & 0 & \xi_1 \\ \xi_2 & -\xi_1 & 0 \end{pmatrix}.$$

Denote by $\Lambda_1(\mathbf{x},\boldsymbol{\xi}), \Lambda_2(\mathbf{x},\boldsymbol{\xi})$ the positive eigenvalues of the algebraic problem

$$r(\boldsymbol{\xi})\mu(\mathbf{x})^{-1}r(\boldsymbol{\xi})\mathbf{h} = \Lambda\varepsilon(\mathbf{x})\mathbf{h}, \qquad \mathbf{h}\in\mathbb{C}^3,\ \mathbf{x}\in\Omega,\ \boldsymbol{\xi}\in\mathbb{R}^3;$$

the third eigenvalue is equal to zero. We put

$$\Gamma(\varepsilon,\mu,\Omega)=\frac{1}{24\pi^3}\int_\Omega\int_{|\boldsymbol{\xi}|=1}\left(\Lambda_1(\mathbf{x},\boldsymbol{\xi})^{-3/2}+\Lambda_2(\mathbf{x},\boldsymbol{\xi})^{-3/2}\right)dS(\boldsymbol{\xi})d\mathbf{x},$$

where dS denotes the area element on the unit sphere. Then the eigenfrequencies m_k satisfy the asymptotic formula

$$\lim_{k\to\infty}k^{-1}m_k^3=\Gamma(\varepsilon,\mu,\Omega). \tag{4.1}$$

Under some rather general assumptions, this formula remains valid also in the nonsmooth case. Its justification needs a careful study of the structure of the space $\Phi(\tau,\varepsilon)$. First of all, A. Alekseev and M. Birman [**65, 66**] established a general geometric scheme (in the sense of the geometry of Hilbert spaces), showing that the problem in a filled resonator can always be reduced to the case of the empty one. For resonators with smooth boundary this led to the formula (4.1) for any measurable and bounded matrices $\varepsilon(\mathbf{x}),\mu(\mathbf{x})$, having the bounded inverse. The case of the nonsmooth boundary turned out to be more difficult. Further progress was achieved on the basis of the results on the analytic structure of vector fields belonging to the space $\Phi(\tau,\varepsilon)$. This analysis was developed by M. Birman and M. Solomyak in the papers [**90, 93, 97, 117**]. Namely, it was shown that in domains with Lipschitz boundary, and also in domains with screens, any vector field $\mathbf{u}\in\Phi(\tau,\mathbf{1})$ can be decomposed into the sum of a term from the Sobolev space $H^1_0(\Omega;\mathbb{C}^3)$ and another term which is the gradient of a weak solution of the Poisson equation $-\Delta w=f$, $w|_{\partial\Omega}=0$, with some $f\in L_2(\Omega)$. This fact allowed these authors to justify (in [**91**]) the formula (4.1) for an empty resonator with Lipschitz boundary.

Recently it was shown by M. Birman and N. Filonov [**160**] that the existence of the above decomposition alone, without any explicit requirements about $\partial\Omega$, already implies the formula (4.1). This is a crucial result, since it reduces the problem of the calculation of spectral asymptotics for the Maxwell operator to the problem (of Real Analysis) of describing singularities of vector fields of a certain class. In particular, this led to the proof of (4.1) for a filled resonator with Lipschitz boundary, under the same assumptions about $\varepsilon(\mathbf{x}),\mu(\mathbf{x})$ as in [**65, 66**], that is, measurability and boundedness of the matrices and their inverses. Some technical tools, which are necessary for the proof of this result, were developed by M. Birman, A. Alekseev, and N. Filonov in [**157**].

A programme for studying spectral properties of the "nonsmooth" Maxwell operator was initiated by M. Birman in the early seventies, though his first papers on the subject are [**65**] and [**66**], joint with his student A. Alekseev. The papers [**157**] and [**160**] can be considered as concluding the long series of results devoted to the realization of this programme.

References

[1*] V. Buslaev, M. Solomyak, D. Yafaev, *On the scientific work of Mikhail Shlëmovich Birman*, Differential Operators and Spectral Theory (V. Buslaev, M. Solomyak, D. Yafaev, eds.), Amer. Math. Soc. Transl. Ser. 2, vol. 189, Amer. Math. Soc., Providence, RI, 1999, pp. 1–15.

[2*] V. Buslaev, M. Solomyak, D. Yafaev (eds.), *List of publications of M. Sh. Birman*, Differential Operators and Spectral Theory, Amer. Math. Soc. Transl. Ser. 2, vol. 189, Amer. Math. Soc., Providence, RI, 1999, pp. 17–26.

[3*] V. S. Buslaev, A. M. Vershik, I. M. Gelfand, et al., *Mikhail Shlemovich Birman (on occasion of his 70-th birthday)*, (Russian) Uspekhi Matem. Nauk **55** (2000), no. 1, 204–207; English transl., Russian Math. Surveys **55** (2000), no. 1, 201–205.

[4*] V. S. Buslaev, M. Z. Solomyak, D. R. Yafaev, *Mikhail Shlemovich Birman (on occasion of his 75-th birthday)*, Algebra i Analiz **16** (2004), no. 1, 5–14; English transl., St. Petersburg Math. J. **16** (2005), no. 1, 1–8.

Department of Mathematics, The Weizmann Institute of Science, Rehovot 76100, Israel

E-mail address: solom@wisdom.weizmann.ac.il

Department of Physics, St. Petersburg State University, Ul'yanovskaya 3, Petrodvorets, St. Petersburg, 198504, RUSSIA

E-mail address: suslina@list.ru

Amer. Math. Soc. Transl.
(2) Vol. **225**, 2008

Continuation of the List of Publications of M. Sh. Birman

[130] M. Sh. Birman, A. Laptev, *The negative discrete spectrum of a two-dimensional Schrödinger operator*, Comm. Pure Appl. Math. **49** (1996), no. 9, 967–997. MR 1399202 (97i:35131)

[131] M. Sh. Birman, T. A. Suslina, *The two-dimensional periodic magnetic Hamiltonian is absolutely continuous*, (Russian) Algebra i Analiz **9** (1997), no. 1, 32–48; English transl., St. Petersburg Math. J. **9** (1998), no. 1, 21–32. MR1458417 (98g:47038)

[138] M. Sh. Birman, T. A. Suslina, *Absolute continuity of the two-dimensional periodic magnetic Hamiltonian with discontinuous vector-valued potential*, (Russian) Algebra i Analiz **10** (1998), no. 4, 1–36; English transl., St. Petersburg Math. J. **10** (1999), no. 4, 579–601. MR 1654063 (99k:81060)

[139] M. Sh. Birman, T. A. Suslina, *Two-dimensional periodic Pauli operator. The effective masses at the lower edge of the spectrum*, Mathematical results in quantum mechanics (Prague, 1998), Oper. Theory Adv. Appl., vol. 108, Birkhäuser, Basel, 1999, pp. 13–31. MR 1708785 (2000g:81049)

[140] M. Sh. Birman, T. A. Suslina, *The periodic Dirac operator is absolutely continuous*, Integral Equations Operator Theory **34** (1999), no. 4, 377–395. MR 1702229 (2000h:47068)

[141] M. Sh. Birman, T. A. Suslina, *Periodic magnetic Hamiltonian with variable metric. The problem of absolute continuity*, (Russian) Algebra i Analiz **11** (1999), no. 2, 1–40; English transl., St. Petersburg Math. J. **11** (2000), no. 2, 203–232. MR 1702587 (2000i:35026)

[142] M. Sh. Birman, R. G. Shterenberg, T. A. Suslina, *Absolute continuity of the spectrum of a two-dimensional Schrödinger operator with potential supported on a periodic system of curves*, (Russian) Algebra i Analiz **12** (2000), no. 6, 140–177; English transl., St. Petersburg Math. J. **12** (2001), no. 6, 983–1012. MR 1816514 (2002k:35227)

[143] M. Sh. Birman, T. A. Suslina, *On the absolute continuity of the periodic Schrödinger and Dirac operators with magnetic potential*, Differential equations and mathematical physics (Birmingham, AL, 1999), AMS/IP Stud. Adv. Math. **16**, Amer. Math. Soc., Providence, RI, 2000, pp. 41–49. MR 1764740

[144] M. Sh. Birman, A. Laptev, T. A. Suslina, *The discrete spectrum of a two-dimensional second-order periodic elliptic operator perturbed by a decreasing potential.* I. *A semi-infinite gap*, (Russian) Algebra i Analiz **12** (2000), no. 4, 36-78; English transl., St. Petersburg Math. J. **12** (2001), no. 4, 535–567. MR 1793617 (2003b:47078)

[145] M. Sh. Birman, T. A. Suslina, *Threshold effects near the lower edge of the spectrum for periodic differential operators of mathematical physics*, Systems, approximation, singular integral operators, and related topics (Bordeaux, 2000), Oper. Theory Adv. Appl., vol. 129, Birkhäuser, Basel, 2001, pp. 71–107. MR 1882692 (2003f:35220)

[146] M. Sh. Birman, M. Z. Solomyak, *On the negative discrete spectrum of a periodic elliptic operator in a waveguide-type domain, perturbed by a decaying potential*, J. Anal. Math. **83** (2001), 337–391. MR 1828497 (2002k:35226)

[147] B. A. Amosov, M. Sh. Birman, M. I. Vishik, et al., *Mikhail Semenovich Agranovich (on the occasion of his seventieth birthday)*, (Russian) Uspekhi Mat. Nauk **56** (2001), no. 4(340), 163–168; English transl., Russian Math. Surveys **56** (2001), no. 4, 777–784. MR 1861459 (2002h:01026)

[148] M. Sh. Birman, T. A. Suslina, *Absolute continuity of the spectrum of the periodic operator of elasticity theory for constant shear modulus*, Nonlinear problems in mathematical physics and related topics, II (in honor of prof. O. A. Ladyzhenskaya), Int. Math. Ser. (N. Y.), 2, Kluwer/Plenum, New York, 2002, pp. 69–74. MR 1971990 (2004c:35296)

[149] A. A. Arkhipova, M. Sh. Birman, V. S. Buslaev, et al., *On the jubilee of Olga Aleksandrovna Ladyzhenskaya*, (Russian) Boundary Value Problems of Mathematical Physics and Related Topics of Function Theory. 32. Zap. Nauchn. Sem. S.-Peterburg. Otdel. Mat. Inst. Steklov. (POMI) **288** (2002), 5–13; English transl., J. Math. Sci. (N. Y.) **123** (2004), no. 6, 4523–4526. MR 1923542

[150] M. Sh. Birman, *On the homogenization procedure for periodic operators in a neighborhood of an edge of an internal gap*, (Russian) Algebra i Analiz **15** (2003), no. 4, 61–71; English transl., St. Petersburg Math. J. **15** (2004), no. 4, 507–513. MR 2068979 (2006i:35010)

[151] M. Sh. Birman, T. A. Suslina, *Periodic second-order differential operators. Threshold properties and homogenization*, (Russian) Algebra i Analiz **15** (2003), no. 5, 1–108; English transl., St. Petersburg Math. J. **15** (2004), no. 5, 639–714. MR 2068790 (2005k:47097)

[152] M. Sh. Birman, M. Z. Solomyak, *Double operator integrals in a Hilbert space*, Integral Equations Operator Theory **47** (2003), no. 2, 131–168. MR 2002663 (2004f:47029)

[153] M. Sh. Birman, T. A. Suslina, *Homogenization of a multidimensional periodic elliptic operator in a neighborhood of an edge of an inner gap*, (Russian) Boundary Value Problems of Mathematical Physics and Related Topics of Function Theory. 36. Zap. Nauchn. Sem. S.-Peterburg. Otdel. Mat. Inst. Steklov. (POMI) **318** (2004), 60–74; English transl., J. Math. Sci. (N. Y.) **136** (2006), no. 2, 3682–3690. MR 2120232 (2005j:35006)

[154] V. I. Arnold, M. Sh. Birman, A. M. Vershik, et al., *Olga Aleksandrovna Ladyzhenskaya*, (Russian) Uspekhi Mat. Nauk **59** (2004), no. 3(357), 151–152; English transl., Russian Math. Surveys **59** (2004), no. 3, 553–555. MR 2116538

[155] M. Sh. Birman, T. A. Suslina, *Threshold approximations with corrector for the resolvent of a factorized selfadjoint operator family*, (Russian) Algebra i Analiz **17** (2005), no. 5, 69–90; English transl., St. Petersburg Math. J. **17** (2006), no. 5, 745–762. MR 2241423

[156] M. Sh. Birman, T. A. Suslina, *Homogenization with corrector term for periodic elliptic differential operators*, (Russian) Algebra i Analiz **17** (2005), no. 6, 1–104; English transl., St. Petersburg Math. J. **17** (2006), no. 6, 897–973. MR 2202045 (2006k:35011)

[157] A. B. Alekseev, M. Sh. Birman, N. D. Filonov, *Spectrum asymptotics for one "nonsmooth" variational problem with solvable constraint*, (Russian) Algebra i Analiz **18** (2006), no. 5, 1–22; English transl., St. Petersburg. Math. J. **18** (2007), no. 5, 681–697. MR 2301038 (2008b:35193)

[158] M. Sh. Birman, T. A. Suslina, *Homogenization with corrector for periodic differential operators. Approximation of solutions in the Sobolev class $H^1(\mathbb{R}^d)$*, (Russian) Algebra i Analiz **18** (2006), no. 6, 1–130; English transl., St. Petersburg Math. J. **18** (2007), no. 6, 857–955. MR 2307356

[159] M. Sh. Birman, T. A. Suslina, *Homogenization of the stationary periodic Maxwell system in the case of constant permeability*, (Russian) Funktsional. Anal. i Prilozhen. **41** (2007), no. 2, 3–23; English transl., Funct. Anal. Appl. **41** (2007), no. 2, 81–98. MR 2345036

[160] M. Sh. Birman, N. D. Filonov, *Weyl asymptotics of the spectrum of the Maxwell operator with non-smooth coefficients in Lipschitz domains*, Nonlinear equations and spectral theory, Amer. Math. Soc. Transl. Ser. 2, v. 220 (2007), pp. 27–44. MR 2343605

[161] M. Sh. Birman, T. A. Suslina, *Homogenization of periodic differential operators as a spectral threshold effect*, Proceedings of XV International Congress on Mathematical Physics, 2008, to appear.

[162] M. Sh. Birman, T. A. Suslina, *Operator error estimates for homogenization of nonstationary periodic equations*, (Russian) Algebra i Analiz **20** (2008), no. 6, 30–107.

[163] M. Sh. Birman, T. A. Suslina, *The limit absorption principle and homogenization procedure for periodic elliptic operators*, (Russian) Funktsional. Anal. i Prilozhen. **42** (2008), no. 4.

Amer. Math. Soc. Transl.
(2) Vol. **225**, 2008

Perturbations of the Continuous Spectrum of a Singular Elliptic Operator by Varying the Boundary and the Boundary Conditions

M. Sh. Birman

ABSTRACT. This paper was published in Russian (Vestnik Leningrad. Univ. **1**, no. 1 (1962), 22–55) and until now it was never translated into English. Many approaches and technical tools which later became classical in spectral and scattering theory appeared in this paper for the first time. This urged us to publish the paper in this volume. The translation was prepared by M. Solomyak. Some minor editorial changes were made. This especially concerns the proof of Theorem 5.2, which was only outlined in the original text. All the changes are approved by the author.

The translation is supplemented with short translator's comments. The references to these comments that are not quoted in the main body of the paper are indicated by asterisks in the references list.

Editors

0. Introduction

During the last few years several papers have been published, whose authors investigated the essential spectrum of an operator of a singular boundary-value problem. It was proved that this spectrum is determined by the data in an arbitrarily small vicinity of the set of singularities, and that it is nonsensitive with respect to other (regular) data of the problem. We recall that the essential spectrum of a selfadjoint operator in a Hilbert space consists of its continuous spectrum and of the eigenvalues of infinite multiplicity.

Apparently, Povzner [**29**] was the first who studied this question. He considered the Schrödinger operator

$$-\Delta u + q(x)u \tag{0.1}$$

in the exterior Ω of a bounded three-dimensional domain Ω_0, with the boundary condition

$$\frac{\partial u}{\partial n} + \sigma(x)u\big|_{\Gamma} = 0, \qquad \Gamma = \partial\Omega. \tag{0.2}$$

Povzner showed that the essential spectrum of the corresponding selfadjoint operator in $L_2(\Omega)$ is stable with respect to the change of the boundary Γ or of the

2000 *Mathematics Subject Classification.* Primary 35P25; Secondary 47A40, 47A55, 47F05.

function $\sigma(x)$. His approach was based upon a well-known Weyl theorem stating that perturbation of a selfadjoint operator by a compact one does not change the essential spectrum. Using the tools of potential theory, Povzner carried out a detailed study of the difference of resolvent kernels of two problems of the type (0.2) for the operator (0.1). He proved that this difference belongs to the class $L_2(\Omega\times\Omega)$. Povzner's analysis substantially depends on the form of the operator (0.1), and even on the dimension. In this connection we have the problem of obtaining results of this nature by using less special means and in a more general setting.

Let us discuss this in more detail, restricting ourselves to the operator (0.1) with the boundary condition (0.2) or

$$u\big|_{\Gamma}=0. \tag{0.3}$$

Clearly, it is sufficient to compare the operator A generated by (0.1) in the whole space, and the operator A_1 that corresponds to the orthogonal sum of the problems (0.1), (0.2) (or (0.1), (0.3)) in the spaces $L_2(\Omega)$ and $L_2(\Omega_0)$. Note that only the first term may contribute to the essential spectrum of A_1. The operators A, A_1 can be treated as two different selfadjoint extensions of the same symmetric operator A^0 whose domain consists of functions vanishing near Γ and outside some ball. In this way the problem considered is included in the general scheme of extension theory.[1] Then, for the problems in dimension one, the desired result follows immediately from the fact that A^0 has finite deficiency indices, and hence the difference of the resolvents of the operators A and A_1 is a finite rank operator. However, in higher dimensions this is not true anymore, and new arguments are necessary. This obstacle was overcome independently and by using different approaches by Glazman [**14**] and Birman [**3**]. Avoiding studying the difference of the resolvents, Glazman showed that for every point of the essential spectrum of the operator (0.1), there exists a singular sequence of functions, in the sense of Weyl, whose terms vanish in an arbitrary (but fixed) bounded domain. This immediately yields coincidence of the essential spectra of both operators A, A_1. This approach, developed by Glazman for the operator (0.1), admits various generalizations.

In the paper [**3**] Birman considered the semibounded second-order elliptic operators of the form

$$\Lambda u=-\sum_{i,j=1}^{m}\frac{\partial}{\partial x_i}\left(a_{ij}(x)\frac{\partial u}{\partial x_j}\right)+a(x)u \qquad (a(x)\geq 1). \tag{0.4}$$

He studied the difference of inverse operators by using ideas of general extension theory for symmetric operators in a Hilbert space. This theory was developed by Krein [**19**], Vishik [**36**], and Birman [**2, 4**]. On the basis of the results of [**19, 2**], the question of compactness of the difference of inverse operators was reduced in [**3**] to obtaining compactness criteria in embedding theorems [**35**] for unbounded domains, but not for the whole Sobolev space: such criteria are needed only for its subspace formed by the solutions of the homogeneous equation $\Lambda u=0$. Such criteria were established in [**3**]. In the same paper, similar problems were solved for elliptic operators with degeneration of ellipticity on the boundary or in interior points of the domain.

[1] The above-described method of such an inclusion ('splitting of an operator') was introduced by Glazman. It was applied systematically in his papers [**10, 11, 12, 13**]. See also [**15**], where other works by this author are listed.

The essential spectrum of singular operators was also studied in the papers [**37, 38**] by Wolf. In [**37**] he, actually following Glazman's approach, investigated the degenerate elliptic high-order nonselfadjoint operators in bounded domains. In [**38**] Wolf proved that the essential spectrum of the Dirichlet problem for the nonselfadjoint operator

$$-\Delta u + \sum_{i=1}^{3} a_i(x) \frac{\partial u}{\partial x_i} + a(x)u$$

coincides with that of the operator $-\Delta u$ in the whole space, provided that the functions a_i, a belong to $L_2(E_3)$ and are bounded.

Note that the latter result is actually not new. It follows from the possibility of including the condition (0.3) in the definition of the operator $-\Delta u$ and from the well-known fact that adding lower-order terms satisfying the above restrictions does not affect the essential spectrum.

New possibilities for studying the continuous spectrum of singular boundary value problems are provided by recent results of Rosenblum [**30**] and Kato [**17, 18**]. They significantly improve Weyl's theorem. The Rosenblum–Kato theorem states that, up to unitary equivalence, perturbation of a selfadjoint operator by a trace class operator preserves its absolutely continuous component (see Subsection 1.4 for the precise formulation). For an application of this result to the perturbation of the boundary and of the boundary conditions, one needs to show that the difference of some (large enough) powers of the resolvents is trace class. We succeeded in doing this, under some rather wide assumptions, by improving and generalizing the approach suggested in [**3**]. The approach developed by Glazman evidently does not work here.

In this paper we present the corresponding results. At the same time, we reprove the results of [**3**], where the proofs were only outlined. In the course of our exposition, we remove some redundant conditions imposed in [**3**]. In particular, we need no restrictions on the growth of the coefficients $a_{ij}(x)$ at infinity. We mostly concentrate on the case where the singular point lies at infinity. Operators with the degeneration of ellipticity on the boundary are also analyzed. The generalization of the approach to other types of singularities is obvious.

In Section 1, we present the necessary information from operator theory in Hilbert space, mostly without proofs. Section 2 is devoted to including the problems considered in the framework of operator theory. In Section 3, we study the difference of the resolvents of two nonidentical boundary-value problems for the differential expression (0.4), for the case of domains with compact boundary Γ. We evaluate the rate of decrease of the eigenvalues of such operators. The obtained estimates are order-sharp, and we believe that they are of independent interest. Note that the results of this section also remain meaningful for the regular boundary-value problems in bounded domains (see Subsection 3.4). The results of Section 3 allow one to apply the Rosenblum–Kato theorem in the two-dimensional case. In Section 4 similar estimates are obtained for the difference of integer powers of the resolvents. This is necessary for the analysis of multi-dimensional problems. A large part of the results obtained in Section 3 can be extended to domains with noncompact boundary and to degenerate elliptic operators. We do this in Section 5. Our main results on the spectrum are formulated in Section 6. They are immediate consequences of the estimates established in Sections 3, 4, and 5.

In the exposition we always try to replace direct analytic calculations, such as integration by parts in unbounded domains, etc., by arguments of an operator-theoretic nature. This makes it possible to avoid various additional restrictions regarding the data, in particular, on the smoothness of the boundary and on the growth of the coefficients.

A short communication on the main results of the paper was published in [**7**]. Some of the results are published here for the first time.

1. Auxiliary material from operator theory

1.1. Here we present formulations of some facts from the extension theory of operators in Hilbert space. Let H be a separable Hilbert space, and let S be a positive definite symmetric operator with the dense domain $D(S)$. In what follows, $D(K)$ and $R(K)$ always stand for the domain and the range of an operator K, and K^* stands for the adjoint operator.

Let $\widetilde{S}_0$ be the Friedrichs extension of S (see [**9, 27, 34**]), and let $\widetilde{S}$ be some other positive definite selfadjoint extension of S. Along with the operator $\widetilde{S}$, let us consider its quadratic form

$$\widetilde{S}[f,f] = (\widetilde{S}^{1/2}f, \widetilde{S}^{1/2}f)$$

defined on the domain $D[\widetilde{S}] = D(\widetilde{S}^{1/2})$. The set $D[\widetilde{S}]$ is a complete Hilbert space with respect to the scalar product

$$\widetilde{S}[f,g] = (\widetilde{S}^{1/2}f, \widetilde{S}^{1/2}g).$$

The operator domain $D(\widetilde{S})$ is dense in this Hilbert space. It was shown by Krein [**19**] that always $D[\widetilde{S}] \supseteq D[\widetilde{S}_0]$, and

$$\widetilde{S}[f,f] = \widetilde{S}_0[f,f] \qquad (f \in D(\widetilde{S}_0)).$$

Thus, $D[\widetilde{S}_0]$ is a (closed) subspace in $D[\widetilde{S}]$. It is important to characterize the orthogonal complement of $D[\widetilde{S}_0]$ in $D[\widetilde{S}]$. Let U be the set of all solutions of the equation $S^*u = 0$. Clearly, U is a subspace in H, and $U = H \ominus R(S)$. The following lemma is due to Krein [**19**].

LEMMA 1.1. *The set $D[\widetilde{S}] \cap U$ is the orthogonal complement of $D[\widetilde{S}_0]$ in the Hilbert space $D[\widetilde{S}]$. In particular, if $f \in D[\widetilde{S}_0]$ and $u \in D[\widetilde{S}] \cap U$, then*

$$\widetilde{S}[f,u] = 0.$$

Now we characterize the same set $D[\widetilde{S}] \cap U$ in other terms. Let us introduce the operator

$$B = \widetilde{S}^{-1} - \widetilde{S}_0^{-1}.$$

Evidently, B is a bounded and selfadjoint operator in H, $B \geq 0$, and $Bh = 0$ for $h \in R(S)$. Let U_1 stand for the orthogonal complement in H of the kernel $\{h \in H : Bh = 0\}$. It follows from the construction that $U_1 \subseteq U$. The operator B is positive on U_1 and has on this subspace the selfadjoint inverse operator $G = B^{-1}$, which, in general, is unbounded. The next Lemma 1.2, which plays an important role in the further exposition, was established by the author [**2, 4**]. In fact, in these papers a more complicated situation was analyzed: arbitrary selfadjoint lower semibounded extensions of S were considered (see Theorem 3 and Corollary 2 in [**4**]). For the positive definite extensions, the proof given in [**4**] becomes much simpler. For this special case, Lemma 1.2 can be also derived from Krein's result; see formula (5.5) in [**19**].

LEMMA 1.2. *The operator G is positive definite on U_1. The domain $D[G]$ of its quadratic form $G[u, u]$ coincides with the set $D[\widetilde{S}] \cap U$ and, therefore, the space $D[\widetilde{S}]$ decomposes into the orthogonal sum*

$$D[\widetilde{S}] = D[\widetilde{S}_0] \oplus D[G].$$

In particular, note that for $u \in D[G]$ we have $\widetilde{S}[u, u] = G[u, u]$. Now, Lemma 1.2 and the variational principle yield the following result.

LEMMA 1.3. *One can obtain the spectral decomposition of the operator G by considering the Rayleigh quotient*

$$\frac{\widetilde{S}[u, u]}{(u, u)}, \qquad u \in D[\widetilde{S}] \cap U.$$

If any subset of $D[\widetilde{S}] \cap U$, which is bounded in $D[\widetilde{S}]$, turns out to be pre-compact in H, then the operator B is compact. In this case, the nonzero eigenvalues of B are equal to the consecutive maxima of the Rayleigh quotient[2]

$$\frac{(u, u)}{\widetilde{S}[u, u]}, \qquad u \in D[\widetilde{S}] \cap U. \tag{1.1}$$

In the conclusion of this subsection, we recall some elementary facts about the relations between positive definite quadratic forms and selfadjoint operators. For details, see the paper [**9**] and the book [**27**].

Let $\widetilde{Q}[f, f]$ be a positive definite quadratic form in H, with the domain $D[\widetilde{Q}]$:

$$\widetilde{Q}[f, f] \geq \gamma^2(f, f), \qquad \forall f \in D[\widetilde{Q}] \qquad (\gamma^2 > 0).$$

The quadratic form $\widetilde{Q}[f, f]$ is said to be closed in H if $D[\widetilde{Q}]$ is a complete Hilbert space with respect to the scalar product, generated by the corresponding sesquilinear form $\widetilde{Q}[f, g]$. Any closed and positive definite quadratic form generates a selfadjoint operator in the following way. For an arbitrary element $h \in H$ there exists a unique element $g \in D[\widetilde{Q}]$ such that

$$\widetilde{Q}[g, f] = (h, f), \qquad \forall f \in D[\widetilde{Q}]. \tag{1.2}$$

If $h_1 \neq h_2$, the corresponding elements g_1, g_2 also differ. The operator $h = \widetilde{Q}g$ is positive definite and selfadjoint in H, and its domain $D(\widetilde{Q})$ is dense in the space $D[\widetilde{Q}]$. Further, $D[\widetilde{Q}] = D(\widetilde{Q}^{1/2})$ and $\widetilde{Q}[f_1, f_2] = (\widetilde{Q}^{1/2} f_1, \widetilde{Q}^{1/2} f_2)$. Note that it is sufficient to check relation (1.2) only for the elements f from some dense subset of $D[\widetilde{Q}]$.

If the quadratic form $\widetilde{Q}[f, f]$ is not closed, but the elements of the completion of $D[\widetilde{Q}]$ in the corresponding metric can be treated as elements of H, then after this completion the quadratic form $\widetilde{Q}[f, f]$ becomes closed in H. In this case, we say that the quadratic form $\widetilde{Q}[f, f]$ is closable. By an appropriate shift, this scheme extends to the case of quadratic forms that are semibounded from below.

---

[2]In this connection, the author often calls these maxima *the spectrum* of the Rayleigh quotient (or 'ratio') (1.1) – *translator's remark.*

1.2. In this subsection we consider an elementary generalization of the variational principle for eigenvalues. Let A be a selfadjoint operator in H, and let $E_\lambda(A)$ be the family of its spectral projections, normalized by the condition of continuity from the left. We define the total multiplicity of the spectrum of A to the left of a point $\nu \in \mathbb{R}$ as the dimension of the subspace $E_\nu(A)H$. Similarly, the total multiplicity of the spectrum of A to the right of ν is the dimension of $(E - E_{\nu+0}(A))H$; here E is the identity operator. Finally, the total multiplicity of the spectrum of A outside a segment $[\nu_1, \nu_2]$ is defined as the sum of total multiplicities to the left of ν_1 and to the right of ν_2.

LEMMA 1.4. *Let A and $A_r, r = 1, \dots, l$, be bounded selfadjoint operators in H, and let $\Phi(\xi_1, \dots, \xi_l)$ be a real-valued function of real arguments that is nondecreasing in each ξ_r. Suppose that*

$$(Au, u) \le \Phi\left((A_1 u, u), \dots, (A_l u, u)\right), \qquad \forall u \in H. \tag{1.3}$$

Let the total multiplicity m_r of the spectrum of A_r to the right of a point ν_r be finite for each r. Then the total multiplicity m of the spectrum of A to the right of the point $\nu = \Phi(\nu_1, \dots, \nu_l)$ is finite. Moreover,

$$m \le \sum_r m_r.$$

PROOF. Suppose that $m > \sum_r m_r$. Then in the subspace $(E - E_{\nu+0}(A))H$ there exists an element u_0 such that $\|u_0\| = 1$ and u_0 is orthogonal to all the subspaces $(E - E_{\nu_r+0}(A_r))H$, $r = 1, \dots, l$. It follows that $(A_r u_0, u_0) \le \nu_r$ and, by (1.3),

$$(Au_0, u_0) \le \Phi\left((A_1 u_0, u_0), \dots, (A_l u_0, u_0)\right) \le \Phi(\nu_1, \dots, \nu_l) = \nu.$$

However, this is impossible, since $u_0 \in (E - E_{\nu+0}(A))H$, and hence,

$$(Au_0, u_0) > \nu(u_0, u_0) = \nu.$$

□

It is not difficult to extend this lemma to the case of semibounded operators. We leave this to the reader.

Now we present another version of this statement.

LEMMA 1.5. *Let A and $A_r, r = 1, \dots, l$, be bounded selfadjoint operators in H, and let $\Phi(\xi_1, \dots, \xi_l) \ge 0$ be a real-valued function of nonnegative arguments that is nondecreasing in each ξ_r. Suppose that*

$$|(Au, u)| \le \Phi\left(|(A_1 u, u)|, \dots, |(A_l u, u)|\right), \qquad \forall u \in H.$$

If the total multiplicity M_r of the spectrum of A_r outside a segment $[-\nu_r, \nu_r]$ is finite for each r, then the total multiplicity M of the spectrum of A outside the segment $[-\nu, \nu]$, where $\nu = \Phi(\nu_1, \dots, \nu_l)$, is finite. Moreover,

$$M \le 2 \sum_r M_r.$$

PROOF. Let us show, for instance, that $\dim E_{-\nu}(A)H \le M_1 + \dots + M_l$. In the opposite case, we can find an element $u_0 \in E_{-\nu}(A)H$ such that $\|u_0\| = 1$ and $\|A_r u_0\| \le \nu_r$ for all $r = 1, \dots, l$. Then $|(A_r u_0, u_0)| \le \|A_r u_0\| \|u_0\| \le \nu_r$ and

$$|(Au_0, u_0)| \le \Phi\left(|(A_1 u_0, u_0)|, \dots, |(A_l u_0, u_0)|\right) \le \Phi(\nu_1, \dots, \nu_l) = \nu.$$

This is impossible, since $u_0 \in E_{-\nu}(A)H$, which implies $(Au_0, u_0) < -\nu$. The total multiplicity to the right of ν is estimated in the same way. □

1.3. Here we introduce a certain class of operators in H. Let A be a compact, selfadjoint operator, and let $\lambda_n(A)$ be its eigenvalues, enumerated in decreasing order of their absolute values, taking multiplicities into account.

DEFINITION 1.6. A compact, selfadjoint operator A in H belongs to class T_α, $\alpha > 0$, if for some $c > 0$ the estimate

$$|\lambda_n(A)| \le cn^{-\alpha}, \qquad \forall n \in \mathbb{N}, \tag{1.4}$$

is satisfied.

Let us list some elementary properties of class T_α.

1) Let $N_A(\lambda)$ stand for the total multiplicity of the spectrum of a compact, selfadjoint operator A outside the segment $[-\lambda, \lambda]$. It is easy to see that the estimate (1.4) is equivalent to the inequality

$$N_A(\lambda) \le c_0 \lambda^{-1/\alpha}, \qquad \forall \lambda > 0,$$

with some $c_0 > 0$.

2) If an operator A is selfadjoint in H, operators A_r, $r = 1, \dots, l$, belong to class T_α, and

$$|(Au, u)| \le \sum_{r=1}^{l} |(A_r u, u)|, \qquad \forall u \in H,$$

then $A \in T_\alpha$. Indeed, we can conclude from Lemma 1.5 (with $\Phi = \xi_1 + \ldots + \xi_l$) that

$$N_A(\lambda) \le 2 \sum_{r=1}^{l} N_{A_r}(\lambda/l) \le 2 \left(\sum_{r=1}^{l} c_{0,r} \right) (\lambda/l)^{-1/\alpha} = c_0 \lambda^{-1/\alpha}.$$

3) If $A_r \in T_\alpha$ and β_r are real numbers, $r = 1, \dots, l$, then

$$A := \sum_{r=1}^{l} \beta_r A_r \in T_\alpha.$$

This immediately follows from Definition 1.6 and statement 2).

4) An operator A belongs to class T_α if and only if the same is true for its positive and negative components, A^+ and A^-. This is evident.

5) Let $A \in T_\alpha$, and suppose that a bounded operator K is such that the product $B = AK$ is a selfadjoint operator. Then $B \in T_\alpha$. Indeed, the properties 3) and 4) allow one to assume that $A \ge 0$ and $\|K\| = 1$. We have

$$\begin{aligned} |(Bu, u)| = |(AKu, u)| &\le \sqrt{(Au, u)}\sqrt{(AKu, Ku)} \\ &\le \frac{1}{2}(Au, u) + \frac{1}{2}(AKu, Ku), \end{aligned}$$

and, by property 2), it is sufficient to prove that $K^*AK \in T_\alpha$. The spectrum of the latter operator is given by the consecutive maxima of the Rayleigh quotient

$$\frac{(AKu, Ku)}{(u, u)}.$$

Since

$$\frac{(AKu, Ku)}{(u, u)} \le \frac{(AKu, Ku)}{\|Ku\|^2} = \frac{(Av, v)}{(v, v)}, \qquad v = Ku,$$

we derive the desired result directly from the minimax principle. The same result holds for the product KA.

1.4. In this subsection we present the formulation of the Rosenblum–Kato theorem. We start with the necessary definitions.

a) *Trace class operators* (see e.g. Krein [**20**]). A selfadjoint operator is said to be trace class if it is compact and its eigenvalues form an absolutely convergent series. Let B be such an operator, and let $\{\phi_k\}$ be some complete orthonormal system in H. Then the sum

$$\sum_k (B\phi_k, \phi_k) \tag{1.5}$$

is finite and does not depend on the choice of the system $\{\phi_k\}$. Moreover, it is equal to the sum of all eigenvalues (the trace) of B. If a selfadjoint operator B is nonnegative, and for some complete orthonormal system $\{\phi_k\}$ the sum (1.5) is finite, then B is trace class. Note that operators of the class T_α with $\alpha > 1$ meet this property.

b) *Absolutely continuous part of an operator* (see e.g. Kato [**17**] or Kuroda [**21**]). Let C be a selfadjoint operator in H, and let E_λ be its resolution of identity (the family of its spectral projections). Consider the set $H_a(C)$ of all elements $u \in H$ such that the function $(E_\lambda u, u)$ is absolutely continuous. The set $H_a(C)$ is a reducing subspace for C. The part C_a of the operator C, acting in $H_a(C)$, is called its absolutely continuous part.

c) **Rosenblum–Kato theorem;** see [**30, 17, 18**]. The formulation given below is due to Kato [**18**].

THEOREM 1.7. *Suppose that A is a selfadjoint operator, $B = B^*$ is trace class, and $C = A + B$. Then the absolutely continuous parts of the operators A and C are unitarily equivalent.*

Unitary equivalence means that there exists an isometric operator W that maps $H_a(A)$ onto the whole of $H_a(C)$ and is such that

$$D(C_a) = WD(A_a); \qquad WA_af = C_aWf, \ \forall f \in D(A_\alpha).$$

In the papers [**30, 17, 18**] an explicit expression for the operator W is given (see also formula (7.1) in the Appendix to the present paper).

In conclusion, let us note that the notion of unitary equivalence of the operators A_a and C_a makes sense also in the case where A and C act in different Hilbert spaces.

2. Notation. Setting of the boundary value problems

In this section we introduce the notation and notions that will be used in the course of our exposition.

2.1. Below E_m stands for the m-dimensional Euclidean space, $x = (x_1, \dots, x_m)$ denotes its generic point, and dx is the volume element. Let Ω be a domain in E_m (i.e., an open and connected set). Let Γ stand for its boundary and $\overline{\Omega} = \Omega \cup \Gamma$ for its closure. We always impose one of the following conditions on the structure of Γ.

1) The boundary $\Gamma = \partial\Omega$ consists of a finite number of isolated, C^2-smooth $(m-1)$-dimensional manifolds without boundary. If this is the case, we simply say that Γ is *smooth*.

2) A more general class of domains is formed by the domains *with piecewise-smooth boundary of the class* C^2. By this we mean restrictions that guarantee the possibility of extending any function $u \in C^2(\overline{\Omega})$ to the whole of E_m in such a way that the extended function also belongs to C^2. The rigorous definition of this class of domains can be found, for instance, in [**1**].

Sometimes for these two types of boundaries we use the term *surface* (respectively, smooth or piecewise-smooth). We denote by ds the area element on Γ. Note that if Γ is compact, it may consist only of a finite number of connected components.

If a domain Ω is bounded and its boundary meets one of the above conditions, then the Sobolev embedding theorems [**35**] for the functions having generalized second-order derivatives are satisfied in Ω.

2.2. Here we introduce some function classes.

Let $u(x)$ be a function in Ω, and let F be a subset of Ω. We say that $u(x)$ vanishes near F if $u(x) = 0$ in the intersection of Ω with some neighborhood of F. We say that $u(x)$ vanishes at infinity if $u(x) = 0$ outside some large ball. As usual, $C^r(\overline{\Omega})$, $r = 0, 1, \dots$, stands for the set of all functions that have continuous derivatives up to the order r in $\overline{\Omega}$, and

$$C^\infty(\overline{\Omega}) = \bigcap_1^\infty C^r(\overline{\Omega}).$$

Further, $C_0^r(\Omega)$ refers to the set of compactly supported functions from $C^r(\overline{\Omega})$ (i.e., functions vanishing near Γ and at infinity). We also need the class $M^r(\Omega)$ which consists of all functions from $C^{r-1}(\overline{\Omega})$ that have all possible generalized derivatives (in the sense of Sobolev) of order r, and these derivatives are measurable functions which are bounded on any compact subset of $\overline{\Omega}$; for $r = 0$, only the latter requirement survives. Unless otherwise stipulated, we interpret derivatives in the generalized (Sobolev) sense, and use standard notation for them. When it is nonambiguous, we use the following shortened notation for the first-order derivatives:

$$\frac{\partial u}{\partial x_j} = u_j, \qquad j = 1, \dots, m.$$

The space $L_2(\Omega)$ is always the basic Hilbert space for us. Respectively, the definitions of $\|u\|$ and (u, v), reserved for this space, are

$$\|u\|^2 = \int_\Omega |u(x)|^2 dx, \qquad (u, v) = \int_\Omega u(x)\overline{v(x)} dx.$$

In other cases, we use an additional subindex.

Let ω be a bounded domain. The Sobolev space $W_2^{(l)}(\omega)$ ([**35**]; see also [**34**]) consists of all functions whose derivatives up to the order l lie in $L_2(\omega)$. The norm in $W_2^{(l)}(\omega)$ is defined by the equality

$$\|u\|^2_{W_2^{(l)}(\omega)} = \sum_{s=0}^{l} \|D^s u\|^2_{L_2(\omega)},$$

where $D^s u$ represents all possible derivatives of order s.

Now, suppose that the boundary $\Gamma_\omega = \partial\omega$ is smooth. We need the Sobolev space of order $1/2$ of functions defined on Γ_ω. This space $W_2^{1/2}(\Gamma_\omega)$ was introduced and studied in the papers [**31, 32**]. We will not give here the corresponding definitions. We need only the following fact: the space $W_2^{1/2}(\Gamma_\omega)$ can be identified with the set of traces on Γ_ω of all functions from $W_2^{(1)}(\omega)$, or with that of normal derivatives of all functions from $W_2^{(2)}(\omega)$.

We say that a function $u(x)$ belongs to the class $W_{2,\text{fin}}^{(l)}(\Omega)$ (with l integer), if $u \in W_2^{(l)}(\omega)$ in any bounded subdomain $\omega \subset \Omega$.

Finally, we introduce one more Hilbert space. Let $\Omega \subset E_m$ be an unbounded domain; the possibility $\Omega = E_m$ is not excluded. On the set of all functions $u \in W_{2,\text{fin}}^{(1)}(\Omega)$, vanishing at infinity, consider the metric that is generated by the quadratic form

$$\int_\Omega \left(a_{ij}(x) \frac{\partial u}{\partial x_i} \frac{\partial \overline{u}}{\partial x_j} + a(x)|u|^2 \right) dx. \tag{2.1}$$

As usual, we drop the symbol of summation from 1 to m over the repeated indices i, j. We suppose that $a_{ij} \in M^1(\Omega)$, $a \in M^0(\Omega)$, $a(x) \geq 1$, and the quadratic form

$$a_{ij}(x)\xi_i\overline{\xi}_j, \qquad \xi_i \in \mathbb{C},\ i = 1, \dots, m, \tag{2.2}$$

is positive definite at each point $x \in \Omega$. By completing the original set of functions in the metric of the quadratic form (2.1), we obtain a Hilbert space that we denote by $W_2^{(1)}(\Omega; a_{ij}, a)$. All its elements are functions, at least of the class $W_{2,\text{fin}}^{(1)}(\Omega)$. The squared norm of an element $u \in W_2^{(1)}(\Omega; a_{ij}, a)$ is given by the integral (2.1), and the scalar product is defined as

$$\int_\Omega (a_{ij} u_i \overline{v_j} + a(x) u \overline{v})\, dx.$$

In what follows, the letter c, often with an index, stands for a positive constant whose value is indifferent. Sometimes, different constants are denoted by the same symbol.

2.3. We study the spectral properties of the selfadjoint operator in $L_2(\Omega)$, generated by an elliptic differential expression

$$\Lambda g = -\frac{\partial}{\partial x_i} \left(a_{ij}(x) \frac{\partial g}{\partial x_j} \right) + a(x) g, \tag{2.3}$$

complemented by a system of boundary conditions. Now, we provide accurate definitions of the notions involved.

It is convenient to assume that the coefficients in (2.3) are defined on the whole of E_m. The coefficient $a(x)$ is supposed to be bounded from below. By adding an appropriate constant to $a(x)$ (which causes only a shift of the spectrum of the corresponding operator), we can always assume that $a(x) \geq 1$; the only exclusion will be made in Subsection 6.3. Further, we always assume that $a_{ij} \in M^1(E_m)$, $a \in M^0(E_m)$, and the quadratic form (2.2) is positive definite uniformly in any ball.[3]

Now we describe the boundary conditions we will be dealing with. Let γ and δ be two nonintersecting subsets of the boundary $\Gamma = \partial\Omega$ such that $\Gamma = \gamma \cup \delta$. We always assume that γ is bounded; the case $\delta = \Gamma$ is not excluded. Let $\sigma(x)$ be a measurable and bounded real-valued function on γ. Below, n stands for the

[3] This agreement is violated in Subsection 5.2 and in some statements of Section 6, where we allow the quadratic form (2.2) to degenerate at some points. This will always be stated explicitly.

direction of the outer normal vector to Γ. Along with the differential expression (2.3) in Ω, we consider the boundary conditions

$$g|_\delta = 0, \tag{2.4}$$

$$\frac{\partial g}{\partial \nu} + \sigma(x)g|_\gamma = 0. \tag{2.5}$$

Here $\frac{\partial g}{\partial \nu}$ stands for the co-normal derivative:

$$\frac{\partial g}{\partial \nu} = a_{ij}\cos(n, x_i)g_j.$$

We aim at interpreting the boundary conditions (2.4), (2.5), and also the differential expression (2.3) itself, in some generalized sense. It is convenient to define them, using the corresponding quadratic form.

On the set $D[\widetilde{S}]$ of all functions from $W_2^{(1)}(\Omega; a_{ij}, a)$ that satisfy the boundary condition (2.4) (in the sense of the embedding theorems), let us consider the quadratic form

$$\widetilde{S}[g, g] = \int_\Omega \left(a_{ij}g_i\overline{g_j} + a|g|^2\right) dx + \int_\gamma \sigma|g|^2 ds. \tag{2.6}$$

Choose a bounded subset ω of Ω, whose boundary is piecewise-smooth and contains the set γ. For any $\varepsilon > 0$ the following inequality is satisfied (see [**35**]):

$$\int_\gamma |g|^2 ds \le \varepsilon \int_\omega |\nabla g|^2 dx + c(\varepsilon) \int_\omega |g|^2 dx. \tag{2.7}$$

It follows that the quadratic form (2.6) is bounded from below in $L_2(\Omega)$. Replacing $a(x)$ by $a(x) + C$, with an appropriate constant C, we can assume that

$$\widetilde{S}[g, g] \ge (g, g). \tag{2.8}$$

This condition is always assumed to be satisfied. The inequalities (2.7) and (2.8) show that on the set $D[\widetilde{S}]$ the quadratic forms (2.6) and (2.1) are metrically equivalent:

$$c_1\widetilde{S}[g, g] \le \|g\|^2_{W_2^{(1)}(\Omega; a_{ij}, a)} \le c_2\widetilde{S}[g, g] \qquad (0 < c_1 \le c_2 < \infty). \tag{2.9}$$

From here it follows that the quadratic form $\widetilde{S}[g, g]$ is closed and, therefore, it generates a selfadjoint and positive definite operator $\widetilde{S}$ in $L_2(\Omega)$. By what was stated in Subsection 1.1, the operator domain of $\widetilde{S}$ consists of elements $g \in D[\widetilde{S}]$ such that for some $h \in L_2(\Omega)$ the identity

$$\widetilde{S}[g, f] = (h, f), \qquad \forall f \in D[\widetilde{S}], \tag{2.10}$$

is fulfilled. If $g \in C_0^2(\Omega)$, one can integrate by parts in the expression $\widetilde{S}[g, f]$, which reduces it to the form (h, f), where $h = \Lambda g$. Moreover, it is easy to see that any element $g \in D[\widetilde{S}]$, which satisfies the identity (2.10), belongs to the class $W_2^{(2)}(\omega)$ in any bounded domain $\omega \Subset \Omega$. The function $h(x)$ appearing in (2.10) necessarily coincides with the one given by the expression (2.3). This allows one to link the operator $\widetilde{S}$ to the differential expression (2.3). Further, the boundary condition (2.4) is evidently fulfilled for the functions $g \in D(\widetilde{S})$. The condition (2.5) is a natural condition in the sense of the Calculus of Variations. This condition arises automatically when passing from the quadratic form to the operator. It is also possible to show that this condition is fulfilled in a certain 'weak' sense. One can find the related material, e.g., in the books [**35, 27**].

Later on, we often speak about the selfadjoint operator $\widetilde{S}$ in $L_2(\Omega)$, which corresponds to the differential expression (2.3) and the boundary conditions (2.4), (2.5). In all such cases we mean the above-described scheme. Note that the operators $\widetilde{S}$, which correspond to different conditions of the type (2.4), (2.5), are selfadjoint extensions of the same symmetric operator S^0. This operator is defined by the differential expression (2.3) on functions $g \in C_0^2(\Omega)$. Varying the boundary conditions corresponds to the passage from a selfadjoint, positive definite extension of S^0 to another one. Here we keep in mind changing of the subsets γ, δ, which form a partition of Γ, or changing of the function $\sigma(x)$ on γ. Incidentally, the subset γ has to remain bounded. We also assume that the condition (2.8) remains valid, but, in fact, this is not a restriction.

It is easy to show that the Friedrichs extension of S^0 corresponds to the Dirichlet boundary condition, that is, to the case where $\delta = \Gamma$.

3. Domains with compact boundary, I

3.1. If Γ is compact, then it consists of a finite number of isolated connected components. Thus, Ω is the exterior domain with respect to a finite family of compact, piecewise-smooth surfaces. The possibility that $\Omega = E_m$ is not excluded.

Let $\widetilde{S}$ be the operator of some boundary value problem (2.4), (2.5) for the differential expression (2.3) in the domain Ω. By this we mean the operator defined by the procedure described in Subsection 2.3. Choose a smooth closed surface $\Gamma_0 \subset E_m$ that contains Γ inside. Let $\Omega_1 \subset E_m$ be the exterior domain with respect to Γ_0, and denote $\Omega_2 = \Omega \setminus \overline{\Omega_1}$. Consider the quadratic form $\widetilde{S}[g,g]$ of the operator $\widetilde{S}$ on the set of all functions from $D[\widetilde{S}]$ vanishing on Γ_0. This new quadratic form, which we denote by $\widetilde{S}_0[g,g]$, is evidently closed in $L_2(\Omega)$, and hence, it generates a selfadjoint operator in this space. This operator $\widetilde{S}_0$ corresponds to a 'split' boundary value problem for the differential expression (2.3). Namely, in the domain Ω_2 we have the same conditions (2.4) and (2.5), with the additional Dirichlet condition on Γ_0. In the domain Ω_1 we have only the latter condition. The operator $\widetilde{S}_0$ is the direct orthogonal sum of the operators that correspond to these two boundary value problems in $L_2(\Omega_2)$ and in $L_2(\Omega_1)$ respectively.

In this way, we have two selfadjoint operators, $\widetilde{S}$ and $\widetilde{S}_0$, which we are going to compare. Clearly, the restriction of each of them onto the domain $D = D(\widetilde{S}) \cap D(\widetilde{S}_0)$ yields the same symmetric and positive definite operator, which we denote by S. Its domain $D(S) = D$ is dense in $L_2(\Omega)$, since it contains the set of functions from $C_0^2(\Omega)$ vanishing also near the surface Γ_0. Thus, both $\widetilde{S}$ and $\widetilde{S}_0$ can be considered as selfadjoint extensions of the same operator S. Now, let $\beta(x)$ be a cut-off function in Ω, i.e., $\beta \in C^\infty(\Omega)$, $0 \le \beta(x) \le 1$, $\beta(x) = 0$ in some neighborhood of Γ_0, and $\beta(x) = 1$ outside some wider neighborhood of Γ_0. We assume that both of these neighborhoods lie inside Ω.

We need the following simple fact.

LEMMA 3.1. *If $\phi(x)$ lies in one of the sets $D(\widetilde{S}_0), D(\widetilde{S})$, and if $\beta(x)$ is a cut-off function in Ω, then $\beta\phi \in D(S)$.*

PROOF. First of all, we note that $\phi \in D[\widetilde{S}]$ implies $\beta\phi \in D[\widetilde{S}_0]$. Indeed, we have $\beta\phi|_{\Gamma_0} = 0$. Hence, it is sufficient to show that $\beta\phi$ can be approximated in the norm of $W_2^{(1)}(\Omega; a_{ij}, a)$ by functions from $D(\widetilde{S})$ vanishing at infinity. For ϕ such an approximation is certainly possible; let $\{\phi_n\}$ be the corresponding approximating sequence. Then it is easy to see that the sequence $\{\beta\phi_n\}$ approximates the function $\beta\phi$.

Further, suppose that $\phi \in D(\widetilde{S}_0)$. Let us show that $\beta\phi \in D(\widetilde{S})$. We have to verify that for any function $g \in D[\widetilde{S}]$ vanishing at infinity, and some $h \in L_2(\Omega)$, the equality

$$\widetilde{S}[\beta\phi, g] = (h, g)$$

is fulfilled. We have $\phi \in D(\widetilde{S}_0)$ and $\beta\phi \in D[\widetilde{S}_0]$. Hence, putting $\widetilde{S}_0\phi = h_0$, we find that

$$\widetilde{S}[\beta\phi, g] = \int_\Omega \left(a_{ij}(\beta\phi)_i\overline{g}_j + a\beta\phi\overline{g}\right) dx + \int_\gamma \sigma\beta\phi\overline{g}ds$$

$$= \int_\Omega \left(a_{ij}\phi_i(\overline{\beta g})_j + a\phi\overline{\beta g}\right) dx + \int_\gamma \sigma\phi\overline{\beta g}ds + \int_\Omega a_{ij}(\phi\beta_i\overline{g}_j - \phi_i\beta_j\overline{g})dx$$

$$= \widetilde{S}_0[\phi, \beta g] - \int_\Omega \left(a_{ij}\phi_i\beta_j + \frac{\partial}{\partial x_j}(a_{ij}\phi\beta_i)\right)\overline{g}dx = (h, g),$$

where

$$h = \beta h_0 - \frac{\partial}{\partial x_j}(a_{ij}\phi\beta_i) - a_{ij}\phi_i\beta_j \in L_2(\Omega).$$

This argument also shows that $\beta\phi \in D(\widetilde{S}_0)$. Thus, $\beta\phi \in D(S)$. The case $\phi \in D(\widetilde{S})$ is settled similarly. □

LEMMA 3.2. *The operator $\widetilde{S}_0$ is the Friedrichs extension of S.*

PROOF. We have to check that any function $\phi \in D(\widetilde{S}_0)$ admits an approximation in the space $W_2^{(1)}(\Omega; a_{ij}, a)$ by functions from $D(S)$. Let $\beta_n(x)$ be a sequence of cut-off functions such that $\beta_n(x) \neq 1$ only in a neighborhood of Γ_0 of width $1/n$. Take a function $\phi \in D(\widetilde{S}_0)$. Then the functions $\beta_n\phi$ lie in $D(S)$ and converge to ϕ in $W_2^{(1)}(\Omega; a_{ij}, a)$. This was shown by Sobolev; see [**35**], Section 12 (lemma to the uniqueness theorem for the Dirichlet boundary value problem). □

3.2. The next statement is the main result of Section 3.

THEOREM 3.3. *Let $\widetilde{S}$ and $\widetilde{S}_0$ be the operators defined in Subsection 3.1. Then the operator $B = \widetilde{S}^{-1} - \widetilde{S}_0^{-1}$ belongs to class T_α, where $\alpha = 2(m-1)^{-1}$.*

PROOF. Lemma 3.2 allows one to apply the results of Subsection 1.1, and, in particular, Lemmas 1.1 and 1.3. Therefore, the nonzero spectrum of B coincides with the spectrum of the Rayleigh quotient

$$I(u) = \frac{\int_\Omega |u|^2 dx}{\int_\Omega \left(a_{ij}u_i\overline{u}_j + a|u|^2\right) dx + \int_\gamma \sigma|u|^2 ds} \tag{3.1}$$

considered on the set of all functions $u \in D[\widetilde{S}]$, which are orthogonal in the $\widetilde{S}$-metric to the set $D(\widetilde{S}_0)$:

$$\widetilde{S}[u, g] \equiv \int_\Omega \left(a_{ij}u_i\overline{g}_j + au\overline{g}\right) dx + \int_\gamma \sigma u\overline{g}ds = 0, \qquad \forall g \in D(\widetilde{S}_0). \tag{3.2}$$

We start by proving the following lemma, which plays an important role in studying the ratio (3.1).

LEMMA 3.4. *Let ω be some neighborhood of Γ_0, such that $\overline{\omega} \subset \Omega$. Then the functions $u \in D[\widetilde{S}]$, subject to condition* (3.2), *satisfy the inequality*

$$\int_{\Omega} |u|^2 dx \leq c \int_{\omega} |u|^2 dx. \tag{3.3}$$

The constant c depends on Ω and on ω, but does not depend on u.

Proof of Lemma 3.4. Let ω_0 be a narrower neighborhood of Γ_0, so that $\omega_0 \Subset \omega$, and let $\beta(x)$ be a cut-off function such that $\beta(x) = 0$ in ω_0 and $\beta(x) = 1$ outside ω. Suppose that $u(x)$ meets the conditions of this lemma, and take $\phi = \widetilde{S}^{-1}u$. By Lemma 3.1, $\beta\phi \in D(S) \subset D[\widetilde{S}_0]$, and hence, by virtue of (3.2),

$$(u, \widetilde{S}(\beta\phi)) = \widetilde{S}[u, \beta\phi] = 0.$$

From here we find, putting $\alpha(x) = 1 - \beta(x)$:

$$\|u\|^2 = (u, \widetilde{S}\phi) = (u, \widetilde{S}\phi) - (u, \widetilde{S}(\beta\phi)) = (u, \widetilde{S}(\alpha\phi)). \tag{3.4}$$

Since $\widetilde{S}(\alpha\phi) = \alpha u - 2a_{ij}\alpha_i\phi_j - \phi\frac{\partial}{\partial x_i}(a_{ij}\alpha_j)$, we conclude from (3.4) and from the positive definiteness of the operator $\widetilde{S}$ that

$$\begin{aligned}\int_{\Omega} |u|^2 dx &\leq \int_{\Omega} \alpha |u|^2 dx + c_1 \|u\|_{L_2(\omega)} \|\phi\|_{W_2^{(1)}(\omega)} \\ &\leq c_2 \|u\|_{L_2(\omega)} \left(\|u\| + \sqrt{\widetilde{S}[\phi, \phi]} \right) \leq c_3 \|u\| \|u\|_{L_2(\omega)}.\end{aligned}$$

The last inequality immediately leads to (3.3). This completes the proof of Lemma 3.4.

Now we return to the proof of Theorem 3.3. It follows from (3.3) and the inequality (2.9) that the Rayleigh quotient (3.1) can be estimated as follows:

$$I(u) \leq c_0 I_0(u) := c_0 \frac{\int_{\omega} |u|^2 dx}{\int_{\omega} (a_{ij} u_i \overline{u}_j + a|u|^2) dx}. \tag{3.5}$$

The spectrum of the Rayleigh quotient $I_0(u)$ is discrete. This follows from the compactness of the corresponding embedding operator; see, e.g., [**35, 34**]. By the variational (minimax) principle, the same is true for the spectrum of the ratio (3.1). Thus, B is a compact operator.

3.3. Here we prove that $B \in T_\alpha$, $\alpha = 2(m-1)^{-1}$. To this end, we study the spectrum of the Rayleigh quotient $I_0(u)$ in more detail. The next lemma is the core of our argument.

LEMMA 3.5. *Let $\omega' \subset E_m$ be a bounded domain with smooth boundary Γ', and let us introduce the scalar product*

$$[f, g] = \int_{\omega'} (a_{ij} f_i \overline{g}_j + a f \overline{g}) dx \tag{3.6}$$

in the space $W_2^{(1)}(\omega')$. Let $U_2^{(1)}(\omega')$ stand for the subspace of those $u \in W_2^{(1)}(\omega')$ that are orthogonal to all $g \in C_0^2(\omega')$:

$$[u, g] = 0, \qquad \forall g \in C_0^2(\omega'). \tag{3.7}$$

Then, in the space $U_2^{(1)}(\omega')$, *the quadratic form*

$$\int_{\omega'} |u|^2 dx$$

generates a compact operator of the class T_α, $\alpha = 2(m-1)^{-1}$.

Proof of Lemma 3.5. We are dealing with the spectrum of the Rayleigh quotient of the same type as $I_0(u)$, that is, of the ratio

$$\frac{\int_{\omega'} |u|^2 dx}{\int_{\omega'} (a_{ij} u_i \overline{u}_j + a|u|^2) dx} \tag{3.8}$$

considered on the set of functions subject to condition (3.7). This condition means that $u(x)$ is a weak solution of the homogeneous equation $\Lambda u = 0$. If the coefficients $a_{ij}(x)$ and $a(x)$ are smooth enough, then, as is well known, any weak solution of this equation is a classical solution. However, we impose no smoothness conditions on the coefficients.

Choose a covering of the closed domain $\overline{\omega'}$ by a finite number of closed subdomains $\overline{\omega_k}$ with smooth boundary (in general, mutually overlapping). This can be done in such a way that each $\overline{\omega_k}$ is C^2-diffeomorphic to the unit ball. Since

$$\int_{\omega'} |u|^2 dx \le \sum_k \int_{\omega_k} |u|^2 dx,$$

it is sufficient to prove the statement of the lemma for each quadratic form $\int_{\omega_k} |u|^2 dx$ separately; here we rely upon statement 2) in Subsection 1.3. Now, it follows from the variational principle that it suffices to consider the spectrum of the ratio (3.8) in which the domain ω' is replaced by any of the subdomains ω_k. Moreover, we relax the condition (3.7), replacing the set $C_0^2(\omega')$ by a narrower set $C_0^2(\omega_k)$. Then the consecutive maxima of the ratio may only increase. By mapping the domain ω_k onto the unit ball, we transform the expression (3.6) (for ω_k) into an expression of the same type for the ball, with some other coefficients. Finally, the quadratic form $\int_{\omega_k} |u|^2 dx$ is controlled by the similar integral over the ball. All this allows us to conclude that it is sufficient to prove the result for the case where ω' is the unit ball.

Our next step is to show that the desired result for an arbitrary operator (2.3) will follow from its special case where the operator is the Laplacian. Consider the harmonic function $v \in W_2^{(1)}(\omega')$ that satisfies the boundary condition $v|_{\Gamma'} = u|_{\Gamma'}$. It follows from the variational description of u and v as solutions of the corresponding homogeneous equations that $u \mapsto v$ is a one-to-one mapping of the space $U_2^{(1)}(\omega')$ onto the set of all harmonic functions from $W_2^{(1)}(\omega')$. Note that

$$[u,u] \ge c \int_{\omega'} (|\nabla u|^2 + |u|^2) dx, \qquad c > 0,$$

and that v is the minimizer for the integral $\int_{\omega'} |\nabla f|^2 dx$ under the condition $f|_{\Gamma'} = u|_{\Gamma'}$. This yields

$$[u,u] \ge c \int_{\omega'} (|\nabla v|^2 + |u|^2) dx \qquad (\Delta v = 0,\ v|_{\Gamma'} = u|_{\Gamma'}). \tag{3.9}$$

Now, we show that

$$\int_{\omega'} |u|^2 dx \le c_1 \int_{\omega'} |v|^2 dx \le c_2 \int_{\omega'} |u|^2 dx. \tag{3.10}$$

For definiteness, we check the first inequality. Let a function $\phi(x)$ solve the equation $\Lambda\phi = u$ in ω', under the boundary condition $\phi|_{\Gamma'} = 0$. Then, according to Ladyzhenskaya's result [**25**], we have $\phi \in W_2^{(2)}(\omega')$ and

$$\|\phi\|_{W_2^{(2)}(\omega')} \leq c_3\|u\|_{L_2(\omega')}. \tag{3.11}$$

It was shown by Slobodetskij [**32**] that

$$\frac{\partial\phi}{\partial\nu}|_{\Gamma'} \in W_2^{(1/2)}(\Gamma')$$

and that the norm of $\frac{\partial\phi}{\partial\nu}|_{\Gamma'}$ in this space is controlled by the norm of ϕ in $W_2^{(2)}(\omega')$. In addition, it is possible to construct a function $\psi \in W_2^{(2)}(\omega')$ such that

$$\psi|_{\Gamma'} = 0, \ \frac{\partial\psi}{\partial n}|_{\Gamma'} = \frac{\partial\phi}{\partial\nu}|_{\Gamma'}$$

and

$$\|\psi\|_{W_2^{(2)}(\omega')} \leq c_4\|\frac{\partial\phi}{\partial\nu}\|_{W_2^{(1/2)}(\Gamma')}. \tag{3.12}$$

We have

$$\int_{\omega'} |u|^2 dx = \int_{\omega'} u\overline{\Lambda\phi} dx = [u, \phi] - \int_{\Gamma'} u\overline{\frac{\partial\phi}{\partial\nu}} ds. \tag{3.13}$$

Here the term $[u, \phi]$ vanishes due to condition (3.7). Continuing this calculation, we find that

$$\int_{\omega'} |u|^2 dx = -\int_{\Gamma'} v\overline{\frac{\partial\psi}{\partial n}} ds = -\int_{\omega'} v\overline{\Delta\psi} dx \leq c_5\|v\|_{L_2(\omega')}\|\psi\|_{W_2^{(2)}(\omega')}. \tag{3.14}$$

The inequalities (3.12) and (3.11), combined with the fact that the norm of $\frac{\partial\phi}{\partial\nu}$ in the space $W_2^{(1/2)}(\Gamma')$ is controlled by $\|\phi\|_{W_2^{(2)}(\omega')}$, yield the estimate

$$\int_{\omega'} |u|^2 dx \leq c_6\|v\|_{L_2(\omega')}\|u\|_{L_2(\omega')}.$$

Thus, the first inequality in (3.10) is justified. Proof of the second one is similar.

The inequalities (3.9) and (3.10) show us that it suffices to consider the Rayleigh quotient

$$\frac{\int_{\omega'} |v|^2 dx}{\int_{\omega'}(|\nabla v|^2 + |v|^2) dx} \tag{3.15}$$

on harmonic functions. It is a little bit more convenient to analyze the inverse ratio. More exactly, we consider the successive minima of the quadratic form

$$\int_{\omega'} |\nabla v|^2 dx \tag{3.16}$$

under the conditions

$$\int_{\omega'} |v|^2 dx = 1, \qquad \Delta v = 0. \tag{3.17}$$

The term $\int_{\omega'} |v|^2 dx$ in (3.16) is dropped, since it does not affect the rate of growth of the eigenvalues. Expansion of an arbitrary harmonic function in ω' into the series over the system of m-dimensional spherical functions reduces both the quadratic form in (3.16) and the integral $\int_{\omega'} |v|^2 dx$ to the diagonal form. This means that the m-dimensional spherical functions form a complete system of the eigenfunctions of the problem considered. Let v_S be such a function of order S, i.e., $v_S = r^S Y_S$, where Y_S is a surface spherical function. Then

$$\int_{\omega'} |\nabla v_S|^2 dx = \int_{\Gamma'} v_S \frac{\partial \overline{v_S}}{\partial r} ds = S \int_{\Gamma'} |Y_S|^2 ds,$$

$$\int_{\omega'} |v_S|^2 dx = \int_0^1 r^{2S+m-1} dr \int_{\Gamma'} |Y_S|^2 ds = (2S+m)^{-1} \int_{\Gamma'} |Y_S|^2 ds.$$

This shows that the successive minima of the quadratic form (3.16) under the conditions (3.17) are given by the numbers $S(2S+m)$. The multiplicity of each such eigenvalue is equal to

$$\frac{(2S+m-2)(S+m-3)!}{(m-2)!S!}$$

(see [**8**]). For large values of S this quantity behaves asymptotically as $\frac{2}{(m-2)!} S^{m-2}$. This implies that the total number of eigenvalues smaller than a number $\mu > 0$ can be estimated as $O(\mu^{(m-1)/2})$. Returning to the ratio (3.15) and using statement 1) in Subsection 1.3, we complete the proof of Lemma 3.5.

Now we are in a position to finish the proof of Theorem 3.3. We choose a neighborhood ω in Lemma 3.4 in such a way that its boundary is smooth. Evidently, each of the domains $\omega' = \omega \cap \Omega_1$ and $\omega'' = \omega \cap \Omega_2$ also has a smooth boundary. By statement 2) in Subsection 1.3, studying the spectrum of the Rayleigh quotient $I_0(u)$ can be replaced by investigating similar ratios for the domains ω' and ω''. Consider, for instance, the ratio (3.8). Evidently, condition (3.7) is weaker than condition (3.2). Thus, Lemma 3.5 applies and yields the desired result. □

3.4. Theorem 3.3 also allows one to understand how the spectral properties of the inverse operator are affected by adding a new component of the boundary, with the Dirichlet boundary condition imposed on this component. In this connection, we supplement Theorem 3.3 by some additional remarks.

REMARK 3.6. The result, and also the proof of Theorem 3.3, remains true if Γ_0 is taken as the union of a finite family of disjoint compact and smooth surfaces lying inside the domain Ω.

If the components of Γ_0 are only piecewise-smooth, then the above proof of Theorem 3.3 (more exactly, that of Lemma 3.5) fails, though it is possible that the fact itself remains true. At the moment, we can only state that the following, weaker statement holds in this case.

REMARK 3.7. Suppose that under the assumptions of Remark 3.6 the components of Γ_0 are only piecewise-smooth. Then the statement of Theorem 3.3 is valid with $\alpha = 2m^{-1}$ in place of $\alpha = 2(m-1)^{-1}$.

Indeed, let us return to analyzing the spectrum of the Rayleigh quotient $I_0(u)$ appearing in (3.5). We remove the restriction (3.2) and replace the condition $u \in D[\widetilde{S}]$ by the weaker condition $u \in W_2^{(1)}(\omega)$. Then the successive maxima of the ratio $I_0(u)$ may only increase. Under these new restrictions, the spectrum of $I_0(u)$ coincides with the sequence λ_n^{-1}, where λ_n are the eigenvalues of the Neumann problem for the operator (2.3) in ω. Using the classical asymptotic formula for such eigenvalues in a bounded domain, we justify the statement of Remark 3.7.

For $m = 2$, Theorem 3.3, combined with the Rosenblum–Kato theorem, allows one to draw some important conclusions in spectral theory (see Section 6). For $m > 2$, similar conclusions become possible based on the results of Section 4, where we will investigate the difference of the integer powers of the inverse operators. Still, we believe that Theorem 3.3 and its analogues are of independent interest. In this connection, we note that this theorem gives no direct answer to our original question: how varying the boundary conditions affects the properties of the inverse operator. The answer is given by the following statement.

THEOREM 3.8. *Let Ω be an exterior domain with respect to a finite family of closed, piecewise-smooth compact surfaces, and let the selfadjoint operators $\widetilde{S}_1$ and $\widetilde{S}_2$, acting in $L_2(\Omega)$, correspond to two different boundary-value problems of the type* (2.4), (2.5) *for the same differential expression* (2.3) *in Ω. Then*

$$\widetilde{S}_1^{-1} - \widetilde{S}_2^{-1} \in T_\alpha, \qquad \alpha = 2m^{-1}.$$

If the boundary Γ is smooth, a similar inclusion is valid with $\alpha = 2(m-1)^{-1}$.

The first statement is a part of the more general Theorem 4.1; see the next section. The second statement is easy to prove using Lemma 3.5 and taking into account Remark 4.5, also in Section 4.

Note that the second statement of Theorem 3.8, as well as Theorem 3.3, remains meaningful even for bounded domains. Namely, it shows that the operator $\widetilde{S}_1^{-1} - \widetilde{S}_2^{-1}$ belongs to the same class T_α as the inverse to the operator of the Dirichlet (or Neumann, etc.) boundary-value problem in a bounded domain of dimension $m-1$. *The index $\alpha = 2(m-1)^{-1}$ is sharp.* Indeed, consider the operators $\widetilde{S}_1$ and $\widetilde{S}_2$ of two boundary-value problems for the Laplacian in the ball ω. Namely, let $\widetilde{S}_1$ correspond to the Dirichlet condition on Γ, and $\widetilde{S}_2$ correspond to the condition

$$\frac{\partial v}{\partial r} + v|_\Gamma = 0.$$

By Lemma 1.3, the spectrum of the operator $\widetilde{S}_1^{-1} - \widetilde{S}_2^{-1}$ coincides with that of the Rayleigh quotient

$$\frac{\int_\omega |v|^2 dx}{\int_\omega |\nabla v|^2 dx + \int_\Gamma |v|^2 ds},$$

considered on the set of all harmonic functions from $W_2^{(1)}(\omega)$. Using spherical functions in the same way as was done when proving Lemma 3.5, we see that the eigenvalues in this case decay exactly as $n^{-\alpha}$ with $\alpha = 2(m-1)^{-1}$.

3.5. In this last subsection of Section 3 we investigate the following question: how the inverse operator is affected by a change of the coefficients in the differential expression (2.3) in some bounded domain. We always assume that these changes preserve all the conditions imposed in Subsection 2.3. This question is easy to solve by using Theorem 3.3. Indeed, let the operators $\widetilde{S}$ and $\widetilde{S}_0$ be the same as in this theorem. We choose the surface Γ_0 in such a way that the domain where the coefficients are changed lies inside Γ_0. Let $\widetilde{R}$ be the operator, corresponding to the same boundary-value problem (2.4), (2.5), but for the differential expression (2.3) with the new coefficients. Starting from $\widetilde{R}$ and the surface Γ_0, we construct the operator $\widetilde{R}_0$ in the same way as the operator $\widetilde{S}_0$ was constructed in Subsection 3.1. It is important that $\widetilde{R}_0^{-1}$ and $\widetilde{S}_0^{-1}$ differ only by their parts acting in $L_2(\Omega_2)$, since in $L_2(\Omega_1)$ both operators $\widetilde{R}_0$ and $\widetilde{S}_0$ correspond to the Dirichlet problem for the same differential expression. In the space $L_2(\Omega_2)$ the operators $\widetilde{R}_0^{-1}$ and $\widetilde{S}_0^{-1}$ belong to class $T_{2/m}$, which follows from the well-known asymptotic behavior of their eigenvalues. By Theorem 3.3, the passage from $\widetilde{S}^{-1}$ to $\widetilde{S}_0^{-1}$, or from $\widetilde{R}^{-1}$ to $\widetilde{R}_0^{-1}$ corresponds to a perturbation of class T_α with $\alpha = 2(m-1)^{-1}$. So, we conclude that $\widetilde{S}^{-1} - \widetilde{R}^{-1} \in T_{2/m}$. In general, this result is the best possible. However, it can be strengthened if the coefficients $a_{ij}(x)$ are the same for both operators, and only $a(x)$ varies. We now show that in this case

$$\widetilde{S}^{-1} - \widetilde{R}^{-1} \in T_{4/m}.$$

Indeed, suppose that in the differential expression (2.3) the coefficient $a(x)$ is replaced by $b(x) = a(x) + \rho(x)$, where $b(x) \geq 1$ and $\rho(x) = 0$ outside some bounded subdomain $\Omega_2 \subset \Omega$. Without loss of generality, we may assume that the boundary $\Gamma = \partial\Omega$ is a part of $\partial\Omega_2$. The operator $(\widetilde{R} - \widetilde{S})u = \rho u$ is bounded, and the operator $\widetilde{S}^{-1} - \widetilde{R}^{-1}$ can be represented as

$$\widetilde{S}^{-1} - \widetilde{R}^{-1} = \widetilde{S}^{-1}(\widetilde{R} - \widetilde{S})\widetilde{R}^{-1} = \widetilde{S}^{-1}\rho\widetilde{S}^{-1}(\widetilde{S}\widetilde{R}^{-1}).$$

Here the last factor is evidently a bounded operator. Hence, by statement 5) in Subsection 1.3, it is sufficient to study the spectrum of $\widetilde{S}^{-1}\rho\widetilde{S}^{-1}$. Here we can assume that $\rho \geq 0$; then the result for general ρ will follow from statement 3) in Subsection 1.3. For $\rho \geq 0$ the spectrum of $\widetilde{S}^{-1}\rho\widetilde{S}^{-1}$ coincides with that of the Rayleigh quotient

$$\frac{\left(\widetilde{S}^{-1}\rho\widetilde{S}^{-1}h, h\right)}{\|h\|^2}, \qquad h \in L_2(\Omega).$$

Putting here $\widetilde{S}^{-1}h = u$, we arrive at the ratio

$$\frac{\int_{\Omega_2} \rho|u|^2 dx}{\|\widetilde{S}u\|^2}, \qquad u \in D(\widetilde{S}). \tag{3.18}$$

Let $\alpha_0(x)$ be a nonnegative, infinitely differentiable function such that $\alpha_0(x) = 1$ in Ω_2 and $\alpha_0(x) = 0$ outside some wider bounded subdomain $\Omega_0 \subset \Omega$. It is easy to check that

$$\|\widetilde{S}(\alpha_0 u)\| \leq c\|\widetilde{S}u\|, \qquad \forall u \in D(\widetilde{S}). \tag{3.19}$$

Actually, this is a special case of a more general estimate; see inequality (4.9) (for $k = 2$) in Section 4.

The estimate (3.19) shows that the ratio (3.18) is controlled by

$$\frac{\int_{\Omega_0} |\alpha_0 u|^2 dx}{\|\widetilde{S}(\alpha_0 u)\|^2}.$$

Letting here $\alpha_0 u = v$, we arrive at the ratio

$$\frac{\int_{\Omega_0} |v|^2 dx}{\|\widetilde{L}v\|^2}. \tag{3.20}$$

Here $\widetilde{L}$ is the operator of the boundary-value problem in Ω_0, corresponding to the differential expression (2.3), the boundary conditions (2.4) and (2.5) on Γ and the Dirichlet condition on the rest of $\partial\Omega_0$. Next, we relax the restrictions on the class of functions v in (3.20), keeping only the condition $v \in D(\widetilde{L})$. Then the successive maxima of the ratio (3.20) coincide with the spectrum of the selfadjoint operator $\widetilde{L}^{-2}$ in $L_2(\Omega_0)$. Using the well-known asymptotic formula for the eigenvalues of a regular boundary-value problem in a bounded domain, we find that $\widetilde{L}^{-1} \in T_{2/m}$, and hence $\widetilde{L}^{-2} \in T_{4/m}$.

Thus, we have proved the following theorem.

THEOREM 3.9. *Let the domain Ω be the same as in Theorem 3.8, and let a selfadjoint operator $\widetilde{S}$ in $L_2(\Omega)$ correspond to the differential expression (2.3) and the boundary conditions (2.4) and (2.5). Let another selfadjoint operator $\widetilde{R}$ in $L_2(\Omega)$ correspond to the differential expression*

$$Mg = -\frac{\partial}{\partial x_i}\left(b_{ij}(x)\frac{\partial g}{\partial x_j}\right) + b(x)g$$

of type (2.3), and to the same boundary conditions[4] *(2.4) and (2.5). Then, if $b_{ij}(x) = a_{ij}(x)$ and $b(x) = a(x)$ outside some bounded subdomain in Ω, then $\widetilde{S}^{-1} - \widetilde{R}^{-1} \in T_{2/m}$. If $b_{ij}(x) = a_{ij}(x)$ in the whole of Ω, then $\widetilde{S}^{-1} - \widetilde{R}^{-1} \in T_{4/m}$.*

4. Domains with compact boundary, II

4.1. The results of Section 3 are not sufficient for applying the Rosenblum–Kato theorem in dimensions greater than 2. In this connection, one has to study operators of the type $\widetilde{S}_1^{-k} - \widetilde{S}_2^{-k}$, with an integer k. We start from the case where the boundary conditions are varied. Namely, we will prove the following theorem, which generalizes the first statement of Theorem 3.8.

THEOREM 4.1. *Let the assumptions of Theorem 3.8 be satisfied. Suppose, in addition, that in a neighborhood $\widehat{\omega}$ of the boundary Γ the coefficients $a_{ij}(x)$ and $a(x)$ belong to the classes M^r with $r = 2k-1$ and $r = 2k-2$ respectively, for some $k \in \mathbb{N}$. Then*

$$\widetilde{S}_1^{-k} - \widetilde{S}_2^{-k} \in T_{2k/m}. \tag{4.1}$$

Note that the boundary Γ is supposed to be piecewise-smooth, as it was in Theorem 3.8. At the same time, we need the additional smoothness of the coefficients. Apparently, in the case of a smooth boundary, the exponent α can be improved in the spirit of Theorem 3.8. However, such an improvement is useless for applications to spectral theory.

As in Section 3, the proof is based on including the problem into the general scheme of extension theory for symmetric operators. Technical problems arising in the course of the proof will be solved in a series of lemmas.

Define the operator Q as the restriction of any of the operators $\widetilde{S}_1^k$ and $\widetilde{S}_2^k$ onto the set

$$D(Q) = D(\widetilde{S}_1^k) \cap D(\widetilde{S}_2^k).$$

Let us show that the operator Q is symmetric, positive definite, and closable. To this end, it is sufficient to check whether the set $D(Q)$ is dense in $L_2(\Omega)$. This is evident if the coefficients $a_{ij}(x)$ and $a(x)$ are smooth enough in the whole of Ω. In the general case, density is implied by the following lemma.

[4]This means that the sets γ and δ, and also the coefficient $\sigma(x)$, remain the same. However, the expression for the co-normal derivative changes, along with the coefficients $a_{ij}(x)$.

LEMMA 4.2. *Let a nonnegative function* $\beta \in C^\infty(E_m)$, $0 \le \beta(x) \le 1$, *be equal to* 1 *outside some neighborhood* ω *of the boundary* Γ *and vanish in some other neighborhood* ω_0 *of* Γ. *We assume that* $\omega_0 \Subset \omega \Subset \widehat{\omega}$. *Let* $u \in D(\widetilde{S}_1^k)$. *Then* $v := \beta u \in D(Q)$.

PROOF. By the results of Ladyzhenskaya, [**24, 25, 26**], and also those of Guseva [**16**] and Slobodetskij [**33**], the function $u(x)$ belongs to the class $W_2^{(2k)}(\omega_1)$, where $\omega_1 = (\omega \setminus \overline{\omega_0}) \cap \Omega$. Moreover, we have

$$\|u\|_{W_2^{(2k)}(\omega_1)} \le c\|\widetilde{S}_1^k u\|. \tag{4.2}$$

Note that in the papers listed the authors discuss operators in the bounded domains with smooth boundary. However, the results are of a 'local' nature, so that they apply to the case of any strictly interior subdomain of Ω, in which the coefficients are smooth enough. Thus, the estimate (4.2) is justified.

Our next task is to show that the expression $\Lambda^k v$ is well defined. Here the differential expression Λ is understood in the generalized (Sobolev) sense. We also establish the equality

$$\Lambda^r v = \beta \Lambda^r u + M_r u, \qquad r = 1, \dots, k, \tag{4.3}$$

where $M_r u$ is a differential expression of order $2r - 1$ with smooth coefficients vanishing outside ω_1. For $r = 1$ the equality (4.3) follows directly from the fact that $u \in W_2^{(2)}$ in any strictly interior and bounded subdomain of Ω. Now we prove (4.3) for $r+1$, assuming that it is satisfied for some r. We have to check that for an arbitrary function $g \in C_0^2(\Omega)$, the following equality holds:

$$\int_\Omega \Lambda^r v \, \Lambda \overline{g} dx = \int_\Omega h \overline{g} dx, \tag{4.4}$$

where $h = \beta \Lambda^{r+1} u + M_{r+1} u$. By (4.3), we have

$$\begin{aligned}\int_\Omega \Lambda^r v \overline{\Lambda g} dx &= \int_\Omega \beta \Lambda^r u \overline{\Lambda g} dx + \int_\Omega M_r u \overline{\Lambda g} dx \\ &= \int_\Omega \Lambda^r u \overline{\Lambda(\beta g)} dx + 2\int_\Omega \Lambda^r u \cdot a_{ij} \beta_i \overline{g_j} dx + \int_\Omega \Lambda^r u \cdot \overline{g} \frac{\partial}{\partial x_i}(a_{ij}\beta_j) dx \\ &\quad + \int_\Omega M_r u \overline{\Lambda g} dx.\end{aligned}$$

Here only the first term in the rightmost side needs some care. According to the definition of the differential expression Λ in the generalized sense, this term is equal to

$$\int_\Omega \beta \Lambda^{r+1} u \overline{g} dx.$$

In all other terms we actually integrate only over the subdomain ω_1. Integration by parts is possible, since $u \in W_2^{(2k)}(\omega_1)$. Integrals over the boundary do not appear, and the sum of all three integrals reduces to the form

$$\int_\Omega M_{r+1} u \overline{g} dx.$$

It is clear that the differential expression M_{r+1} satisfies all the required properties. Thus, the equality (4.4), and hence also (4.3), is justified.

Now we show that

$$\Lambda^r v \in W_2^{(1)}(\Omega; a_{ij}, a), \qquad r = 0, 1, \dots, k-1. \tag{4.5}$$

The equality (4.3) yields the finiteness of integral (2.1) for the functions $\Lambda^r v$. Further, let $\{\phi_n\}$ be a sequence of functions, vanishing at infinity and approximating $\Lambda^r u$ in $W_2^{(1)}(\Omega; a_{ij}, a)$. Then the functions $\psi_n = \beta\phi_n + M_r u$ also vanish at infinity, and it follows from (4.3) that they approximate $\Lambda^r v$ in the same metric.

Finally, let us show that $v \in D(\widetilde{S}_1^k)$ and $v \in D(\widetilde{S}_2^k)$. For definiteness, we concentrate on the second inclusion. The relations (4.5) and $\Lambda^r v|_\Gamma = 0$ imply that $\Lambda^r v \in D[\widetilde{S}_2]$ for all $r = 0, 1, \dots, k-1$. If $g(x)$ is an arbitrary function from $D[\widetilde{S}_2]$ that vanishes at infinity, then

$$\widetilde{S}_2[\Lambda^r v, g] = \int_\Omega \left(a_{ij}(\Lambda^r v)_i \bar{g}_j + a\Lambda^r v\bar{g}\right) dx, \qquad r = 0, 1, \dots, k-1.$$

Integration by parts reduces the right-hand side of this integral to the form $\int_\Omega h\bar{g}dx$, with some $h \in L_2(\Omega)$. This integration by parts is justified by the relation (4.3) and the estimate (4.2). The equality obtained means that $\Lambda^r v \in D(\widetilde{S}_2)$ for all $r = 0, 1, \dots, k-1$. □

4.2. In order to complete the proof of Theorem 4.1, we need a few more steps. First of all, note that the set $D(Q)$ is dense in $L_2(\Omega)$. Indeed, let us choose a sequence of neighborhoods ω and ω_0, as those used in the proof of Lemma 4.2, and shrinking to the surface Γ. Let $\{\beta(x)\}$ be the sequence of the corresponding cut-off functions. Then for any $u \in D(\widetilde{S}_1^k)$ the functions $\beta(x)u(x)$ lie in $D(Q)$ and approximate u in $L_2(\Omega)$.

Now, let us consider the operators $\widetilde{S}_1^k$ and $\widetilde{S}_2^k$ as two selfadjoint and positive definite extensions of the symmetric operator Q. For brevity, we denote $\widetilde{S}_1^k = \widetilde{Q}_1$, $\widetilde{S}_2^k = \widetilde{Q}_2$, and introduce the Friedrichs extension $\widetilde{Q}_0$ of Q. It is sufficient for us to show that the operators

$$D_1 = \widetilde{Q}_1^{-1} - \widetilde{Q}_0^{-1}, \qquad D_2 = \widetilde{Q}_2^{-1} - \widetilde{Q}_0^{-1}$$

belong to the class $T_{2k/m}$. Both operators are studied in the same way. For definiteness, we consider D_1. By Lemma 1.3, its nonzero spectrum coincides with that of the Rayleigh quotient

$$\frac{\int_\Omega |u|^2 dx}{\left(\widetilde{S}_1^{k/2}u, \widetilde{S}_1^{k/2}u\right)} \tag{4.6}$$

considered for $u \in U \cap D(\widetilde{S}_1^{k/2})$, where U is the set of all solutions of the equation $Q^*u = 0$. To study this ratio, we use the same scheme as that employed in Section 3 for the proof of Remark 3.7. We need two more lemmas.

LEMMA 4.3. *Let ω be a neighborhood of Γ, and let $\omega_2 = \omega \cap U$. Then the inequality*

$$\int_\Omega |u|^2 dx \le c_0 \int_{\omega_2} |u|^2 dx \tag{4.7}$$

holds for any function $u \in U \cap D(\widetilde{S}_1^{k/2})$. The constant factor c_0 depends on Ω and ω, but does not depend on the function $u(x)$.

PROOF. Let $u \in U \cap D(\widetilde{S}_1^{k/2})$, and let $\phi(x)$ be the solution of the equation $\widetilde{Q}_1\phi = u$. Take the same cut-off function $\beta(x)$ as in Lemma 4.2. Then, by this lemma, $\beta\phi \in D(Q)$, and

$$(\widetilde{Q}_1(\beta\phi), u) = (Q(\beta\phi), u) = (\beta\phi, Q^*u) = 0. \tag{4.8}$$

Take $\alpha(x) = 1 - \beta(x) \geq 0$. Then, by (4.8) and (4.3), we get

$$\begin{aligned}\|u\|^2 = (\widetilde{Q}_1\phi, u) &= (\widetilde{Q}_1(\phi), u) - (\widetilde{Q}_1(\beta\phi), u) = (\Lambda^k\phi, u) - (\Lambda^k(\beta\phi), u) \\ &= (\alpha\Lambda^k\phi, u) - (M_k\phi, u) = \int_\Omega \alpha|u|^2 dx - \int_\Omega M_k\phi\overline{u}dx.\end{aligned}$$

In the estimation below we adopt the notation introduced in the course of the proof of Lemma 4.2. We have

$$\|u\|^2 \leq \|u\|^2_{L_2(\omega_2)} + \|u\|_{L_2(\omega_1)}\|\phi\|_{W_2^{(2k-1)}(\omega_1)} \leq \|u\|_{L_2(\omega_2)}\left(\|u\| + \|\phi\|_{W_2^{(2k-1)}(\omega_1)}\right),$$

whence, by (4.2),

$$\|u\|^2 \leq c_1\|u\|_{L_2(\omega_2)}\|u\|.$$

□

Our next lemma will allow us to reduce the study of the ratio (4.6) to an investigation of a similar ratio in a bounded domain.

LEMMA 4.4. *Let ω and $\omega_3 \supset \overline{\omega}$ be two neighborhoods of Γ, and let $\alpha_1 \in C^\infty(E_m)$ be a function such that $0 \leq \alpha_1(x) \leq 1$, $\alpha_1(x) = 0$ outside ω_3, and $\alpha_1(x) = 1$ in ω. Then for any function $\phi \in D(\widetilde{S}_1^{k/2})$, the inequality*

$$\|\widetilde{S}_1^{k/2}(\alpha_1\phi)\| \leq c_2\|\widetilde{S}_1^{k/2}\phi\| \tag{4.9}$$

holds,[5] *with a constant c_2 depending on Ω, ω_3, ω, and on k, but not on ϕ.*

PROOF. Denote $\omega_4 = (\omega_3 \setminus \overline{\omega}) \cap \Omega$, and note that ω_4 is a domain of the same type as the domain ω_1 appearing in the course of the proof of Lemma 4.2. Hence, the inequality (4.2) holds in ω_4. The equality (4.3) also holds, with $\beta(x)$ replaced by $\beta_1(x) = 1 - \alpha_1(x)$. We may assume that the coefficients in the differential expressions M_r vanish outside ω_4.

For k odd and k even our argument is different. We start with the case of k even, $k = 2\nu$. From (4.3) and (4.2) we derive that

$$\begin{aligned}\|\widetilde{S}_1^\nu(\alpha_1\phi)\| &= \|\Lambda^\nu(\alpha_1\phi)\| \leq \|\alpha_1\Lambda^\nu\phi\| + \|M_\nu\phi\| \\ &\leq \|\Lambda^\nu\phi\| + c_3\|\phi\|_{W_2^{(2\nu-1)}(\omega_4)} \leq c_2\|\widetilde{S}_1^\nu\phi\|,\end{aligned}$$

and we are done.

Now, let $k = 2\nu + 1$ be an odd number. We have

$$\|\widetilde{S}_1^{k/2}(\alpha_1\phi)\|^2 = \int_\Omega \left(a_{ij}\frac{\partial}{\partial x_i}(\Lambda^\nu\alpha_1\phi) \cdot \frac{\partial}{\partial x_j}(\overline{\Lambda^\nu\alpha_1\phi}) + a|\Lambda^\nu(\alpha_1\phi)|^2\right) dx.$$

---

[5] We assume that ω_3 is contained in the vicinity of Γ in which the coefficients $a_{ij}(x)$ and $a(x)$ are smooth enough. For $k = 1, 2$ this requirement is redundant.

By using equality (4.3) for $r = \nu$ and estimating the integrals, we obtain

$$\begin{aligned}\|\widetilde{S}_1^{k/2}(\alpha_1\phi)\|^2 \le& \int_\Omega \left(a_{ij}\frac{\partial}{\partial x_i}(\alpha_1\Lambda^\nu\phi)\cdot\frac{\partial}{\partial x_j}(\overline{\alpha_1\Lambda^\nu\phi}) + a\alpha_1^2|\Lambda^\nu\phi|^2\right)dx \\ &+c_4\|\phi\|^2_{W_2^{(2\nu)}(\omega_4)} + c_5\|\Lambda^\nu\phi\|^2_{W_2^{(1)}(\omega_4)} \\ \le& \int_\Omega \alpha_1^2\left(a_{ij}\frac{\partial}{\partial x_i}(\Lambda^\nu\phi)\cdot\frac{\partial}{\partial x_j}(\overline{\Lambda^\nu\phi}) + a|\Lambda^\nu\phi|^2\right)dx \\ &+c_6\|\Lambda^\nu\phi\|^2_{W_2^{(1)}(\omega_4)} + c_4\|\phi\|^2_{W_2^{(2\nu)}(\omega_4)}.\end{aligned}$$

By (4.2), the last term in the right-hand side is controlled by $\|\Lambda^\nu\phi\|^2$. Hence, we arrive at the inequality

$$\|\widetilde{S}_1^{k/2}(\alpha_1\phi)\| \le c\|\Lambda^\nu\phi\|_{W_2^{(1)}(\Omega;a_{ij},a)}.$$

This is equivalent to the desired estimate (4.9). □

4.3. Now it is easy to complete the proof of Theorem 4.1. We assume that the neighborhood ω of Γ, appearing in the conditions of Lemma 4.4, is the same as in Lemma 4.3; the domain ω_3 and the function $\alpha_1(x)$ are the same as in Lemma 4.4, and $\omega_5 = \omega_3 \cap \Omega$. Then the inequalities (4.7) and (4.9), in combination with the variational principle, show that the ratio (4.6) can be replaced by

$$\frac{\int_{\omega_5}|\alpha_1 u|^2 dx}{\left\|\widetilde{S}_1^{k/2}(\alpha_1 u)\right\|^2}. \tag{4.10}$$

Consider a selfadjoint operator $\widetilde{L}$ in $L_2(\omega_5)$, which is generated by the differential expression Λ and by the boundary conditions that we now describe. On Γ these boundary conditions are the same as those for the operator $\widetilde{S}_1$. On the part of $\partial\omega_5$ lying inside Ω we impose the Dirichlet boundary condition. By the same reasoning as used for the proof of Lemma 4.2, we conclude that the function $g = \alpha_1 u$ lies in $D(\widetilde{L}^{k/2})$, and the ratio (4.10) can be written as

$$\frac{\|g\|^2_{L_2(\omega_5)}}{\left\|\widetilde{L}^{k/2}g\right\|^2_{L_2(\omega_5)}}. \tag{4.11}$$

The successive maxima of this ratio may only increase, if we subject the functions g in (4.11) to a single restriction, namely, $g \in D\left(\widetilde{L}^{k/2}\right)$. Under this restriction, the spectrum of the Rayleigh quotient (4.11) coincides with that of the operator $\widetilde{L}^{-k}$. Now the inclusion $\widetilde{L}^{-1} \in T_{2/m}$ yields $\widetilde{L}^{-k} \in T_{2k/m}$. The proof of Theorem 4.1 is complete.

REMARK 4.5. If $k = 1$, the above proof does not need Lemma 4.4. Based on Lemma 4.3, we reduce the problem to studying the ratio

$$\frac{\int_{\omega_2}|u|^2dx}{\int_{\omega_2}(a_{ij}u_i\overline{u_j} + a|u|^2)dx}$$

under the condition of the same type as in (4.6). If the boundary Γ is smooth, Lemma 3.5 immediately implies the second statement of Theorem 3.8.

4.4. Here we discuss the effect of varying the boundary. Relying on Theorem 4.1, we may restrict ourselves to the case of the Dirichlet boundary condition. Further, it is sufficient to compare the operator of the Dirichlet boundary value problem in Ω with the operator generated by the differential expression (2.3) in the whole of E_m. The following theorem solves the problem.

THEOREM 4.6. *Let Ω be an exterior domain with respect to a finite family of closed, piecewise-smooth surfaces. Let $\Gamma = \partial\Omega$ and $\Omega_0 = E_m \setminus \overline{\Omega}$. Suppose that in a neighborhood of Γ the coefficients $a_{ij}(x)$ and $a(x)$ of the differential expression* (2.3) *belong to the classes M^r with $r = 2k-1$ and $r = 2k-2$ respectively, for some $k \in \mathbb{N}$. Let $\widetilde{S}$ stand for the operator generated by the differential expression* (2.3) *in E_m, and let $\widetilde{S}_0$ be the orthogonal sum in $L_2(E_m)$ of the operators generated by the same differential expression* (2.3) *in $L_2(\Omega)$ and in $L_2(\Omega_0)$ with the Dirichlet boundary conditions on Γ. Then*

$$\widetilde{S}^{-k} - \widetilde{S}_0^{-k} \in T_{2k/m}. \tag{4.12}$$

The proof is quite similar to the proof of Theorem 4.1, and we only outline it. The operators $\widetilde{S}^k$ and $\widetilde{S}_0^k$ are considered as extensions of the operator Q, defined on the set $D(Q) = D(\widetilde{S}^k) \cap D(\widetilde{S}_0^k)$ by the equality $Qg = \widetilde{S}^k g = \widetilde{S}_0^k g$. An auxiliary statement, which is an analogue of Lemma 4.2, can be obtained in the same way as its prototype. After this is done, an investigation of the spectrum of the operator (4.12) is reduced to the study of the ratios

$$\frac{\int_\Omega |u|^2 dx}{\|\widetilde{S}^{k/2} u\|^2}, \qquad \frac{\int_\Omega |u|^2 dx}{\|\widetilde{S}_0^{k/2} u\|^2},$$

for $u \in U^{(1)} \cap D(\widetilde{S}^{k/2})$ and $u \in U^{(1)} \cap D(\widetilde{S}_0^{k/2})$, respectively. Here $U^{(1)}$ stands for the set of all solutions of the equation $Q^* u = 0$. For the proof, one needs the appropriate analogues of Lemmas 4.3 and 4.4. Their formulations are clear, and their proofs remain basically the same. After this is achieved, the same argument as in Subsection 4.3 allows one to finish the proof.

5. Domains with noncompact boundary. Degeneration of ellipticity

5.1. A certain part of the results of the two previous sections can be easily extended to the case where the boundary $\Gamma = \partial\Omega$ is noncompact. First of all, we mention the case where *an isolated component of Γ is changed (or added), or where the boundary condition varies on such a component. Then Theorems* 3.3, 3.8, 4.1, *and* 4.6, *with some obvious changes in their formulations, remain valid.* Moreover, their proofs do not require any modification.

The situation becomes more complicated if the changes concern an unbounded component of Γ. For definiteness, below we discuss an analogue of Theorem 3.3. As in this theorem, we assume that the boundary is piecewise-smooth. Let $\widetilde{S}$ be the operator of some boundary-value problem of the type (2.4), (2.5) for the differential expression (2.3) in Ω. The meaning of this is the same as in Subsection 2.3. Let $\Gamma' \subset E_m$ be a smooth closed surface and let $\Gamma_0 = \Gamma' \cap \Omega$. We assume that Γ_0 divides Ω into two parts Ω_1 and Ω_2, and that the set $\overline{\Omega_2}$ is compact and its boundary contains the set γ. The surface Γ' can be chosen in such a way that both boundaries $\partial\Omega_1$ and $\partial\Omega_2$ are piecewise-smooth and the distance between Γ_0 and γ is positive. In addition, one can assume that Γ_0 has nonempty intersections only with a finite number of components of Γ. On both sides of Γ_0 we impose the additional Dirichlet boundary condition and define the operator $\widetilde{S}_0$ in the same way as was done in Subsection 3.1. This operator corresponds to the orthogonal sum of two boundary-value problems acting in $L_2(\Omega_1)$ and $L_2(\Omega_2)$. Further, in a neighborhood of Γ_0 we construct a cut-off function $\beta(x)$. We assume that the distance between γ and the set where $\beta(x) \neq 1$ is positive. Then it is easy to extend Lemmas 3.1 and 3.2 to this new situation. We only note that the term

$$\int_\Omega a_{ij}\phi\beta_i\overline{g_j}dx,$$

appearing in the course of the proof of Lemma 3.1, after integration by parts, reduces to the form

$$-\int_\Omega \frac{\partial}{\partial x_j}(a_{ij}\phi\beta_i)\cdot\overline{g}dx + \int_\Gamma \phi\frac{\partial\beta}{\partial\nu}\overline{g}ds. \tag{5.1}$$

In a similar calculation in Subsection 3.1, the integral over Γ was equal to zero due to the condition $\frac{\partial\beta}{\partial\nu}|_\Gamma = 0$. Now the latter equality is violated on the part of Γ that is close to $\Gamma\cap\Gamma_0$. However, this part lies in δ, which means that there $\phi = 0$. Therefore, the integral over Γ in (5.1) vanishes, as before.

Lemma 3.4 also remains true in our situation. For the proof, we only need to specify the set ω in inequality (3.3). Namely, we take ω as the intersection of Ω with some, narrow enough, vicinity of Γ_0. As a result, we reduce our task to the study of the Rayleigh quotient

$$I_0(u) = \frac{\int_\omega |u|^2dx}{\int_\omega(a_{ij}u_i\overline{u_j} + a|u|^2)dx} \tag{5.2}$$

under the conditions $u \in D[\widetilde{S}]$ and

$$\int_\omega (a_{ij}u_i\overline{g_j} + au\overline{g})dx = 0, \tag{5.3}$$

where $g \in D[\widetilde{S}_0]$ is an arbitrary function such that $g(x) = 0$ outside ω. However, an attempt to investigate the ratio (5.2) with the help of Lemma 3.5 fails, since the domains $\omega' = \Omega_1\cap\omega$ and $\omega'' = \Omega_2\cap\omega$ have boundaries that are only piecewise-smooth (rather than smooth), even if Γ is smooth. Still, the argument used in Section 3 for justifying Remark 3.7 does work, and we come to the following theorem.

THEOREM 5.1. *Let $\widetilde{S}$ and $\widetilde{S}_0$ be the operators defined above in this section. Then $B := \widetilde{S}^{-1} - \widetilde{S}_0^{-1} \in T_{2/m}$.*

In spite of the above-mentioned obstacles, under the assumptions of Theorem 5.1, the class $T_{2/m}$ can be replaced by T_α with $\alpha = 2(m-1)^{-1}$. This refinement is important for applications in spectral theory only if $m = 2$. Below we restrict ourselves to this case, in view of its greater simplicity.

THEOREM 5.2. *Let $m = 2$. Then, under the assumptions of Theorem 5.1, we have $B \in T_2$.*

PROOF. We may assume that Γ_0 intersects the curve $\Gamma = \partial\Omega$ in the points where the latter is smooth, and that these curves are orthogonal to each other at the points of intersection. Fix one of these corner points. By an appropriate change of coordinates, one can reduce the general case to the situation where the parts of Γ and Γ_0 near this point are some intervals parallel to the coordinate axes.

Clearly, it is sufficient to estimate the eigenvalues of the ratio (5.2) for the particular case where in (5.2) and (5.3) the sub-domain ω is replaced by a small square ω_0', adjoining the corner point.

Along with the differential expression (2.3), let us consider the similar expression Λ^0 in which all the coefficients a_{ij} with $i \neq j$ are replaced by zero. In our case $(m = 2)$ this means that

$$\Lambda^0 g = -\frac{\partial}{\partial x_1}\left(a_{11}\frac{\partial g}{\partial x_1}\right) - \frac{\partial}{\partial x_2}\left(a_{22}\frac{\partial g}{\partial x_2}\right) + ag. \tag{5.4}$$

Let $[f, g]^0$ stand for the scalar product of the form (3.6) generated by the differential expression (5.4), and let $\frac{\partial}{\partial \nu^0}$ denote the corresponding co-normal derivative. Define a function v as the solution of the boundary value problem $\Lambda^0 v = 0$ in ω_0', $v|_{\partial\omega_0'} = u|_{\partial\omega_0'}$. Let us show that the two-sided inequality (3.10) is satisfied for u and this new function v. Again, we check only the first inequality in (3.10). To this end, consider the function ϕ satisfying the equation $\Lambda\phi = u$ in ω_0' and the boundary condition $\phi|_{\partial\omega_0'} = 0$. Then, by the result of **[25]**, $\phi \in W_2^{(2)}(\omega_0')$ and the estimate (3.11) is valid, in spite of the fact that the boundary of ω_0' is nonsmooth. Note that, due to the boundary condition, we have

$$\frac{\partial \phi}{\partial \nu}|_{\partial\omega_0'} = \frac{\partial \phi}{\partial \nu^0}|_{\partial\omega_0'}.$$

Now, a calculation parallel to (3.13), (3.14) gives:

$$\int_{\omega_0'} |u|^2 dx = \int_{\omega_0'} u\overline{\Lambda\phi} dx = [u, \phi] - \int_{\partial\omega_0'} u\frac{\overline{\partial\phi}}{\partial\nu} ds = -\int_{\partial\omega_0'} u\frac{\overline{\partial\phi}}{\partial\nu} ds$$

and further,

$$-\int_{\partial\omega_0'} u\frac{\overline{\partial\phi}}{\partial\nu} ds = -\int_{\partial\omega_0'} v\frac{\overline{\partial\phi}}{\partial\nu^0} ds = [v, \phi]^0 - \int_{\partial\omega_0'} v\frac{\overline{\partial\phi}}{\partial\nu^0} ds = \int_{\omega_0'} v\overline{\Lambda_0\phi} dx.$$

Note that $[u, \phi] = [v, \phi]^0 = 0$, in view of the conditions $\Lambda u = \Lambda^0 v = 0$. Now we conclude the proof of (3.10) exactly as was done in Lemma 3.5. So, we managed to avoid using the class $W_2^{(1/2)}(\partial\omega_0')$.

Absence of the off-diagonal terms in (5.4) allows one to apply the symmetry principle for extending $v(x)$ across the boundary Γ. More exactly, let Γ' be the side of the square ω_0' that lies on Γ. Denote by ω_1' the square obtained from ω_0' by reflection in Γ', and by $\widetilde{\omega}'$ the rectangle $\omega_0' \cup \omega_1' \cup \Gamma'$. Since $v \in D[\widetilde{S}]$, the function $v(x)$ vanishes on Γ'. We extend $v(x)$ from ω_0' to $\widetilde{\omega}'$ as an odd function and the coefficients $a_{11}(x)$, $a_{22}(x)$, and $a(x)$ as even functions. Then $v \in W_2^{(1)}(\widetilde{\omega}')$, and the extended coefficients belong to the classes M^1 and M^0 respectively. It is also easy to see that the condition of type (5.3) remains valid:

$$\int_{\widetilde{\omega}'} (a_{11}v_1\overline{g_1} + a_{22}v_2\overline{g_2} + av\overline{g})dx = 0, \qquad g \in C_0^2(\widetilde{\omega}'). \tag{5.5}$$

All this allows one to reduce the problem to studying the ratio of type (5.2) (with $a_{12} = a_{21} = 0$) for the domain $\widetilde{\omega}'$ under the condition (5.5). The corner point now lies on the smooth part of $\partial\widetilde{\omega}'$. Thus, Lemma 3.5 applies and this finishes the proof. □

It is easy to see that for domains with noncompact boundary an analogue of the first statement in Theorem 3.8 also remains true. However, some additional technical obstacles arise when one attempts to extend Theorems 4.1 and 4.6 to this case. We will not discuss this matter here, and in the next Section 6 we restrict ourselves to applications of Theorems 5.1 and 5.2, whose proofs were given above.

5.2. Now, let us assume that the ellipticity condition is violated on some compact subset of Ω. For definiteness, we consider the case where the quadratic form

$$a_{ij}(x)\xi_i\overline{\xi_j} \tag{5.6}$$

degenerates on a subset of Γ. Then the essential spectrum may appear even if Ω is a bounded domain (see [**28**]). Below we discuss this case in more detail. However, our argument can be easily extended to a wide class of problems in unbounded domains.

So, let $\Omega \subset E_m$ be a bounded domain with piecewise-smooth boundary Γ, and let an elliptic differential expression (2.3) be defined in Ω. We assume that the quadratic form (5.6) degenerates on a subset $\Gamma^0 \subset \Gamma$; the case $\Gamma^0 = \Gamma$ is not excluded. We also assume that the set $\Gamma \setminus \Gamma^0$ splits into the union of two disjoint subsets, $\Gamma \setminus \Gamma^0 = \gamma \cup \delta$, and that the distance between γ and Γ^0 is positive. Let $\sigma(x)$ be a bounded measurable function on γ. We consider the quadratic form

$$\widetilde{S}[g,g] = \int_\Omega \left(a_{ij}g_i\overline{g_j} + a|g|^2\right) + \int_\gamma \sigma|g|^2 ds. \tag{5.7}$$

Originally, we define it on the set of all functions $g \in W_2^{(1)}(\Omega)$ that are equal to zero on δ and vanish near Γ^0, and then we take the closure of this quadratic form. The closed quadratic form (5.7) is bounded from below and generates a selfadjoint operator $\widetilde{S}$ in $L_2(\Omega)$ which corresponds to the boundary-value problem (2.4), (2.5) for the differential expression (2.3) in Ω. It is easy to understand that in further construction the role of the set Γ^0 is the same as the one played by the point at infinity for domains with noncompact boundary. If a smooth surface Γ_0 divides Ω into two parts, and $\Gamma_0 \cap \Gamma \subset \delta$, then the operator $\widetilde{S}_0$ can be constructed in the same way as was done in Subsections 3.1 and 5.1. The argument of Subsection 5.1 also works in this new situation, and we arrive at the following statement.

THEOREM 5.3. *The statements of Theorems* 5.1 *and* 5.2 *remain valid for the operators* $\widetilde{S}$ *and* $\widetilde{S}_0$ *defined in this subsection.*

If $\Gamma^0 = \Gamma$, then in Theorems 5.1 and 5.2 any closed smooth surface, lying inside Ω, can be taken as Γ_0. This allows one to analyze the case where a new isolated component of $\partial\Omega$ is added, provided that the quadratic form (5.6) does not degenerate on this component. In addition, Theorems 3.3, 3.8, 4.1, and 4.6 extend to this case in the way described in Subsection 5.1.

6. Theorems on the spectrum

6.1. The results of this section are immediate consequences of theorems proved in Sections 3, 4, and 5 in combination with the Weyl and the Rosenblum–Kato theorems.

THEOREM 6.1. *Let* $\Omega \subset E_m$ *be a domain with piecewise-smooth boundary, and let* $\widetilde{S}$ *be the operator in* $L_2(\Omega)$, *corresponding to the boundary-value problem* (2.4),

(2.5) *for the elliptic differential expression* (2.3) *in* Ω. *Then varying a compact portion of the boundary* $\partial\Omega$ *or replacing conditions* (2.4), (2.5) *by other conditions of the same type does not affect the essential spectrum of* $\widetilde{S}$.

PROOF. The result easily follows from Theorem 5.1 and Weyl's theorem. Indeed, the surface Γ_0 (see Subsection 5.1) can be chosen in such a way that all changes in the data lie in the domain Ω_2. On the other hand, by Theorem 5.1, the operator $\widetilde{S}^{-1} - \widetilde{S}_0^{-1}$ is compact, and hence the essential spectra of $\widetilde{S}^{-1}$ and $\widetilde{S}_0^{-1}$ are the same. Then the essential spectra of the operators $\widetilde{S}$ and $\widetilde{S}_0$ also coincide. Further, the part of $\widetilde{S}_0$ acting in the bounded domain Ω_2 has a discrete spectrum. This shows that the essential spectrum of the operator $\widetilde{S}$ coincides with that of the operator corresponding to the Dirichlet problem in Ω_1. Therefore, it does not depend on the shape of the domain Ω_2 and on the boundary conditions on $\partial\Omega_2$. □

A short version of this proof was published in the author's paper [**3**], where some additional restrictions on the behavior of the coefficients $a_{ij}(x)$ as $|x| \to \infty$ were imposed. As was mentioned in the Introduction, similar results for the operator $-\Delta u + q(x)u$ in the exterior of a bounded domain in E_3 were obtained by Povzner [**29**] and Glazman [**14**]. Note that these authors did not assume that the potential $q(x)$ is bounded from below.

A similar result, which is given below, is implied by Theorem 5.3. A short communication on this result was published in [**3**]. The case where the quadratic form (5.6) degenerates in an interior point of Ω was also analyzed there.

THEOREM 6.2. *Let* $\Omega \subset E_m$ *be a bounded domain with piecewise-smooth boundary* Γ, *and let the elliptic differential expression* (2.3) *degenerate on a portion* Γ^0 *of the boundary* Γ. *Let* $\widetilde{S}$ *be the operator in* $L_2(\Omega)$ *defined by the differential expression* (2.3) *and the boundary conditions* (2.4), (2.5); *we assume that the subset* $\gamma \subset \Gamma$, *on which the condition* (2.5) *is imposed, lies at a positive distance from* Γ^0. *Then varying a portion of the boundary* Γ *lying at a positive distance from* Γ^0, *or changing the boundary conditions on such a portion does not affect the essential spectrum of* $\widetilde{S}$.

More advanced results on the spectrum can be obtained on the basis of the Rosenblum–Kato theorem.

THEOREM 6.3. *Let the conditions of Theorem* 6.1 *or Theorem* 6.2 *be fulfilled for* $m = 2$. *Then the absolutely continuous part of the operator* $\widetilde{S}$ *is stable, within the unitary equivalence.*

PROOF. For definiteness, we give the proof under the assumptions of Theorem 6.1. By Theorem 5.2, the operator $\widetilde{S}^{-1} - \widetilde{S}_0^{-1}$ is trace class. By the Rosenblum–Kato theorem, the absolutely continuous parts of the operators $\widetilde{S}^{-1}$ and $\widetilde{S}_0^{-1}$ are unitarily equivalent, and hence the same is true for $\widetilde{S}$ and $\widetilde{S}_0$. The absolutely continuous part of $\widetilde{S}_0$ is nothing but that of the operator corresponding to the Dirichlet boundary value problem in Ω_1. This part does not depend on the shape of the domain Ω_2 and on the boundary conditions on $\partial\Omega_2$. □

COROLLARY 6.4. *Let* $m = 2$, *and suppose that the coefficients* $a_{ij}(x)$ *and* $a(x)$ *in the differential expression* (2.3) *are changed in a bounded subdomain of* Ω. *Under the assumptions of Theorem 6.2, we also suppose that this subdomain lies on a positive distance from* Γ^0. *Then the absolutely continuous part of the operator* $\widetilde{S}$ *is stable with respect to such a change, within the unitary equivalence.*

Indeed, Theorem 6.3 shows that the boundary of Ω can be altered in such a way that the set on which the coefficients are subject to changes lies outside Ω.

6.2. We obtain an analogue of Theorem 6.3 for $m > 2$ by applying Theorems 4.1 and 4.6. Here we are forced to restrict ourselves by the domains with compact boundary. The operators (4.1) and (4.12), appearing in the formulation of these theorems, are trace class when $2k > m$. This fact, combined with the Rosenblum–Kato theorem, immediately yields the following basic result.

THEOREM 6.5. *Let* $\Omega \subset E_m$, $m > 2$, *be an exterior domain with respect to a finite family of closed, piecewise-smooth surfaces. Let* $\widetilde{S}$ *be the selfadjoint operator in* $L_2(\Omega)$, *corresponding to a boundary-value problem* (2.4), (2.5) *for an elliptic differential expression* (2.3). *Assume that in some neighborhood of the boundary* $\Gamma = \partial\Omega$ *the coefficients* $a_{ij}(x)$ *and* $a(x)$ *belong to the classes* M^{2k-1} *and* M^{2k} *respectively, where* $k = [m/2] + 1$. *Then varying the boundary* Γ *(such that the above assumptions about the coefficients are fulfilled also near the new boundary), or varying the boundary conditions of type* (2.4), (2.5) *does not affect the absolutely continuous part of the operator, within the unitary equivalence.*

COROLLARY 6.6. *Under the assumptions of Theorem* 6.5, *changing the coefficients* $a_{ij}(x)$ *and* $a(x)$ *in a bounded subdomain of* Ω *does not affect (within the unitary equivalence) the absolutely continuous part of the operator* $\widetilde{S}$, *provided that in some neighborhood of this subdomain the coefficients satisfy the smoothness assumptions of Theorem* 6.5.

The proof is evident.

Below we point out an important particular case where these additional smoothness conditions are unnecessary. Namely, instead of the general operator (2.3), let us consider the Schrödinger operator

$$-\Delta u + q(x)u \tag{6.1}$$

in E_3. If $q \notin M^2$, we can change the potential q in an arbitrary bounded subdomain to make it smooth enough. Theorem 3.9 shows that under this procedure the inverse operator is perturbed by a trace class operator. Applying Theorem 6.5 and returning to the original potential $q(x)$, we arrive at the following statement.

THEOREM 6.7. *Let* $\Omega \subset E_3$ *be an exterior domain with respect to a finite family of closed, piecewise-smooth surfaces, and let* $\widetilde{S}$ *be the selfadjoint operator in* $L_2(\Omega)$, *corresponding to a boundary-value problem* (2.4), (2.5) *for the Schrödinger operator* (6.1), *with the measurable potential* $q(x) \geq 1$ *which is bounded in each ball. Then varying the boundary* Γ, *or the potential* $q(x)$, *or the boundary conditions of type* (2.4), (2.5) *in a bounded subdomain of* Ω, *does not affect the absolutely continuous part of the operator, within the unitary equivalence.*

Let us add one more remark.

REMARK 6.8. The results of this subsection extend to the domains with noncompact boundary Γ, provided that the change of the boundary or the boundary conditions concerns only a compact and isolated component of Γ. A similar statement also holds for the problems with the degeneration of ellipticity at the boundary, provided that the changes concern only an isolated component of the boundary on which there is no degeneration.

This follows from the fact that Theorems 3.3, 3.8, 4.1, and 4.6 extend to the situation considered. The possibility of such an extension was mentioned at the beginning of Section 5,

6.3. Finally, we discuss a problem of a somewhat different nature. Unlike our former assumptions, we consider the case where $a(x) \geq 0$, admitting the possibility that $a(x) \equiv 0$. Let the operator $\widetilde{S}$ correspond to the differential expression (2.3) and the boundary conditions (2.4), (2.5) in a domain Ω with piecewise-smooth boundary, in general, unbounded. If $\sigma(x) \geq 0$, the operator $\widetilde{S}$ is evidently nonnegative. Otherwise, its spectrum may have a nonempty negative part. By Theorem 6.1, this part is always discrete. The next theorem makes this statement more precise.

THEOREM 6.9. *The negative spectrum of the operator $\widetilde{S}$ can consist only of a finite number of isolated eigenvalues of finite multiplicity.*

PROOF. The variational argument shows that it suffices to give the proof for the case where $a(x) = 0$ and $\sigma(x) \leq 0$. Then the quadratic form $\widetilde{S}[g, g]$ is

$$\widetilde{S}[g, g] = \int_{\Omega} a_{ij} g_i \overline{g_j} dx + \int_{\gamma} \sigma |g|^2 ds, \qquad g|_{\delta} = 0.$$

Along with $\widetilde{S}$, let us consider the positive operator $\widetilde{S}_1$ of the same type as $\widetilde{S}$ but with $\sigma(x) = 1$. Then

$$\widetilde{S}_1[g, g] = \int_{\Omega} a_{ij} g_i \overline{g_j} dx + \int_{\gamma} |g|^2 ds,$$

$$\widetilde{S}[g, g] = \widetilde{S}_1[g, g] - \int_{\gamma} \widetilde{\sigma} |g|^2 ds,$$

where $\widetilde{\sigma}(x) = 1 - \sigma(x) \geq 1$. A result known in perturbation theory (see Lemma 3 and the consequent corollary in [**5**] for a short exposition, and Theorem 1.4 in [**6**] for the details) shows that it suffices to check the following property: any set of functions $g \in D[\widetilde{S}]$, bounded in the metric $\widetilde{S}_1[g, g]$, is pre-compact in the metric $\int_{\gamma} \widetilde{\sigma} |g|^2 ds$. Since the set γ is supposed to be bounded, the required property holds by virtue of the compactness of the embedding $W_2^{(1)}(\Omega)$ into $L_2(\gamma)$; see [**35**]. □

7. Appendix. On the existence of the wave operators

The Rosenblum–Kato theorem (see Subsection 1.4) is closely connected with the existence of the so-called wave operators (see [**30, 17, 18**] and [**21**]). Namely, under the assumptions of this theorem the strong limits exist,

$$W_{\mp}(C, A) = \text{s-}\lim_{t \to \pm\infty} e^{itC} e^{-itA} P_a, \tag{7.1}$$

where P_a is the orthogonal projection onto the subspace $H_a(A)$. Moreover, each of the wave operators $W_{\pm}$ maps $H_a(A)$ onto $H_a(C)$ isometrically and establishes the unitary equivalence of the operators A_a and C_a. However, the Rosenblum–Kato theorem says nothing about the case where the inverse operators A^{-1}, C^{-1}, or some other functions of A and C meet the conditions of this theorem, rather than the operators themselves. At the same time, the existence of the limits (7.1) is important for many applications, in particular, in quantum scattering theory. In this connection, Kuroda [**22, 23**] got some refinement of the Rosenblum–Kato conditions of the existence of the wave operators. However, Kuroda's results require coincidence of the domains of the operators, or, at least, of their quadratic forms. Such results cannot be applied to the problems where the boundary $\Gamma = \partial\Omega$ is varied.

Recently the author proved the following theorem. For operators with bounded inverse it contains the results by Kuroda.

THEOREM 7.1. *Let the selfadjoint operators A and C have bounded inverses, and assume that the operator $C^{-1} - A^{-1}$ is trace class. Then the strong limits* (7.1) *do exist, and, moreover,*

$$W_{\pm}(C, A) = W_{\mp}(C^{-1}, A^{-1}).$$

The same is true for the operators $W_{\pm}(A, C)$.

This theorem can be applied under the assumptions of Theorem 6.3 and allows one to prove the existence of the wave operators in the problem of perturbation of the boundary and the boundary conditions in the domains $\Omega \subset E_2$. For $m > 2$ the problem still remains open.

Added in proof. Recently this problem was solved by the author, who proved the following theorem. The strong limits (7.1) exist if the operators A and C are positive definite, the difference $C^{-1} - A^{-1}$ is compact, and for some integer number $n \geq 1$ the operator $C^{-n} - A^{-n}$ is trace class. This theorem yields the desired result for an arbitrary m.

TRANSLATOR'S COMMENTS

1. To Subsection 1.3. The classes T_α, or, more precisely, their analogues for not necessarily selfadjoint operators, are nothing but the 'weak Schatten ideals' $\mathfrak{S}_{p,w}$, $p = \alpha^{-1}$ (other notation $\mathfrak{S}_{p,\infty}$, Σ_p).

Their systematic theory (but only for $p > 1$ when they can be equipped by a norm) was established later, in the book [43*], as a part of a more general theory. The spaces $\mathfrak{S}_{p,w}$ for all $p > 0$ are quite natural from the point of view of interpolation theory of normed and quasi-normed spaces; see Section 1.19.7 in the book [45*]. In both mentioned books, Birman's paper is not cited.

2. To Lemma 1.3. This is an important statement that has numerous applications in the spectral theory of boundary-value problems. It states that the nonzero spectra of two certain operators, acting in different Hilbert spaces and having a rather different nature, do coincide. As is mentioned in Subsection 1.1, this is a particular case of a more general statement established in [**4**]. Up to some extent, by its character this lemma is close to the famous Birman–Schwinger principle that was discovered by the author later, in the paper [**5**] (detailed exposition in [**6**]).

3. To Subsection 2.1. At present, the most relevant reference is Section 6.3 in the book [44*]. The Calderón–Stein extension theorem was yet unknown to the author when the work on the paper was in progress. Now it can be said that in dimensions $m \geq 3$ the assumption $\partial\Omega \in \mathrm{Lip}$ is sufficient for the validity of all the results in the paper in which $\partial\Omega$ is supposed to be piecewise-smooth. Such boundaries for $m = 2$ are discussed in Theorem 5.2 only. There it is sufficient to suppose that each component of Γ consists of a finite number of C^2-smooth simple curves, intersecting transversely at each corner point.

4. To Lemma 3.5. Much later, the author returned to this class of problems. For domains with smooth boundary he, in cooperation with M. Solomyak, found not only the estimates but also the asymptotics for the operators generated by the variational problems considered on the space of solutions of homogeneous elliptic equations. See their papers [41*, 42*].

5. To Theorem 7.1. In more detail, this theorem, and also the result mentioned as *added in proof*, appeared in the papers [39*, 40*]. This was the first version of the famous result for which the name 'Invariance principle' was coined later. Later on, statements of this nature were repeatedly strengthened and generalized by the author and by Kato.

References

[1] Babich V. M., *On the extension of functions,* (Russian) Uspehi Mat. Nauk (N.S.) **8**, no. 2(54), (1953), 111–113.

[2] Birman M. Sh., *On the theory of selfadjoint extensions of positive definite operators,* (Russian) Dokl. Akad. Nauk SSSR **91**, no. 2, (1953), 189–191.

[3] Birman M. Sh., *On the spectrum of singular boundary problems for elliptic differential equations,* (Russian) Dokl. Akad. Nauk SSSR **97**, no. 1, (1954), 5–7.

[4] Birman M. Sh., *On the theory of selfadjoint extensions of positive definite operators,* (Russian) Mat. Sb. (N.S.) **38(80)**, no. 4 (1956), 431–450.

[5] Birman M. Sh., *Perturbations of quadratic forms and the spectrum of singular boundary value problems,* (Russian) Dokl. Akad. Nauk SSSR **125**, no. 3, (1959), 471–474.

[6] Birman M. Sh., *On the spectrum of singular boundary-value problems,* (Russian) Mat. Sb. (N.S.) **55(97)**, no. 2, (1961), 125–174. English translation in: Eleven Papers on Analysis, AMS Transl. **53**, 23–80, AMS, Providence, R.I., 1966.

[7] Birman M. Sh., *Perturbation of the spectrum of a singular elliptic operator under variation of the boundary and boundary conditions,* (Russian) Dokl. Akad. Nauk SSSR **137**, no. 4, (1961), 761–763.

[8] Erdelyi A., Magnus F., Oberhettinger F., and Tricomi F., *Higher transcendental functions. Vols. I, II,* McGraw-Hill, New York, 1953. xxvi+302, xvii+396 pp.

[9] Friedrichs K., *Spektraltheorie halbbeschränkter Operatoren und Anwendung auf die Spektralzerlegung von Differentialoperatoren,* (German) Math. Ann. **109**, no. 1 (1934), 685–713.

[10] Glazman I. M., *On the theory of singular differential operators,* (Russian) Uspehi Mat. Nauk (N.S.) **5** no. 6(40), (1950), 102–135. English transl. in: Amer. Math. Soc. Translation 1953, (1953), no. 96, 43 pp.

[11] Glazman I. M., *On the spectrum of linear differential operators,* (Russian) Dokl. Akad. Nauk SSSR **80**, no. 2, (1951), 153–156.

[12] Glazman I. M., *On the character of the spectrum of one-dimensional singular boundary problems,* (Russian) Dokl. Akad. Nauk SSSR (N.S.) **87**, no. 1, (1952), 5–8.

[13] Glazman I. M., *On the character of the spectrum of multidimensional singular boundary problems,* (Russian) Dokl. Akad. Nauk SSSR (N.S.) **87**, no. 2, (1952), 171–174.

[14] Glazman I. M., *On an application of the method of decomposition to multidimensional singular boundary problems,* (Russian) Mat. Sb. N.S. **35(77)**, no. 2, (1954), 231–246.

[15] Glazman I. M., *Direct methods of the qualitative spectral analysis of singular differential operators,* (Russian) Ph.D. thesis, Kharkov State University, 1958.

[16] Guseva O. B., *On boundary problems for strongly elliptic systems,* (Russian) Dokl. Akad. Nauk SSSR (N.S.) **102**, no. 6, (1955), 1069–1072.

[17] Kato T., *On finite-dimensional perturbations of self-adjoint operators,* J. Math. Soc. Japan **9**, 2, (1957), 239–249.

[18] Kato T., *Perturbation of continuous spectra by trace class operators,* Proc. Japan Acad. **33**, 5, (1957), 260–264.

[19] Krein M. G., *The theory of self-adjoint extensions of semi-bounded Hermitian transformations and its applications. I,* (Russian) Math. Sb. (N.S.) **20(62)**, no. 3, (1947), 431–495.

[20] Krein M. G., *On the trace formula in perturbation theory,* (Russian) Mat. Sb. (N.S.) **33(75)**, no. 3, (1953), 597–626.

[21] Kuroda S. T., *On the existence and the unitarity property of the scattering operator,* Nuovo Cimento **12**, 5, (1959), 431–454.

[22] Kuroda S. T., *Perturbation of continuous spectra by unbounded operators. I,* J. Math. Soc. Japan **11**, 3, (1959), 246–262.

[23] Kuroda S. T., *Perturbation of continuous spectra by unbounded operators. II,* J. Math. Soc. Japan **12**, 3, (1960), 243–257.

[24] Ladyzhenskaya O. A., *On the closure of the elliptic operator,* (Russian) Dokl. Akad. Nauk SSSR (N.S.) **79**, no. 5, (1951), 723–725.

[25] Ladyzhenskaya O. A., *A simple proof of the solvability of the fundamental boundary problems and of a problem on eigenvalues for linear elliptic equations,* (Russian) Vestnik Leningrad. Univ. **10**, no. 11, (1955), 23–29.

[26] Ladyzhenskaya O. A., *The mixed problem for a hyperbolic equation,* (Russian) Gosudarstv. Izdat. Tehn.-Teor. Lit., Moscow, 1953. 279 pp.

[27] Mihlin S. G., *The problem of the minimum of a quadratic functional,* (Russian) Gosudarstv. Izdat. Tehn.-Teor. Lit., Moscow, 1952. 216 pp. English transl. in: Holden-Day Series in Mathematical Physics, Holden-Day, Inc., San Francisco, 1965, ix+155 pp.

[28] Mihlin S. G., *Degenerate elliptic equations,* (Russian) Vestnik Leningrad. Univ. **9**, no. 8, (1954), 19–48.

[29] Povzner A. Ya., *On the expansion of arbitrary functions in characteristic functions of the operator* $-\Delta u + cu$, (Russian) Mat. Sb. (N.S.) **32(74)**, no. 1, (1953), 109–156.

[30] Rosenblum M., *Perturbation of the continuous spectrum and unitary equivalence,* Pacific J. Math. **7**, 1, (1957), 997–1010.

[31] Slobodetskij L. N. and Babich V. M., *On boundedness of the Dirichlet integrals,* (Russian) Dokl. Akad. Nauk SSSR (N.S.) **106**, no. 4, (1956), 604–606.

[32] Slobodetskij L. N., *Generalized Sobolev spaces and their application to boundary problems for partial differential equations,* (Russian) Leningrad. Gos. Ped. Inst. Uchen. Zap. **197**, 1958, 54–112.

[33] Slobodetskij L. N., *Estimates in* L_2 *for solutions of linear elliptic and parabolic systems. I,* (Russian) Vestnik Leningrad. Univ. **15** no. 7, (1960), 28–47.

[34] Smirnov V. I., *A course in higher mathematics. Vol.* 5, (Russian) Fizmatgiz, Moscow, 1959. 655 pp. English transl: Pergamon Press, Oxford; Addison-Wesley, 1964, xiv+635 pp.

[35] Sobolev S. L., *Some applications of functional analysis in mathematical physics,* (Russian) Izdat. Leningrad. Gos. Univ., Leningrad, 1950. 255 pp. English translation: Translations of Mathematical Monographs, Vol. 7, American Mathematical Society, Providence, R.I., 1963, vii+239 pp.

[36] Vishik M. I., *On general boundary problems for elliptic differential equations,* (Russian) Trudy Moskov. Mat. Ob. **1**, (1952), 187–246.

[37] Wolf F., *On the essential spectrum of partial differential boundary problems,* Comm. Pure Appl. Math. **12**, 2, (1959), 211–228.

[38] Wolf F., *On the invariance of the essential spectrum under a change of boundary conditions of partial differential operators,* Nederl. Akad. Wetensch. Proc. Ser. A 62 = Indag. Math. **21**, 2, (1959), 142–147.

[39*] Birman M. Sh., *On existence conditions for wave operators,* Dokl. Akad. Nauk SSSR **143**, no. 3, (1962), 506–509. English transl., Soviet Math. Dokl. **3**, (1962), 740-744.

[40*] Birman M. Sh., *Existence conditions for wave operators,* Izv. Akad. Nauk SSSR Ser. Mat. **27**, no. 4, (1963), 883–906.

[41*] Birman M. Sh. and Solomyak M. Z., *Asymptotic behavior of the spectrum of variational problems on solutions of elliptic equations,* Sibirsk. Mat. Zh. **20**, no. 1, (1979), 3–22. English transl., Siberian Math. J. **20**, (1979), 1-15.

[42*] Birman M. Sh. and Solomyak M. Z., *Asymptotic behavior of the spectrum of variational problems on solutions of elliptic equations in unbounded domains,* (Russian) Funktsional. Anal. i Prilozhen. **14**, no. 4, (1980), 27-35. English transl., Functional Anal. Appl. **14**, (1981), 267-274.

[43*] Gohberg I. C. and Krein M. G., *Introduction to the theory of linear non-selfadjoint operators in Hilbert space*, (Russian), Izdat. 'Nauka', Moscow 1965. 448 pp. English translation: Amer. Math. Soc., Providence, 1969, xv+378 pp.

[44*] Stein E. M., *Singular integrals and differentiability properties of functions*, Princeton University Press, Princeton, N.J., 1970, xiv+290 pp.

[45*] Triebel H., *Interpolation theory, function spaces, differential operators*, North-Holland Publishing Co., Amsterdam-New York, 1978, 528 pp.

Department of Physics, St.Petersburg State University, 3, Ul'yanovskaya, Petrodvorets, St.Petersburg, 198504 Russia

E-mail address: mbirman@list.ru

Amer. Math. Soc. Transl.
(2) Vol. **225**, 2008

Asymptotic Behavior of the Eigenfunctions of the Many-particle Schrödinger Operator. I. One-dimensional Particles

V. S. Buslaev and S. B. Levin

To Mikhail Shlëmovich Birman on his 80th birthday

ABSTRACT. We study the three-body nonrelativistic quantum scattering problem with rapidly decreasing at infinity pair potentials for the case of one-dimensional particles. We consider the solution $\chi(\mathbf{z},\mathbf{q})$ of the corresponding Schrödinger equation that can be interpreted as the scattered plane wave $e^{i\langle \mathbf{z},\mathbf{q}\rangle}$ and find its asymptotic behavior as $|\mathbf{z}| \to \infty$. More precisely, we explicitly construct a function $\chi_0(\mathbf{z},\mathbf{q})$ that determines the asymptotic behavior of the solution up to a diverging circle wave with a smooth amplitude coefficient. The method is based on analogies between the quantum scattering problem and the diffraction of the plane wave by a system of half-transparent infinite screens. We believe that the formalism can be useful also for studying many-dimensional particle scattering and for the case of long range pair potentials.

1. Introduction

In this paper we deal with a nonrelativistic Hamiltonian H for n particles,

(1)
$$H\psi = -\triangle\psi + \frac{1}{2}\sum_{i,j=1}^{n} v(z_i - z_j)\psi, \quad \psi = \psi(\mathbf{z}) \in \mathbb{C}, \ \mathbf{z} = (z_1, ..., z_n) \in \mathbb{R}^{nd}, \ z_i \in \mathbb{R}^d,$$

where $\triangle$ is the Laplace operator on $\mathbb{R}^{nd}$, d is the number of degrees of freedom of a single particle, v is a given two-particle interaction. It is assumed that $v(z)$, $z \in \mathbb{R}^d$, is a smooth real-valued function vanishing as $z \to \infty$. From the formula for the operator H it is clear that the masses of all particles are assumed to be equal and all the pair potentials are assumed to be identical.

There are several approaches to the spectral theory of such operators and to the description of the corresponding scattering processes. However, for the numerical computations of parameters that are important for applications, only a stationary technique based on integral equations for the eigenfunctions can be used. Unfortunately, so far there is no known mathematically satisfactory regular stationary

2000 *Mathematics Subject Classification.* Primary 81U10.

approach to many-particle systems with the pair potential slowly decreasing at infinity. Such an approach has not been satisfactorily developed even for the three-particle systems with the Coulomb type potential. Notice that the case of rapidly decreasing at infinity pair potentials for multidimensional particles, $d > 1$, can be treated as stationary in the framework of Faddeev's theory; see [**1**].

In [**2**] the problem was linked to the study of the asymptotic behavior at infinity in the configuration space $\mathbb{R}^{nd}$ of the eigenfunctions which correspond to the continuous spectrum of the operator H. Although it would be sufficient to find this asymptotic behavior only heuristically, it was not described with completeness that would guarantee a rigorous treatment of the problem. Moreover, even for the Hamiltonians with a rapidly decreasing pair potential, a complete description of the asymptotic behavior of eigenfunctions has not been derived. It is worth mentioning that some additional difficulties appear in the case of one-dimensional particles where the character of the asymptotic behavior becomes essentially different.

Let us describe now the content of this paper more formally. We consider the system of three one-dimensional particles with pair potential that rapidly decreases at infinity (and even has finite support). We separate the motion of the center of mass that technically means that the Hamiltonian is restricted to the set of functions defined on the subspace $\Gamma = \{\mathbf{z} : z_1 + z_2 + z_3 = 0\}$. In this case $n = 3$, $d = 1$, so that $\dim \Gamma = 2$. After such a reduction the operator H can be described by the original formula (1) in which, however, $\triangle$ has to be considered now as the Laplace-Beltrami operator on the Euclidean space Γ. We assume further that $v(z) \geq 0$. Under this assumption the two-particle subsystems and the operator H do not have bound states, so that the spectrum of the Hamiltonian is purely continuous (of infinite multiplicity) covering the semiaxis $E > 0$. These simplifications are not essential if we take into account the main goals pursued in this work.

We consider solutions $\chi(\mathbf{z}, \mathbf{q})$ which we call "scattered plane waves" and which can be characterized by their asymptotic behavior as $|\mathbf{z}| \to \infty$:

$$\chi(\mathbf{z}, \mathbf{q}) = e^{i\langle \mathbf{z}, \mathbf{q}\rangle} + f(\hat{\mathbf{z}}, \mathbf{q}) \frac{e^{i|\mathbf{z}||\mathbf{q}|}}{|\mathbf{z}|^{1/2}} + o\left(\frac{1}{|\mathbf{z}|^{1/2}}\right), \tag{2}$$

where $\hat{\mathbf{z}} = \frac{\mathbf{z}}{|\mathbf{z}|} \in S^1$. In the last formula the so-called scattering amplitude f is a singular distribution in the variable $\hat{\mathbf{z}}$, and the asymptotic behavior is understood in a weak sense for functions of $\hat{\mathbf{z}}$ as $|\mathbf{z}| \to \infty$. Such solutions as $|\mathbf{q}|^2 = E$ have to generate the infinitesimal eigenspace of the operator H that corresponds to the eigenvalue $E > 0$.

Our goal is to separate explicitly the term $\chi_0(\mathbf{z}, \mathbf{q})$ that is responsible for the singularities of the amplitude so that the correction $\chi_1 = \chi - \chi_0$ behaves asymptotically as the diverging circle wave with a smooth amplitude $g(\hat{\mathbf{z}}, \mathbf{q})$:

$$\chi_1(\mathbf{z}, \mathbf{q}) = g(\hat{\mathbf{z}}, \mathbf{q}) \frac{e^{i|\mathbf{z}||\mathbf{q}|}}{|\mathbf{z}|^{1/2}} + o\left(\frac{1}{|\mathbf{z}|^{1/2}}\right). \tag{3}$$

It is obvious that the term $\chi_0(\mathbf{z}, \mathbf{q})$ is not defined uniquely.

Such asymptotic formulas have already been considered earlier; see [**3**]–[**4**]. However, they were obtained only for the vectors $\hat{\mathbf{z}}$ and $\mathbf{q}$ that do not belong to some angle neighborhoods of six rays that lie on the straight lines $l_j = \{\mathbf{z} : z_i = z_k;\ i \neq k \neq j;\ i, j, k = 1, 2, 3\}$. We will retain this assumption with respect to $\mathbf{q}$, and propose asymptotic formulas that are uniform with respect to $\hat{\mathbf{z}}$.

In this paper the asymptotic formulas are obtained only heuristically. In future publications we are planning to generalize the formulas to the case of long range potentials such that for $z \to \infty$, $v(z) = \frac{\alpha}{|z|} + O(\frac{1}{|z|^N})$, where N is sufficiently large, $\alpha = const$. Of course, for such slowly decreasing potentials, formulas (2) and (3) have to be appropriately modified. In principle, possessing a sufficiently precise asymptotic description of the solutions $\chi(\mathbf{z}, \mathbf{q})$ at infinity, even obtained heuristically, for a rigorous characterization of the solutions one can construct an integral equation that is closer to the ordinary Lippmann–Schwinger equation than to Faddeev's system. We postpone the discussion of the details for subsequent publications.

Let us make brief remarks about the structure of the paper. The next section contains the list of the main formulas for the function $\chi_0(\mathbf{z}, \mathbf{q})$. These formulas describe the function in terms of solutions for the Hamiltonians of two-particle subsystems. The motivation has an essentially geometrical and ray character and was first proposed in the already-mentioned paper [**3**]. In fact, the motivation is more complicated than the formulas themselves, and we discuss it in the next section. Trying to characterize it briefly we can say that if one observes the supports of pair potentials from a large distance, the supports are reduced to screens of zero thickness. Therefore, the problem asymptotically transforms to the scattering of a plane wave by the system of three intersecting screens whose reflection and transmission coefficients are inherited from the one-dimensional scattering by pair potentials. In turn, the diffraction of a plane wave by such a system of screens can be explicitly described in terms of geometrical optics. The explicit formulas we imply here represent the solution up to a diverging circle wave with a smooth scattering amplitude. The last sentence is not absolutely correct since the geometric-optical formulas for the field have jumps on two certain rays going from the origin to infinity. Modifications of the ray formulas that allow us to eliminate these jumps are considered in the last section of the paper. It is worth mentioning that the contents of Sections 3 and 4 differ from the contents of the corresponding Sections of [**3**] only methodically. However, these methodical distinctions become very essential for future constructions. In any case, the contents of Sections 3 and 4 are needed to present a closed description of the main formulas of Section 2.

The authors would like to thank S. L. Yakovlev for fruitful discussions. The work was partially supported by RFBR , grant 08-01-00209, by grant 60866/2007 of the Swedish Institute and by grant 629-2002-8331 of the Swedish Research Council.

2. Main formulas

2.1. Geometry of the model. The Hamiltonian H of the system of three one-dimensional particles is defined on the set of complex-valued functions on Γ. Here Γ is a hyperplane in $\mathbf{R}^3$ whose points $\mathbf{z} = (z_1, z_2, z_3) \in \mathbf{R}^3$ satisfy the condition $z_1 + z_2 + z_3 = 0$. On Γ we have a scalar product induced by the standard scalar product on $\mathbf{R}^3$: $\langle \mathbf{z}, \mathbf{z}' \rangle = z_1 z_1' + z_2 z_2' + z_3 z_3'$.

On Γ, we consider the three straight lines $l_j = \{z : z_i = z_k,\ i \neq k \neq j;\ i, j, k = 1, 2, 3\}$, such that the coordinates of two particles coincide on each of them. Introduce on Γ three unit vectors $\mathbf{l}_j$, $j = 1, 2, 3$, that belong to the straight lines l_j and are oriented in the directions of growth of the corresponding variables z_j:

$$\mathbf{l}_1 = \sqrt{\frac{2}{3}}\left(1, -\frac{1}{2}, -\frac{1}{2}\right), \quad \mathbf{l}_2 = \sqrt{\frac{2}{3}}\left(-\frac{1}{2}, 1, -\frac{1}{2}\right), \quad \mathbf{l}_3 = \sqrt{\frac{2}{3}}\left(-\frac{1}{2}, -\frac{1}{2}, 1\right).$$

Further, we introduce three other unit vectors orthogonal to $\mathbf{l}_j$:

$$\mathbf{k}_1 = \frac{1}{\sqrt{2}}(0, -1, 1), \qquad \mathbf{k}_2 = \frac{1}{\sqrt{2}}(1, 0, -1), \qquad \mathbf{k}_3 = \frac{1}{\sqrt{2}}(-1, 1, 0). \tag{4}$$

Consider the vector $\mathbf{z} \in \Gamma$, $\mathbf{z} = (z_1, z_2, z_3)$, and write it as $\mathbf{z} = y_j \mathbf{l}_j + x_j \mathbf{k}_j$. It is obvious that

$$x_1 = \frac{1}{\sqrt{2}}(z_3 - z_2), \qquad x_2 = \frac{1}{\sqrt{2}}(z_1 - z_3), \qquad x_3 = \frac{1}{\sqrt{2}}(z_2 - z_1), \tag{5}$$

and

$$y_j = \sqrt{\frac{3}{2}} z_j, \qquad j = 1, 2, 3.$$

Traditionally the pairs of coordinates (x_j, y_j), $j = 1, 2, 3$, are called the Jacobi coordinates.

Consider now two vectors $\mathbf{z}$ and $\mathbf{q}$ and denote the coordinates of the second one by (k_j, p_j), $j = 1, 2, 3$. Then

$$\langle \mathbf{q}, \mathbf{z} \rangle = q_1 z_1 + q_2 z_2 + q_3 z_3 = \frac{2}{3}(x_1 k_1 + x_2 k_2 + x_3 k_3) = y_j p_j + x_j k_j, \quad j = 1, 2, 3. \tag{6}$$

As a consequence, for each j one obtains the following representation for the Laplace–Beltrami operator on Γ:

$$\Delta = \frac{\partial^2}{\partial y_j^2} + \frac{\partial^2}{\partial x_j^2}.$$

The main three-particle Hamiltonian H has the structure

$$H = -\Delta + v(x_1) + v(x_2) + v(x_3).$$

Note that compared to the original formula (1) we slightly changed the notation for the pair potential; namely we introduced a new scale in the arguments of the potential. In what follows we will use this latter notation.

We will also use the operators

$$H_j = -\Delta + v(x_j) = -\frac{\partial^2}{\partial y_j^2} + h_j, \quad h_j = -\frac{\partial^2}{\partial x_j^2} + v(x_j). \tag{7}$$

It is clear that the equation for the eigenfunctions of the operator H_j admits a separation of variables: the product $\chi = \psi(x_j, k) e^{ipy_j}$ satisfies the equation $H_j \chi = E\chi$, $E = k^2 + p^2$. We denoted here by $\psi(x, k)$ a solution of the equation

$$h\psi = k^2 \psi. \tag{8}$$

The notation $\psi(x, k)$ will be specified later on.

The lines l_j define six sectors on the plane Γ. The internal part of a certain one consists of the vectors (z_1, z_2, z_3) whose coordinates satisfy the condition $z_{j_1} > z_{j_2} > z_{j_3}$, where $\sigma = (j_1, j_2, j_3)$ is a permutation of the numbers (123). With each sector, we associate the corresponding permutation σ and write $\lambda = \lambda_\sigma$ (see Figure 1).

Let the group S_3 of permutations acts on Γ in the following way:

$$(\sigma, \mathbf{z}) \to \sigma\mathbf{z} = (z_{j_1}, z_{j_2}, z_{j_3}), \qquad \sigma = (j_1, j_2, j_3), \quad \mathbf{z} = (z_1, z_2, z_3).$$

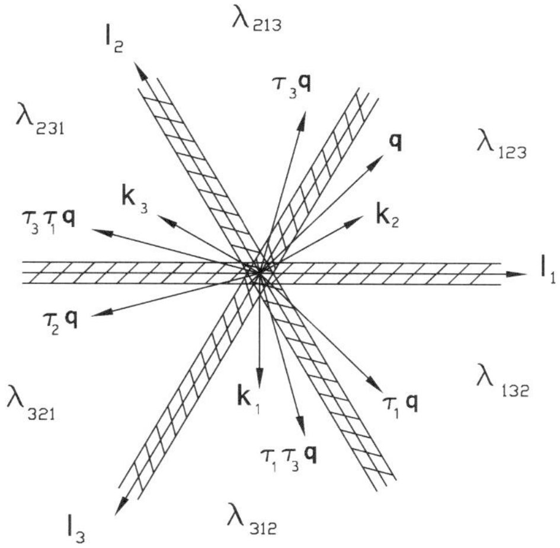

FIGURE 1

The group contains 6 elements. A permutation can be identical, or a transposition of two elements, or a composition of two transpositions, where some of the compositions coincide. We introduce the following notation for the transpositions: $\tau_1 = (132)$, $\tau_2 = (321)$, $\tau_3 = (213)$, and we notice that $\tau_i^2 = I$, $i = 1, 2, 3$. The action of the transposition on Γ will be denoted by the same symbol τ_j, $j = 1, 2, 3$. It corresponds to the reflection with respect to the line l_j, $j = 1, 2, 3$. The composition of two transpositions generates a rotation. In particular, the following equalities hold: $\tau_1\tau_2 = \tau_3\tau_1 = \tau_2\tau_3$, $\tau_2\tau_1 = \tau_1\tau_3 = \tau_3\tau_2$.

Six elements τ of the group S_3 generate six vectors $\tau\mathbf{q}$. If $\mathbf{q} \in \lambda_\sigma$, then $\tau\mathbf{q} \in \lambda_{\tau\sigma}$.

In what follows we will use two tables of the scalar products $\langle \sigma\mathbf{q}, \mathbf{k}_j \rangle$ and $\langle \sigma\mathbf{q}, \mathbf{l}_j \rangle$.

j \ σ	e	τ_1	τ_2	τ_3	$\tau_3\tau_1$	$\tau_1\tau_3$
1	k_1	$-k_1$	$-k_3$	$-k_2$	k_3	k_2
2	k_2	$-k_3$	$-k_2$	$-k_1$	k_1	k_3
3	k_3	$-k_2$	$-k_1$	$-k_3$	k_2	k_1

Table of scalar products $\langle \sigma\mathbf{q}, \mathbf{k}_j \rangle$

j \ σ	e	τ_1	τ_2	τ_3	$\tau_3\tau_1$	$\tau_1\tau_3$
1	p_1	p_1	p_3	p_2	p_3	p_2
2	p_2	p_3	p_2	p_1	p_1	p_3
3	p_3	p_2	p_1	p_3	p_2	p_1

Table of scalar products $\langle \sigma\mathbf{q}, \mathbf{l}_j \rangle$

Considering the solution $\chi(\mathbf{z}, \mathbf{q})$, we assume that $\mathbf{q} \in \lambda_{123}$. This is not a limitation since $\chi(\mathbf{z}, \mathbf{q}) = \chi(\sigma\mathbf{z}, \sigma\mathbf{q})$, $\sigma \in S_3$.

2.2. Setting of the problem. In this work we study solutions $\chi = \chi(\mathbf{z}, \mathbf{q})$ of the equation

$$-\Delta\chi + V\chi = \mathbf{q}^2\chi, \quad V = v(x_1) + v(x_2) + v(x_3), \qquad \mathbf{z}, \mathbf{q} \in \Gamma.$$

We assume that the pair potential $v(x)$ is a real-valued function sufficiently rapidly vanishing at infinity. To simplify and make more transparent the exposition of the results we assume, moreover, that the potential $v(x)$ is an even nonnegative function. This function is equal to zero outside of the interval $|x| < \frac{b}{2}$. We mentioned already that these assumptions, which at first glance can be viewed as too restrictive, do not make the problem trivial and do not contradict the goals of the paper. The total three-particle potential V viewed as a function on Γ vanishes outside the union of the strips $|x_j| < \frac{b}{2}$, $j = 1, 2, 3$; see Figure 1.

As was already mentioned in the Introduction, the solution χ, which is called the "scattered plane wave", can be characterized by its behavior at infinity ($|\mathbf{z}| \to \infty$):

$$\chi \sim e^{\langle \mathbf{q}, \mathbf{z} \rangle} + f(\hat{\mathbf{z}}, \mathbf{q}) \frac{e^{i|\mathbf{q}||\mathbf{z}|}}{|\mathbf{z}|^{\frac{1}{2}}}, \qquad \hat{\mathbf{z}} = \frac{\mathbf{z}}{|\mathbf{z}|}.$$

This behavior has to be treated in a suitable "weak" sense. The vector $\mathbf{q} \neq 0$ in this setting has to be regarded as a fixed parameter.

In this work we explicitly describe a "singular part" of the solution χ, the function χ_0, such that for $|\mathbf{z}| \to \infty$ the following formula takes place:

$$\chi_1 = \chi - \chi_0 \sim g(\hat{\mathbf{z}}, \mathbf{q}) \frac{e^{i|\mathbf{q}||\mathbf{z}|}}{|\mathbf{z}|^{\frac{1}{2}}}.$$

Here g is a continuous (even smooth) function of $\mathbf{z}$.

As we mentioned above the function χ_0 can be obtained with the help of heuristic ideas. Indeed, the fact that the function χ_0 correctly retrieves the singular part of the scattering amplitude has to be rigorously proved separately.

The function χ_0 will be explicitly described in terms of the eigenfunctions of the operators H_j, $j = 1, 2, 3$. The eigenfunctions $\chi_j(\mathbf{z}, \mathbf{q})$ of H_j that are analogs of the scattered plane waves can be obtained by the separation of variables:

$$\chi_j(\mathbf{z}, \mathbf{q}) = \psi(x_j, k_j) e^{ip_j y_j}, \quad E = q^2 = k_j^2 + p_j^2. \tag{9}$$

In formula (9) the Jacobi coordinates are used for the description of the position vector $\mathbf{z}$, $\mathbf{z} \sim (x_j, y_j)$ and for the wave vector $\mathbf{q}$, $\mathbf{q} \sim (k_j, p_j)$.

In the last formula the function $\psi(x, k)$ is a solution of the two-particle Schrödinger equation on the axis (8). Now we have to specify this solution by its asymptotic behavior as $x \to \infty$. If $kx > 0$, then

$$\psi(x, k) = s(k) e^{ikx}. \tag{10}$$

If $kx < 0$, then

$$\psi(x, k) = e^{ikx} + r(k) e^{-ikx}. \tag{11}$$

Here s and r are undefined coefficients which have to be determined when we construct the solution $\psi(x, k)$. They are so-called transmission and reflection coefficients. Notice that under our assumptions, $s(k) = s(|k|)$ and $r(k) = r(|k|)$.

2.3. Formulas for $\hat{\chi}$. Here we start to describe the main formulas that will characterize (after a certain modification) the field χ_0. The field that is constructed here will differ a little from the final asymptotical field χ_0, and will be denoted by $\hat{\chi}$. In the next section we will suggest a ray (trajectory) interpretation of these formulas and then they will become very natural.

Consider six rays l_j^t, $t = +, -$, $l_j^+ = \{\mathbf{z} \in l_j : z_j \geq 0\}$, $l_j^- = \{\mathbf{z} \in l_j : z_j \leq 0\}$. They are naturally oriented. These rays bound six sectors λ_σ. In each sector there exists exactly one vector of the form $\mathbf{q}_\sigma = \sigma\mathbf{q}$, $\sigma \in S_3$. Consider now the new six sectors that are bounded by the neighboring rays directed along the vectors $\mathbf{q}_\sigma$. They can be associated with the rays $l_j^\pm$ which belong to the sectors. Let us denote the sector corresponding to $l_j^\pm$ by $K_j^\pm$. In what follows we describe the field $\hat{\chi}$ in the sectors $K_j^\pm$, $j = 1, 2, 3$.

Sector K_1^+: $\hat{\chi} = \psi_1^+$,

$$\psi_1^+(\mathbf{z}, \mathbf{q}) = \chi_1(\mathbf{z}, \mathbf{q}) s_2 s_3.$$

We use here the following notation: $s_j = s(k_j)$, $\quad r_j = r(k_j)$.

Sector K_3^-: $\hat{\chi} = \psi_3^-$,

$$\psi_3^-(\mathbf{z}, \mathbf{q}) = \chi_3(\mathbf{z}, \mathbf{q}) s_1 s_2.$$

Sector K_2^+: $\hat{\chi} = \psi_2^+$,

$$\psi_2^+(\mathbf{z}, \mathbf{q}) = \chi_2(\mathbf{z}, \mathbf{q}) s_1 + \chi_2(\mathbf{z}, \tau_3\mathbf{q}) s_2 r_3.$$

Sector K_2^-: $\hat{\chi} = \psi_2^-$,

$$\psi_2^-(\mathbf{z}, \mathbf{q}) = \chi_2(\mathbf{z}, \mathbf{q}) s_3 + \chi_2(\mathbf{z}, \tau_1\mathbf{q}) s_2 r_1.$$

Sector K_1^-: $\hat{\chi} = \psi_1^-$,

$$\psi_1^-(\mathbf{z}, \mathbf{q}) = \chi_1(\mathbf{z}, \mathbf{q}) + \chi_1(\mathbf{z}, \tau_2\mathbf{q}) r_2 s_1 + \chi_1(\mathbf{z}, \tau_3\tau_1\mathbf{q}) r_2 r_1 + \chi_1(\mathbf{z}, \tau_3\mathbf{q}) r_3.$$

Sector K_3^+: $\hat{\chi} = \psi_3^+$,

$$\psi_3^+(\mathbf{z}, \mathbf{q}) = \chi_3(\mathbf{z}, \mathbf{q}) + \chi_3(\mathbf{z}, \tau_2\mathbf{q}) r_2 s_3 + \chi_3(\mathbf{z}, \tau_1\tau_3\mathbf{q}) r_2 r_3 + \chi_3(\mathbf{z}, \tau_1\mathbf{q}) r_1.$$

The total field $\hat{\chi}$ is defined by the formula

$$\hat{\chi} = \theta_1^+\psi_1^+ + \theta_3^-\psi_3^- + \theta_2^+\psi_2^+ + \theta_2^-\psi_2^- + \theta_1^-\psi_1^- + \theta_3^+\psi_3^+.$$

We used here the notation $\theta_j^{(\pm)}$ for the indicator function of the corresponding sector $K_j^\pm$,

$$\theta_1^+ + \theta_3^- + \theta_2^+ + \theta_2^- + \theta_1^- + \theta_3^+ = 1.$$

The last formula for $\hat{\chi}$ does not define the field on the boundaries of the sectors. In the next section we will see that on all boundaries except two directed along the vectors $\tau_3\tau_1 q$ and $\tau_1\tau_3 q$, the constructed field is smooth. In Section 4, in neighborhoods of two mentioned rays, the field will be modified and after that will become a smooth function everywhere.

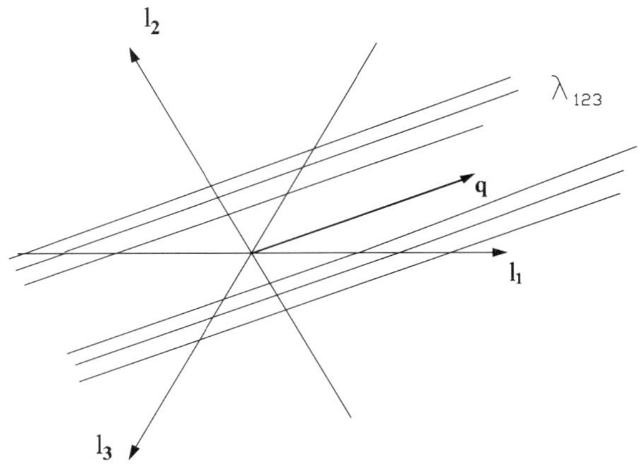

FIGURE 2

3. Ray-optical motivation

If a point $\mathbf{z} \in \Gamma$ is located outside of the supports of the pair potentials, $|x_j| \geq \frac{b}{2}$, $j = 1, 2, 3$, the solutions ψ_j can be reduced, in accordance with formulas (10)–(11), to linear combinations of the exponents. Any sector $K_j^{\pm}$ has a nontrivial intersection with two distinct sectors λ_σ. The reduction is described by different formulas on these intersections, since on different λ_σ the variable x_j has different signs. After such reductions the function $\hat{\chi}$ is replaced by a function that has a clear ray interpretation. We will denote this function by $\chi^{(ray)}$.

Note also that the continuity of the wave fields will be controlled only for those $\mathbf{z}$ that are not too small, so that the supports of different pair potentials do not intersect. Since we are interested here only in the asymptotic behavior of the field as $|\mathbf{z}| \to \infty$, the lack of information about the field near the origin does not play any role.

Sector λ_{123}.

From the inside $\lambda_{123}^{(1)} = K_1^+ \cap \lambda_{123}$, where $x_1 \leq -\frac{b}{2}$, the solution ψ_1^+ is reduced to an exponent and consequently the solution $\chi^{(ray)}$ becomes a plane wave

$$\psi_1^+ = e^{i\langle \mathbf{q}, \mathbf{z}\rangle} s_1 s_2 s_3.$$

In the semisector $\lambda_{123}^{(3)} = K_3^- \cap \lambda_{123}$, where $x_3 \leq -\frac{b}{2}$, it is similarly reduced to

$$\psi_3^- = e^{i\langle \mathbf{q}, \mathbf{z}\rangle} s_1 s_2 s_3.$$

Therefore, everywhere in λ_{123} outside of the supports of the potentials we have

$$\chi_{123}^{(ray)} = e^{i\langle \mathbf{q}, \mathbf{z}\rangle} s_1 s_2 s_3. \tag{12}$$

This implies that the combination of the solutions

$$\chi_{123} = \theta_{123}^{(1)} \psi_1^+ + \theta_{123}^{(3)} \psi_3^-$$

is a smooth function in the entire sector λ_{123}.

The plane wave (12) can be interpreted as a wave field that corresponds to the field of rays

$$\mathbf{z} = \mathbf{q}t + \mathbf{z}_0, \qquad \mathbf{z}_0 \perp \mathbf{q}.$$

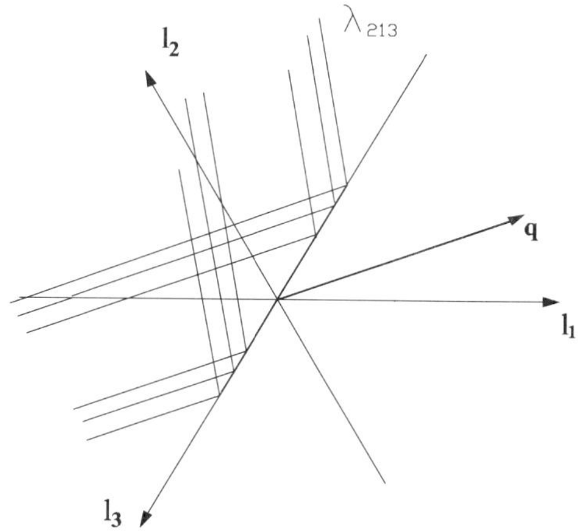

FIGURE 3

These rays coming from the "initial" sector λ_{321} into the sector λ_{123} cross successively the screens l_3, l_2, l_1 or the screens l_1, l_2, l_3. Each crossing is accompanied by the multiplication of the corresponding wave field, i.e., the plane wave $e^{i\langle \mathbf{q},\mathbf{z}\rangle}$, by the corresponding transmission coefficient.

Sector λ_{213}.

In the semisector $\lambda_{213}^{(3)} = K_3^- \cap \lambda_{213}$, where $x_3 \geq \frac{b}{2}$, the solution ψ_3^- is reduced to the sum of plane waves

$$\psi_3^- = \left(e^{i\langle \mathbf{q},\mathbf{z}\rangle} + r_3 e^{i\langle \tau_3\mathbf{q},\mathbf{z}\rangle}\right) s_1 s_2.$$

In the complementary semisector $\lambda_{213}^{(2)} = K_2^+ \cap \lambda_{213}$, where $x_2 \geq \frac{b}{2}$, we have

$$\psi_2^+ = e^{i\langle \mathbf{q},\mathbf{z}\rangle} s_2 s_1 + e^{i\langle \tau_3\mathbf{q},\mathbf{z}\rangle} s_1 s_2 r_3.$$

Thus in λ_{213} outside of the supports of the pair potentials

$$\chi_{213}^{(ray)} = e^{i\langle \mathbf{q},\mathbf{z}\rangle} s_2 s_1 + e^{i\langle \tau_3\mathbf{q},\mathbf{z}\rangle} s_1 s_2 r_3. \tag{13}$$

This implies that the combination of the solutions

$$\chi_{213} = \theta_{213}^{(3)}\psi_3^- + \theta_{213}^{(2)}\psi_2^+$$

is a smooth function in the sector λ_{213}.

The field (13) can be interpreted as the sum of the wave fields corresponding to the ray fields as shown in Figure 3.

We see that in the sector $\lambda_{213}^{(3)}$ there are two ray fields: the field described earlier and the field of rays reflected off the screen l_3.

If rays intersect a screen, the field is accompanied by the corresponding coefficient s, and if they are reflected off the screen, the field is accompanied by the coefficient r. Although the fates of the rays coming into the sector $\lambda_{213}^{(2)}$ and the fates of the rays penetrating into the sector $\lambda_{213}^{(3)}$ differ, the corresponding wave fields are represented analytically by the same expression.

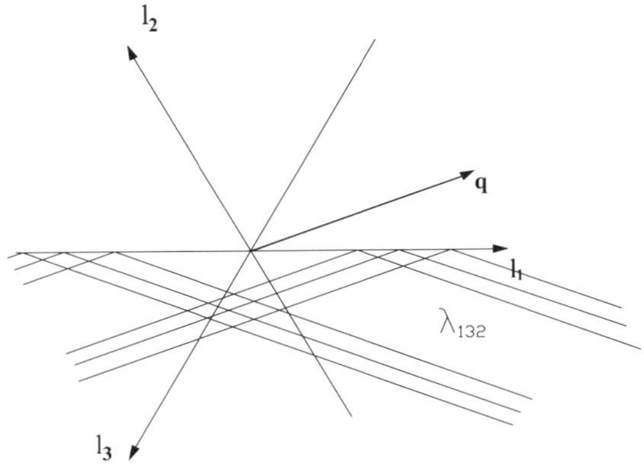

FIGURE 4

Sector λ_{132}.

In this sector the expression

$$\chi_{132} = \theta_{132}^{(1)}\psi_1^+ + \theta_{132}^{(2)}\psi_2^-$$

is a smooth function. Outside of the supports of the potentials,

$$\chi_{132}^{(ray)} = e^{i\langle \mathbf{q},\mathbf{z}\rangle} s_2 s_3 + e^{i\langle \tau_1\mathbf{q},\mathbf{z}\rangle} s_2 s_3 r_1. \tag{14}$$

We will not discuss here the structure of the corresponding fields of rays and refer to Figure 4.

Sector λ_{321}.

In the semisector $\lambda_{321}^{(1)} = K_1^- \cap \lambda_{321}$ as $x_1 \geq \frac{b}{2}$, the solution ψ_1^- takes the form

$$\psi_1^- = \left(e^{i\langle \mathbf{q},\mathbf{z}\rangle} + e^{i\langle \tau_1\mathbf{q},\mathbf{z}\rangle} r_1\right) + e^{i\langle \tau_2\mathbf{q},\mathbf{z}\rangle} s_1 r_2 s_3 + \left(e^{i\langle \tau_3\tau_1\mathbf{q},\mathbf{z}\rangle} + r_3 e^{i\langle \tau_2\mathbf{q},\mathbf{z}\rangle}\right) r_2 r_1 + \left(e^{i\langle \tau_3\mathbf{q},\mathbf{z}\rangle} + e^{i\langle \tau_1\tau_3\mathbf{q},\mathbf{z}\rangle} r_2\right) r_3.$$

In the semisector $\lambda_{321}^{(3)} = K_3^+ \cap \lambda_{321}$, where $x_3 \geq \frac{b}{2}$, the solution ψ_3^+ is reduced to the function

$$\psi_3^+ = \left(e^{i\langle \mathbf{q},\mathbf{z}\rangle} + e^{i\langle \tau_3\mathbf{q},\mathbf{z}\rangle} r_3\right) + e^{i\langle \tau_2\mathbf{q},\mathbf{z}\rangle} s_3 r_2 s_1 + \left(e^{i\langle \tau_1\tau_3\mathbf{q},\mathbf{z}\rangle} + r_1 e^{i\langle \tau_2\mathbf{q},\mathbf{z}\rangle}\right) r_2 r_3 + \left(e^{i\langle \tau_1\mathbf{q},\mathbf{z}\rangle} + e^{i\langle \tau_3\tau_1\mathbf{q},\mathbf{z}\rangle} r_2\right) r_1.$$

Since these two expressions are identical,

$$\chi_{321} = \theta_{321}^{(1)}\psi_1^- + \theta_{321}^{(3)}\psi_3^+$$

is a smooth function in the sector λ_{321}. The corresponding fields of rays are shown in Figures 5, 6.

The field $\chi_{321}^{(ray)}$ is described in this case by the formula

$$\chi_{321}^{(ray)} = e^{i\langle \mathbf{q},\mathbf{z}\rangle} + e^{i\langle \tau_3\mathbf{q},\mathbf{z}\rangle} r_3 + e^{i\langle \tau_1\mathbf{q},\mathbf{z}\rangle} r_1 + e^{i\langle \tau_1\tau_3\mathbf{q},\mathbf{z}\rangle} r_2 r_3 + e^{i\langle \tau_3\tau_1\mathbf{q},\mathbf{z}\rangle} r_2 r_1 + e^{i\langle \tau_2\mathbf{q},\mathbf{z}\rangle} s_3 r_2 s_1 + e^{i\langle \tau_2\mathbf{q},\mathbf{z}\rangle} r_1 r_2 r_3.$$

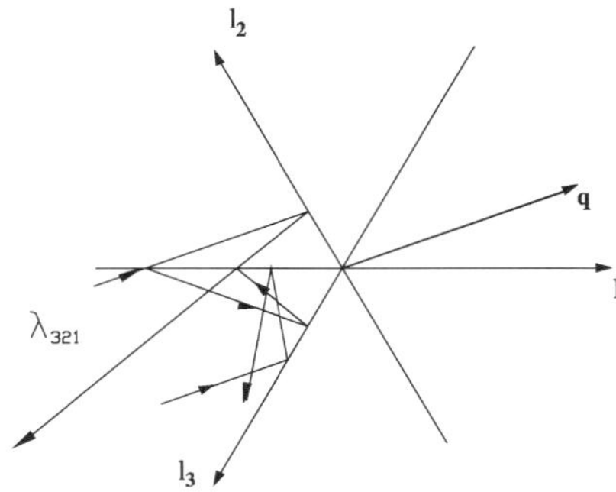

FIGURE 5

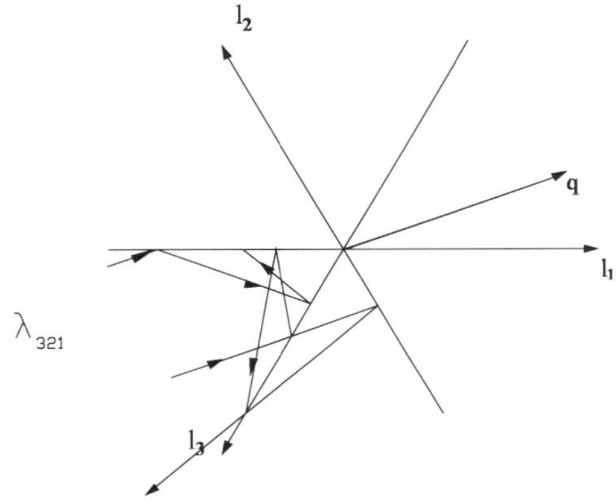

FIGURE 6

Sector λ_{231}.

In the semisector $\lambda_{231}^{(2)} = K_2^+ \cap \lambda_{231}$, where $x_2 \leq -\frac{b}{2}$, the solution ψ_2^+ takes the form

$$\psi_2^+ = \left(e^{i\langle \mathbf{q},\mathbf{z}\rangle} + e^{i\langle \tau_2\mathbf{q},\mathbf{z}\rangle} r_2\right) s_1 + \left(e^{i\langle \tau_3\mathbf{q},\mathbf{z}\rangle} + e^{i\langle \tau_2\tau_3\mathbf{q},\mathbf{z}\rangle} r_1\right) s_2 r_3.$$

In the semisector $\lambda_{231}^{(1)} = K_1^- \cap \lambda_{231}$. as $x_1 \leq -\frac{b}{2}$, the following expression for the solution ψ_2^+ can be obtained:

$$\psi_1^- = e^{i\langle \mathbf{q},\mathbf{z}\rangle} s_1 + \left(e^{i\langle \tau_2\mathbf{q},\mathbf{z}\rangle} + r_3 e^{i\langle \tau_1\tau_2\mathbf{q},\mathbf{z}\rangle}\right) r_2 s_1 + e^{i\langle \tau_3\tau_1\mathbf{q},\mathbf{z}\rangle} s_3 r_2 r_1 + e^{i\langle \tau_3\mathbf{q},\mathbf{z}\rangle} s_2 r_3.$$

It is easy to see that these two wave fields do not coincide. Therefore, the sum in the sector λ_{231},

$$\chi_{231} = \theta_{231}^{(2)} \psi_2^+ + \theta_{231}^{(1)} \psi_1^-,$$

has a jump along the ray $\tau_2\tau_3\mathbf{q}$. As has already been explained, we assume that the vector $\mathbf{q}$ and, consequently, the vector $\tau_2\tau_3\mathbf{q}$ belong to the interiors of the corresponding sectors. The fields of the rays are shown in Figures 7, 8.

The corresponding field $\chi_{231}^{(ray)}$ has the form

$$\chi_{231}^{(ray)} = \chi_{231}^{(c)} + \chi_{231}^{(j)},$$

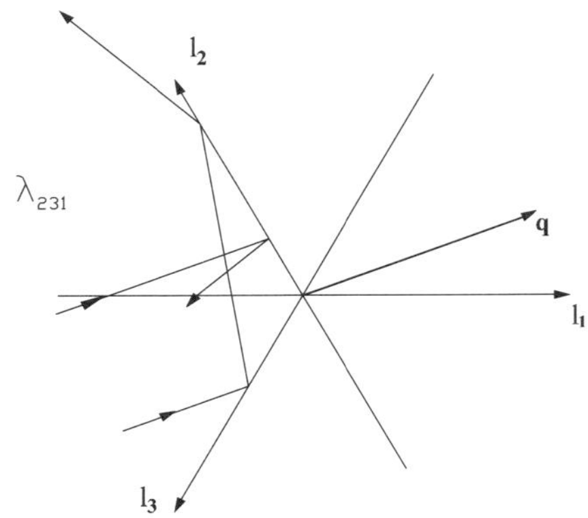

FIGURE 7

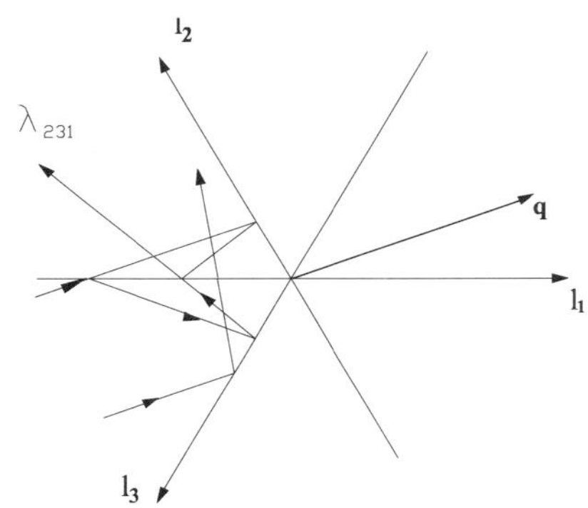

FIGURE 8

where $\chi_{231}^{(c)}$ is the continuous part of the field $\chi_{231}^{(ray)}$:

$$\chi_{231}^{(c)} = e^{i\langle \mathbf{q},\mathbf{z}\rangle} s_1 + e^{i\langle \tau_3\mathbf{q},\mathbf{z}\rangle} r_2 s_1 + e^{i\langle \tau_3\mathbf{q},\mathbf{z}\rangle} s_2 r_3, \tag{15}$$

and $\chi_{231}^{(j)}$ is the discontinuous part:

$$\begin{gathered} \chi_{231}^{(j)} = e^{i\langle \tau_2\tau_3\mathbf{q},\mathbf{z}\rangle} \left(\theta_{231}^{(2)} R_1 + \theta_{231}^{(1)} R_2\right), \\ R_1 = r_1 s_2 r_3, \qquad R_2 = r_3 r_2 s_1 + s_3 r_2 r_1. \end{gathered} \tag{16}$$

Sector λ_{312}.

In the semisector $\lambda_{312}^{(2)} = K_2^- \cap \lambda_{312}$, where $x_2 \leq -\frac{b}{2}$, the solution ψ_2^- looks as follows:

$$\psi_2^- = \left(e^{i\langle \mathbf{q},\mathbf{z}\rangle} + r_2 e^{i\langle \tau_2\mathbf{q},\mathbf{z}\rangle}\right) s_3 + \left(e^{i\langle \tau_1\mathbf{q},\mathbf{z}\rangle} + r_3 e^{i\langle \tau_2\tau_1\mathbf{q},\mathbf{z}\rangle}\right) s_2 r_1.$$

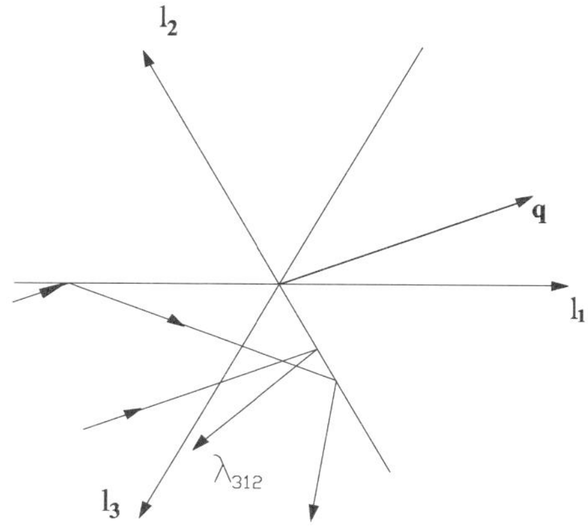

FIGURE 9

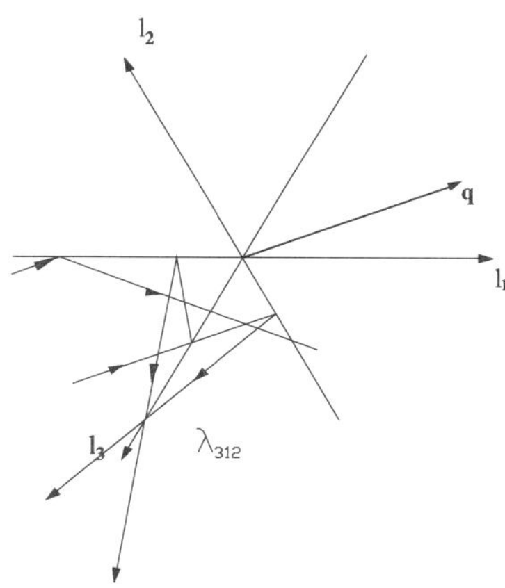

FIGURE 10

In the semisector $\lambda_{312}^{(3)} = K_3^+ \cap \lambda_{312}$, where $x_3 \leq -\frac{b}{2}$, the following expression for ψ_3^+ can be easily obtained:

$$\psi_3^+ = e^{i\langle \mathbf{q},\mathbf{z}\rangle} s_3 + \left(e^{i\langle \tau_2\mathbf{q},\mathbf{z}\rangle} + r_1 e^{i\langle \tau_3\tau_2\mathbf{q},\mathbf{z}\rangle}\right) r_2 s_3 + e^{i\langle \tau_1\tau_3\mathbf{q},\mathbf{z}\rangle} s_1 r_2 r_3 + e^{i\langle \tau_1\mathbf{q},\mathbf{z}\rangle} s_2 r_1.$$

The fields obtained in the semisectors do not coincide. Again, it means that the sum of the fields in the sector λ_{312},

$$\chi_{312} = \theta_{312}^{(2)}\psi_2^- + \theta_{312}^{(3)}\psi_3^+,$$

has a jump along the ray $\tau_1\tau_3\mathbf{q}$. The ray fields are shown in Figures 9, 10.

The corresponding field $\chi_{312}^{(ray)}$ has the form

$$\chi_{312}^{(ray)} = \chi_{312}^{(c)} + \chi_{312}^{(j)},$$

where $\chi_{312}^{(c)}$ is the continuous part of the field $\chi_{312}^{(ray)}$,

$$\chi_{312}^{(c)} = e^{i\langle \mathbf{q},\mathbf{z}\rangle} s_3 + e^{i\langle \tau_2\mathbf{q},\mathbf{z}\rangle} r_2 s_3 + e^{i\langle \tau_1\mathbf{q},\mathbf{z}\rangle} s_2 r_1,$$

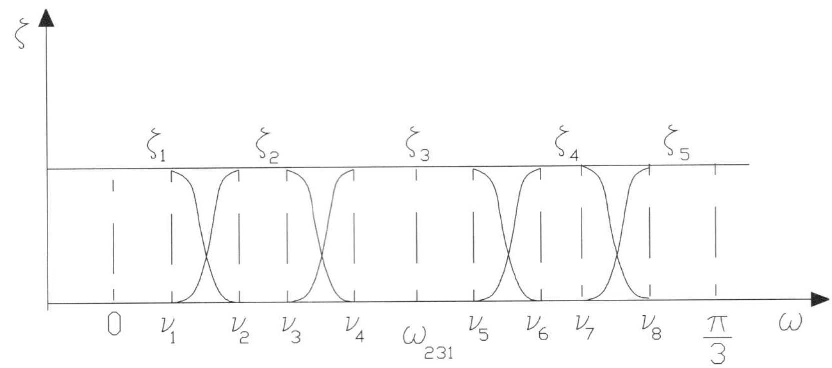

FIGURE 11

and $\chi_{312}^{(j)}$ is the discontinuous part:

$$\chi_{312}^{(j)} = e^{i\langle\tau_2\tau_1\mathbf{q},\mathbf{z}\rangle}\left(\theta_{312}^{(2)}R_1 + \theta_{312}^{(3)}R_2\right).$$

Concluding this section, let us note that the ray fields shown in Figures 2–10 exhaust all possible ray (trajectory) fields with different fates. The ray fields differ by the sequences of intersections of the screens and reflections off them arising from the field $\chi^{(in)}(\mathbf{z},\mathbf{q}) = e^{i\langle\mathbf{z},\mathbf{q}\rangle}$, $\mathbf{q} \in \lambda_{123}$, that fall onto the system of the screens. Therefore we can claim that the described ray field is complete in this approximation. Indirectly this implies the completeness of the description of the field $\hat{\chi}$.

4. Regularization of the jumps

4.1. Notation. In this section we consider the field in the sectors λ_{231} and λ_{312} where the function $\hat{\chi}$ has jumps. As was shown in Section 3, these jumps are localized on the rays directed along the vectors $\tau_2\tau_3\mathbf{q} = \tau_1\tau_2\mathbf{q} = \tau_3\tau_1\mathbf{q} \in \lambda_{231}$ and $\tau_2\tau_1\mathbf{q} = \tau_3\tau_2\mathbf{q} = \tau_1\tau_3\mathbf{q} \in \lambda_{312}$.

On the plane Γ, consider the polar coordinates and denote the polar angle by ω. In the two sectors under consideration, it is convenient to orient the variable ω differently. Referring to Figure 1, let us orient ω in the sector λ_{231} clockwise, and in the sector λ_{312} counterclockwise. Let the angle variable be zero on the ray l_1^- at λ_{231}, and let it be zero on the ray l_3^+ at λ_{312}. Denote the angle variable corresponding to the ray of jump in λ_{231} by ω_{231}, and in λ_{312} by ω_{312}.

Consider now the sector λ_{231}. The variable ω runs over the interval $[0,\pi/3]$. On the interval $(0,\pi/3)$, we fix the eight points ν_i, $i = 1,\dots,8$, so that $\nu_i < \nu_{i+1}$ and $\nu_4 < \omega_{231} < \nu_5$. Consider the open covering of $[0,\pi/3]$ by the subintervals $[0,\nu_2)$, (ν_1,ν_4), (ν_3,ν_6), (ν_5,ν_8), $(\nu_7,\pi/3]$. Introduce a partition of unity corresponding to this covering: $\zeta_1+\zeta_2+\zeta_3+\zeta_4+\zeta_5 = 1$; see Figure 11. Denote also $\widetilde{\zeta} = \zeta_2+\zeta_3+\zeta_4$.

In the sector λ_{231} the discontinuous field $\hat{\chi}$ can be described by the following formula:

$$\hat{\chi} = \psi_1^-(\mathbf{x},\mathbf{q})\zeta_1(\omega) + \psi_{231}^{(ray)}\widetilde{\zeta}(\omega) + \psi_2^+(\mathbf{x},\mathbf{q})\zeta_5(\omega).$$

Recall that

$$\psi_{231}^{(ray)} = \psi_{231}^{(c)} + \psi_{231}^{(j)}, \quad \psi_{231}^{(j)} = e^{i\langle\tau_2\tau_3\mathbf{q},\mathbf{z}\rangle}\left(R_1\theta_{231}^{(2)} + R_2\theta_{231}^{(1)}\right).$$

Since the vector $\mathbf{q}$ does not approach any of the bounding screens we have

$$\psi_{231}^{(c)} = \chi_{231}^{(c)},$$

where the function $\chi_{231}^{(c)}$ was defined in (15).

Let us write down a similar representation for the sector λ_{312}.

The variable ω again runs over the interval $[0, \pi/3]$. In complete analogy with the case of the sector λ_{231} we introduce the points μ_i, the corresponding open covering of the interval $[0, \pi/3]$, and a partition of unity corresponding to this covering: $\eta_1 + \eta_2 + \eta_3 + \eta_4 + \eta_5 = 1$, $\widetilde{\eta} = \eta_2 + \eta_3 + \eta_4$.

Let us consider the field $\hat{\chi}$ on λ_{312} as the sum:

$$\hat{\chi} = \psi_1^-(\mathbf{z}, \mathbf{q})\eta_1(\omega) + \psi_{312}^{(ray)}\widetilde{\eta}(\omega) + \psi_2^+(\mathbf{z}, \mathbf{q})\eta_5(\omega).$$

Here

$$\psi_{312}^{(ray)} = \psi_{312}^{(c)} + \psi_{312}^{(j)}, \quad \psi_{312}^{(j)} = e^{i\langle \tau_2\tau_1\mathbf{q}, \mathbf{z}\rangle}\left(R_1\theta_{312}^{(2)} + R_2\theta_{312}^{(3)}\right).$$

Further we will show how to modify and transform the discontinuous components of the function $\hat{\chi}$ to the smooth functions.

4.2. Modification of the discontinuous component. Let us start once again with the sector λ_{231}. We will replace the discontinuous components of the asymptotic solution by smooth exact solutions of the equation $(-\triangle - E)\chi = 0$ that asymptotically, as $|\mathbf{z}| \to \infty$, behave like the discontinuous functions $\psi_{231}^{(j)}$ and additionally contain diverging circle waves with smooth amplitudes. This procedure is well known and was studied in connection with the Sommerfeld problem about the diffraction of a plane wave by a screen.

Practically the modification procedure is nothing else but a simple replacement of the discontinuous indicator functions $\theta_{231}^{(1)}$ and $\theta_{231}^{(2)}$ by some smooth functions $\Phi_{231}^{(1)}$ and $\Phi_{231}^{(2)}$. To describe these smooth functions consider the following expression:

$$\Phi(\mathbf{z}, \mathbf{q}, \mathbf{q}^{\perp}) = \Phi\left((|\mathbf{z}||\mathbf{q}| - \langle \mathbf{z}, \mathbf{q}\rangle)^{1/2} sgn\langle \mathbf{z}, \mathbf{q}^{\perp}\rangle\right).$$

Here $(|\mathbf{z}||\mathbf{q}| - \langle \mathbf{z}, \mathbf{q}\rangle)^{1/2} \geq 0$, and $\mathbf{q}^{\perp}$ is a nonzero vector orthogonal to $\mathbf{q}$, $|\mathbf{q}| \neq 0$; the function $\Phi(\mathbf{z}, \mathbf{q}, \mathbf{q}^{\perp})$ depends only on the orientation of $\mathbf{q}^{\perp}$. The function Φ is defined by the integral

$$\Phi(\alpha) = \frac{e^{-i\pi/4}}{\sqrt{\pi}} \int_{-\infty}^{\alpha} e^{it^2} dt, \qquad \alpha \in \mathbf{R}.$$

Notice that

$$\Phi(\alpha) \to 1, \quad \alpha \to +\infty; \qquad \Phi(\alpha) \to 0, \quad \alpha \to -\infty.$$

It is known that the function $e^{i\langle \mathbf{z}, \mathbf{q}\rangle}\Phi(\mathbf{z}, \mathbf{q}, \mathbf{q}^{\perp})$ satisfies the equation

$$(-\triangle - E)\chi = 0.$$

Further introduce the function $\Phi_{\sigma}^{(i)}$:

$$\theta_{231}^{(2)} \Longrightarrow \Phi_{231}^{(2)}(\mathbf{z}) = \Phi(\mathbf{z}, \tau_2\tau_3\mathbf{q}, (\tau_2\tau_3\mathbf{q})^{\perp}),$$

assuming that the pair $\left(\tau_2\tau_3\mathbf{q}, (\tau_2\tau_3\mathbf{q})^{\perp}\right)$ has the same orientation as the pair $(\mathbf{l}_2, \mathbf{k}_2)$.

Analogously,

$$\theta_{231}^{(1)} \Longrightarrow \Phi_{231}^{(1)}(\mathbf{z}) = \Phi(\mathbf{z}, \tau_3\tau_1\mathbf{q}, -(\tau_3\tau_1\mathbf{q})^{\perp}).$$

We can propose similar formulas also for the sector λ_{312}:

$$\theta^{(2)}_{312} \Longrightarrow \Phi^{(2)}_{312}(\mathbf{z}) = \Phi(\mathbf{z}, \tau_2\tau_1\mathbf{q}, (\tau_2\tau_1\mathbf{q})^{\perp}).$$

We assume here that the pair $\left(\tau_2\tau_1\mathbf{q}, (\tau_2\tau_1\mathbf{q})^{\perp}\right)$ has the same orientation as the pair $(-\mathbf{l}_2, \mathbf{k}_2)$. Finally,

$$\theta^{(3)}_{312} \Longrightarrow \Phi^{(3)}_{312}(\mathbf{z}) = \Phi(\mathbf{z}, \tau_3\tau_2\mathbf{q}, -(\tau_3\tau_2\mathbf{q})^{\perp}).$$

Now we have to describe how to match the discontinuous and the smooth components of the solutions. For that it is useful to investigate more precisely the behavior of the integral $\Phi(\alpha)$ as $|\alpha| \to \infty$. It is easy to see that

$$\Phi(\alpha) \underset{\alpha\to+\infty}{\sim} 1 + \frac{e^{-i\pi/4}}{\sqrt{\pi}}\frac{1}{2i}e^{i\alpha^2}\left(\frac{1}{\alpha} + \frac{\xi_1}{\alpha^2} + \dots\right),$$

$$\Phi(\alpha) \underset{\alpha\to-\infty}{\sim} \frac{e^{-i\pi/4}}{\sqrt{\pi}}\frac{1}{2i}e^{i\alpha^2}\left(\frac{1}{|\alpha|} + \frac{\xi_1}{\alpha^2} + \dots\right).$$

4.3. Matching of wave fields. The discontinuous component of the field $\psi^{(j)}_{231}$ can be replaced by the following smooth function:

$$\psi^{(sm)}_{231} = e^{i\langle\tau_2\tau_3\mathbf{q},\mathbf{z}\rangle}\left\{R_2\left[\zeta_2 + \zeta_3\Phi^{(1)}_{231}\right] + R_1\left[\zeta_3\Phi^{(2)}_{231} + \zeta_4\right]\right\}.$$

The functions ζ_i, $i = 2, 3, 4$ were defined in Subsection 4.1.

It is important to see how two expressions in the brackets multiplied by R_1, R_2 unify in the answer. The machinery is the same for both terms and literally repeats at the other sector. Let us make short comments with respect to the sum $\zeta_2+\zeta_3\Phi^{(1)}_{231}$. As $|\mathbf{z}| \to \infty$, on the interval (ν_3, ν_4) both terms in the leading order are equal to 1, which means that the sum has the same leading order. The additional terms introduced by the integral $\Phi^{(1)}_{231}$ contribute only in diverging waves with smooth amplitudes.

Now let us write down a similar expression for the field $\psi^{(sm)}_{312}$ that smooths the discontinuous field $\psi^{(j)}_{312}$. The discontinuous function $\psi^{(j)}_{312}$ has to be replaced now by the smooth function $\psi^{(sm)}_{312}$,

$$\psi^{(sm)}_{312} = e^{i\langle\tau_2\tau_1\mathbf{q},\mathbf{z}\rangle}\left\{R_2\left[\eta_2 + \eta_3\Phi^{(3)}_{312}\right] + R_1\left[\eta_3\Phi^{(2)}_{312} + \eta_4\right]\right\}.$$

To combine the obtained formulas in the final expression let us recall that in all sectors λ_σ there were introduced the functions χ_σ that describe in these sectors the field $\hat{\chi}$:

$$\hat{\chi} = \chi_{123} + \chi_{213} + \chi_{132} + \chi_{321} + \chi_{231} + \chi_{312}.$$

Here each term contains as a factor the indicator function of the corresponding sectors. All terms except the last two are continuous and they even smooth the corresponding sectors. They combine into smooth functions in the neighborhoods of the boundaries of the sectors. The fields corresponding to the last two sectors χ_σ were modified. After that the discontinuous fields become continuous. With such a modification the field $\hat{\chi}$ becomes the smooth field χ_0. Denote the smoothened components of the field by $\chi^{(sm)}_{231}$ and $\chi^{(sm)}_{312}$. Thus,

$$\chi_0 = \chi_{123} + \chi_{213} + \chi_{132} + \chi_{321} + \chi^{(sm)}_{231} + \chi^{(sm)}_{312}. \tag{17}$$

Let us write down the formulas for the last two fields. Following the computations of the last section we have

$$\chi_{231}^{(sm)} = \psi_1^- \zeta_1 + \psi_2^+ \zeta_5 + \psi_{231}^{(c)} \widetilde{\zeta} + \psi_{231}^{(sm)}.$$

Explicit formulas for $\psi_{231}^{(sm)}$ are given in the present subsection.

Similarly,

$$\chi_{312}^{(sm)} = \psi_3^+ \eta_1 + \psi_2^- \eta_5 + \psi_{312}^{(c)} \widetilde{\eta} + \psi_{312}^{(sm)}.$$

In conclusion we formulate a theorem that will be rigorously proved in subsequent publications. As above, denote the "scattered plane wave" by $\chi(\mathbf{z}, \mathbf{q})$ and recall that $\mathbf{q} \in \lambda_{123}$, $\mathbf{q} \neq 0$.

THEOREM 4.1. *The function $\chi_0(\mathbf{z}, \mathbf{q})$ such that the difference $\chi(\mathbf{z}, \mathbf{q}) - \chi_0(\mathbf{z}, \mathbf{q})$ is, asymptotically, as $|\mathbf{z}| \to \infty$, a diverging circle wave with a continuous amplitude, can be described by formula* (17).

References

[1] L. D. Faddeev, *Mathematical questions in the quantum theory of scattering for a system of three particles.* Trudy Mat. Inst. Steklov. **69** (1963), 122 pp. (Russian)

[2] S. P. Merkuriev, *Coordinate asymptotic behavior of* (3 → 3) *wave functions for a system of three charged particles.* Teoret. Mat. Fiz. **32** (1977), no. 2, 187–207. (Russian)

[3] V. S. Buslaev, S. P. Merkuriev, and S. P. Salikov, *Diffraction characteristics of scattering in a quantum system of three one-dimensional particles.* Scattering theory. Theory of oscillations, Probl. Mat. Fiz., Leningrad. Univ., Leningrad, **9**, (1979), 14–30. (Russian)

[4] ———, *Description of pair potentials for which the scattering in the quantum system of three one-dimensional particles is free of diffractional effects.* Zap. Nauchn. Sem. Leningrad. Otdel. Mat. Inst. Steklov. (LOMI) **84**, (1979), 16–22. (Russian)

V. A. FOCK INSTITUTE FOR PHYSICS, ST. PETERSBURG STATE UNIVERSITY, 198504 ST. PETERSBURG, RUSSIA

Amer. Math. Soc. Transl.
(2) Vol. **225**, 2008

Spectral Asymptotics of the Maxwell Operator on Lipschitz Manifolds with Boundary

M. N. Demchenko and N. D. Filonov

To Mikhail Shlëmovich Birman on his 80th birthday

ABSTRACT. The Maxwell operator is considered on a compact Lipschitz manifold with boundary. The Weyl asymptotics for its spectrum is proved.

1. Introduction

1.1. In the early twentieth century, H. Weyl developed a general technique of calculation of spectral asymptotics for numerous operators of mathematical physics. The problem of an electromagnetic resonator $\Omega \subset \mathbb{R}^3$ with a perfectly conducting boundary $\partial\Omega$ was one of the examples. Frequencies of electromagnetic oscillations correspond to the spectrum of the Maxwell operator, which acts on a pair $\{E, H\}$ by the formula

$$\mathcal{M}\{E, H\} = \{i\varepsilon^{-1} \operatorname{curl} H, -i\mu^{-1} \operatorname{curl} E\}, \tag{1.1}$$

where E, H denote electric and magnetic components of the field. The divergence free conditions

$$\operatorname{div}(\varepsilon E) = \operatorname{div}(\mu H) = 0 \tag{1.2}$$

and the boundary conditions

$$E_\tau \mid_{\partial\Omega} = (\mu H)_\nu \mid_{\partial\Omega} = 0 \tag{1.3}$$

are supposed to be fulfilled. Here $\varepsilon(x)$, $\mu(x)$ are matrix-valued functions describing the dielectric permittivity and the magnetic permeability of the medium which fills the resonator Ω. Indices τ and ν denote the tangential and normal components of a vector on $\partial\Omega$. The principal term of the spectral asymptotics for an empty resonator with smooth boundary was calculated by H. Weyl [**13**] in 1912:

$$m_k \sim \left(\frac{\operatorname{mes}\Omega}{3\pi^2}\right)^{-1/3} k^{1/3}, \quad k \to \infty, \tag{1.4}$$

2000 *Mathematics Subject Classification.* Primary 35Q60; Secondary 35P20.
Key words and phrases. Spectral asymptotics, Maxwell operator, Lipschitz manifolds.
The second author was supported by the EPSRC grant GR/T25552/01.

where m_k are the positive eigenvalues of the Maxwell operator (with $\varepsilon = \mu = \mathbb{1}$) enumerated in increasing order.

The problem becomes considerably more difficult when the boundary $\partial\Omega$ or ε or μ becomes nonsmooth. Only in 1976 did A. B. Alexeev and M. Sh. Birman [**2**] justify the asymptotics of the spectrum of the Maxwell operator for nonsmooth bounded ε, μ and a smooth boundary $\partial\Omega$. In this case, the asymptotic formula becomes more complicated than (1.4); see (4.3), (5.2) below. In 1987, M. Sh. Birman and M. Z. Solomyak [**8**] proved (1.4) for the case of a Lipschitz domain and $\varepsilon = \mu = \mathbb{1}$. Note that in this situation the Weyl variational technique cannot be applied directly. Moreover, even the proper definition of a selfadjoint Maxwell operator for nonsmooth $\partial\Omega$ meets certain obstacles (see [**7**] for the details).

In [**4, 3**], the methods of [**2**] and [**8**] were combined and the Weyl asymptotics was obtained for the case of nonsmooth (bounded, measurable and positive definite) coefficients ε, μ and a Lipschitz boundary $\partial\Omega$. The purpose of this paper is to extend the latter result to the case of a Lipschitz arbitrary-dimensional manifolds with boundary (see Theorem 4.3 below). The Maxwell operator on manifolds was considered by various authors, but to the best of our knowledge, the question of spectral asymptotics in non-smooth situations has not been studied before.

1.2. In [**14**], H. Weyl used the language of differential forms in order to extend the stationary Maxwell system in Euclidean space to the case of an arbitrary dimension. This system was extended to smooth Riemannian manifolds in [**12**]. In [**14, 12**], the electric and magnetic fields $\tilde{E}$ and $\tilde{H}$ are treated as k- and $(k+1)$-forms, while the coefficients $\tilde{\varepsilon}$ and $\tilde{\mu}$ are considered as linear transformations on k- and $(k+1)$-forms respectively. In this context the Maxwell operator acts in the following way:

$$\tilde{\mathcal{M}}\{\tilde{E}, \tilde{H}\} = \{-i\tilde{\varepsilon}^{-1}(-1)^{k(n-k-1)} * d * \tilde{H}, -i\tilde{\mu}^{-1}d\tilde{E}\},$$

where $*$ denotes the Hodge operator. We propose a different generalization of the Maxwell operator, which does not require a Riemannian structure on the manifold. In our case E is a k-form, H is an $(n-k-1)$-form, ε is a linear operator from the space of k-forms to the space of $(n-k)$-forms, and μ is a linear operator from the space of $(n-k-1)$-forms to the space of $(k+1)$-forms. Then the operator M acts as follows:

$$M\{E, H\} = \{(-1)^{k+1}\varepsilon^{-1}dH, \mu^{-1}dE\}.$$

If the manifold is equipped with some metric tensor, then one can easily construct tensors $\tilde{\varepsilon}$, $\tilde{\mu}$ and the operator $\tilde{M}$ corresponding to given ε and μ. The operators M and $\tilde{M}$ will be unitarily equivalent regardless of the metric tensor. Therefore we find it more reasonable to work with M, which does not require any Riemannian structure. Our approach turns out to be considerably less demanding to the smoothness of the manifold.

Note that domains in Euclidean space with boundary singularities such as screens and conical points can also be considered as Lipschitz manifolds (see Section 5.2 below). Thus, the main result of this paper (Theorem 4.3) yields the Weyl asymptotics for such domains. This result has been previously known [**4**, Section 5] only under some additional restrictions.

We use the following strategy of the proof. We reduce the question of the spectral asymptotics on a manifold to the same question for a Euclidean domain by means of localization. The asymptotics for Euclidean domains of arbitrary dimension can be proved by following the approach of [**4**] (where the three-dimensional case was considered); the proof of this will be published separately in [**10**]. Note that most of the ideas used for reducing the case of a manifold to the case of a Euclidean domain are borrowed from [**8**].

In Section 2 we recall some facts from the theory of Lipschitz manifolds. The Maxwell operator on a manifold is defined in Section 3. In Section 4 we state the main result. In Section 5, for further reference we state a result (Theorem 5.1) on the Weyl asymptotics for the Maxwell operator in Euclidean domains of arbitrary dimension and discuss some particular cases; the proof of this result will be published in [**10**]. In Section 6 we reformulate the problem of the Weyl asymptotics for the Maxwell operator on a manifold in terms of a quotient of quadratic forms. Section 7 contains a few lemmas necessary for the reduction to a Euclidean domain. Finally, in Section 8 we derive the main result from Theorem 5.1.

The authors thank A. Pushnitski for useful discussions.

2. Lipschitz manifolds

2.1. Differential forms. In this section we recall the basic concepts of Lipschitz manifold theory (see, e.g., [**11**]). We denote $\mathbb{R}^n_+ = \{x = (x^1, \dots, x^n) \in \mathbb{R}^n : x^n \geq 0\}$. We say that $U \subset \mathbb{R}^n_+$ is open if it is open in the relative topology of $\mathbb{R}^n_+$. Let $U \subset \mathbb{R}^n_+$ be open; a map $\phi : U \to \mathbb{R}^m$ is said to be *locally Lipschitz* if its restriction on every compact $K \subset U$ is Lipschitz. The bijective map $\phi : U \to V$ is called *locally bi-Lipschitz*, if ϕ and ϕ^{-1} are locally Lipschitz. An n-dimensional topological manifold X with boundary ∂X is called a *Lipschitz manifold* if it has an atlas $\{V_\alpha, \alpha\}$ such that for any two charts $\alpha : V_\alpha \to \mathbb{R}^n_+$ and $\gamma : V_\gamma \to \mathbb{R}^n_+$ the corresponding transition map $\gamma \circ \alpha^{-1}$ is locally bi-Lipschitz.

For an open set $U \subset \mathbb{R}^n_+$ and $1 \leq p \leq \infty$ we define $\mathcal{L}^k_p(U)$ as a class of k-forms, having coefficients in a standard basis from $L_{p,loc}(U)$. Lipschitz maps are almost everywhere differentiable; hence for a locally bi-Lipschitz map $\phi : U \to V$, $U, V \subset \mathbb{R}^n_+$, one can define the pullback map ϕ^* of differential forms as usual. The pullback map ϕ^* preserves the classes $\mathcal{L}^k_p : \phi^* \mathcal{L}^k_p(V) = \mathcal{L}^k_p(U)$.

Let X be a Lipschitz manifold and $\{V_\alpha, \alpha\}$ be an atlas on X. Let $\{\omega_\alpha\}$ be a set of complex-valued differential forms defined on sets V_α. We call $\omega = \{\omega_\alpha\}$ a differential form on X if for any two charts (V_α, α) and (V_γ, γ) the equality $(\alpha \circ \gamma^{-1})^* \omega_\alpha = \omega_\gamma$ holds. Let

$$\mathcal{L}^k_p(X) = \{k\text{-forms } \omega = \{\omega_\alpha\} \,:\, \omega_\alpha \in \mathcal{L}^k_p(\alpha(V_\alpha)) \; \forall \alpha\}.$$

It is clear that

$$\omega \in \mathcal{L}^k_p(X), \quad \theta \in \mathcal{L}^l_q(X) \quad \Rightarrow \quad \omega \wedge \theta \in \mathcal{L}^{k+l}_r(X), \quad 1/p + 1/q = 1/r.$$

We also denote

$$\mathcal{W}^k(X) = \{\omega \in \mathcal{L}^k_2(X) \,:\, d\omega \in \mathcal{L}^{k+1}_2(X)\}.$$

Note that the space $\mathcal{W}^0$ is exactly the Sobolev space of scalar functions $H^1(X)$, while $\mathcal{W}^n = \mathcal{L}^n_2$.

For a compact oriented manifold X and $\omega \in \mathcal{L}^n_1(X)$ the integral $\int_X \omega$ is defined as usual and is finite.

2.2. Boundary conditions. Let X be a compact oriented manifold with boundary. Let $j : \partial X \to X$ be the natural embedding and ω be a differential form on X. We need to define the equality $j^*\omega = 0$. In order to motivate the following definition, note that, for a smooth manifold X and a smooth k-form ω, the equality $j^*\omega = 0$ is equivalent to the identity $\int_{\partial X} j^*(\omega \wedge \theta) = 0$ for all $(n-k-1)$-forms θ on X. Having in mind Stokes' formula, we introduce

DEFINITION 2.1. Let $\omega \in \mathcal{W}^k$, $k < n$. The equality $j^*\omega = 0$ means that

$$\int_X \omega \wedge d\theta = (-1)^{k+1} \int_X d\omega \wedge \theta \quad \forall \theta \in \mathcal{W}^{n-k-1}.$$

For a form $\omega \in \mathcal{W}^n = \mathcal{L}_2^n$ the condition $j^*\omega = 0$ is always supposed to be satisfied.

REMARK 2.2. It is easy to check that $j^*\omega = 0$ implies $j^*(d\omega) = 0$..

2.3. Coefficients. Fix a standard basis in $\mathbb{R}^n$ and consider a linear operator $\beta : \mathcal{F}^k(n) \to \mathcal{F}^{n-k}(n)$, where $\mathcal{F}^k(n)$ is the space of all antisymmetric k-linear complex-valued functions ω in $\mathbb{R}^n$. We say that $\beta \in M_+^k(n)$ if $\overline{\beta\omega} = \beta\overline{\omega}$ and the function $b(\omega)$ specified by the equality $\omega \wedge \beta\overline{\omega} = b(\omega) dx^1 \wedge \cdots \wedge dx^n$ is positive for any $\omega \neq 0$. For any $\beta \in M_+^k(n)$, one has the identity

$$\omega \wedge \beta\theta = \theta \wedge \beta\omega \quad \omega, \theta \in \mathcal{F}^k(n).$$

The proof of this identity is analogous to the proof of the elementary statement from linear algebra that if the quadratic form of a matrix is real valued, then this matrix is Hermitian. For further reference, we rewrite the above identity as

$$\omega \wedge \beta\theta = (-1)^{k(n-k)} \beta\omega \wedge \theta, \quad \omega, \theta \in \mathcal{F}^k(n). \tag{2.1}$$

Next, $\beta \in M_+^k(n)$ has an inverse operator

$$\beta^{-1} : \mathcal{F}^{n-k}(n) \to \mathcal{F}^k(n),$$

and $(-1)^{k(n-k)}\beta^{-1} \in M_+^{n-k}(n)$. Now we extend this definition to Lipschitz manifolds.

DEFINITION 2.3. A linear map $\beta : \mathcal{L}_2^k(X) \to \mathcal{L}_2^{n-k}(X)$ belongs to the class $M_+^k(X)$ if for every chart (V_α, α) there is a function $\beta_\alpha \in L_\infty(\alpha(V_\alpha), M_+^k(n))$ satisfying

$$(\beta\omega)_\alpha = \beta_\alpha \omega_\alpha \quad \forall \omega \in \mathcal{L}_2^k(X),$$

$$(-1)^{k(n-k)} \beta_\alpha^{-1} \in L_\infty(\alpha(V_\alpha), M_+^{n-k}(n)).$$

REMARK 2.4. $\beta \in M_+^k(X) \Leftrightarrow (-1)^{k(n-k)}\beta^{-1} \in M_+^{n-k}(X)$.

From now on we suppose that for every $k = 0, 1, \ldots, n$, some $\beta_k \in M_+^k(X)$ is chosen. We define $\mathcal{L}_2^k(\beta_k)$ as a Hilbert space $\mathcal{L}_2^k$ with the inner product

$$(\omega, \theta)_{\beta_k} = \int_X \omega \wedge \beta_k \overline{\theta}. \tag{2.2}$$

2.4. The operator d_k. Let us introduce the operator

$$d_k : \mathcal{L}_2^k(\beta_k) \to \mathcal{L}_2^{k+1}(\beta_{k+1}), \quad d_k\omega = d\omega, \quad \operatorname{Dom} d_k = \{\omega \in \mathcal{W}^k : j^*\omega = 0\}.$$

It is easy to see that d_k is densely defined.

LEMMA 2.5. *The operator d_k is closed. The adjoint operator d_k^* acts as*

$$d_k^*\theta = (-1)^{k+1}\beta_k^{-1}d(\beta_{k+1}\theta), \quad \operatorname{Dom} d_k^* = \{\theta \in \mathcal{L}_2^{k+1} : d(\beta_{k+1}\theta) \in \mathcal{L}_2^{n-k}\}.$$

PROOF. The space $\operatorname{Dom} d_k$ is complete with respect to the scalar product $\int d\omega \wedge \beta_{k+1}(\overline{d\omega}) + \int \omega \wedge \beta_k\overline{\omega}$, which is an immediate consequence of [**11**]. It follows that d_k is closed. Now let $\theta \in \operatorname{Dom} d_k^*$ and $\eta = d_k^*\theta$, i.e.,

$$\int_X d\omega \wedge \beta_{k+1}\overline{\theta} = \int_X \omega \wedge \beta_k\overline{\eta} \qquad \forall \omega \in \mathcal{W}^k, \ j^*\omega = 0. \tag{2.3}$$

Then $(-1)^{k+1}d(\beta_{k+1}\theta) = \beta_k\eta$. On the other hand, if $\theta \in \mathcal{L}_2^{k+1}$ and $d(\beta_{k+1}\theta) \in \mathcal{L}_2^{n-k}$, then the equality (2.3) holds by virtue of Definition 2.1. □

3. The Maxwell operator

3.1. Function spaces. Here we introduce a number of spaces of k-forms. First of all, let

$$G^k = \{d\varphi : \varphi \in \mathcal{W}^{k-1}\}, \quad G_0^k = \{d\varphi : \varphi \in \mathcal{W}^{k-1}, j^*\varphi = 0\}$$

for $k \geq 1$, and suppose $G_0^0 = G^0 = \{0\}$. Furthermore,

$$J^k(\beta_k) = \{\omega \in \mathcal{L}_2^k : d(\beta_k\omega) = 0\}, \quad J_0^k(\beta_k) = \{\omega \in J^k(\beta_k) : j^*(\beta_k\omega) = 0\}.$$

Clearly $J_0^0 = J^0 = \mathcal{L}_2^0$. Recalling the definition (2.2) of the inner product in $\mathcal{L}_2^k(\beta_k)$, it is not difficult to see that the orthogonal complement to G_0^k is J^k, and the orthogonal complement to G^k is J_0^k:

$$\mathcal{L}_2^k(\beta_k) = G_0^k \oplus_{\beta_k} J^k = G^k \oplus_{\beta_k} J_0^k. \tag{3.1}$$

This orthogonal decomposition appears in several branches of the theory of PDE and differential geometry and is attributed to Weyl, Helmholtz, and Hodge.

In what follows we suppose $k \leq n-1$. Introduce the Hilbert space

$$F^k(\beta_k) = \{\omega \in \mathcal{W}^k : d(\beta_k\omega) \in \mathcal{L}_2^{n-k+1}\}$$

with the norm

$$\|\omega\|_{F^k}^2 = \int_X d\omega \wedge \beta_{k+1}(\overline{d\omega}) + \beta_{k-1}^{-1}d(\beta_k\omega) \wedge d(\overline{\beta_k\omega}) + \omega \wedge \beta_k\overline{\omega} \tag{3.2}$$

(in the case $k = 0$ the second term vanishes), and its subspaces of "electric" and "magnetic" fields

$$F_{el}^k(\beta_k) = \{\omega \in F^k(\beta_k) : j^*\omega = 0\}, \quad F_m^k(\beta_k) = \{\omega \in F^k(\beta_k) : j^*(\beta_k\omega) = 0\}.$$

REMARK 3.1. For $k = 0$ the space F_{el}^0 coincides with the usual Sobolev space $\mathring{H}^1(X)$, and $F_m^0 = F^0$ coincides with $H^1(X)$.

Furthermore, we set

$$\Phi_{el}^k(\beta_k) = F_{el}^k(\beta_k) \cap J^k(\beta_k), \quad E_{el}^k(\beta_k) = F_{el}^k(\beta_k) \cap G_0^k,$$

$$\Phi_m^k(\beta_k) = F_m^k(\beta_k) \cap J_0^k(\beta_k), \quad E_m^k(\beta_k) = F_m^k(\beta_k) \cap G^k.$$

Obviously,

$$F^k_{el} = \Phi^k_{el} \oplus_{\beta_k} E^k_{el}, \quad F^k_m = \Phi^k_m \oplus_{\beta_k} E^k_m, \tag{3.3}$$
$$\Phi^0_{el} = \mathring{H}^1(X), \quad \Phi^0_m = H^1(X), \quad E^0_{el} = E^0_m = \{0\}.$$

Finally, we set

(3.4)

$$K^k_{el}(\beta_k) = \{\omega \in \Phi^k_{el}(\beta_k) : d\omega = 0\} = \{\omega \in \mathcal{L}^k_2 : d\omega = 0,\ d(\beta_k\omega) = 0,\ j^*\omega = 0\}.$$

In Section 7 we prove that the space $K^k_{el}(\beta_k)$ is finite dimensional (and may be trivial); see Corollary 7.7. For the operator d_k, introduced in Section 2.4, we have

$$\ker d_k = G^k_0 \oplus_{\beta_k} K^k_{el}(\beta_k). \tag{3.5}$$

REMARK 3.2. Recall (see Remark 2.4) that

$$\beta_k \in M^k_+(X) \Leftrightarrow (-1)^{k(n-k)}\beta_k^{-1} \in M^{n-k}_+(X).$$

In what follows we shall work with form spaces depending on $(-1)^{k(n-k)}\beta_k^{-1}$. In order to simplify our notation, we shall write $J^{n-k}(\beta_k^{-1})$, etc. instead of $J^{n-k}((-1)^{k(n-k)}\beta_k^{-1})$, etc.

3.2. The Maxwell operator: definition. Our purpose is to construct a self-adjoint first-order differential operator from blocks of type βd. Obviously, the spaces G^k and G^k_0 of exact forms will be in the kernels of the corresponding blocks. Therefore, we shall work with their orthogonal complements J^k_0, J^k (see (3.1)).

For $k \le n-1$ introduce the operator (see Remark 3.2 for the space in the r.h.s.)

$$R_k(\beta_k, \beta_{k+1}) : J^k(\beta_k) \to J^{n-k-1}_0(\beta_{k+1}^{-1}),$$
$$R_k\omega = \beta_{k+1} d\omega, \quad \mathrm{Dom}\, R_k = \Phi^k_{el}(\beta_k).$$

The operator R_k is closed, and for its kernel we have $\ker R_k = K^k_{el}(\beta_k)$. The description of the adjoint operator can be obtained similarly to the proof of Lemma 2.5:

$$R_k(\beta_k, \beta_{k+1})^* : J^{n-k-1}_0(\beta_{k+1}^{-1}) \to J^k(\beta_k),$$
$$R_k^*\theta = (-1)^{k+1}\beta_k^{-1} d\theta, \quad \mathrm{Dom}\, R_k^* = \Phi^{n-k-1}_m(\beta_{k+1}^{-1}).$$

DEFINITION 3.3. The Maxwell operator $M_k(\beta_k, \beta_{k+1})$, $k = 0, 1, ..., n-1$, is the blockwise operator in the Hilbert space $J^k(\beta_k) \oplus J^{n-k-1}_0(\beta_{k+1}^{-1})$ given by

$$M_k(\beta_k, \beta_{k+1}) = \begin{pmatrix} 0 & R_k^* \\ R_k & 0 \end{pmatrix},$$
$$M_k \begin{pmatrix} \omega \\ \theta \end{pmatrix} = \begin{pmatrix} (-1)^{k+1}\beta_k^{-1} d\theta \\ \beta_{k+1} d\omega \end{pmatrix},$$
$$\mathrm{Dom}\, M_k(\beta_k, \beta_{k+1}) = \Phi^k_{el}(\beta_k) \oplus \Phi^{n-k-1}_m(\beta_{k+1}^{-1}).$$

The operator $M_k(\beta_k, \beta_{k+1})$ is selfadjoint and its kernel is finite-dimensional. Later on, we will show that the spectrum $\sigma(M_k)$ is discrete. By the identity

$$\begin{pmatrix} I & 0 \\ 0 & -I \end{pmatrix} \begin{pmatrix} 0 & R_k^* \\ R_k & 0 \end{pmatrix} = - \begin{pmatrix} 0 & R_k^* \\ R_k & 0 \end{pmatrix} \begin{pmatrix} I & 0 \\ 0 & -I \end{pmatrix},$$

the operators M_k and $-M_k$ are unitarily equivalent and so the spectrum $\sigma(M_k)$ is symmetric with respect to zero.

For a selfadjoint operator A with a discrete spectrum, we denote by $N(\lambda, A)$ the number of eigenvalues of A (enumerated with multiplicities taken into account) in the interval $(0, \lambda)$. The main object of our research is the counting function $N(\lambda, M_k(\beta_k, \beta_{k+1}))$. The main result (see Theorem 4.3) is the asymptotics of this counting function as $\lambda \to \infty$.

REMARK 3.4. It is easy to see that

$$M_k^2 = \begin{pmatrix} R_k^* R_k & 0 \\ 0 & R_k R_k^* \end{pmatrix}.$$

The operators $R_k^* R_k$ and $R_k R_k^*$ are unitarily equivalent on the orthogonal complements to their kernels; therefore

$$N(\lambda, M_k(\beta_k, \beta_{k+1})) = N(\lambda^2, R_k(\beta_k, \beta_{k+1})^* R_k(\beta_k, \beta_{k+1})). \tag{3.6}$$

Note that the zero eigenvalue of $R_k^* R_k$ is not counted in the r.h.s.

4. Asymptotics of the spectrum

The leading term of the asymptotics of $N(\lambda, M_k)$ as $\lambda \to \infty$ involves a fairly complicated expression. In order to write down this expression, we need to consider some auxiliary objects, which is done in Section 4.1. The asymptotics of $N(\lambda, M_k)$ is stated in Section 4.2.

4.1. Algebraic problem. To every point $x \in X$ we relate an auxiliary finite-dimensional spectral problem. Fix a point x in a chart (V_α, α) and a real 1-form $\xi \in \mathcal{F}^1(n)$. Let us consider the algebraic problem

$$\xi \wedge \beta_{k+1,\alpha}(x)(h \wedge \xi) = \Lambda \beta_{k,\alpha}(x) h, \tag{4.1}$$

where $h \in \mathcal{F}^k(n)$ and $\Lambda \in \mathbb{C}$ are unknown. After multiplying (4.1) by $\overline{h}$ on the left, we get

$$(\overline{h} \wedge \xi) \wedge \beta_{k+1,\alpha}(h \wedge \xi) = \Lambda \overline{h} \wedge \beta_{k,\alpha} h,$$

whence all eigenvalues Λ are nonnegative by virtue of the positiveness of β_k and β_{k+1}. Let $0 < \Lambda_1(x, \xi) \leq \Lambda_2(x, \xi) \leq \dots$ be the nonzero eigenvalues of the problem (4.1) enumerated with multiplicities taken into account. We introduce the counting function $N_{k,\alpha}(\Lambda; x, \xi) = \#\{j : \Lambda_j(x, \xi) < \Lambda\}$. Further, let us define

$$\nu_{k,\alpha}(\Lambda; x) = (2\pi)^{-n} \int_{\mathbb{R}^n} N_{k,\alpha}(\Lambda; x, \xi) d^n \xi; \tag{4.2}$$

here we use the standard identification of real 1-forms and vectors in $\mathbb{R}^n$. Thus we have a collection of functions $\{\nu_{k,\alpha}(\Lambda)\}$ on the sets V_α. These functions behave as n-forms under the coordinate transformations, which is the content of the following

LEMMA 4.1. *For all $\Lambda > 0$ the collection $\{\nu_{k,\alpha}(\Lambda)\}$ (see (4.2)) defines an n-form $\nu_k(\Lambda)$ on the manifold X, and $\nu_k(\Lambda) \in \mathcal{L}_\infty^n(X)$.*

PROOF. Let (V_α, α), (V_γ, γ) be two intersecting charts, and $\phi = \alpha \circ \gamma^{-1}$ be a transition map. We need to verify that

$$\nu_{k,\alpha}(\Lambda; x) = |\partial\phi/\partial x| \, \nu_{k,\gamma}(\Lambda; \phi^{-1}(x)),$$

where $|\partial\phi/\partial x|$ is the corresponding Jacobian. If the parameters ξ_α and ξ_γ are related by the rule of transformation of 1-forms, i.e. $\xi_\gamma = \phi^* \xi_\alpha$, then the solutions of (4.1) are related by $h_{j,\gamma} = \phi^* h_{j,\alpha}$, $\Lambda_{j,\gamma} = \Lambda_{j,\alpha}$, and so

$$N_{k,\alpha}(\Lambda; x, \xi_\alpha) = N_{k,\gamma}(\Lambda; \phi^{-1}(x), \xi_\gamma)$$

(in the case of a smooth manifold X we could say that $N_k(\Lambda; x, \xi)$ is a scalar function on the cotangent bundle). Therefore,

$$\begin{aligned}\nu_{k,\alpha}(\Lambda; x) &= (2\pi)^{-n} \int_{\mathbb{R}^n} N_{k,\alpha}(\Lambda; x, \xi_\alpha)\, d^n\xi_\alpha \\ &= (2\pi)^{-n} \int_{\mathbb{R}^n} N_{k,\gamma}(\Lambda; \phi^{-1}(x), \xi_\gamma)|\partial\phi/\partial x| d^n\xi_\gamma = |\partial\phi/\partial x|\, \nu_{k,\gamma}(\Lambda; \phi^{-1}(x)).\end{aligned}$$

It is easy to see that the forms $\nu_{k,\alpha}(\Lambda)$ are measurable and bounded, so $\nu_k(\Lambda) \in \mathcal{L}^n_\infty(X)$. □

4.2. Formulation of the result. First of all, we show that the forms $\nu_k(\Lambda)$ are homogeneous in Λ of order $n/2$. Again, we fix a point x in the chart (V_α, α). The eigenvalues Λ_j of the problem (4.1) are homogeneous in ξ of order two, $\Lambda_j(x, r\xi) = r^2\Lambda_j(x, \xi)$ for all $r > 0$. Therefore,

$$\begin{aligned}\int_{\mathbb{R}^n} N_{k,\alpha}(\Lambda; x, \xi)\, d^n\xi &= \sum_j \operatorname{mes}\{\xi : \Lambda_j(x, \xi) < \Lambda\} \\ &= \sum_j \int_{|\xi|=1} \int_0^{(\Lambda/\Lambda_j(x,\xi))^{1/2}} r^{n-1} dr\, dS(\xi) = \frac{\Lambda^{n/2}}{n} \int_{|\xi|=1} \sum_j \Lambda_j(x, \xi)^{-n/2} dS(\xi),\end{aligned}$$

where $dS(\xi)$ is the Lebesgue measure on the unit sphere. Thus, we have proved the following

LEMMA 4.2. *The form $\nu_k(\Lambda)$ defined by (4.2) is homogeneous in Λ of order $n/2$, $\nu_k(\Lambda) = \Lambda^{n/2}\nu_k(1)$. The form $\nu_k = \nu_k(1) \in \mathcal{L}^n_\infty(X)$ is given by*

$$\nu_k(x) = \frac{1}{n(2\pi)^n} \int_{|\xi|=1} \sum_j \Lambda_j(x, \xi)^{-n/2} dS(\xi)$$

in the local coordinates.

Now we are ready to state our main result.

THEOREM 4.3. *Let X be a compact oriented Lipschitz manifold with boundary, $\beta_k \in M^k_+(X)$, $\beta_{k+1} \in M^{k+1}_+(X)$. Let $M_k(\beta_k, \beta_{k+1})$ be the Maxwell operator defined in Section 3.2. Then for the counting function of the positive eigenvalues of $M_k(\beta_k, \beta_{k+1})$, the asymptotics*

$$N(\lambda, M_k(\beta_k, \beta_{k+1})) \sim \int_X \nu_k(\lambda^2) = \lambda^n \int_X \nu_k, \quad \lambda \to \infty, \tag{4.3}$$

holds true, where $\nu_k(\Lambda)$ and ν_k are the forms defined above.

REMARK 4.4. Here and in what follows the symbol $\sim$ means that the ratio of the quantities in the left and right hand sides tends to 1.

For the positive eigenvalues $0 < m_1 \le m_2 \le \cdots$ of $M_k(\beta_k, \beta_{k+1})$ this asymptotic formula means that the limit

$$\lim_{j\to\infty} j^{-1/n} m_j = \left(\int_X \nu_k\right)^{-1/n}$$

exists.

REMARK 4.5. The statement of Theorem 4.3 is valid also in the case when X is a manifold without boundary. In this case, the definition of the Maxwell operator is simpler: there are no boundary conditions, and some function spaces coincide,

$$G_0^k = G^k, \quad J_0^k = J^k, \quad F_{el}^k = F_m^k = F^k, \quad \Phi_{el}^k = \Phi_m^k, \quad E_{el}^k = E_m^k.$$

The proof of Theorem 4.3 is the same as in the case of a nonempty boundary.

5. Domains in Euclidean space

In this section we consider as an example a particular case of the Lipschitz manifold X. We will consider $X = \overline{\Omega}$, where $\Omega \subset \mathbb{R}^n$ is a bounded domain with Lipschitz boundary (in a neighborhood of every point of the boundary $\partial\Omega$ there is a local coordinate system in $\mathbb{R}^n$, such that the domain in this system is the subgraph of a Lipschitz function).

THEOREM 5.1. *Let $\Omega \subset \mathbb{R}^n$ be a bounded domain with Lipschitz boundary, $\beta_k \in M_+^k(\overline{\Omega})$, $\beta_{k+1} \in M_+^{k+1}(\overline{\Omega})$, and let $M_k(\beta_k, \beta_{k+1})$ be the Maxwell operator. Then for the counting function $N(\lambda, M_k(\beta_k, \beta_{k+1}))$ the asymptotics* (4.3) *holds true.*

This theorem is proved in [**4, 3**] for the case $n = 3$, $k = 1$ (see also Section 5.2 below). The proof for an arbitrary dimension n and an arbitrary order k uses the same ideas and will be published in [**10**].

5.1. Scalar problems. Let us discuss the case $k = 0$. It is natural to identify $\beta_0(x)$ with a scalar positive function, and $\beta_1(x)$ with an $(n \times n)$-matrix-valued positive definite function; $\beta_0, \beta_1 \in L_\infty(\Omega)$. The operator $R_0^* R_0$ coincides with the operator

$$A_0 = -\beta_0(x)^{-1} \operatorname{div}(\beta_1(x)\nabla \,\cdot)$$

of the Dirichlet problem in the weighted space $L_2(\Omega, \beta_0(x)d^n x)$. The problem (4.1) reads as follows:

$$\langle \beta_1(x)\xi, \xi\rangle h = \Lambda\beta_0(x)h, \quad h \in \mathbb{C} \setminus \{0\},$$

where ξ is identified with a vector in $\mathbb{R}^n$. Thus, there exists only one eigenvalue $\Lambda(x,\xi) = \beta_0(x)^{-1}\langle \beta_1(x)\xi, \xi\rangle$. Therefore,

$$\nu_0(\Lambda, x) = (2\pi)^{-n} \operatorname{mes}\{\xi : \Lambda(x,\xi) < \Lambda\} = \frac{\kappa_n \Lambda^{n/2}\beta_0(x)^{n/2}}{(2\pi)^n(\det \beta_1(x))^{1/2}},$$

where κ_n is the volume of the unit ball in $\mathbb{R}^n$. Taking into account Remark 3.4, the asymptotic formula (4.3) can be recast as

$$N(\lambda, A_0) \sim \frac{\lambda^{n/2}\kappa_n}{(2\pi)^n}\int_\Omega \frac{\beta_0(x)^{n/2} d^n x}{(\det \beta_1(x))^{1/2}}, \quad \lambda \to \infty. \tag{5.1}$$

In a similar manner, if we consider the case $k = n - 1$ and the operator $R_{n-1}R_{n-1}^*$, we get the Neumann problem for a second-order equation. The asymptotics of the type (5.1) is also valid in this case. These results for domains in the Euclidean space are well known; see for example [**5**] (where the asymptotics for the Dirichlet problem is proved not only for Lipschitz domains, but for arbitrary bounded open sets). Note also that the Weyl asymptotics of the spectrum of the scalar problems on the Lipschitz manifold can be easily deduced from the corresponding results for the domains in the Euclidean space by the so-called "Dirichlet-Neumann bracketing".

5.2. The Maxwell operator. Let $n = 3$, $k = 1$, and let $\Omega \subset \mathbb{R}^3$ be a bounded domain with Lipschitz boundary. It is natural to identify 1-forms ω, θ with vector-valued functions E, H, and the coefficients β_1, β_2 with (3×3)-matrix-valued functions; also, let $\varepsilon = \beta_1$, $\mu = \beta_2^{-1}$. The function spaces J and Φ can be written in the following way:

$$J^1(\varepsilon) = \{E \in L_2(\Omega, \mathbb{C}^3) : \operatorname{div}(\varepsilon E) = 0\},$$
$$J_0^1(\mu) = \{H \in L_2(\Omega, \mathbb{C}^3) : \operatorname{div}(\mu H) = 0,\ (\mu H)_\nu \mid_{\partial\Omega} = 0\},$$
$$\Phi_{el}^1(\varepsilon) = \{E \in L_2(\Omega, \mathbb{C}^3) : \operatorname{curl} E \in L_2(\Omega, \mathbb{C}^3),\ \operatorname{div}(\varepsilon E) = 0,\ E_\tau \mid_{\partial\Omega} = 0\},$$
$$\Phi_m^1(\mu) = \{H \in L_2(\Omega, \mathbb{C}^3) : \operatorname{curl} H \in L_2(\Omega, \mathbb{C}^3),\ \operatorname{div}(\mu H) = 0,\ (\mu H)_\nu \mid_{\partial\Omega} = 0\}.$$

Thus, conditions (1.2), (1.3) are provided by the definition of the spaces $\Phi_{el}^1(\varepsilon)$, $\Phi_m^1(\mu)$. The Maxwell operator acts in the space $J^1(\varepsilon) \oplus J_0^1(\mu)$ on the domain $\Phi_{el}^1(\varepsilon) \oplus \Phi_m^1(\mu)$ by the formula

$$M_1 \begin{pmatrix} E \\ H \end{pmatrix} = \begin{pmatrix} \varepsilon^{-1} \operatorname{curl} H \\ \mu^{-1} \operatorname{curl} E \end{pmatrix}.$$

The Maxwell operator $\mathcal{M}$, defined in the papers [**6, 7**] and corresponding to the formulas (1.1) – (1.3), differs from the Maxwell operator M_1 by the presence of the imaginary unit:

$$\mathcal{M} \begin{pmatrix} E \\ H \end{pmatrix} = \begin{pmatrix} i\varepsilon^{-1} \operatorname{curl} H \\ -i\mu^{-1} \operatorname{curl} E \end{pmatrix}.$$

Clearly, the operators M_1 and $\mathcal{M}$ are unitarily equivalent. In this case, the asymptotics of Theorem 4.3 becomes
(5.2)

$$N(\lambda, \mathcal{M}) \sim \frac{\lambda^3}{24\pi^3} \int_\Omega \int_{|\xi|=1} \left(\Lambda_1(x,\xi)^{-3/2} + \Lambda_2(x,\xi)^{-3/2}\right) dS(\xi) d^3x, \quad \lambda \to \infty,$$

where $\Lambda_{1,2}$ are the nonzero eigenvalues of the problem

$$\left[\xi, \mu(x)^{-1}[h, \xi]\right] = \Lambda \varepsilon(x) h, \quad \xi, h \in \mathbb{C}^3, \quad \Lambda \in \mathbb{C},$$

and $[\,.\,,\,.\,]$ is the vector product in $\mathbb{R}^3$. Note that the asymptotic formula (5.2) is the main result of [**4**].

In fact, Theorem 4.3 allows one to take X to be a more "exotic" set than a domain with Lipschitz boundary. The following two examples illustrate this point.

Example 1. Let us consider the unit ball $\{x = (r, \varphi, \theta) : r < 1\}$ with the cut along the half-disc $\{x = (r, \varphi, \theta) : r < 1, \varphi = 0, \theta \in [0, \pi]\}$, where (r, φ, θ) are the spherical coordinates in $\mathbb{R}^3$. Denote by Ω the completion of the described domain with respect to its intrinsic metric. Thus we distinguish two sides of the cut, $\varphi = 0$ and $\varphi = 2\pi$.

The map $(r, \varphi, \theta) \mapsto (r, \varphi/2, \theta)$ is bi-Lipschitz and one-to-one from Ω to a closed half-ball, which makes Ω a Lipschitz manifold. In the same way arbitrary *domains with screens* (see [**9**]) are Lipschitz manifolds. The asymptotics of the spectrum of the Maxwell operator in domains with screens was obtained in [**9**] for $\varepsilon = \mu = \mathbb{1}$ and in [**4, 3**] for arbitrary ε, μ. However, in these papers the boundary outside the "edge" of the screen (i.e. the screens themselves and the "exterior" part of the boundary) are C^2-smooth. Our result includes the "Lipschitz screens" as well.

Example 2. Let $\Omega = \{x = rs : r \in [0, 1],\ s \in \overline{\Xi}\}$, where Ξ is an open simply connected subset of the unit sphere with smooth boundary. Once again, it is easy to introduce the Lipschitz manifold structure on Ω; nevertheless the boundary $\partial\Omega$ can be non-Lipschitz near the point 0.

6. Variational formulation

6.1. Quadratic forms. In order to study the asymptotics of the eigenvalues of a selfadjoint operator it is convenient to deal with the spectrum of the quotient of quadratic forms. Let a and b be two positive quadratic forms in a Hilbert space $\mathcal{H}$. Let the form a define an inner product in $\mathcal{H}$, and let the form b correspond to a compact selfadjoint operator B with respect to this inner product. We consider the quotient

$$b[x]/a[x], \qquad x \in \mathcal{H}, \tag{6.1}$$

and denote by $n(\lambda, (6.1))$ the counting function of the sequence of maxima of the quotient (6.1). This is equivalent to defining $n(\lambda, (6.1)) = \#\{j : b_j > \lambda\}$, where b_j are the eigenvalues of the operator B enumerated with multiplicities taken into account. The following statement is proved in [**8**] in a particular case. Almost the same proof applies in the general situation.

LEMMA 6.1. *Let $\mathcal{H}_1$, $\mathcal{H}_2$ be two Hilbert spaces, and let $A : \mathcal{H}_1 \to \mathcal{H}_2$ be a densely defined closed operator. We assume that the embedding* $\operatorname{Dom} A \subset \mathcal{H}_1$ *is compact, where* $\operatorname{Dom} A$ *is considered with respect to the graph norm of A. Consider the quotients*

$$\|y\|^2_{\mathcal{H}_1}/\|Ay\|^2_{\mathcal{H}_2}, \quad y \in \operatorname{Dom} A, \quad y \perp \ker A, \tag{6.2}$$

$$\|Ax\|^2_{\mathcal{H}_2}/\|A^*Ax\|^2_{\mathcal{H}_1}, \quad x \in \operatorname{Dom}(A^*A), \quad x \perp \ker A. \tag{6.3}$$

Then $n(\lambda, (6.2)) = n(\lambda, (6.3))$.

PROOF. We have the polar representation $A = U|A|$ for the operator A, where U is an isometric isomorphism of the space $\mathcal{H}_1 \ominus \ker A$ into $\mathcal{H}_2 \ominus \ker A^*$. We have

$$\operatorname{Dom}|A| = \operatorname{Dom} A, \quad \ker|A| = \ker A, \quad A^* = |A|U^* \quad \text{and} \quad A^*A = |A|^2.$$

For $x \in \operatorname{Dom} A$, $x \perp \ker A$, we put $y = |A|x$. Then $\|y\|_{\mathcal{H}_1} = \|Ax\|_{\mathcal{H}_2}$. Further, $x \in \operatorname{Dom}(A^*A)$ if and only if $y \in \operatorname{Dom}|A| = \operatorname{Dom} A$, and

$$\|A^*Ax\|_{\mathcal{H}_1} = \||A|y\|_{\mathcal{H}_1} = \|Ay\|_{\mathcal{H}_2}.$$

Thus, the spectra of the problems (6.2) and (6.3) coincide. □

6.2. Variational quotients. Instead of M_k, we will deal with the operator $R_k^*R_k$; see Remark 3.4. It is more convenient to consider the inverse operator in the orthogonal complement to $\ker R_k = K^k_{el}$. The inverse operator corresponds to the quotient

$$\frac{\|\omega\|^2_{\mathcal{L}^k_2(\beta_k)}}{\|R_k(\beta_k, \beta_{k+1})\omega\|^2_{\mathcal{L}_2^{n-k-1}(\beta^{-1}_{k+1})}} = \frac{\int_X \omega \wedge \beta_k \overline{\omega}}{\int_X d\omega \wedge \beta_{k+1}(\overline{d\omega})}, \tag{6.4}$$

$$\omega \in \Phi^k_{el}(\beta_k), \qquad \int_X \omega \wedge \beta_k \overline{\eta} = 0 \ \forall \eta \in K^k_{el}.$$

It is evident that

$$N(\lambda^2, R_k(\beta_k, \beta_{k+1})^* R_k(\beta_k, \beta_{k+1})) = n(\lambda^{-2}, (6.4)). \tag{6.5}$$

The analysis of the quotient (6.4) is complicated by the fact that the elements $\omega \in \Phi^k_{el}$ do not admit multiplication by cut-off functions (the condition $d(\beta_k\omega) = 0$ is lost after multiplication). In order to overcome this problem, we will consider the quotient

(6.6)
$$\frac{\int_X \theta \wedge \beta_k \overline{\theta}}{\int_X d\theta \wedge \beta_{k+1}(\overline{d\theta}) + \beta_{k-1}^{-1}(d(\beta_k\theta)) \wedge d(\overline{\beta_k\theta})}, \qquad \theta \in F^k_{el}(\beta_k), \int_X \theta \wedge \beta_k\overline{\eta} = 0 \ \forall \eta \in K^k_{el}$$

(if $k = 0$ the second term in the denominator must be omitted). The restriction of the quotient (6.6) onto the subspace Φ^k_{el} coincides with the quotient (6.4), and the space F^k_{el} is invariant with respect to multiplication by Lipschitz cut-off functions.

The key to our proof is a recurrence relation (6.8) for the counting function $n(\lambda, (6.4))$. In order to state this relation, let us display the quotient (6.4) with the change of index $k \mapsto k-1$:

$$\text{(6.7)} \qquad \frac{\int_X \omega \wedge \beta_{k-1}\overline{\omega}}{\int_X d\omega \wedge \beta_k(\overline{d\omega})}, \qquad \omega \in \Phi^{k-1}_{el}(\beta_{k-1}), \int_X \omega \wedge \beta_{k-1}\overline{\eta} = 0 \ \forall \eta \in K^{k-1}_{el}$$

(we assume $k \geq 1$). The proof of the following theorem is based on the orthogonal decomposition $F^k_{el} = \Phi^k_{el} \oplus_{\beta_k} E^k_{el}$ (see (3.3)).

THEOREM 6.2. *In the notation introduced above we have* $n(\lambda, (6.6)) = n(\lambda, (6.4))$ *when* $k = 0$, *and*

$$\text{(6.8)} \qquad n(\lambda, (6.6)) = n(\lambda, (6.4)) + n(\lambda, (6.7))$$

when $k \geq 1$.

PROOF. The statement is clear if $k = 0$. Suppose $k \geq 1$. The terms in the decomposition $F^k_{el} = \Phi^k_{el} \oplus E^k_{el}$ are orthogonal both with respect to the form in the numerator of (6.6) and with respect to the form in the denominator of (6.6). The restriction of the quotient (6.6) onto the subspace Φ^k_{el} coincides with the quotient (6.4), and the restriction onto E^k_{el} coincides with the quotient

$$\text{(6.9)} \qquad \frac{\int_X d\psi \wedge \beta_k(\overline{d\psi})}{\int_X \beta_{k-1}^{-1}(d(\beta_k d\psi)) \wedge d(\overline{\beta_k d\psi})}, \qquad \psi \in \mathcal{W}^{k-1}, \quad j^*\psi = 0,$$
$$d(\beta_k d\psi) \in \mathcal{L}_2^{n-k+1}, \qquad \int_X \psi \wedge \beta_{k-1}\overline{\eta} = 0 \ \forall \eta \in \mathcal{W}^{k-1} : d\eta = 0.$$

Therefore,

$$n(\lambda, (6.6)) = n(\lambda, (6.4)) + n(\lambda, (6.9)).$$

It remains to prove that $n(\lambda, (6.9)) = n(\lambda, (6.7))$. For this purpose we use Lemma 6.1 with

$$\mathcal{H}_1 = \mathcal{L}_2^{k-1}(\beta_{k-1}), \quad \mathcal{H}_2 = \mathcal{L}_2^k(\beta_k), \quad A = d_{k-1}.$$

By (3.5), we have $\mathcal{H}_1 \ominus \ker A = J^{k-1} \ominus K^{k-1}_{el}$, and so the quotient (6.7) coincides with the quotient (6.2). Furthermore, Lemma 2.5 implies

$$\begin{aligned} \operatorname{Dom} A^*A &= \{\psi \in \operatorname{Dom} d_{k-1} : d\psi \in \operatorname{Dom} d^*_{k-1}\} \\ &= \{\psi \in \mathcal{W}^{k-1} : j^*\psi = 0, \ d(\beta_k d\psi) \in \mathcal{L}_2^{n-k+1}\}, \end{aligned}$$

and $A^*A\psi = (-1)^k \beta_{k-1}^{-1} d(\beta_k d\psi)$. So the quotients (6.9) and (6.3) coincide. □

6.3. The strategy of the proof of the main result. First of all, from Theorems 6.2 and 5.1 we easily obtain the asymptotics for the quotient (6.6) in the case of a domain in Euclidean space:

LEMMA 6.3. *Let $\Omega \subset \mathbb{R}^n$ be a bounded domain with Lipschitz boundary. Then*

$$n(\lambda, (6.6)) \sim \lambda^{-n/2} \int_\Omega \nu_k(\beta_k, \beta_{k+1}) + \nu_{k-1}(\beta_{k-1}, \beta_k), \quad \lambda \to +0, \tag{6.10}$$

where we assume $\nu_{-1} = 0$.

PROOF. Theorem 5.1 and formulas (3.6) and (6.5) yield

$$n(\lambda, (6.4)) \sim \lambda^{-n/2} \int_\Omega \nu_k(\beta_k, \beta_{k+1}), \quad \lambda \to +0.$$

Since the quotient (6.7) is identical to the quotient (6.4) after the replacement of k by $k-1$, we have

$$n(\lambda, (6.7)) \sim \lambda^{-n/2} \int_\Omega \nu_{k-1}(\beta_{k-1}, \beta_k), \quad \lambda \to +0.$$

A reference to Theorem 6.2 completes the proof. □

In the following two sections, we extend the asymptotic formula (6.10) to the case of a Lipschitz manifold. In the proof of this, we use the fact that the quotient (6.6) is amenable to the usual technique of analysis on manifolds: changes of variables and multiplication by cut-off functions.

Finally, in Section 8.3 we derive the asymptotics for the quotient (6.4) on a Lipschitz manifold by induction in k. Here we again use the recurrence relation (6.8). Note that the idea to consider quotients of type (6.6) instead of ones of type (6.4) is borrowed from [**8**].

7. Compactness of the embedding $F^k_{el} \subset \mathcal{L}^k_2$

7.1. Covering lemmas. The usual compactness argument yields the following

LEMMA 7.1. *Let X be a compact Lipschitz manifold. Then there exists a finite collection of open sets $\{U_j\}_{j=1}^m$, such that $X = \bigcup_{j=1}^m U_j$, the closure of each set U_j lies in a single chart $\overline{U}_j \subset V_{\alpha_j}$, and the image $\alpha_j(U_j) \subset \mathbb{R}^n_+$ is a ball or a half-ball.*

The following fact is also well known.

LEMMA 7.2. *a) Let $M \subset \mathbb{R}^n$ be an open set. For all $\varepsilon > 0$ there exists an open set $Y \subset M$, such that $\partial Y \in C^\infty$ and* $\mathrm{mes}(M \setminus Y) < \varepsilon$.

b) Let $N \subset V \subset \mathbb{R}^n_+$, where N is a closed set and V is an open set. Then for all $\varepsilon > 0$ there exists an open set Z, such that $N \subset Z \subset V$, $\partial Z \in C^\infty$ and $\mathrm{mes}(Z \setminus N) < \varepsilon$.

LEMMA 7.3. *Let X be a compact Lipschitz manifold. Then there exists a natural number m, which depends only on X, such that for all $\rho \in \mathcal{L}^n_\infty(X)$ and all $\sigma > 0$ there exists an open subset Y and open subsets $Z_1, \dots, Z_m$ of X with the following properties:*

- $X = Y \cup \left(\bigcup_{j=1}^m Z_j\right)$;
- $Y \cap \partial X = \emptyset$;

- *Y is Lipschitz homeomorphic to an open possibly disconnected bounded set in $\mathbb{R}^n$ with a smooth boundary;*
- *for each Z_j there exists an open possibly disconnected bounded set in $\mathbb{R}^n_+$ with a smooth boundary, such that Z_j is Lipschitz homeomorphic to this set;*
- $\left|\int_{Z_j} \rho\right| < \sigma$ *for all j.*

PROOF. Let $\{U_j\}_{j=1}^m$ be a covering of X from Lemma 7.1, $\overline{U}_j \subset V_{\alpha_j}$. In each chart (V_{α_j}, α_j) the n-form ρ can be represented by a bounded function $\rho_j \in L_\infty(\alpha_j(V_{\alpha_j}))$. Denote $\hat{\rho} := \max_j \|\rho_j\|_{L_\infty}$. We choose an open set $Y_1 \subset U_1 \setminus \partial X$ in such a way that $\partial(\alpha_1(Y_1)) \in C^\infty$ and $\operatorname{mes} \alpha_1(U_1 \setminus Y_1) < \sigma/(2\hat{\rho})$. Then we choose an open set Z_1, $\overline{U}_1 \setminus Y_1 \subset Z_1 \subset V_{\alpha_1}$, in such a way that $\partial(\alpha_1(Z_1)) \in C^\infty$ and $\operatorname{mes} \alpha_1(Z_1) < \sigma/\hat{\rho}$. The existence of the sets Y_1, Z_1 is guaranteed by Lemma 7.2. Then we repeat this procedure m times: at the $(l+1)$-st step we construct open sets Y_{l+1}, Z_{l+1} such that

$$Y_{l+1} \subset U_{l+1} \setminus \bigcup_{j=1}^{l} \overline{Y}_j \setminus \partial X, \quad \partial(\alpha_{l+1}(Y_{l+1})) \in C^\infty, \quad \operatorname{mes} \alpha_{l+1}\left(U_{l+1} \setminus \bigcup_{j=1}^{l+1} Y_j\right) < \frac{\sigma}{2\hat{\rho}},$$

$$\overline{U}_{l+1} \setminus \bigcup_{j=1}^{l+1} Y_j \subset Z_{l+1} \subset V_{\alpha_{l+1}}, \quad \partial(\alpha_{l+1}(Z_{l+1})) \in C^\infty, \quad \operatorname{mes} \alpha_{l+1}(Z_{l+1}) < \sigma/\hat{\rho}.$$

Finally, we set $Y = \bigcup_{j=1}^m Y_j$. The sets $Y, Z_1, \dots, Z_m$ satisfy all the requirements of the lemma. □

7.2. Compactness of the embedding $F_{el}^k \subset \mathcal{L}_2^k$. We denote by $\operatorname{Lip}(X)$ the space of all scalar Lipschitz functions on X. Clearly, if $\zeta \in \operatorname{Lip}(X)$, then $d\zeta \in \mathcal{L}_\infty^1(X)$.

LEMMA 7.4. *Let Ξ be an open subset of X such that $\overline{\Xi}$ is a Lipschitz submanifold of X. Let $\omega \in F_{el}^k(X)$, $\zeta \in \operatorname{Lip}(X)$, $\operatorname{supp} \zeta \subset \Xi$. Then $\zeta\omega \in F_{el}^k(\overline{\Xi})$.*

PROOF. Evidently, $\zeta\omega \in \mathcal{L}_2^k(\overline{\Xi})$, $d(\zeta\omega) \in \mathcal{L}_2^{k+1}(\overline{\Xi})$ and

$$d(\beta_k(\zeta\omega)) = \zeta d(\beta_k\omega) + d\zeta \wedge \beta_k\omega \in \mathcal{L}_2^{n-k+1}(\overline{\Xi}).$$

It remains to check the boundary conditions. Let $\theta \in \mathcal{W}^{n-k-1}(\overline{\Xi})$. The conditions $j^*\omega = 0$, $\operatorname{supp}\zeta \subset \Xi$ imply $\int_\Xi \omega \wedge d(\zeta\theta) = (-1)^{k+1} \int_\Xi d\omega \wedge \zeta\theta$. Therefore,

$$\begin{aligned} \int_\Xi \zeta\omega \wedge d\theta &= \int_\Xi \omega \wedge d(\zeta\theta) - \int_\Xi \omega \wedge d\zeta \wedge \theta \\ &= (-1)^{k+1} \int_\Xi d\omega \wedge \zeta\theta - \int_\Xi \omega \wedge d\zeta \wedge \theta = (-1)^{k+1} \int_\Xi d(\zeta\omega) \wedge \theta. \end{aligned}$$

□

LEMMA 7.5. *The embedding $F_{el}^k(\beta_k) \subset \mathcal{L}_2^k(\beta_k)$ is compact.*

PROOF. Let $\{U_j\}_{j=1}^m$ be a covering of X from Lemma 7.1, and let $\{\zeta_j\}_{j=1}^m$ be a subordinate partition of unity. Due to Lemma 7.4 we can represent the embedding operator $I_{F_{el}^k \to \mathcal{L}_2^k}$ as

$$I_{F_{el}^k(X) \to \mathcal{L}_2^k(X)} = \sum_{j=1}^m I_{F_{el}^k(\overline{U}_j) \to \mathcal{L}_2^k(X)}(\zeta_j \,\cdot\,).$$

It is not difficult to check that the spaces $K^k_{el}(\overline{U}_j)$ are trivial. Together with Lemma 6.3 this yields the compactness of the embedding operators $I_{F^k_{el}(\overline{U}_j)\to\mathcal{L}^k_2(\overline{U}_j)}$. Therefore, the operator $I_{F^k_{el}(X)\to\mathcal{L}^k_2(X)}$ is also compact. □

COROLLARY 7.6. *The spectrum of the Maxwell operator is discrete.*

PROOF. The compactness of the embedding $F^k_{el}(\beta_k)\subset\mathcal{L}^k_2(\beta_k)$ immediately yields the compactness of the embedding

$$\operatorname{Dom} R_k(\beta_k,\beta_{k+1})=\Phi^k_{el}(\beta_k)\subset\mathcal{L}^k_2(\beta_k)$$

(since $\Phi^k_{el}(\beta_k)$ is a subspace of $F^k_{el}(\beta_k)$). It follows that the spectrum of the operator $R^*_k R_k$ is discrete. It remains to refer to Remark 3.4. □

COROLLARY 7.7. *The space $K^k_{el}(\beta_k)$ is finite-dimensional.*

PROOF. By the definition (3.4) of the space $K^k_{el}(\beta_k)$, the norm $\|\cdot\|_{F^k_{el}(\beta_k)}$ (see (3.2)) coincides with the norm $\|\cdot\|_{\mathcal{L}^k_2(\beta_k)}$ on $K^k_{el}(\beta_k)$. Due to the compactness of the embedding $F^k_{el}(\beta_k)\subset\mathcal{L}^k_2(\beta_k)$, we get the required statement. □

7.3. Transformation of (6.6)**.** Below we transform the variational quotient (6.6) to a more convenient form. If we add the term $\int_X\theta\wedge\beta_k\overline{\theta}$ to the denominator of (6.6), we obtain the quotient

$$\frac{\int_X\theta\wedge\beta_k\overline{\theta}}{\int_X d\theta\wedge\beta_{k+1}(\overline{d\theta})+\beta^{-1}_{k-1}(d(\beta_k\theta))\wedge d(\overline{\beta_k\theta})+\theta\wedge\beta_k\overline{\theta}},\tag{7.1}$$

$$\theta\in F^k_{el}(\beta_k),\qquad\int_X\theta\wedge\beta_k\overline{\eta}=0\quad\forall\eta\in K^k_{el}.$$

In terms of the consecutive maxima $\{b_j\}$, this corresponds to the change $b_j\mapsto b_j/(1+b_j)$. Therefore,

$$n(\lambda,(6.6))\sim n(\lambda,(7.1)),\qquad\lambda\to+0.$$

Now let us drop the orthogonality conditions in (7.1). We arrive at the variational quotient

$$\frac{\int_X\theta\wedge\beta_k\overline{\theta}}{\int_X d\theta\wedge\beta_{k+1}(\overline{d\theta})+\beta^{-1}_{k-1}(d(\beta_k\theta))\wedge d(\overline{\beta_k\theta})+\theta\wedge\beta_k\overline{\theta}},\qquad\theta\in F^k_{el}(\beta_k).\tag{7.2}$$

By the inequalities

$$n(\lambda,(7.1))\le n(\lambda,(7.2))\le n(\lambda,(7.1))+\dim K^k_{el}(\beta_k)$$

and Corollary 7.7, the leading terms of the asymptotics of $n(\lambda,(7.1))$ and $n(\lambda,(7.2))$ coincide. Thus, the limit $\lim_{\lambda\to0}\lambda^{n/2}n(\lambda,(7.2))$ exists if and only if the limit $\lim_{\lambda\to0}\lambda^{n/2}n(\lambda,(6.6))$ does, and if these limits do exist, they coincide:

$$\lim_{\lambda\to0}\lambda^{n/2}n(\lambda,(6.6))=\lim_{\lambda\to0}\lambda^{n/2}n(\lambda,(7.2)).\tag{7.3}$$

8. Proof of Theorem 4.3

8.1. The space F_0^k. For a subspace $\Psi \subset F_{el}^k(X)$, let us denote

$$\delta(\Psi) = \liminf_{\lambda \to 0} \lambda^{n/2} n(\lambda, (7.2), \Psi), \quad \Delta(\Psi) = \limsup_{\lambda \to 0} \lambda^{n/2} n(\lambda, (7.2), \Psi),$$

where $n(\lambda, (7.2), \Psi)$ is the counting function of the sequence of the maxima of the quotient (7.2), considered on the space Ψ. Let $F_0^k(X)$ be the closure of the set $\{\omega \in F^k(\beta_k) : \operatorname{supp} \omega \cap \partial X = \emptyset\}$ in $F^k(X)$. It is clear that $F_0^k(X) \subset F_{el}^k(X)$.

LEMMA 8.1. *Let $X = \bigcup_{j=0}^m \Xi_j$, where Ξ_j are open sets, $\overline{\Xi}_j$ are Lipschitz submanifolds, and $\Xi_0 \cap \partial X = \emptyset$. Then*

$$\delta(F_{el}^k(X))^{2/(n+2)} \le \delta(F_0^k(\Xi_0))^{2/(n+2)} + \sum_{j=1}^m \Delta(F_{el}^k(\Xi_j))^{2/(n+2)}, \tag{8.1}$$

$$\Delta(F_{el}^k(X))^{2/(n+2)} \le \Delta(F_0^k(\Xi_0))^{2/(n+2)} + \sum_{j=1}^m \Delta(F_{el}^k(\Xi_j))^{2/(n+2)}. \tag{8.2}$$

Unfortunately, the restricted volume of the present paper does not allow us to give the proof of this lemma. This proof follows the same steps as the proof of the estimate from above of the asymptotic functional for the quotient (12) in the paper [**8**].

Next, we show that in the case of the domain in Euclidean space the asymptotics of the spectrum of the quotient (7.2) does not change if we consider it on the space $F_0^k(\Omega)$ instead of $F_{el}^k(\Omega)$.

LEMMA 8.2. *Let $\Omega \subset \mathbb{R}^n$ be a bounded domain with Lipschitz boundary. Then*

$$\begin{aligned} \exists \lim_{\lambda \to 0} \lambda^{n/2} n(\lambda, (7.2), F_0^k(\Omega)) &= \lim_{\lambda \to 0} \lambda^{n/2} n(\lambda, (7.2), F_{el}^k(\Omega)) \\ &= \int_\Omega \nu_k(\beta_k, \beta_{k+1}) + \nu_{k-1}(\beta_{k-1}, \beta_k). \end{aligned} \tag{8.3}$$

PROOF. The second equality is due to Lemma 6.3 and formula (7.3). It is also clear that $n(\lambda, (7.2), F_0^k(\Omega)) \le n(\lambda, (7.2), F_{el}^k(\Omega))$. So it suffices to prove that $\delta(F_0^k(\Omega)) \ge \delta(F_{el}^k(\Omega))$.

Let $\sigma > 0$ and $\Xi_0, \Xi_1 \subset \Omega$ be open sets with Lipschitz boundaries such that $\Omega = \Xi_0 \cup \Xi_1$, $\overline{\Xi}_0 \cap \partial\Omega = \emptyset$ and $\operatorname{mes} \Xi_1 < \sigma$. Applying inequality (8.1) with $m = 1$ at the first step and Lemma 6.3 at the second step, we get

$$\begin{aligned} \delta(F_{el}^k(\Omega))^{2/(n+2)} &\le \delta(F_0^k(\Xi_0))^{2/(n+2)} + \Delta(F_{el}^k(\Xi_1))^{2/(n+2)} \\ &\le \delta(F_0^k(\Omega))^{2/(n+2)} + \left(\int_{\Xi_1} \nu_k + \nu_{k-1} \right)^{2/(n+2)}. \end{aligned}$$

The last term in the right hand side can be made arbitrarily small by the choice of σ, so $\delta(F_{el}^k(\Omega)) \le \delta(F_0^k(\Omega))$. □

REMARK 8.3. If Ω is a manifold which is Lipschitz homeomorphic to a domain in $\mathbb{R}^n$ with a Lipschitz boundary, then formula (8.3) also holds true. Indeed, both the variational quotient (7.2) and the asymptotic coefficient in the r.h.s. of (8.3) are expressed in terms of integrals of n-forms, and therefore are invariant with respect to changes of variables.

8.2. The spectrum of (6.6) on the manifold.

THEOREM 8.4. *Let X be a compact oriented Lipschitz manifold with boundary. Then*

$$n(\lambda, (6.6)) \sim \lambda^{-n/2} \int_X \nu_k(\beta_k, \beta_{k+1}) + \nu_{k-1}(\beta_{k-1}, \beta_k), \quad \lambda \to +0,$$

where we assume $\nu_{-1} = 0$.

PROOF. Due to (7.3) it suffices to consider the quotient (7.2). Let $\sigma > 0$, and let $Y, Z_1, \ldots, Z_m$ be a covering of X guaranteed by Lemma 7.3 with $\rho = \nu_k + \nu_{k-1}$.

First we obtain a lower bound for $\delta(F^k_{el}(X))$. The minimax principle yields

$$n\left(\lambda, (7.2), F^k_{el}(X)\right) \geq n\left(\lambda, (7.2), F^k_0(Y)\right)$$

and therefore $\delta(F^k_{el}(X)) \geq \delta(F^k_0(Y))$. Next, due to Lemma 8.2 and Remark 8.3, we have $\delta(F^k_0(Y)) = \int_Y \nu_k + \nu_{k-1}$. Further, by Lemma 7.3, we have

$$\int_Y \nu_k + \nu_{k-1} \geq \int_X (\nu_k + \nu_{k-1}) - m\sigma.$$

As σ is arbitrary, this chain of reasoning yields

$$\delta(F^k_{el}(X)) \geq \int_X \nu_k + \nu_{k-1}.$$

Next, we obtain an estimate from above for $\Delta(F^k_{el}(X))$. In order to do this, we use inequality (8.2) with $\Xi_j = Z_j$ for $j = 1, \ldots, m$, and $\Xi_0 = Y$:

$$\Delta(F^k_{el}(X))^{2/(n+2)} \leq \Delta(F^k_0(Y))^{2/(n+2)} + \sum_{j=1}^m \Delta(F^k_{el}(Z_j))^{2/(n+2)}.$$

Further, again by Lemma 8.2, Remark 8.3 and Lemma 7.3,

$$\Delta(F^k_0(Y)) = \int_Y \nu_k + \nu_{k-1} \leq \int_X \nu_k + \nu_{k-1}, \quad \Delta(F^k_{el}(Z_j)) = \int_{Z_j} \nu_k + \nu_{k-1} \leq \sigma,$$

whence

$$\Delta(F^k_{el}(X))^{2/(n+2)} \leq \left(\int_X \nu_k + \nu_{k-1}\right)^{2/(n+2)} + m\sigma^{2/(n+2)}.$$

Again, as σ is arbitrarily small, this implies $\Delta(F^k_{el}(X)) \leq \int_X \nu_k + \nu_{k-1}$. □

8.3. Proof of Theorem 4.3. By virtue of formulas (3.6) and (6.5) it suffices to show that

$$n(\lambda, (6.4)) \sim \lambda^{-n/2} \int_X \nu_k(\beta_k, \beta_{k+1}), \quad \lambda \to +0. \tag{8.4}$$

Recall that the spectral asymptotics for the variational quotient (6.6) has already been proven (see Theorem 8.4), and that the variational quotient (6.4) differs from (6.6) only by the choice of the function space on which it is considered. We establish (8.4) by induction in k using the recurrence relation (6.8) and Theorem 8.4. In order to use our induction argument, we need to assume that $\beta_j \in M^j_+(X)$ are defined for all j. Since the manifold X is oriented, the spaces $M^j_+(X)$ are not empty, and so we can choose some β_j for all j.

For $k = 0$ formula (8.4) is immediate by Theorem 8.4, as $\Phi^0_{el}(X) = F^0_{el}(X)$.

Now suppose that (8.4) holds true for $k-1$, which means

$$n(\lambda, (6.7)) \sim \lambda^{-n/2} \int_X \nu_{k-1}(\beta_{k-1}, \beta_k), \quad \lambda \to +0.$$

Then by the recurrence relation (6.8) and Theorem 8.4 we get

$$n(\lambda, (6.4)) = n(\lambda, (6.6)) - n(\lambda, (6.7)) \sim \lambda^{-n/2} \int_X \nu_k(\beta_k, \beta_{k+1}), \quad \lambda \to +0.$$

The induction argument is complete. □

References

[1] Alekseev A. B. *Spectral asymptotics of elliptic boundary value problems with solvable constraints*, Ph.D. thesis, Leningrad State University (1977). (Russian)

[2] Alekseev A. B.; Birman M. Sh. *Asymptotic behavior of the spectrum of elliptic boundary value problems with solvable constraints*, Dokl. Akad. Nauk SSSR **230**, no. 3 (1976), 505–507. English transl., Sov. Math., Dokl. **17** (1976), 1319–1322 (1977).

[3] Alekseev A. B.; Birman M. Sh.; Filonov N. D. *Spectrum asymptotics for one "nonsmooth" variational problem with solvable constraint*, Algebra i Analiz **18**, no. 5 (2006), 1–22. English transl., St. Petersburg. Math. J. **18**, no. 5 (2007), 681–697.

[4] Birman M. Sh.; Filonov N. D. *Weyl asymptotics of the spectrum of the Maxwell operator with non-smooth coefficients in Lipschitz domains*, Nonlinear equations and spectral theory, Amer. Math. Soc. Transl., Ser. 2, v. 220 (2007), 27–44.

[5] Birman M. Sh.; Solomyak M. Z. *Quantitative analysis in Sobolev imbedding theorems and applications to spectral theory*, Tenth Mathematical School (Summer School, Kaciveli/Nalchik, 1972) , 5–189. Izdanie Inst. Mat. Akad. Nauk Ukrain. SSR, Kiev, 1974. English transl., American Mathematical Society Translations, Series 2, **114**. AMS, Providence, R.I. (1980).

[6] Birman M. Sh.; Solomyak M. Z. *The Maxwell operator in domains with a nonsmooth boundary*, Sibirsk. Mat. Zh. **28**, no. 1 (1987), 23–36. English transl., Sib. Math. J. **28** (1987), 12–24.

[7] Birman M. Sh.; Solomyak M. Z. *L_2-theory of the Maxwell operator in arbitrary domains* Uspehi Mat. Nauk **42**, no. 6(258) (1987), 61–76. English transl., Russ. Math. Surv. **42**, no. 6 (1987), 75–96.

[8] Birman M. Sh.; Solomyak M. Z. *Weyl asymptotics of the spectrum of the Maxwell operator for domains with a Lipschitz boundary*, Vestnik Leningrad. Univ. Mat. Mekh. Astronom. no. 3 (1987), 23–28. English transl., Vestn. Leningr. Univ., Math. **20**, no. 3 (1987), 15–21.

[9] Birman M. Sh.; Solomyak M. Z. *Principal singularities of the electric component of an electromagnetic field in regions with screens*, Algebra i Analiz **5**, no. 1 (1993), 143–159. English transl., St. Petersburg. Math. J. **5**, no. 1 (1994), 125–139.

[10] M. Demchenko; N. Filonov, *The Weyl asymptotics of the spectrum of the Maxwell operator in arbitrary-dimensional Lipschitz domains*, to appear

[11] Goldshtein, V. M.; Kuz′minov, V. I.; Shvedov, I. A. *Differential forms on a Lipschitz manifold*, Sibirsk. Mat. Zh. **23** (1982), no. 2, 16–30. English transl., Siberian Math J. **23** (1982), no. 2, 151–161.

[12] N. Weck, *Maxwell's Boundary Value Problem on Riemannian Manifolds with Nonsmooth Boundaries*, J. Math. Anal. and Appl., **46** (1974), 410–437.

[13] H. Weyl, *Über das Spectrum der Hohlraumstrahlung*, J. Reine Angew. Math., **141** (1912), 163–181.

[14] H. Weyl, *Die natürlichen Randwertaufgaben im Außenraum für Strahlungsfelder beliebiger Dimension und beliebiger Ranges*, Math. Z., **56** (1952), 105–119.

St. Petersburg Branch, V. A. Steklov Institute of Mathematics, 27 Fontanka, St. Petersburg 191023, Russia

E-mail address: demchenko@pdmi.ras.ru

E-mail address: filonov@pdmi.ras.ru

Amer. Math. Soc. Transl.
(2) Vol. **225**, 2008

Spectral Inequalities for Schrödinger Operators with Surface Potentials

Rupert L. Frank and Ari Laptev

To Mikhail Shlëmovich Birman on his 80th birthday

ABSTRACT. We prove sharp Lieb–Thirring inequalities for Schrödinger operators with potentials supported on a hyperplane and we show how these estimates are related to Lieb–Thirring inequalities for relativistic Schrödinger operators.

1. Introduction

The Cwikel–Lieb–Rozenblum and the Lieb–Thirring inequalities estimate the number and moments of the eigenvalues of the Schrödinger operators $-\Delta - V$ in $L_2(\mathbb{R}^N)$ in terms of an integral of the potential V. They state that the bound

$$\operatorname{tr}(-\Delta - V)_-^\gamma \le L_{\gamma,N} \int_{\mathbb{R}^N} V(x)_+^{\gamma+N/2}\,dx \tag{1.1}$$

holds with a constant $L_{\gamma,N}$ independent of V iff $\gamma \ge 1/2$ for $N=1$, $\gamma > 0$ for $N=2$ and $\gamma \ge 0$ for $N \ge 3$. Here and below, $t_\pm := \max\{0, \pm t\}$ denotes the positive and negative parts of a real number, a real-valued function or a selfadjoint operator t. In particular, the problem of finding the optimal value of the constant $L_{\gamma,N}$ has attracted a lot of attention recently. We refer to the review articles [**H2, LW2**] for background information, references and applications of (1.1).

The purpose of the present paper is twofold. First, we would like to find an analog of inequality (1.1) for Schrödinger operators with singular potentials $V(x) = v(x_1, \dots, x_d)\delta(x_N)$, $d := N-1$, supported on a hyperplane. It turns out that such an inequality is indeed valid, provided the integral on the right hand side of (1.1) is replaced by

$$\int_{\mathbb{R}^d} v(x_1, \dots, x_d)_+^{2\gamma+d}\,dx_1 \dots dx_d\,.$$

We determine the complete range of γ's for which the resulting inequality holds. Moreover, we find the sharp values of the constants for $\gamma \ge 3/2$ by using the method

2000 *Mathematics Subject Classification.* Primary 35J10.

This work has been supported by DAAD grant D/06/49117 (R. F.).

of 'lifting with respect to the dimension'. This provides yet another example of the power and flexibility of this method, which was used by Laptev and Weidl [**LW1**] to obtain the sharp constants in (1.1) for $\gamma \geq 3/2$.

The second purpose of this paper is to point out a relation between the Schrödinger operator $-\Delta - v(x_1, \ldots, x_d)\delta(x_N)$ in $L_2(\mathbb{R}^N)$ and the relativistic Schrödinger operator $\sqrt{-\Delta} - v(x_1, \ldots, x_d)$ in $L_2(\mathbb{R}^d)$. Note that the space dimension $d = N - 1$ of the relativistic operator differs from that of the nonrelativistic operator. The basic idea is to relate the eigenfunctions of the Schrödinger operator with singular potential to the eigenfunctions of a nonlinear eigenvalue problem involving the relativistic Schrödinger operator. This construction is essentially the Poisson extension and is implicit in several earlier works, e.g., in [**CL, FS, CS**]. In our context the connection between the two operators becomes useful when combined with a monotonicity argument in the spirit of the Birman–Schwinger principle. It allows us both to prove the singular analog of inequality (1.1) and to (slightly) improve upon the known constants in Lieb–Thirring inequalities for relativistic Schrödinger operators.

2. Schrödinger operators with surface potentials

2.1. Main results. In this section we consider the operator

$$H(v)\,u = -\Delta u \quad \text{in } \mathbb{R}^{d+1}_+ := \{(x,y):\ x \in \mathbb{R}^d, y > 0\} \tag{2.1}$$

together with boundary conditions of the third type

$$\frac{\partial u}{\partial \nu} - vu = 0 \quad \text{on } \mathbb{R}^d \times \{0\}\,. \tag{2.2}$$

Here $\partial/\partial\nu = -\partial/\partial y$ denotes the (exterior) normal derivative and v a real-valued function on $\mathbb{R}^d$. If v is form-compact with respect to $\sqrt{-\Delta}$ in $L_2(\mathbb{R}^d)$, then $H(v)$ can be defined as a selfadjoint operator in $L_2(\mathbb{R}^{d+1}_+)$ by means of the quadratic form

$$\iint_{\mathbb{R}^{d+1}_+} |\nabla u|^2\,dx\,dy - \int_{\mathbb{R}^d} v(x)|u(x,0)|^2\,dx, \quad u \in H^1(\mathbb{R}^{d+1}_+)\,. \tag{2.3}$$

The negative spectrum of $H(v)$ consists of eigenvalues with finite multiplicities. We shall prove

THEOREM 2.1 (**Lieb–Thirring inequalities for surface potentials**). *The inequality*

$$\operatorname{tr}[H(v)]_-^\gamma \leq S_{\gamma,d} \int_{\mathbb{R}^d} v(x)_+^{2\gamma+d}\,dx \tag{2.4}$$

holds for all $0 \leq v \in L_{2\gamma+d}(\mathbb{R}^d)$ *iff*

$$\gamma > 0 \quad \text{if } d = 1, \qquad \text{and} \qquad \gamma \geq 0 \quad \text{if } d \geq 2\,. \tag{2.5}$$

Inequality (2.4) reflects the correct order of growth in the strong coupling limit, as can be seen from the Weyl-type asymptotics

$$\lim_{\alpha\to\infty} \alpha^{-2\gamma-d} \operatorname{tr}[H(\alpha\,v)]_-^\gamma = L^{\mathrm{cl}}_{\gamma,d} \int_{\mathbb{R}^d} v(x)_+^{2\gamma+d}\,dx\,, \tag{2.6}$$

with

$$L^{\mathrm{cl}}_{\gamma,d} := 2^{-d}\pi^{-d/2}\frac{\Gamma(\gamma+1)}{\Gamma(\gamma+d/2+1)}\,. \tag{2.7}$$

Since relation (2.6) is not completely standard we comment on its proof in Remark 3.7 below.

Our second result concerns the constants in the bounds of Theorem 2.1. Denoting by $S_{\gamma,d}$ the *sharp* constant in (2.4) we infer from (2.6) that

$$S_{\gamma,d} \geq L^{\mathrm{cl}}_{\gamma,d}\,. \tag{2.8}$$

We shall prove that for sufficiently large values of γ one actually has equality.

THEOREM 2.2 (**Sharp constants**). *Let $d \geq 1$ and $\gamma \geq 3/2$. Then the sharp constant in (2.4) is $S_{\gamma,d} = L^{\mathrm{cl}}_{\gamma,d}$.*

We prove this theorem in Subsection 2.2. Besides the sharp constants for $\gamma \geq 3/2$ our method yields explicit and tight bounds for $S_{\gamma,d}$ for arbitrary γ. In particular, we prove

$$\begin{aligned} S_{\gamma,d} &\leq \frac{\pi}{\sqrt{3}}\, L^{\mathrm{cl}}_{\gamma,d} && \text{if } 1 \leq \gamma < 3/2 \text{ and } d \geq 1,\\ S_{\gamma,1} &\leq 2\, L^{\mathrm{cl}}_{\gamma,1} && \text{if } 1/2 \leq \gamma < 1 \text{ and } d = 1,\\ S_{\gamma,d} &\leq \frac{2\pi}{\sqrt{3}}\, L^{\mathrm{cl}}_{\gamma,d} && \text{if } 1/2 \leq \gamma < 1 \text{ and } d \geq 2; \end{aligned} \tag{2.9}$$

see Remark 2.4. Moreover, the constants in the estimates for the number of negative eigenvalues satisfy

$$S_{0,2} \leq 6.04\, L^{\mathrm{cl}}_{0,2} \qquad \text{if } \gamma = 0 \text{ and } d = 2, \tag{2.10}$$

$$S_{0,3} \leq 6.08\, L^{\mathrm{cl}}_{0,3} \qquad \text{if } \gamma = 0 \text{ and } d = 3, \tag{2.11}$$

$$S_{0,d} \leq 10.332\, L^{\mathrm{cl}}_{0,d} \qquad \text{if } \gamma = 0 \text{ and } d \geq 4; \tag{2.12}$$

see Remark 2.4 and Subsection 3.3. The upper bounds (2.10), (2.11) can be supplemented by the lower bounds

$$S_{0,2} \geq 4\, L^{\mathrm{cl}}_{0,2} \qquad \text{if } \gamma = 0 \text{ and } d = 2, \tag{2.13}$$

$$S_{0,3} \geq 3\, L^{\mathrm{cl}}_{0,3} \qquad \text{if } \gamma = 0 \text{ and } d = 3; \tag{2.14}$$

see Subsection 3.3. In particular, for $d = 2$ the upper bound (2.10) is off by at most a factor 1.51.

2.2. Lifting with respect to dimension. In this subsection we use an argument in the spirit of Laptev and Weidl [**LW1**] to prove

THEOREM 2.3. *Let $d \geq 1$, $\gamma \geq 3/2$ and $\tau \geq 0$. Then*

$$\operatorname{tr}\,[H(v) + \tau]_-^\gamma \leq L^{\mathrm{cl}}_{\gamma,d} \int_{\mathbb{R}^d} \big(v(x)_+^2 - \tau\big)_+^{\gamma+d/2}\, dx \tag{2.15}$$

with $L^{\mathrm{cl}}_{\gamma,d}$ defined in (2.7).

Choosing $\tau = 0$ and recalling (2.6) we obtain Theorem 2.2.

PROOF. We shall prove Theorem 2.3 by induction over d. It is convenient to reflect the dependence on d in the notation of the quadratic form, so we write

$$h_d(v)[u] := \iint_{\mathbb{R}^{d+1}_+} |\nabla u|^2\, dx\, dy - \int_{\mathbb{R}^d} v(x) |u(x,0)|^2\, dx$$

and $H_d(v)$ for the associated operator. Note that this operator is also well defined for $d = 0$ and v a nonnegative real number. Indeed, in this case one has $H_0(v)u =$

$-u''$ and $u'(0) = -vu(0)$ for $u \in \operatorname{dom} H_0(v)$, and one easily finds that $H_0(v)$ has one negative eigenvalue, namely $-v_+^2$. Hence

$$\operatorname{tr}_{L_2(\mathbb{R}_+)} [H_0(v) + \tau]_-^\gamma = \left(v_+^2 - \tau\right)_+^\gamma ,$$

which is the analog of (2.15) for $d = 0$ and all $\gamma \geq 0$.

Now we fix $d \geq 1$ and assume that the assertion is already proved for all smaller dimensions. We write $x = (x_1, x')$ with $x_1 \in \mathbb{R}$, $x' \in \mathbb{R}^{d-1}$ and note that

$$H_d(v) + \tau \geq -\frac{d^2}{dx_1^2} \otimes 1_{L_2(\mathbb{R}_+^d)} - [H_{d-1}(v(x_1, \cdot)) + \tau]_-$$

with the identification $L_2(\mathbb{R}_+^{d+1}) = L_2(\mathbb{R}) \otimes L_2(\mathbb{R}_+^d)$. Hence the variational principle and the operator-valued Lieb–Thirring inequality from [**LW1**] yield for all $\gamma \geq 3/2$,

$$\text{(2.16)} \quad \operatorname{tr}_{L_2(\mathbb{R}_+^{d+1})} [H_d(v) + \tau]_-^\gamma \leq L_{\gamma,1}^{\text{cl}} \int_{-\infty}^{\infty} \operatorname{tr}_{L_2(\mathbb{R}_+^d)} [H_{d-1}(v(x_1, \cdot)) + \tau]_-^{\gamma+1/2} \, dx_1 .$$

By the induction hypothesis, the right hand side is bounded from above by

$$L_{\gamma,1}^{\text{cl}} L_{\gamma+1/2,d-1}^{\text{cl}} \int_{-\infty}^{\infty} \left(\int_{\mathbb{R}^{d-1}} (v(x_1, x')_+^2 - \tau)_+^{\gamma+d/2} \, dx' \right) dx_1$$
$$= L_{\gamma,d}^{\text{cl}} \int_{\mathbb{R}^d} (v(x)_+^2 - \tau)_+^{\gamma+d/2} \, dx ,$$

which establishes the assertion for dimension d and completes the proof of Theorem 2.3. □

REMARK 2.4. The above approach can be used to prove inequality (2.4) for $\gamma \geq 1/2$ and to obtain bounds (2.9) for the sharp constants $S_{\gamma,d}$. Indeed, according to [**HLW**] and [**DLL**] the operator-valued inequality (2.16) holds with an additional factor of $\pi/\sqrt{3}$ on the right hand side if $\gamma \geq 1$, with an additional factor of 2 if $1/2 \leq \gamma < 1$ and $d = 1$ and with an additional factor of $2\pi/\sqrt{3}$ if $\gamma \geq 1/2$ and $d \geq 2$.

Similarly, one can use the operator-valued inequality from [**FLS2**] (see also [**H1**]) to prove (2.4) for $\gamma \geq 0$ and $d \geq 3$ and to obtain (2.12). Extending the original proof of Lieb and Thirring [**LT**] to the operator-valued case would yield (2.4) for $\gamma > 0$ and $d = 2$. However, we do not know how to prove (2.4) with the operator-valued approach for $0 < \gamma < 1/2$ if $d = 1$ and for $\gamma = 0$ if $d = 2$. We give a proof based on a different idea in Subsection 3.3 below.

2.3. Additional remarks.

2.3.1. *Magnetic fields.* Let $A \in L_{2,\text{loc}}(\overline{\mathbb{R}_+^{d+1}}, \mathbb{R}^{d+1})$ and let the operator $H(A, v)$ be defined through the closure of the quadratic form

$$\iint_{\mathbb{R}_+^{d+1}} |(-i\nabla - A)u|^2 \, dx \, dy - \int_{\mathbb{R}^d} v(x)|u(x,0)|^2 \, dx, \quad u \in C_0^\infty(\overline{\mathbb{R}_+^{d+1}}) .$$

By a similar argument as in [**LW1**] one can prove that Theorem 2.3 remains true, with the same constant, if $H(v)$ is replaced by $H(A, v)$. More generally, all the inequalities sketched in Remark 2.4 remain true. The argument behind Theorem 2.1, however, allows only those A that are independent of y and orthogonal to the y-direction; see Remark 3.5. To obtain the analog of (2.4) for the complete range of γ's given in (2.5) one can rely upon an abstract operator-theoretic argument; see [**R**] for $\gamma = 0$ and [**F**] for $\gamma > 0$.

2.3.2. *Leaky graph Hamiltonians.* In the previous subsections we studied eigenvalues of the Laplacian $H(v)$ on the half-space with a perturbation by boundary conditions. This problem is essentially equivalent to the study of eigenvalues of the Schrödinger operator

$$\tilde{H}(v) = -\Delta - v(x)\delta(y)$$

in the whole space $\mathbb{R}^{d+1}$ with a potential supported on a hyperplane. The precise definition of the operator $\tilde{H}(v)$ in $L_2(\mathbb{R}^{d+1})$ is given via the quadratic form

$$\iint_{\mathbb{R}^{d+1}} |\nabla u|^2\, dx\, dy - \int_{\mathbb{R}^d} v(x)|u(x,0)|^2\, dx, \quad u \in H^1(\mathbb{R}^{d+1})\,.$$

The decomposition of a function into an even and an odd part with respect to the variable y induces an orthogonal decomposition of the space $L_2(\mathbb{R}^{d+1})$, which reduces the operator $\tilde{H}(v)$. The part of $\tilde{H}(v)$ on odd functions is unitarily equivalent to the Dirichlet Laplacian on $\mathbb{R}^{d+1}_+$, whereas the part on even functions is unitarily equivalent to $H(\frac{1}{2}v)$. Hence

$$\operatorname{tr}\left[\tilde{H}(v)\right]_-^\gamma = \operatorname{tr}\left[H(\tfrac{1}{2}v)\right]_-^\gamma\,,$$

and we obtain immediately the analogs of Theorems 2.1 and 2.2.

THEOREM 2.5. *The inequality*

$$\operatorname{tr}\left[\tilde{H}(v)\right]_-^\gamma \le \tilde{S}_{\gamma,d} \int_{\mathbb{R}^d} v(x)_+^{2\gamma+d}\, dx \tag{2.17}$$

holds for all $0 \le v \in L_{2\gamma+d}(\mathbb{R}^d)$ *iff*

$$\gamma > 0 \quad \text{if } d = 1, \qquad \text{and} \qquad \gamma \ge 0 \quad \text{if } d \ge 2\,. \tag{2.18}$$

THEOREM 2.6. *Let* $d \ge 1$ *and* $\gamma \ge 3/2$. *Then the sharp constant in* (2.17) *is* $\tilde{S}_{\gamma,d} = 2^{-2\gamma-d} L^{\mathrm{cl}}_{\gamma,d}$.

2.3.3. *Complex-valued surface potentials.* In applications one often encounters the boundary-value problem (2.1), (2.2) with a *complex-valued* function v, which leads to nonreal eigenvalues. If v is sufficiently regular, the quadratic form (2.3) generates an m-sectorial operator, which we continue to denote by $H(v)$. We denote by $\lambda_j(v)$, $j = 1, 2, \ldots$, the (at most countably many) eigenvalues of $H(v)$ in the cut plane $\mathbb{C} \setminus [0, \infty)$, repeated according to their algebraic multiplicities. Following the approach suggested in [**FLLS**] one obtains

THEOREM 2.7. *Let* $d \ge 1$ *and* $\gamma \ge 1$.

(1) *For eigenvalues with negative real part,*

$$\sum_{j:\ \operatorname{Re}\lambda_j(v)<0} (-\operatorname{Re}\lambda_j(v))^\gamma \le S_{\gamma,d} \int_{\mathbb{R}^d} (\operatorname{Re} v(x))_-^{2\gamma+d}\, dx\,.$$

(2) *If* $\kappa > 0$, *then for eigenvalues outside the cone* $\{|\operatorname{Im} z| < \kappa \operatorname{Re} z\}$,

$$\sum_{j:\ |\operatorname{Im}\lambda_j(v)| \ge \kappa \operatorname{Re}\lambda_j(v)} |\lambda_j(v)|^\gamma \le 2^{1+\gamma+d/2}\left(1 + \frac{2}{\kappa}\right)^{2\gamma+d} S_{\gamma,d} \int_{\mathbb{R}^d} |v(x)|^{2\gamma+d}\, dx\,.$$

Here $S_{\gamma,d}$ *is the constant from* (2.4).

2.3.4. *Waveguides.* Let $\omega \subset \mathbb{R}^d$ be a domain of finite measure and put $\Omega := \omega \times \mathbb{R}_+$ and $\Gamma := (\partial\omega) \times \mathbb{R}_+$. The quadratic form (2.3), restricted to $\{u \in H^1(\Omega) : u = 0 \text{ on } \Gamma\}$, defines a selfadjoint operator $H_\omega(v)$ in $L_2(\Omega)$, which corresponds to Dirichlet boundary conditions on Γ and boundary conditions of the third type on $\omega \times \{0\}$. By the variational principle, Theorem 2.1 implies that

$$\operatorname{tr}[H_\omega(v)]_-^\gamma \le S_{\gamma,d} \int_\omega v(x)_-^{2\gamma+d}\,dx$$

for $\gamma > 0$ if $d = 1$ and $\gamma \ge 0$ if $d \ge 2$ with the constant $S_{\gamma,d}$ from (2.4). In particular, $S_{\gamma,d} = L^{\mathrm{cl}}_{\gamma,d}$ for $\gamma \ge 3/2$. We now show that in the special case where $v \equiv v_0$ is a constant, the estimate with the semi-classical constant holds already for $\gamma \ge 1$.

THEOREM 2.8. *Let $\omega \subset \mathbb{R}^d$ be a domain of finite measure and $v \equiv v_0 > 0$ a constant. Then for any $\gamma \ge 1$,*

$$\operatorname{tr}[H_\omega(v)]_-^\gamma \le L^{\mathrm{cl}}_{\gamma,d}\,|\omega|\,v_0^{2\gamma+d}\,. \tag{2.19}$$

PROOF. By separation of variables one has

$$\operatorname{tr}[H_\omega(v)]_-^\gamma = \operatorname{tr}\left[-\Delta_\omega^D - v_0^2\right]_-^\gamma\,,$$

where $-\Delta_\omega^D$ denotes the Dirichlet Laplacian on ω. Therefore the assertion follows from the Berezin-Li-Yau inequality; see [**B, LY**] and also [**L**]. □

The same argument shows that if ω is tiling (in particular, *any* interval ω if $d = 1$), then (2.19) holds for all $\gamma \ge 0$; see [**P**].

3. Relativistic Schrödinger operators

3.1. Statement of the results. In this section we derive a connection between Schrödinger operators $H(v)$ with surface potential in $L_2(\mathbb{R}^{d+1}_+)$ and relativistic Schrödinger operators $\sqrt{-\Delta} - v$ in $L_2(\mathbb{R}^d)$. We begin by recalling Lieb–Thirring and Cwikel–Lieb–Rozenblum inequalities for the latter operator.

PROPOSITION 3.1. *The inequality*

$$\operatorname{tr}\left[\sqrt{-\Delta} - v\right]_-^\gamma \le D_{\gamma,d} \int_{\mathbb{R}^d} v(x)_+^{\gamma+d}\,dx \tag{3.1}$$

holds for all $0 \le v \in L_{\gamma+d}(\mathbb{R}^d)$ iff

$$\gamma > 0 \quad \text{if } d = 1, \qquad \text{and} \qquad \gamma \ge 0 \quad \text{if } d \ge 2\,. \tag{3.2}$$

This result is due to Daubechies [**D**]. The fact that the inequality is not valid for $\gamma = 0$ if $d = 1$ follows from the fact that $\sqrt{-\Delta} - v$ has a negative eigenvalue for any nontrivial $v \ge 0$ if $d = 1$. This can be proved as in [**S**, Prop. 7.4].

The Weyl-type asymptotics in the relativistic case read

$$\lim_{\alpha\to\infty} \alpha^{-\gamma-d} \operatorname{tr}\left[\sqrt{-\Delta} - \alpha\, v\right]_-^\gamma = D^{\mathrm{cl}}_{\gamma,d} \int_{\mathbb{R}^d} v(x)_+^{\gamma+d}\,dx\,, \tag{3.3}$$

with

$$D^{\mathrm{cl}}_{\gamma,d} := 2^{-d}\pi^{-d/2} \frac{\Gamma(\gamma+1)\,\Gamma(d+1)}{\Gamma(\gamma+d+1)\,\Gamma(d/2+1)}\,. \tag{3.4}$$

We denote by $D_{\gamma,d}$ the *sharp* constant in (3.1). In the case $d = 3$, the bound

$$D_{\gamma,3} \le 6.08\, D^{\mathrm{cl}}_{\gamma,3}\,, \quad \gamma \ge 0\,, \tag{3.5}$$

is contained in [**D**]. Similarly one proves that for $d = 2$,

$$D_{\gamma,2} \le 6.04\, D^{\mathrm{cl}}_{\gamma,2}\,, \quad \gamma \ge 0\,. \tag{3.6}$$

We are now in position to state a result which connects relativistic Schrödinger operators and nonrelativistic Schrödinger operators with surface potentials. As usual, we denote by $N(-\tau, T)$ the number of eigenvalues, counting multiplicities, less than $-\tau$ of a selfadjoint, lower semi-bounded operator T, and write $N(T) := N(0, T)$.

THEOREM 3.2. *Assume that v is form-compact with respect to $\sqrt{-\Delta}$. Then for any $\tau \ge 0$ one has*

$$N(-\tau, H(v)) = N(\sqrt{-\Delta + \tau} - v)\,. \tag{3.7}$$

Moreover, for any $\gamma > 0$ and $0 < \rho < 1$ one has

$$\operatorname{tr}\left[\sqrt{-\Delta} - v\right]_-^\gamma \le \operatorname{tr}\left[H(v)\right]_-^{\gamma/2} \le \left(\frac{\rho}{\sqrt{1-\rho^2}}\right)^\gamma \operatorname{tr}\left[\sqrt{-\Delta} - \rho^{-1} v\right]_-^\gamma\,. \tag{3.8}$$

We shall prove this in Subsection 3.2 below, as well as the following

COROLLARY 3.3. *The sharp constants in (2.4) and (3.1) coincide for $\gamma = 0$ and $d \ge 2$,*

$$S_{0,d} = D_{0,d}\,, \tag{3.9}$$

and satisfy for any $\gamma > 0$ and $d \ge 1$,

$$\frac{\gamma^{\gamma/2}\, d^{d/2}}{(\gamma + d)^{(\gamma+d)/2}}\, S_{\gamma/2,d} \le D_{\gamma,d} \le S_{\gamma/2,d}\,. \tag{3.10}$$

We shall use Theorem 3.2 in two directions. In Subsection 3.3 we shall use the known Lieb–Thirring inequalities in the relativistic case to derive the Lieb–Thirring inequalities for surface potentials. In Subsection 3.4 we shall use the estimates on the constants $S_{\gamma,d}$ for surface potentials to improve upon the estimates (3.5) and (3.6) in the relativistic case. We also discuss the connection of our inequality with an inequality by Birman, Koplienko and Solomyak.

3.2. Duality. The following lemma characterizes the negative eigenvalues of the operator $H(v)$ as the values $-\tau$ for which 0 is an eigenvalue of the operator $\sqrt{-\Delta + \tau} - v$.

LEMMA 3.4. *Assume that v is form-compact with respect to $\sqrt{-\Delta}$ and let $\tau > 0$.*
(1) *Let $f \in \ker(\sqrt{-\Delta + \tau} - v)$ and define $u(x,y) := (\exp(-y\sqrt{-\Delta + \tau}) f)(x)$. Then $u \in \ker(H(v) + \tau)$ and $u(x,0) = f(x)$.*
(2) *Let $u \in \ker(H(v) + \tau)$ and define $f(x) := u(x,0)$. Then we have $f \in \ker(\sqrt{-\Delta + \tau} - v)$ and $u(x,y) = (\exp(-y\sqrt{-\Delta + \tau}) f)(x)$.*

The proof of this lemma is straightforward and will be omitted (see [**FS**] for a similar argument). Using a modification of the Birman–Schwinger principle we now give the

PROOF OF THEOREM 3.2. Since the eigenvalues of the operators $\sqrt{-\Delta + t} - v$ are increasing with respect to t, one has for any fixed $\tau \ge 0$,

$$N(\sqrt{-\Delta + \tau} - v) = \#_{\mathrm{m}}\{t > \tau : 0 \text{ is an eigenvalue of } \sqrt{-\Delta + t} - v\}\,.$$

Here $\#_{\rm m}\{\dots\}$ means that the cardinality of $\{\dots\}$ is determined according to multiplicities. By Lemma 3.4, the right hand side coincides with

$$\#_{\rm m}\{t>\tau:\, -t \text{ is an eigenvalue of } H(v)\} = N(-\tau, H(v))\,,$$

as claimed.

To prove (3.8) we note that by the previous argument,

$$\operatorname{tr}[H(v)]_-^{\gamma/2} = \frac{\gamma}{2}\int_0^\infty N(-\tau,H(v))\,\tau^{\gamma/2-1}\,d\tau = \frac{\gamma}{2}\int_0^\infty N(\sqrt{-\Delta+\tau}-v)\,\tau^{\gamma/2-1}\,d\tau\,. \tag{3.11}$$

The elementary inequalities

$$\rho\sqrt{\lambda}+\sqrt{1-\rho^2}\sqrt{\tau}\le\sqrt{\lambda+\tau}\le\sqrt{\lambda}+\sqrt{\tau},\quad \lambda,\tau>0,\, 0<\rho<1\,,$$

imply

$$N(-\sqrt{\tau},\sqrt{-\Delta}-v)\le N(\sqrt{-\Delta+\tau}-v)\le N(-\sqrt{1-\rho^2}\sqrt{\tau},\rho\sqrt{-\Delta}-v)\,.$$

Plugging this into (3.11) we obtain (3.8). □

Proof of Corollary 3.3. Equality (3.9) as well as the second inequality in (3.10) follow immediately from equality (3.7) and the first inequality in (3.8). To prove the first inequality in (3.10) we combine Daubechies' inequality (3.1) with the second inequality in (3.8) to get

$$\operatorname{tr}[H(v)]_-^{\gamma/2}\le D_{\gamma,d}\left(\frac{\rho}{\sqrt{1-\rho^2}}\right)^{\gamma}\rho^{-\gamma-d}\int_{\mathbb{R}^d}v(x)_+^{\gamma+d}\,dx\,.$$

The assertion follows by optimizing over $0<\rho<1$. □

Remark 3.5. The material in this subsection, except for the proof of the second part of Corollary 3.3, is of an abstract nature. If A is a nonnegative operator in a Hilbert space $\mathfrak{H}$ and if B is a selfadjoint operator which is relatively form-compact with respect to A, then define the operator H in $L_2(\mathbb{R}_+,\mathfrak{H})$ by the quadratic form

$$\int_0^\infty\left(\|F'(y)\|_{\mathfrak{H}}^2+\|AF(y)\|_{\mathfrak{H}}^2\right)\,dy-b[F(0)]$$

for $F\in H^1(\mathbb{R}_+,\mathfrak{H})\cap L_2(\mathbb{R}_+,\operatorname{dom}A)$. Here b is the quadratic form of B. Then the argument of this subsection yields

$$N(-\tau,H)=N(\sqrt{A^2+\tau}-B)\,.$$

As an application of this generalization one can extend Theorem 3.2 to relativistic Schrödinger operators with magnetic field or to relativistic Schrödinger operators with a Hardy weight subtracted (see [**FLS1**]).

3.3. Proof of Theorem 2.1. Theorem 2.1 is an immediate consequence of Proposition 3.1 and Theorem 3.2.

Moreover, Daubechies' bounds (3.6), (3.5) yield the upper bounds (2.10), (2.11) for the sharp constants $S_{0,d}$. Similarly, the lower bounds (2.13), (2.14) follow from

$$D_{0,d}\ge\frac{2^{d-1}}{(d-1)^d}\,\Gamma(d+1)\,D_{0,d}^{\rm cl}\,,\qquad d\ge 2. \tag{3.12}$$

(Note that this is only useful for $d\le 7$, since otherwise the factor on the right hand side is less than one and the bound $D_{0,d}\ge D_{0,d}^{\rm cl}$ follows from (3.3).) The lower bound (3.12) can be seen as follows. The definition of $D_{0,d}$ implies that if $\int v^d\,dx < D_{0,d}^{-1}$, then $\sqrt{-\Delta} - v$ is a nonnegative operator. Hence

$$\int_{\mathbb{R}^d} v|u|^2\,dx \le \int_{\mathbb{R}^d} |(-\Delta)^{1/4}u|^2\,dx$$

for all $u \in H^{1/2}(\mathbb{R}^d)$. Choosing $v = \alpha_\varepsilon |u|^{2/(d-1)}$ with a constant α_ε such that $\alpha_\varepsilon^d \int |u|^{2d/(d-1)}\,dx = (D_{0,d}+\varepsilon)^{-1}$ and letting ε tend to zero, we find

$$D_{0,d}^{-1/d}\left(\int_{\mathbb{R}^d} |u|^{2d/(d-1)}\,dx\right)^{(d-1)/d} \le \int_{\mathbb{R}^d} |(-\Delta)^{1/4}u|^2\,dx$$

for all $u \in H^{1/2}(\mathbb{R}^d)$. Hence $D_{0,d}^{-1/d}$ is not greater than the constant in the sharp Sobolev inequality:
(3.13)

$$S_d' \,\|u\|_{2d/(d-1)}^2 \le \left\|(-\Delta)^{1/4}u\right\|^2, \qquad S_d' := \frac{d-1}{2}\,2^{1/d}\,\pi^{(d+1)/2d}\,\Gamma\left(\frac{d+1}{2}\right)^{-1/d};$$

see [**LL**, Thm. 8.4]. Recalling definition (3.4) of $D_{\gamma,d}^{\mathrm{cl}}$ we arrive at (3.12).

REMARK 3.6. Instead of using the 'relativistic' Sobolev inequality (3.13) to prove (3.12) we could have used a similar argument based on the sharp Sobolev trace inequality

$$S_d'\left(\int_{\mathbb{R}^d} |u|^{2d/(d-1)}\,dx\right)^{(d-1)/d} \le \iint_{\mathbb{R}^{d+1}_+} |\nabla u|^2\,dx\,dy \tag{3.14}$$

to directly prove (2.13), (2.14). The constant S_d' in (3.14) is the same as in (3.13); see [**E**]. Indeed, an argument similar to our Lemma 3.4 was used in [**CL**] to derive (3.14) from (3.13).

REMARK 3.7. Weyl-type asymptotics (2.6) can be proved by a bracketing argument, dividing $\mathbb{R}^{d+1}_+$ into domains $Q\times\mathbb{R}_+$ with $Q \subset \mathbb{R}^d$ a small cube. An alternative proof can be based on Theorem 3.2. Indeed, for $\gamma = 0$ the asymptotics (2.6) follows immediately from Theorem 3.2 and (3.3) (which is valid for all smooth v). If $\gamma > 0$, we write as in (3.11),

$$\operatorname{tr}[H(\alpha\, v)]_-^\gamma = \gamma \int_0^\infty N(-\tau, H(\alpha\, v))\,\tau^{\gamma-1}\,d\tau = \gamma\int_0^\infty N(\sqrt{-\Delta+\tau} - \alpha\, v)\,\tau^{\gamma-1}\,d\tau\,.$$

For smooth v one can justify that this is asymptotically equal as $\alpha \to \infty$ to

$$\begin{aligned}
\gamma &\int_0^\infty \iint_{\{(x,\xi)\in\mathbb{R}^d\times\mathbb{R}^d:\ \sqrt{|\xi|^2+\tau}-\alpha\, v(x)<0\}} \frac{dx\,d\xi}{(2\pi)^d}\,\tau^{\gamma-1}\,d\tau \\
&= \gamma L_{0,d}^{\mathrm{cl}} \int_0^\infty \int_{\mathbb{R}^d} \left((\alpha v(x))_+^2 - \tau\right)_+^{d/2}\,dx\,\tau^{\gamma-1}\,d\tau \\
&= \alpha^{2\gamma+d} L_{\gamma,d}^{\mathrm{cl}} \int_{\mathbb{R}^d} v(x)_+^{2\gamma+d}\,dx\,.
\end{aligned}$$

This concludes the sketch of (2.6). We note that by a standard argument based on Theorem 2.1, the asymptotics (2.6) extends to all v for which the right hand side is finite if $\gamma > 0$ and $d = 1$ or if $\gamma \ge 0$ and $d \ge 2$.

3.4. Relation with the BKS inequality. In this subsection we shall use Theorem 3.2 to improve upon known constants for relativistic Schrödinger operators and discuss its relation with an inequality by Birman, Koplienko and Solomyak. We begin with a result pertaining to massive relativistic Schrödinger operators.

REMARK 3.8. Let $d \geq 3$ and $m \geq 0$. Then

$$N(\sqrt{-\Delta + m^2} - m - v) \leq 10.332\, D_{0,d}^{\mathrm{cl}} \int_{\mathbb{R}^d} \left((v(x)+m)_+^2 - m^2\right)_+^{d/2} dx\,. \tag{3.15}$$

This improves upon Daubechies' bound [**D**], who obtains (3.15) with constant $14.14\, D_{0,3}^{\mathrm{cl}}$ for $d = 3$. To prove (3.15) we combine Theorem 3.2 and Remark 2.4 to get

$$\begin{aligned} N(\sqrt{-\Delta + m^2} - m - v) &= N(-m^2, H(v+m)) \\ &\leq 10.332\, L_{0,d}^{\mathrm{cl}} \int_{\mathbb{R}^d} \left((v(x)+m)_+^2 - m^2\right)_+^{d/2} dx\,, \end{aligned}$$

and recall that $L_{0,d}^{\mathrm{cl}} = D_{0,d}^{\mathrm{cl}}$.

We return again to the massless case $m = 0$.

REMARK 3.9. The sharp constants in (3.1) satisfy

$$\begin{aligned} D_{\gamma,2} &\leq \sqrt{3}\pi\, D_{\gamma,2}^{\mathrm{cl}} && \text{if } 2 \leq \gamma < 3 \text{ and } d = 2, \\ D_{\gamma,2} &\leq 4\, D_{\gamma,2}^{\mathrm{cl}} && \text{if } \gamma \geq 3 \text{ and } d = 2, \\ D_{\gamma,3} &\leq \frac{15\pi}{8}\, D_{\gamma,3}^{\mathrm{cl}} && \text{if } \gamma \geq 3 \text{ and } d = 3. \end{aligned} \tag{3.16}$$

This improves upon (3.6) and (3.5). To prove the first inequality in (3.16) we combine (3.10) with (2.9) to get

$$D_{2,2} \leq \frac{\pi}{\sqrt{3}} L_{1,2}^{\mathrm{cl}} = \sqrt{3}\pi\, D_{2,2}^{\mathrm{cl}}\,.$$

By the argument of Aizenman and Lieb [**AL**], this implies $D_{\gamma,2} \leq \sqrt{3}\pi\, D_{\gamma,2}^{\mathrm{cl}}$ for all $\gamma \geq 2$. The other bounds in (3.16) are proved similarly.

In conclusion we would like to recall a result by Birman, Koplienko and Solomyak.

PROPOSITION 3.10. *Let* $0 < s < 1$, $\gamma \geq 1$ *and let* A, B *be nonnegative operators such that* $\operatorname{tr}(A-B)_+^{s\gamma} < \infty$. *Then*

$$\operatorname{tr}(A^s - B^s)_+^\gamma \leq \operatorname{tr}(A-B)_+^{s\gamma}. \tag{3.17}$$

In [**BKS**] this is proved under the additional assumption $A \geq B$, but, as observed in [**LSS**], this assumption can be removed in view of the operator inequality

$$A^s - B^s \leq (B + (A-B)_+)^s - B^s\,.$$

Moreover, [**LSS**] contains an elementary proof of (3.17) in the case $\gamma = 1$. We deduce from (3.17) and (1.1) that

$$\operatorname{tr}(\sqrt{-\Delta} - v)_-^\gamma \leq \operatorname{tr}(-\Delta - v_+^2)_-^{\gamma/2} \leq L_{\gamma/2,d} \int_{\mathbb{R}^d} v(x)_+^{\gamma+d}\, dx, \quad \gamma \geq 1. \tag{3.18}$$

Since the best-known bounds on the constants $L_{\gamma/2,d}$ for $\gamma \geq 1$ coincide with those for $S_{\gamma/2,d}$, Proposition 3.10 yields for $\gamma \geq 1$ the same bounds on $D_{\gamma,d}$ as our Theorem 3.2. In particular, Remark 3.9 can also be derived via (3.18). In contrast, the $\gamma = 0$ result of Remark 3.8 cannot be deduced via (3.18). It is interesting, in our opinion, to understand whether there is a deeper connection between the $s = 1/2$ case of Proposition 3.10 and Theorem 3.2.

References

[AL] M. Aizenman, E. Lieb, *On semiclassical bounds for eigenvalues of Schrödinger operators.* Phys. Lett. A **66** (1978), no. 6, 427–429.

[B] F.A. Berezin, *Covariant and contravariant symbols of operators* [Russian]. English transl: Math. USSR Izv. **6** (1972), 1117–1151.

[BKS] M. S. Birman, L. S. Koplienko, M. Z. Solomyak, *Estimates of the spectrum of a difference between fractional powers of two selfadjoint operators.* Soviet Math. (Iz. VUZ) **19** (1975), no. 3, 1–6.

[CL] E. Carlen, M. Loss, *Competing symmetries of some functionals arising in mathematical physics.* In: Stochastic processes, physics and geometry (Ascona and Locarno, 1988), 277–288, World Sci. Publ., Teaneck, NJ, 1990.

[CS] L. Caffarelli, L. Silvestre, *An extension problem related to the fractional Laplacian.* Comm. Partial Differential Equations **32** (2007), no. 8, 1245–1260.

[D] I. Daubechies, *An uncertainty principle for fermions with generalized kinetic energy.* Comm. Math. Phys. **90** (1983), no. 4, 511–520.

[DLL] J. Dolbeault, A. Laptev, M. Loss, *Lieb–Thirring inequalities with improved constants.* J. Eur. Math. Soc., to appear.

[E] J. Escobar, *Sharp constant in a Sobolev trace inequality.* Indiana Univ. Math. J. **37** (1988), 687–698.

[F] R. L. Frank, *Remarks on eigenvalue estimates and semigroup domination.* In preparation.

[FLLS] R. L. Frank, A. Laptev, E. H. Lieb, R. Seiringer, *Lieb–Thirring inequalities for Schrödinger operators with complex-valued potentials.* Lett.. Math. Phys. **77** (2006), 309–316.

[FLS1] R. L. Frank, E. H. Lieb, R. Seiringer, *Hardy-Lieb–Thirring inequalities for fractional Schrödinger operators.* J. Amer. Math. Soc. **21** (2008), no. 4, 925–950.

[FLS2] R. L. Frank, E. H. Lieb, R. Seiringer, *Number of bound states of Schrödinger operators with matrix-valued potentials.* Lett. Math. Phys. **82** (2007), 107–116.

[FS] R. L. Frank, R. G. Shterenberg, *On the scattering theory of the Laplacian with a periodic boundary condition. II. Additional channels of scattering.* Doc. Math. **9** (2004), 57–77.

[H1] D. Hundertmark, *On the number of bound states for Schrödinger operators with operator-valued potentials.* Ark. Mat. **40** (2002), 73–87.

[H2] D. Hundertmark, *Some bound state problems in quantum mechanics.* In: Spectral theory and mathematical physics: a Festschrift in honor of Barry Simon's 60th birthday, 463–496, Proc. Sympos. Pure Math. **76**, Part 1, Amer. Math. Soc.., Providence, RI, 2007.

[HLW] D. Hundertmark, A. Laptev and T. Weidl, *New bounds on the Lieb–Thirring constants.* Invent. Math., **40** (2000), 693–704.

[L] A. Laptev, *Dirichlet and Neumann eigenvalue problems on domains in Euclidean spaces.* J. Funct. Anal. **151** (1997), no. 2, 531–545.

[LW1] A. Laptev, T. Weidl, *Sharp Lieb–Thirring inequalities in high dimensions.* Acta Math. **184** (2000), no. 1, 87–111.

[LW2] A. Laptev, T. Weidl, *Recent results on Lieb–Thirring inequalities.* Journées "Équations aux Dérivées Partielles" (La Chapelle sur Erdre, 2000), Exp. No. XX, Univ. Nantes, Nantes, 2000.

[LY] P. Li, S.-T. Yau, *On the Schrödinger equation and the eigenvalue problem.* Comm. Math. Phys. **88** (1983), 309–318.

[LL] E. H. Lieb, M. Loss, *Analysis. Second edition.* Graduate Studies in Mathematics **14**, American Mathematical Society, Providence, RI, 2001.

[LSS] E. H. Lieb, H. Siedentop, J. P. Solovej, *Relativistic electrons in classical electromagnetic fields.* J. Stat. Phys. **89** (1997), 37–59.

[LT] E. H. Lieb, W. Thirring, *Inequalities for the moments of the eigenvalues of the Schrödinger Hamiltonian and their relation to Sobolev inequalities.* Studies in Mathematical Physics, 269–303. Princeton University Press, Princeton, NJ, 1976.

[P] G. Pólya, *On the eigenvalues of vibrating membranes.* Proc. London Math. Soc. **11** (1961), 419–433.

[R] G. Rozenblyum, *Domination of semigroups and estimates for eigenvalues.* St. Petersburg Math. J. **12** (2001), no. 5, 831–845.

[S] B. Simon, *Trace ideals and their applications. Second edition.* Mathematical Surveys and Monographs **120**, American Mathematical Society, Providence, RI, 2005.

Department of Mathematics, Princeton University, Fine Hall, Princeton, NJ 08544, USA

E-mail address: `rlfrank@math.princeton.edu`

Department of Mathematics, Imperial College London, London SW7 2AZ, UK and Department of Mathematics, Royal Institute of Technology, 100 44 Stockholm, Sweden

E-mail address: `a.laptev@imperial.ac.uk, laptev@math.kth.se`

Amer. Math. Soc. Transl.
(2) Vol. **225**, 2008

On the Spectrum of the Dirichlet Laplacian in a Narrow Infinite Strip

Leonid Friedlander and Michael Solomyak

To Mikhail Shlëmovich Birman on his 80th birthday

ABSTRACT. This is a continuation of the paper [**3**]. We consider the Dirichlet Laplacian in a family of unbounded domains $\{x \in \mathbb{R},\ 0 < y < \epsilon h(x)\}$. The main assumption is that $x = 0$ is the only point of global maximum of the positive, continuous function $h(x)$. We show that the number of eigenvalues lying below the essential spectrum indefinitely grows as $\epsilon \to 0$, and find the two-term asymptotics in $\epsilon \to 0$ of each eigenvalue and the one-term asymptotics of the corresponding eigenfunction. The asymptotic formulae obtained involve the eigenvalues and eigenfunctions of an auxiliary ODE on $\mathbb{R}$ that depends only on the behavior of $h(x)$ as $x \to 0$.

The proof is based on a detailed study of the resolvent of the operator Δ_ϵ.

1. Introduction

This paper is a continuation of the authors' work [**3**], where we studied the spectrum of the Dirichlet Laplacian $\Delta_{\epsilon,D}$ in a narrow strip

$$\Omega_\epsilon = \{(x,y) : x \in I,\ 0 < y < \epsilon h(x)\}.$$

The main objective in [**3**] was to understand the behavior of the eigenvalues as $\epsilon \to 0$, and the main assumption was that I is a finite segment and the continuous function $h(x)$ has on I a single point of global maximum. We found the two-term asymptotics (in $\epsilon \to 0$) of each eigenvalue, and also the one-term asymptotics of each eigenfunction. Our approach was based upon a careful study of the resolvent

$$\mathfrak{A}_{\epsilon,D}^{-1} := (\Delta_{\epsilon,D} - K\epsilon^{-2})^{-1}$$

with an appropriate choice of the constant K; see (2.6) below. Note that we consider the Laplacian as a positive operator, so that

$$\Delta\psi = -\psi''_{xx} - \psi''_{yy}.$$

Here we apply the same approach to two other problems of a similar nature. One of them concerns the case $I = \mathbb{R}$, and the assumptions about $h(x)$ are basically the same as in [**3**], complemented by a mild additional condition as $|x| \to \infty$. In

2000 *Mathematics Subject Classification.* Primary 35P15.

another problem I is a finite segment (as it was in [**3**]), but the Dirichlet condition at the vertical parts of $\partial\Omega_\epsilon$ is replaced by the Neumann condition. This gives rise to the operator $\Delta_{\epsilon,DN}$ that was not discussed in [**3**]. In both cases, and especially in the second one, the difference with the original problem studied in [**3**] looks minor. However, some technical tools used there no more apply in the new situation, and one has to look for appropriate substitutes.

Note that the case of the Neumann boundary condition on the whole of $\partial\Omega_\epsilon$ is simpler than the case of the Dirichlet condition; see its analysis in [**7**] and [**6**]. The results in [**7**], [**6**] concern a much wider class of domains than those in our paper [**3**].

We believe that our results for unbounded Ω_ϵ, i.e. for $I = \mathbb{R}$, are of some independent interest. They easily extend to the case when I is a half-line; we leave it to the reader. The results for the operator $\Delta_{\epsilon,N}$ are of a more technical character. They are useful, since they allow one to apply the Dirichlet–Neumann bracketing for study of some other problems. In particular, in our problem for $I = \mathbb{R}$ this gives a simple way to obtain the asymptotics of eigenvalues, avoiding an analysis of the resolvent; we show this in Section 6. One more problem concerns the case when $I = \mathbb{R}$ and the function $h(x)$ is periodic (Laplacian in a thin periodic waveguide). Here the spectrum of $\Delta_{\epsilon,D}$ has the band-gap structure, and we study the location of bands and give a bound for their widths; it turns out that they decay exponentially as $\epsilon \to 0$. We present our corresponding results in a separate paper [**4**].

We complement these comments at the end of Section 2, after introducing the necessary notation and formulating some of our main results.

2. The case of a finite segment: Setting of the problem and formulation of main results

2.1. Preliminaries. Let $I = [-a, b]$ be a finite segment and $h(x) > 0$ be a continuous function on I. We assume that

(*i*) $x = 0$ is the only point of global maximum of $h(x)$ on I.

(*ii*) The function $h(x)$ is C^1 on $I \setminus \{0\}$, and in a neighborhood of $x = 0$ it admits an expansion

$$h(x) = \begin{cases} M - c_+ x^m + O\big(x^{m+1}\big), & x > 0, \\ M - c_-|x|^m + O\big(|x|^{m+1}\big), & x < 0, \end{cases} \tag{2.1}$$

where $M, c_\pm > 0$ and $m \geq 1$.

We consider the Laplacian in Ω_ϵ, and we always impose the boundary conditions

$$\psi(x, 0) = \psi(x, \epsilon h(x)) = 0. \tag{2.2}$$

The conditions at $x = -a$ and $x = b$ can be either Dirichlet or Neumann, and we denote the corresponding operators as $\Delta_{\epsilon,D}$ and $\Delta_{\epsilon,DN}$, respectively; in [**3**] the operator $\Delta_{\epsilon,D}$ was denoted as Δ_ϵ. For the sake of brevity, sometimes we speak about the 'D-problem' and the 'DN-problem'.

Denote

$$H^{1,d}(\Omega_\epsilon) = \{\psi \in H^1(\Omega_\epsilon) : \ \psi(x, 0) = \psi(x, \epsilon h(x)) = 0\}.$$

The conditions (2.2) imply

$$\int_0^{\epsilon h(x)} \psi_y'(x,y)^2 dy \geq \frac{\pi^2}{\epsilon^2 h^2(x)} \int_0^{\epsilon h(x)} \psi^2(x,y) dy; \tag{2.3}$$

here ψ is a smooth real-valued function. The nature of the problem allows us to work with such functions only. By (2.3),

$$\int_{\Omega_\epsilon} |\nabla\psi|^2 dxdy \geq \frac{\pi^2}{M^2\epsilon^2} \int_{\Omega_\epsilon} \psi^2 dxdy, \qquad \forall\psi \in H^{1,d}(\Omega_\epsilon). \tag{2.4}$$

It is convenient for us to work with the quadratic form

$$\mathfrak{a}_\epsilon[\psi] = \int_{\Omega_\epsilon} \left(|\nabla\psi|^2 - \frac{\pi^2}{M^2\epsilon^2}\psi^2\right) dxdy, \qquad \psi \in H^{1,d}(\Omega_\epsilon). \tag{2.5}$$

We denote the corresponding operators as $\mathfrak{A}_{\epsilon,D}$ and $\mathfrak{A}_{\epsilon,DN}$, depending on the boundary condition at $x = -a$ and $x = b$. We suppress the subscripts D and DN when our argument applies to both operators. On their respective domains they act as

$$\mathfrak{A}_\epsilon\psi = \Delta\psi - \frac{\pi^2}{M^2\epsilon^2}\psi. \tag{2.6}$$

2.2. Limiting behavior of eigenvalues and eigenfunctions. It turns out that under the conditions (i), (ii) this behavior is determined by the operator on $L^2(\mathbb{R})$ given by

$$\mathbf{H} = -\frac{d^2}{dx^2} + q(x), \qquad q(x) = \begin{cases} 2\pi^2 M^{-3} c_+ x^m, & x > 0, \\ 2\pi^2 M^{-3} c_- |x|^m, & x < 0. \end{cases} \tag{2.7}$$

The spectrum of $\mathbf{H}$ is discrete and consists of simple eigenvalues, which we denote by μ_j. The corresponding eigenfunctions $X_j(x)$, normalized by the conditions $\|X_j\|_{L^2(\mathbb{R})} = 1$, $X_j(x) > 0$ for large $x > 0$, decay as $|x| \to \infty$ superexponentially fast. If $m = 2$ and $c_+ = c_- = c$, then $\mathbf{H}$ turns into the harmonic oscillator.

THEOREM 2.1. *Let I be a finite segment and let $h(x)$ meet the conditions (i) and (ii). Then*

1) The eigenvalues $\lambda_j(\epsilon, D)$, $\lambda_j(\epsilon, DN)$ of the operators $\Delta_{\epsilon,D}$, $\Delta_{\epsilon,DN}$ have the same asymptotic behavior, namely

$$\lim_{\epsilon\to 0} \epsilon^{2\alpha}\left(\lambda_j(\epsilon) - \frac{\pi^2}{M^2\epsilon^2}\right) = \mu_j, \tag{2.8}$$

where

$$\alpha = 2(m+2)^{-1}. \tag{2.9}$$

2) For the normalized eigenfunctions $\Psi_j(\epsilon, D; x, y)$, $\Psi_j(\epsilon, DN; x, y)$ of the operators $\Delta_{\epsilon,D}$, $\Delta_{\epsilon,DN}$ we have, with an appropriate choice of sign:

$$\lim_{\epsilon\to 0} \int_{\Omega_\epsilon} \left(\Psi_j(\epsilon; x, y) - \frac{\sqrt{2}}{\sqrt{\epsilon^{1+\alpha} h(x)}} X_j(x\epsilon^{-\alpha})\right)^2 dxdy = 0.$$

For the D-case this is the result of Theorems 1.1 and 1.4 in [**3**]. For the DN-case the results are new.

Our proof of Theorem 2.1 is based upon the study of the operator family $\mathfrak{A}_\epsilon^{-1}$ as $\epsilon \to 0$. It consists of two steps. Firstly, we reduce the original problem to the one for an auxiliary ordinary differential operator $\mathbf{Q}_\epsilon$ acting on $L^2(I)$. Secondly, we show that the operators $(\epsilon^{2\alpha}\mathbf{Q}_\epsilon)^{-1}$ approach a family of operators on $L^2(\mathbb{R})$ that are unitarily equivalent to, and hence isospectral with, the operator $\mathbf{H}^{-1}$. In the next two subsections we describe these steps and formulate the corresponding results. Their proofs are given in Section 4; the general scheme is explained in Section 3. In Section 5 we show that under mild additional assumptions about the behavior of $h(x)$ as $|x| \to \infty$ this scheme applies also to the case $I = \mathbb{R}$. In the concluding Section 6 we prove that these conditions can be simplified even further if one is interested only in the behavior of the eigenvalues.

2.3. Reduction of dimension. In $L^2(\Omega_\epsilon)$ we take the subspace $\mathfrak{L}_\epsilon$ that consists of the functions

$$\psi(x,y) = \psi_{\epsilon,\chi}(x,y) = \chi(x)\sqrt{\frac{2}{\epsilon h(x)}}\sin\frac{\pi y}{\epsilon h(x)}.$$

The mapping

$$\mathbf{\Pi}_\epsilon : \chi \mapsto \psi_{\epsilon,\chi} \tag{2.10}$$

is an isometric isomorphism of $L^2(I)$ onto $\mathcal{H}_\epsilon$, and we identify any operator $\mathbf{T}$ on $\mathcal{H}_\epsilon$ with the operator $\mathbf{\Pi}_\epsilon^{-1}\mathbf{T}\mathbf{\Pi}_\epsilon$ acting on $L^2(I)$. Let $\chi \in H^1(I)$. A direct computation shows that

$$\mathbf{a}_\epsilon[\psi_{\epsilon,\chi}] = \mathbf{q}_\epsilon[\chi] := \int_I (\chi'(x)^2 + W_\epsilon(x)\chi^2(x))\,dx, \tag{2.11}$$

where

$$W_\epsilon(x) = \frac{\pi^2}{\epsilon^2}\left(\frac{1}{h^2(x)} - \frac{1}{M^2}\right) + \left(\frac{\pi^2}{3} + \frac{1}{4}\right)\frac{h'(x)^2}{h^2(x)}. \tag{2.12}$$

The quadratic form $\mathbf{q}_\epsilon[\chi]$, considered on $H^1(I)$, is positive definite and closed in $L^2(I)$. The same is true for its restriction to $H^{1,0}(I)$. The corresponding selfadjoint operator on $L^2(I)$ acts as

$$\mathbf{Q}_\epsilon\chi = -\chi'' + W_\epsilon(x)\chi, \tag{2.13}$$

with the Dirichlet or the Neumann condition at ∂I. When it is necessary to reflect it in the notation, we denote these operators by $\mathbf{Q}_{\epsilon,D}$ and $\mathbf{Q}_{\epsilon,N}$ respectively.

Below $\mathfrak{L}^\epsilon$ stands for the orthogonal complement of $\mathfrak{L}_\epsilon$ in $L^2(\Omega_\epsilon)$. Given a Hilbert space $\mathfrak{Z}$, we write $\mathbf{0}_\mathfrak{Z}$ for the zero operator on $\mathfrak{Z}$.

THEOREM 2.2. *Under the assumptions of Theorem 2.1 one has*

$$\|\mathfrak{A}_\epsilon^{-1} - \mathbf{Q}_\epsilon^{-1} \oplus \mathbf{0}_{\mathfrak{L}^\epsilon}\| = O(\epsilon^{3\alpha}), \qquad \epsilon \to 0, \tag{2.14}$$

where $\mathfrak{A}_\epsilon$ is either of the operators $\mathfrak{A}_{\epsilon,D}$ and $\mathfrak{A}_{\epsilon,DN}$, and $\mathbf{Q}_\epsilon = \mathbf{Q}_{\epsilon,D}$ in the D-case and $\mathbf{Q}_\epsilon = \mathbf{Q}_{\epsilon,N}$ in the DN-case.

For the D-case this is Theorem 1.2 in [**3**].

2.4. From $\mathbf{Q}_\epsilon$ to $\mathbf{H}$. Along with the operator $\mathbf{H}$ defined in (2.7), let us consider on $L^2(\mathbb{R})$ the operator family

$$\mathbf{H}_\epsilon = -\frac{d^2}{dx^2} + \epsilon^{-2}q(x), \quad \epsilon > 0.$$

In particular $\mathbf{H}_1 = \mathbf{H}$. The substitution $x = t\epsilon^\alpha$ shows that for any $\epsilon > 0$ the operator $\epsilon^{2\alpha}\mathbf{H}_\epsilon$ is unitarily equivalent to $\mathbf{H}$, and hence, $\epsilon^{2\alpha}\mathbf{H}_\epsilon$ is an isospectral family of operators.

THEOREM 2.3. *Let $\mathbf{Q}_\epsilon$ be either of the operators $\mathbf{Q}_{\epsilon,D}$ and $\mathbf{Q}_{\epsilon,N}$. One has*

$$\lim_{\epsilon\to 0} \|(\epsilon^{2\alpha}\mathbf{Q}_\epsilon)^{-1} \oplus \mathbf{0}_{L^2(\mathbb{R}\setminus I)} - (\epsilon^{2\alpha}\mathbf{H}_\epsilon)^{-1}\| = 0.$$

For the D-case this is a reformulation of Theorem 1.3 in [**3**]; see (1.9) therein.

Note that Theorems 2.2 and 2.3 give a stronger result than Theorem 2.1. For instance, they imply that the convergence

$$\epsilon^{-2\alpha}\left(\lambda_j(\epsilon) - \frac{\pi^2}{M^2\epsilon^2}\right)^{-1} \to \mu_j^{-1},$$

cf. (2.8), is uniform in j.

The derivation of Theorem 2.1 from Theorems 2.2 and 2.3 was explained in [**3**], and we do not reproduce it here.

3. Proof of Theorem 2.2: General scheme

For the D-case the proof is given in [**3**], Section 3. For the DN-case the scheme remains the same, and we find it useful to present it in an abstract form.

Let $\mathfrak{H}_\epsilon$, $0 < \epsilon \le \epsilon_0$, be a family of separable Hilbert spaces. Keeping in mind our original problem, it is sufficient to consider them real. Let $\mathcal{H}_\epsilon \subset \mathfrak{H}_\epsilon$ be a family of their (closed) subspaces. We denote by $\mathcal{H}^\epsilon$ the orthogonal complement of $\mathcal{H}_\epsilon$ in $\mathfrak{H}_\epsilon$, so that

$$\mathfrak{H}_\epsilon = \mathcal{H}_\epsilon \oplus \mathcal{H}^\epsilon.$$

Below $\mathbf{P}_\epsilon$ and $\mathbf{P}^\epsilon$ stand for the orthogonal projections in $\mathfrak{H}_\epsilon$ onto the subspaces $\mathcal{H}_\epsilon$ and $\mathcal{H}^\epsilon$ respectively, and for an arbitrary element $\psi \in \mathfrak{H}_\epsilon$ we standardly write

$$\psi_\epsilon = \mathbf{P}_\epsilon\psi, \qquad \psi^\epsilon = \mathbf{P}^\epsilon\psi.$$

For each ϵ, let $\mathfrak{a}_\epsilon[\psi_1, \psi_2]$ be a symmetric bilinear form in $\mathfrak{H}_\epsilon$, defined for ψ_1, ψ_2 lying in a dense domain $\mathfrak{d}_\epsilon$. We suppose that the corresponding quadratic form $\mathfrak{a}[\psi] := \mathfrak{a}_\epsilon[\psi, \psi]$ is nonnegative and closed. We write $\mathfrak{A}_\epsilon$ for the selfadjoint operator on $\mathfrak{H}_\epsilon$, generated by $\mathfrak{a}_\epsilon$.

Suppose that

$$\psi \in \mathfrak{d}_\epsilon \implies \psi_\epsilon \in \mathfrak{d}_\epsilon.$$

Then also $\psi \in \mathfrak{d}_\epsilon \implies \psi^\epsilon \in \mathfrak{d}_\epsilon$ and the sets

$$\mathbf{d}_\epsilon := \{\psi_\epsilon : \psi \in \mathfrak{d}_\epsilon\}, \qquad \mathbf{d}^\epsilon := \{\psi^\epsilon : \psi \in \mathfrak{d}_\epsilon\}$$

are dense in $\mathcal{H}_\epsilon$ and in $\mathcal{H}^\epsilon$, respectively. Indeed, if, say, $\widetilde{\psi} \in \mathcal{H}_\epsilon$ and $(\psi_\epsilon, \widetilde{\psi}) = 0$ for all $\psi_\epsilon \in \mathbf{d}_\epsilon$, then also $(\psi, \widetilde{\psi}) = 0$ for all $\psi \in \mathfrak{d}_\epsilon$, and hence $\widetilde{\psi} = 0$.

Let us consider the families $\mathbf{a}_\epsilon = \mathfrak{a}_\epsilon \restriction \mathbf{d}_\epsilon$ and $\mathbf{a}^\epsilon = \mathfrak{a}_\epsilon \restriction \mathbf{d}^\epsilon$ of quadratic forms in $\mathcal{H}_\epsilon$ and in $\mathcal{H}^\epsilon$. Both $\mathbf{a}_\epsilon$ and $\mathbf{a}^\epsilon$ are closed. We denote by $\mathcal{A}_\epsilon$ and $\mathcal{A}^\epsilon$ the corresponding selfadjoint operators on $\mathcal{H}_\epsilon$ and on $\mathcal{H}^\epsilon$.

The quadratic form $\mathfrak{a}_\epsilon$ decomposes as

$$\mathfrak{a}_\epsilon[\psi] = \mathbf{a}_\epsilon[\psi_\epsilon] + \mathbf{a}^\epsilon[\psi^\epsilon] + 2\mathfrak{a}_\epsilon[\psi_\epsilon, \psi^\epsilon], \qquad \psi \in \mathfrak{d}_\epsilon. \tag{3.1}$$

Now, we make the following assumptions about the behavior of each term in (3.1).

$$\mathbf{a}_\epsilon[\psi_\epsilon] \geq c(\epsilon)\|\psi_\epsilon\|^2, \qquad \forall \psi_\epsilon \in \mathbf{d}_\epsilon, \qquad c(\epsilon) \geq c_0 > 0; \tag{3.2}$$

$$\mathbf{a}^\epsilon[\psi^\epsilon] \geq p(\epsilon)\|\psi^\epsilon\|^2, \qquad \forall \psi^\epsilon \in \mathbf{d}^\epsilon; \tag{3.3}$$

$$p(\epsilon) \to \infty, \qquad c(\epsilon) = O(p(\epsilon)); \tag{3.4}$$

$$|\mathfrak{a}_\epsilon[\psi_\epsilon, \psi^\epsilon]|^2 \leq q^2(\epsilon)\mathbf{a}_\epsilon[\psi_\epsilon]\mathfrak{a}_\epsilon[\psi^\epsilon], \qquad \forall \psi \in \mathfrak{d}_\epsilon, \qquad q(\epsilon) \to 0. \tag{3.5}$$

PROPOSITION 3.1. *Let the conditions* (3.2)–(3.5) *be satisfied. Then for* ϵ *small enough the operator* $\mathfrak{A}_\epsilon$ *is positive definite, and*

$$\|\mathfrak{A}_\epsilon^{-1} - \mathcal{A}_\epsilon^{-1} \oplus \mathbf{0}_{\mathcal{H}^\epsilon}\| \leq p(\epsilon)^{-1} + Cq(\epsilon)c(\epsilon)^{-1}. \tag{3.6}$$

PROOF. Along with $\mathfrak{a}_\epsilon[\psi]$, let us consider its diagonal part, i.e. the quadratic form

$$\widehat{\mathfrak{a}}_\epsilon[\psi] = \mathbf{a}_\epsilon[\psi_\epsilon] + \mathbf{a}^\epsilon[\psi^\epsilon], \qquad \psi \in \mathfrak{d}_\epsilon.$$

By (3.2) and (3.3), $\widehat{\mathfrak{a}}_\epsilon$ is positive definite. It is also closed, since both its components are closed. Let $\widehat{\mathfrak{A}}_\epsilon$ be the corresponding selfadjoint operator on $\mathfrak{H}_\epsilon$. By (3.4), we have for small ϵ:

$$\|\widehat{\mathfrak{A}}_\epsilon^{-1}\| \leq Cc(\epsilon)^{-1},$$

with some constant $C > 0$. Besides, (3.5) implies

$$|\mathfrak{a}_\epsilon[\psi_\epsilon, \psi^\epsilon]| \leq q(\epsilon)\widehat{\mathfrak{a}}_\epsilon[\psi]$$

and hence,

$$|\mathfrak{a}_\epsilon[\psi] - \widehat{\mathfrak{a}}_\epsilon[\psi]| = 2|\mathfrak{a}_\epsilon[\psi_\epsilon, \psi^\epsilon]| \leq 2q(\epsilon)\widehat{\mathfrak{a}}_\epsilon[\psi].$$

For ϵ small enough, so that $q(\epsilon) \leq 1/4$, this implies

$$\tfrac{1}{2}\,\widehat{\mathfrak{a}}_\epsilon[\psi] \leq \mathfrak{a}_\epsilon[\psi] \leq \tfrac{3}{2}\,\widehat{\mathfrak{a}}_\epsilon[\psi], \qquad \forall \psi \in \mathfrak{d}_\epsilon.$$

Hence, for such an ϵ the operator $\mathfrak{A}_\epsilon$ is positive definite (rather than only nonnegative, as was originally assumed), and

$$\|\mathfrak{A}_\epsilon^{-1}\| \leq 2Cc(\epsilon)^{-1}.$$

Note also that (3.3) is equivalent to

$$\|(\mathcal{A}^\epsilon)^{-1}\| \leq p(\epsilon)^{-1}. \tag{3.7}$$

Further, for any $\psi_1, \psi_2 \in \mathfrak{d}_\epsilon$ and ϵ small enough, one has

$$\left|(\mathfrak{A}_\epsilon^{1/2}\psi_1, \mathfrak{A}_\epsilon^{1/2}\psi_2) - (\widehat{\mathfrak{A}}_\epsilon^{1/2}\psi_1, \widehat{\mathfrak{A}}_\epsilon^{1/2}\psi_2)\right| = |\mathfrak{a}_\epsilon[\psi_1, \psi_2] - \widehat{\mathfrak{a}}_\epsilon[\psi_1, \psi_2]|$$

$$= |\mathfrak{a}_\epsilon[\psi_{1,\epsilon}, \psi_2^\epsilon] + \mathfrak{a}_\epsilon[\psi_1^\epsilon, \psi_{2,\epsilon}]| \leq q(\epsilon)\left((\mathbf{a}_\epsilon[\psi_{1,\epsilon}]\mathfrak{a}_\epsilon[\psi_2^\epsilon])^{1/2} + (\mathbf{a}_\epsilon[\psi_{2,\epsilon}]\mathfrak{a}_\epsilon[\psi_1^\epsilon])^{1/2}\right)$$

$$\leq 2q(\epsilon)(\widehat{\mathfrak{a}}_\epsilon[\psi_1]\widehat{\mathfrak{a}}_\epsilon[\psi_2])^{1/2} \leq 2\sqrt{2}\,q(\epsilon)(\widehat{\mathfrak{a}}_\epsilon[\psi_1]\mathfrak{a}_\epsilon[\psi_2])^{1/2}.$$

In the last formula we take $\psi_1 = \widehat{\mathfrak{A}}_\epsilon^{-1}f$, $\psi_2 = \mathfrak{A}_\epsilon^{-1}g$; here $f, g \in \mathfrak{H}_\epsilon$ are arbitrary elements. Then

$$|(\widehat{\mathfrak{A}}_\epsilon^{-1}f, g) - (\mathfrak{A}_\epsilon^{-1}f, g)|$$

$$\leq 2\sqrt{2}\,q(\epsilon)\big((\mathfrak{A}_\epsilon^{-1}g, g)(\widehat{\mathfrak{A}}_\epsilon^{-1}f, f)\big)^{1/2} \leq 4Cq(\epsilon)c(\epsilon)^{-1}\|f\|\|g\|,$$

and therefore

$$\|\mathfrak{A}_\epsilon^{-1} - \widehat{\mathfrak{A}}_\epsilon^{-1}\| \le 4Cq(\epsilon)c(\epsilon)^{-1}. \tag{3.8}$$

Since $\widehat{\mathfrak{A}}_\epsilon^{-1} = \mathcal{A}_\epsilon^{-1} \oplus (\mathcal{A}^\epsilon)^{-1}$, we conclude that

$$\|\widehat{\mathfrak{A}}_\epsilon^{-1} - \mathcal{A}_\epsilon^{-1} \oplus \mathbf{0}_{\mathcal{H}^\epsilon}\| = \|(\mathcal{A}^\epsilon)^{-1}\|.$$

Together with (3.7) and (3.8), this leads to (3.6). □

4. Proof of Theorems 2.2 and 2.3 (the DN-case)

4.1. Preliminaries. To prove Theorem 2.2, we use Proposition 3.1 with $\mathfrak{H}_\epsilon = L^2(\Omega_\epsilon)$ and $\mathcal{H}_\epsilon = \mathcal{L}_\epsilon$. As in Section 2.3, we identify operators $\mathbf{T}$ acting in $\mathcal{L}_\epsilon$ with their images $\mathbf{\Pi}_\epsilon^{-1}\mathbf{T}\mathbf{\Pi}_\epsilon$ acting in $L^2(I)$. Here $\mathbf{\Pi}_\epsilon$ is the isometry given by (2.10). The basic bilinear form is

$$\mathfrak{a}_\epsilon[\psi_1, \psi_2] = \int_{\Omega_\epsilon} \left(\frac{\partial \psi_1}{\partial x}\frac{\partial \psi_2}{\partial x} + \frac{\partial \psi_1}{\partial y}\frac{\partial \psi_2}{\partial y} - \frac{\pi^2}{M^2\epsilon^2}\psi_1\psi_2 \right) dxdy \tag{4.1}$$

defined for $\psi_1, \psi_2 \in \mathfrak{d}_\epsilon := H^{1,d}(\Omega_\epsilon)$. The corresponding quadratic form is given by (2.5). For each $\psi \in \mathbf{d}_\epsilon = H^{1,d}(\Omega_\epsilon) \cap \mathcal{L}_\epsilon$ there exists one and only one $\chi \in H^1(I)$ such that $\psi_\epsilon = \psi_{\epsilon,\chi}$, and then $\mathbf{a}_\epsilon[\psi_\epsilon] = \mathbf{q}_\epsilon[\chi]$. It is immediate that the operator $\mathcal{A}_\epsilon$ is nothing but $\mathbf{Q}_{\epsilon,DN}$, and hence, the proof of Theorem 2.2 for the DN-case reduces to proving the inequalities (3.2)–(3.5) with appropriate constants $c(\epsilon), p(\epsilon)$ and $q(\epsilon)$.

4.2. Proof of (3.2)**.** The inequality (3.2) is the only inequality, the proof of which requires a new argument compared with [**3**]. There are several ways to prove (3.2). We choose a way that is based on a remarkable result due to Birman; see [**1**], [**2**]. This result belongs to the general theory of selfadjoint extensions of symmetric operators.

The inequality (3.2) can be rewritten as

$$\mathbf{q}_\epsilon[\chi] \ge c\epsilon^{-2\alpha}\|\chi\|^2, \qquad \chi \in H^1(I),$$

or equivalently,

$$\|\mathbf{Q}_{\epsilon,N}^{-1}\| \le c^{-1}\epsilon^{2\alpha}. \tag{4.2}$$

A similar inequality for $\mathbf{Q}_{\epsilon,D}^{-1}$ was proved in [**3**], Lemma 2.1. Therefore, it is sufficient to estimate the norm of the operator $\mathbf{Q}_{\epsilon,N}^{-1} - \mathbf{Q}_{\epsilon,D}^{-1}$. For technical reasons, it is more convenient to deal with the operator

$$\mathbf{S}_\epsilon = (\mathbf{Q}_{\epsilon,N} + 1)^{-1} - (\mathbf{Q}_{\epsilon,D} + 1)^{-1}.$$

Its norm can be estimated by using lemma 3.1 in [**2**]. When applied to the operators in question, it yields

$$\|\mathbf{S}_\epsilon\| = \max\left\{\mathcal{R}(u) : \ u \in H^1(I), \ -u'' + (1 + W_\epsilon(x))u = 0\right\}, \tag{4.3}$$

where

$$\mathcal{R}(u) = \frac{\int_I |u|^2 dx}{\int_I (|u'|^2 + (1 + W_\epsilon(x))|u|^2)\, dx}.$$

The weak form of the differential equation from (4.3) is

$$\int_I (u'\phi' + (1 + W_\epsilon(x))u\phi)\, dx = 0, \qquad \forall \phi \in H^{1,0}(I). \tag{4.4}$$

Take $\phi(x) = u(x)\zeta_{ab}(x)$; the function $\zeta_{ab}(x)$ is described below. First, we fix a function $\zeta \in C^\infty(0,2)$ such that

$$\zeta(t) = 1,\ t \le 1;\ \zeta(t) = (2-t)^2,\ t > 3/2;\ 0 \le \zeta(t) \le 1 \text{ everywhere.}$$

Denote $K = \max \zeta'(t)^2/\zeta(t)$, and set

$$\zeta_{ab}(x) = \begin{cases} b^2\zeta(2x/b), & x \in [0,b); \\ a^2\zeta(2|x|/a), & x \in (-a,0]. \end{cases}$$

We have $\zeta_{ab} \in C^\infty(I)$, $\zeta(-a) = \zeta(b) = 0$, and $\zeta'_{ab}(x)^2 \le 4K\zeta_{ab}(x)$. The function $\phi = u\zeta_{ab}$ lies in $H^{1,0}(I)$, and we conclude from (4.4) that

$$2\int_I \left(u'^2 + (1+W_\epsilon(x))u^2\right)\zeta_{ab}dx = -2\int_I u'u\zeta'_{ab}dx$$
$$\le \int_I u'^2\zeta_{ab}dx + \int_I u^2\frac{(\zeta'_{ab})^2}{\zeta_{ab}}dx \le \int_I u'^2\zeta_{ab}dx + 4K\int_{\widetilde{I}} u^2dx,$$

where $\widetilde{I} = (-a,-a/2) \cup (b/2,b)$. From here we derive that

$$\int_I \left(u'^2 + 2(1+W_\epsilon(x))u^2\right)\zeta_{ab}dx \le 4K\int_{\widetilde{I}} u^2dx.$$

Hence,

$$\int_{a/2}^{b/2} u^2dx \le 2K\int_{\widetilde{I}} u^2dx.$$

The conditions on $h(x)$ imply the inequality

$$W_\epsilon(x) \ge \sigma\epsilon^{-2}|x|^m, \qquad \forall\epsilon > 0,\ x \in I,$$

with some $\sigma > 0$; see (2.2) in [**3**]. This yields

$$\mathcal{R}(u) \le (1+2K)\frac{\int_{\widetilde{I}} u^2dx}{\int_{\widetilde{I}} W_\epsilon(x)u^2dx} \le \frac{1+2K}{\sigma(\min(a,b)/2)^m}\epsilon^2.$$

So, $\|\mathbf{S}_\epsilon\| = O(\epsilon^2)$. From the Hilbert resolvent formula we conclude that also

$$\|\mathbf{Q}_{\epsilon,N}^{-1} - \mathbf{Q}_{\epsilon,D}^{-1}\| = O(\epsilon^2). \tag{4.5}$$

By (2.9), $\alpha \le 2/3$ and hence, $\|\mathbf{Q}_{\epsilon,D}^{-1}\| = O(\epsilon^{2\alpha})$ yields the same estimate for $\|\mathbf{Q}_{\epsilon,N}^{-1}\|$. This completes the proof of (4.2) and hence, that of (3.2), with $c(\epsilon) = C\epsilon^{-2\alpha}$.

4.3. Proof of (3.3)**.** This inequality is the easiest to prove. The inclusion $\psi \in \mathcal{H}^\epsilon$ means that

$$\int_0^{\epsilon h(x)} \psi(x,y)\sin\frac{\pi y}{\epsilon h(x)}dy = 0, \qquad \text{a.a. } x \in I. \tag{4.6}$$

In other words, the function $\psi(x,\cdot)$ is orthogonal to the first eigenfunction of the operator $-u''_{yy}$ on the interval $(0,\epsilon h(x))$, with the Dirichlet boundary conditions at its ends. Thus, every function $\psi \in \mathfrak{d}_\epsilon \cap \mathcal{H}^\epsilon$ satisfies an estimate similar to (2.3); its right-hand side contains an additional factor 4. Therefore

$$\|\psi\|^2 \le \frac{M^2\epsilon^2}{3\pi^2}\mathfrak{a}_\epsilon[\psi], \qquad \forall\psi \in \mathbf{d}^\epsilon,$$

which means that (3.3) is satisfied with

$$p(\epsilon) = 3\pi^2M^{-2}\epsilon^{-2}. \tag{4.7}$$

In particular, (4.7) implies (3.4). If $m > 1$ in (2.1), then the symbol O in (3.4) can be replaced by o.

4.4. End of the proof of Theorem 2.2. The proof of (3.5) repeats the proof of the similar inequality in [**3**]. Nevertheless, we outline the argument.

For $\psi \in \mathbf{d}^\epsilon$ we can integrate (4.6) by parts in y and differentiate it in x. This results in the equalities that hold for a.a. $x \in I$:

$$\int_0^{\epsilon h(x)} \psi'_y(x,y) \cos \frac{\pi y}{\epsilon h(x)} dy = 0;$$

$$\int_0^{\epsilon h(x)} \psi'_x(x,y) \sin \frac{\pi y}{\epsilon h(x)} dy = \frac{\pi}{\epsilon} \frac{h'(x)}{h^2(x)} \int_0^{\epsilon h(x)} y\psi(x,y) \cos \frac{\pi y}{\epsilon h(x)} dy.$$

Let $\psi \in \mathfrak{d}_\epsilon$ and $\psi_\epsilon = \psi_{\epsilon,\chi}$ with some $\chi \in H^1(I)$. The off-diagonal part of $\mathfrak{a}_\epsilon[\psi]$ reduces to the form

$$\mathfrak{a}_\epsilon[\psi_\epsilon, \psi^\epsilon] = \int_{\Omega_\epsilon} (\psi_{\epsilon,\chi})'_x (\psi^\epsilon)'_x dxdy \tag{4.8}$$

$$= \frac{\sqrt{2}\pi}{\epsilon^{3/2}} \int_{\Omega_\epsilon} \widetilde{h}(x) \cos \frac{\pi y}{\epsilon h(x)} \left(\phi' \psi_\epsilon - \phi (\psi_\epsilon)'_x\right) y dxdy,$$

where

$$\widetilde{h}(x) = \frac{h'(x)}{h^2(x)}, \qquad \phi(x) = \frac{\chi(x)}{h^{1/2}(x)}.$$

Now the estimate (3.5) with $q(\epsilon) = C\epsilon^\alpha$ follows by the Cauchy-Schwarz inequality.

Applying the estimate (3.6) to the operator $\mathfrak{A}_{\epsilon,DN}$, we conclude that

$$\left\| \mathfrak{A}_\epsilon^{-1} - \mathbf{Q}_\epsilon^{-1} \oplus \mathbf{0}_{\mathfrak{L}^\epsilon} \right\| \le C(\epsilon^2 + \epsilon^{3\alpha});$$

together with $\alpha \le 2/3$, the last estimate implies (2.14). This completes the proof of Theorem 2.2.

4.5. Proof of Theorem 2.3. For the D-case this is an equivalent reformulation of Theorem 1.3 in [**3**]; see (1.10) there. In the DN-case, we apply the inequality (4.5), which yields

$$\|(\epsilon^{2\alpha}\mathbf{Q}_{\epsilon,N})^{-1} \oplus \mathbf{0}_{L^2(\mathbb{R}\setminus I)} - (\epsilon^{2\alpha}\mathbf{H}_\epsilon)^{-1}\|$$

$$\le \|(\epsilon^{2\alpha}\mathbf{Q}_{\epsilon,D})^{-1} \oplus \mathbf{0}_{L^2(\mathbb{R}\setminus I)} - (\epsilon^{2\alpha}\mathbf{H}_\epsilon)^{-1}\| + \epsilon^{-2\alpha}\|\mathbf{Q}_{\epsilon,N}^{-1} - \mathbf{Q}_{\epsilon,D}^{-1}\|.$$

Both terms on the right tend to zero as $\epsilon \to 0$: the first by Theorem 1.3 in [**3**] and the second by (4.5), again keeping in mind that $\alpha < 2/3$. This completes the proof of Theorem 2.3.

5. The case $I = \mathbb{R}$: Behavior of the resolvent

In this and the next section we study the case $I = \mathbb{R}$. The spectrum of the Dirichlet Laplacian Δ_ϵ in Ω_ϵ is now not necessarily discrete, the structure of its essential component $\sigma_{ess}(\Delta_\epsilon)$ depends on the behavior of $h(x)$ at infinity. In this section we study the case when $h(x)$ satisfies conditions (i) and (ii), and also the following additional conditions:

(iii) $\limsup_{|x|\to\infty} h(x) < M$,

(iii') $h'/h \in L^\infty(\mathbb{R})$.

Our goal is to show that under these conditions, analogues of Theorems 2.2 and 2.3 hold. A result similar to Theorem 2.1 then follows automatically, though its formulation becomes a bit more complicated because the operator Δ_ϵ can have a nonempty essential spectrum.

The situation simplifies if one is interested only in the behavior of the eigenvalues. In the next section 6 we will give a short proof of an analogue of statement 1) in Theorem 2.1. For that purpose, the condition (*iii'*) turns out to be not necessary.

5.1. Behavior of the resolvent: formulations and auxiliary results. Let $I = \mathbb{R}$. The quadratic form (2.11) is positive definite and closed on the natural domain

$$\mathbf{d}_\epsilon = \{\chi \in H^1(\mathbb{R}) : \mathbf{q}_\epsilon[\chi] < \infty\}.$$

As before, the corresponding selfadjoint operator $\mathbf{Q}_\epsilon$ is formally given by (2.13). Note also that instead of two operators $\mathfrak{A}_{\epsilon,D}$, $\mathfrak{A}_{\epsilon,DN}$ we have only one operator $\mathfrak{A}_\epsilon$.

THEOREM 5.1. *Let $I = \mathbb{R}$ and let $h(x)$ satisfy the conditions (i)–(iii'). Then the equality* (2.14) *holds.*

For the analogue of Theorem 2.3 we do not need condition (*iii'*).

THEOREM 5.2. *Let $I = \mathbb{R}$ and let $h(x)$ satisfy the conditions (i)–(iii). Then*

$$\lim_{\epsilon \to 0} \|(\epsilon^{2\alpha}\mathbf{Q}_\epsilon)^{-1} - (\epsilon^{2\alpha}\mathbf{H}_\epsilon)^{-1}\| = 0. \tag{5.1}$$

As in Section 4, in the proofs of these theorems we rely upon Proposition 3.1, taking $\mathfrak{H}_\epsilon = L^2(\Omega_\epsilon)$ and $\mathcal{H}_\epsilon = \mathcal{L}_\epsilon$. The basic bilinear form is again given by (4.1). We have $\mathcal{A}_\epsilon = \mathbf{Q}_\epsilon$. The latter operator acts on $L^2(\mathbb{R})$; it is generated by the quadratic form (2.11) defined on its natural domain. The most important thing is to prove (5.1). Indeed, the conditions (*i*)–(*iii*) may lead to a function $W_\epsilon(x)$ that is bounded; then Theorem 2.16 from the book [**8**], which was the main ingredient of our proof of Theorem 1.3 in [**3**], does not apply. We need an appropriate substitute. First, we prove the following lemma, which can be considered as a partial generalization of Theorem 2.16 in [**8**]. By $\mathfrak{C}$ we denote the ideal of all compact operators in the algebra of all bounded operators.

PROPOSITION 5.3. *Let $\mathbf{T} \ge 0$ and $\mathbf{T}_\epsilon \ge 0$, $0 < \epsilon \le \epsilon_0$, be bounded selfadjoint operators in a separable Hilbert space $\mathfrak{H}$, and let $\mathbf{T}_\epsilon \to \mathbf{T}$ strongly as $\epsilon \to 0$. Suppose also that there exists a bounded selfadjoint operator $\mathbf{T}_0$ such that $\mathbf{T}_\epsilon \le \mathbf{T}_0$ for all $\epsilon \le \epsilon_0$, and*

$$\operatorname{dist}(\mathbf{T}_0, \mathfrak{C}) = m, \qquad m \ge 0.$$

Then

$$\limsup_{\epsilon \to 0} \|\mathbf{T}_\epsilon - \mathbf{T}\| \le m.$$

PROOF. Fix $\eta > 0$ and find an operator $\mathbf{S} = \mathbf{S}^* \in \mathfrak{C}$ that satisfies $\|\mathbf{T}_0 - \mathbf{S}\| < m + \eta$. Let $\mathfrak{F}$ be a finite-dimensional subspace in $\mathfrak{H}$, such that $\|\mathbf{S}g\| \le \eta\|g\|$ for any $g \perp \mathfrak{F}$. Then

$$\|\mathbf{T}_0 g\| \le \|(\mathbf{T}_0 - \mathbf{S})g\| + \|\mathbf{S}g\| \le (m + 2\eta)\|g\|, \qquad \forall g \perp \mathfrak{F}.$$

Now, let $h \in \mathfrak{H}$ be an arbitrary element, and let $h = f + g$ where $f \in \mathfrak{F}$ and $g \perp \mathfrak{F}$. Then

$$((\mathbf{T}_\epsilon - \mathbf{T})h, h) = ((\mathbf{T}_\epsilon - \mathbf{T})f, f) + ((\mathbf{T}_\epsilon - \mathbf{T})g, g) + 2\operatorname{Re}((\mathbf{T}_\epsilon - \mathbf{T})f, g).$$

Since the dimension of $\mathfrak{F}$ is finite, strong convergence $\mathbf{T}_\epsilon \to \mathbf{T}$ implies

$$\lim_{\epsilon\to 0} \sup_{f\in\mathfrak{F}} \|(\mathbf{T}_\epsilon - \mathbf{T})f\|/\|f\| = 0.$$

Therefore,

$$|\,((\mathbf{T}_\epsilon - \mathbf{T})h, h)\,| \le |\,((\mathbf{T}_\epsilon - \mathbf{T})g, g)\,| + \eta\|f\|^2 + 2\eta\|f\|\|g\|$$

if ϵ is small enough. Further,

$$-(\mathbf{T}_0 g, g) \le ((\mathbf{T}_\epsilon - \mathbf{T})g, g) \le (\mathbf{T}_0 g, g),$$

since both $\mathbf{T}$ and $\mathbf{T}_\epsilon$ are nonnegative operators. Therefore,

$$|\,((\mathbf{T}_\epsilon - \mathbf{T})g, g)\,| \le (\mathbf{T}_0 g, g) \le (m + 2\eta)\|g\|^2.$$

Finally, this yields

$$|\,((\mathbf{T}_\epsilon - \mathbf{T})h, h)\,| \le (m + 2\eta)\|g\|^2 + 2\eta\|f\|^2 + \eta\|g\|^2 \le (m + 3\eta)\|h\|^2.$$

Since $\eta > 0$ is arbitrary, the statement of the proposition follows. □

Below, the symbol $\mathbf{Z}_V$ denotes the Schrödinger operator on $L^2(\mathbb{R})$ with the nonnegative potential $V(x)$. We define the operator $\mathbf{Z}_V$ via its quadratic form. The next statement is a substitute for proposition 4.1 in [**3**].

PROPOSITION 5.4. *Let $V(x)$ and $V_\epsilon(x)$, $0 < \epsilon \le \epsilon_0$, be nonnegative measurable functions on $\mathbb{R}$, such that $V(x) \to \infty$ as $|x| \to \infty$ and*

(5.2) $$V_\epsilon(x) \to V(x) \text{ as } \epsilon \to 0, \qquad \text{uniformly on compact sets.}$$

Suppose also that

$$V_\epsilon(x) \ge V_\epsilon^\circ(x), \qquad \forall x \in \mathbb{R},\ 0 < \epsilon \le \epsilon_0,$$

where $V_\epsilon^\circ(x)$ is another family of measurable functions on $\mathbb{R}$, which is monotone in ϵ:

(5.3) $$\epsilon_1 > \epsilon_2 \implies V_{\epsilon_1}^\circ(x) \le V_{\epsilon_2}^\circ(x), \qquad \forall x \in \mathbb{R}$$

and

(5.4) $$c(\epsilon) := \liminf_{|x|\to\infty} V_\epsilon^\circ(x) \to \infty \qquad \text{as } \epsilon \to 0.$$

Then

(5.5) $$\|\mathbf{Z}_{V_\epsilon}^{-1} - \mathbf{Z}_V^{-1}\| \to 0, \qquad \epsilon \to 0.$$

PROOF. Denote

$$\mathbf{T} = \mathbf{Z}_V^{-1}, \qquad \mathbf{T}_\epsilon = \mathbf{Z}_{V_\epsilon}^{-1}.$$

Also let $\mathbf{T}_\epsilon^\circ$ be the inverse to the Schrödinger operator with the potential $V_\epsilon^\circ(x)$. The assumption (5.2) implies that $\mathbf{Z}_{V_\epsilon} u \to \mathbf{Z}_V u$ for any $u \in C_0^\infty(\mathbb{R})$, and Theorem 8.1.5 in [**5**] guarantees the strong convergence $\mathbf{T}_\epsilon \to \mathbf{T}$.

Fix $\epsilon^* \in (0, \epsilon_0)$; then for $\epsilon < \epsilon^*$ the conditions of Proposition 5.3 are satisfied with $\mathbf{T}_0 = \mathbf{T}_{\epsilon^*}^\circ$. From this proposition we conclude that

$$\limsup_{\epsilon\to 0} \|\mathbf{T}_\epsilon - \mathbf{T}\| \le 1/c(\epsilon^*).$$

Taking $\epsilon^* \to 0$, we arrive at (5.5). □

5.2. Proof of Theorem 5.2. Introduce the potential

$$V_\epsilon(t) = \epsilon^{2\alpha} W_\epsilon(t\epsilon^\alpha),$$

where $W_\epsilon(x)$ is the function defined in (2.12). The assumption (2.1) implies that

$$V_\epsilon(t) = q(t) + \pi^2 \rho_1(t\epsilon^\alpha) t^{m+1}\epsilon^\alpha + \epsilon^{2\alpha} v(t\epsilon^\alpha),$$

where $\rho_1(x)$ is some function which is bounded on any finite interval $(-a, a)$; cf. proof of Theorem 1.3 in [**3**]. Hence, $V_\epsilon(t) \to q(t)$ uniformly on compact subsets in $\mathbb{R}$.

It follows from the assumptions (i) and (iii) that for $|x| \geq 1$ the function $\pi^2(h^{-2}(x) - M^{-2})$ is bounded below:

$$\pi^2(h^{-2}(x) - M^{-2}) \geq c_0, \qquad |x| \geq 1,$$

with some $c_0 > 0$. On $[-1, 1]$ the inequality $W_\epsilon(x) \geq \sigma\epsilon^{-2}|x|^m$ with some $\sigma > 0$ is fulfilled, which leads to the estimate

$$W_\epsilon(x) \geq \sigma_1 \epsilon^{-2} \min(|x|^m, 1), \qquad \forall x \in \mathbb{R},\ \epsilon > 0.$$

Hence,

$$V_\epsilon(t) \geq \sigma_1 \min(\epsilon^{-2+2\alpha+m\alpha}|t|^m, \epsilon^{-2+2\alpha}) = \sigma_1 \min(|t|^m, \epsilon^{-2+2\alpha}).$$

The conditions (5.3) and (5.4) are satisfied if we take

$$V_\epsilon^\circ(t) = \sigma_1 \min(|t|^m, \epsilon^{-2+2\alpha}).$$

So, Theorem 5.4 applies and yields

$$\|\widehat{\mathbf{Q}}_\epsilon^{-1} - \mathbf{H}^{-1}\| \to 0,$$

where $\widehat{\mathbf{Q}}_\epsilon$ is the Schrödinger operator with the potential $V_\epsilon(t)$. The substitution $t = x\epsilon^{-\alpha}$ leads to (5.1).

5.3. Theorem 5.1: Outline of proof. The argument necessary for proving Theorem 5.1 is quite similar to the one in [**3**], and also in Section 4 of the present paper; we only outline it, concentrating on the few new moments.

First, note that

$$\|(\epsilon^{2\alpha}\mathbf{H}_\epsilon)^{-1}\| = \|\mathbf{H}^{-1}\| = const,$$

and hence (5.1) yields

$$\|\mathbf{Q}_\epsilon^{-1}\| \leq C\epsilon^{2\alpha}.$$

Equivalently, this means that (3.2) is satisfied with $c(\epsilon) = C^{-1}\epsilon^{-2\alpha}$. The condition (3.3) is satisfied with the same $p(\epsilon)$ as in (4.7); its proof does not change. Since $\alpha \leq 2/3$, (3.4) also holds. It only remains to prove (3.5). To this end, we repeat the reasoning in Section 4.4. The only difference is that now for estimating the integral in the right-hand side of (4.8) we need the condition (iii').

6. The case $I = \mathbb{R}$: Behavior of eigenvalues, a simple proof

An analogue of Theorem 2.1 follows from Theorems 5.1 and 5.2 in the same way as Theorem 2.1 itself follows from Theorems 2.2 and 2.3. However, there is a much simpler and independent way to prove an analogue of statement 1) in Theorem 2.1. It uses the Dirichlet–Neumann bracketing, which is possible, since Theorem 2.1 is now at our disposal for both the D and the DN cases. For the proof, the condition (iii') is not needed.

In Theorem 6.1 below, $\nu(\epsilon)$ stands for the bottom of $\sigma_{ess}(\Delta_\epsilon)$, and we take $\nu(\epsilon) = \infty$ if Δ_ϵ has no essential spectrum. We denote by $n(\epsilon)$, $n(\epsilon) \leq \infty$, the number of eigenvalues $\lambda_j(\epsilon) < \nu(\epsilon)$.

THEOREM 6.1. *Let $I = \mathbb{R}$. If $h(x)$ satisfies the conditions (i), (ii) and (iii), then $\nu(\epsilon) \to \infty$ as $\epsilon \to 0$. Moreover, for small values of ϵ the spectrum of Δ_ϵ below $\nu(\epsilon)$ is nonempty, $n(\epsilon) \to \infty$ as $\epsilon \to 0$, and for each $j \in \mathbb{N}$ the equality* (2.8) *holds; μ_j are eigenvalues of the operator* (2.7).

PROOF. Take any segment $\widehat{I} = [-a, a]$. It follows from the conditions (i) and (iii) that

$$M_a := \sup_{x \notin \widehat{I}} h(x) < M.$$

Denote

$$\Omega_{a,\epsilon} = \{(x,y) \in \Omega_\epsilon : x \in \widehat{I}\}, \qquad \Omega'_{a,\epsilon} = \{(x,y) \in \Omega_\epsilon : x \notin \widehat{I}\}.$$

Let $\Delta_{a,\epsilon,D}$, $\Delta'_{a,\epsilon,D}$ be the Dirichlet Laplacians in $\Omega_{a,\epsilon}$, $\Omega'_{a,\epsilon}$ respectively; $\Delta_{a,\epsilon,DN}$ and $\Delta'_{a,\epsilon,DN}$ have a similar meaning. Then

$$\Delta_{a,\epsilon,DN} + \Delta'_{a,\epsilon,DN} < \Delta'_\epsilon < \Delta_{a,\epsilon,D} + \Delta'_{a,\epsilon,D}.$$

For $\Omega'_{a,\epsilon}$ an inequality similar to (2.4), with M_a in place of M, is satisfied; it implies that the spectra of both operators $\Delta'_{a,\epsilon,D}$ and $\Delta'_{a,\epsilon,DN}$ lie above the number $C(\epsilon) := \frac{\pi^2}{M_a^2\epsilon^2}$, which is greater than $\frac{\pi^2}{M^2\epsilon^2}$. It follows that $\nu(\epsilon) \geq C(\epsilon)$, and the eigenvalues $\lambda_j(\epsilon) = \lambda_j(\Delta_\epsilon)$ lying below $C(\epsilon)$ satisfy the two-sided inequality

$$\lambda_j(\Delta_{a,\epsilon,DN}) \leq \lambda_j(\epsilon) \leq \lambda_j(\Delta_{a,\epsilon,D}).$$

The asymptotics (2.8) for the eigenvalues $\lambda_j(\Delta_{a,\epsilon,DN})$, $\lambda_j(\Delta_{a,\epsilon,D})$ implies that the number $\nu(\epsilon)$ grows indefinitely as $\epsilon \to 0$, and that the eigenvalues $\lambda_j(\epsilon)$ have the same asymptotics given by (2.8). □

Another problem of the same nature was analyzed in Section 6.1 of [**3**]. It concerns the case when the segment I is finite but the function $h(x)$ is allowed to vanish at the ends of I.

Acknowledgments

This work was mostly motivated by the questions asked by participants of the workshop in Quantum Graphs, their Spectra and Application (Cambridge, April 2007) after the talk given by the second author. We are grateful to all who asked questions. Especially, we would like to thank Brian Davies and Leonid Parnovski for their suggestion to use Dirichlet–Neumann bracketing in the case of the whole line. The work was mostly done when both authors visited the Isaac Newton Institute for Mathematical Sciences in Cambridge, UK. We acknowledge the hospitality of the Newton Institute. The first author was partially supported by the NSF grant DMS 0648786.

References

[1] M. Sh. Birman, *On the theory of self-adjoint extensions of positive definite operators,* (Russian) Mat. Sb. (N.S.) **38(80)**, no. 4 (1956), 431–450.

[2] M. Sh. Birman, *Perturbations of the continuous spectrum of a singular elliptic operator by varying the boundary and the boundary conditions*, Vestnik Leningrad. Univ. **1**, no. 1 (1962), 22–55. English translation in this volume, pp. 19–53.

[3] L. Friedlander, M. Solomyak, *On the spectrum of the Dirichlet Laplacian in a narrow strip*, Israeli Math. J., in press.

[4] L. Friedlander, M. Solomyak, *On the spectrum of narrow periodic waveguides*, Russ. J. Math. Phys. **15** (2008), no. 2, 238–242.

[5] T. Kato, *Perturbation Theory for Linear Operators*, Second Edition, Springer, Berlin – Heidelberg – New York, 1976.

[6] P. Kuchment, H. Zheng, *Asymptotics of spectra of Neumann Laplacians in thin domains*, Contemp. Math. **327** (2003), 199–213.

[7] J. Rubinstein, M. Schatzman, *Variational Problems on Multiply Connected Thin Strips I: Basic Estimates and Convergence of the Laplacian Spectrum*, Arch. Rational Mech. Anal. **160** (2001), 271–308.

[8] B. Simon, *Trace Ideals and Their Applications*, London Math. Soc. Lecture Notes Series 35, Cambridge University Press, 1979.

University of Arizona, Tucson, Arizona, USA
E-mail address: friedlan@math.arizona.edu

Department of Mathematics, The Weizmann Institute of Science, Rehovot 76100, Israel
E-mail address: solom@wisdom.weizmann.ac.il

Amer. Math. Soc. Transl.
(2) Vol. **225**, 2008

Lyapunov Functions of Periodic Matrix-valued Jacobi Operators

Evgeny Korotyaev and Anton Kutsenko

To Mikhail Shlëmovich Birman on his 80th birthday

ABSTRACT. We consider periodic matrix-valued Jacobi operators. The spectrum of such an operator is absolutely continuous and consists of intervals separated by gaps. We study spectral data, which determine the spectrum of these operators (counting multiplicity). 1) We define the Lyapunov function, which is analytic on an associated Riemann surface. On each sheet the Lyapunov function has the standard properties of the Lyapunov function for the scalar case. We obtain that this function has (real or complex) branch points, which we call resonances. We show that the endpoints of any gap are periodic or anti-periodic eigenvalues or resonances. 2) We prove that the spectral data (part of the j-periodic eigenvalues for some $j \geqslant 2$) determine the spectrum (counting multiplicity) of the Jacobi operator.

1. Introduction and main results

Consider selfadjoint matrix-valued Jacobi operators $\mathcal{J}$ acting on $\ell^2(\mathbb{Z})^m, m \geqslant 1$, and given by

$$(\mathcal{J}y)_n = a_n y_{n+1} + b_n y_n + a_{n-1}^\top y_{n-1}, \quad y_n \in \mathbb{C}^m, \ y = (y_n)_{n\in\mathbb{Z}} \in \ell^2(\mathbb{Z})^m, \tag{1.1}$$

for all $n \in \mathbb{Z}$, where $y_n \in \mathbb{C}^m$ and $a_n, b_n = b_n^\top$ are p-periodic sequences of the $m \times m$ real matrices and $\det a_n \neq 0$ for all $n \in \mathbb{Z}$. It is well known that the spectrum $\sigma(\mathcal{J})$ of $\mathcal{J}$ is absolutely continuous and consists of nondegenerate intervals $[\lambda_{n-1}^+, \lambda_n^-]$, $\lambda_{n-1}^+ < \lambda_n^- \leqslant \lambda_n^+, n = 1, \dots, N < \infty$. These intervals are separated by the gaps $\gamma_n = (\lambda_n^-, \lambda_n^+)$ with length > 0. Introduce the fundamental $m \times m$ matrix-valued solutions $\varphi = (\varphi_n(z))_{n\in\mathbb{Z}}, \vartheta = (\vartheta_n(z))_{n\in\mathbb{Z}}, z \in \mathbb{C}$ of the equation

$$a_n y_{n+1} + b_n y_n + a_{n-1}^\top y_{n-1} = z y_n, \qquad \varphi_0 \equiv \vartheta_1 \equiv 0, \ \varphi_1 \equiv \vartheta_0 \equiv I_m, \tag{1.2}$$

where I_m is the identity $m \times m$ matrix. We define the monodromy $2m \times 2m$ matrix $\mathcal{M}_p$ by

$$\mathcal{M}_p(z) = \begin{pmatrix} \vartheta_p(z) & \varphi_p(z) \\ \vartheta_{p+1}(z) & \varphi_{p+1}(z) \end{pmatrix}. \tag{1.3}$$

2000 *Mathematics Subject Classification.* Primary 47B36.

Introduce the modified monodromy matrix M and the determinant D by

$$M = P_0 \mathcal{M}_p P_0^{-1}, \quad P_0 = \begin{pmatrix} a_0 & 0 \\ 0 & I_m \end{pmatrix}, \tag{1.4}$$

$$D(z,\tau) = \det(M(z) - \tau I_{2m}), \ \tau, z \in \mathbb{C}.$$

Let $\tau_j = \tau_j(z), j \in \mathbb{N}_{2m}$ be the eigenvalues of $M(z)$, where $\mathbb{N}_m = \{1, \ldots, m\}$. An eigenvalue of $M(z)$ is called a *multiplier*.

THEOREM (Lyapunov–Poincaré). *i) The following identities hold:*

$$M^\top J M = J, \quad J = \begin{pmatrix} 0 & I_m \\ -I_m & 0 \end{pmatrix}, \tag{1.5}$$

$$D(z,\tau) = \tau^{2m} D(z, \tau^{-1}), \qquad all \quad z, \tau \in \mathbb{C}, \ \tau \neq 0. \tag{1.6}$$

ii) If $\tau(z)$ is a multiplier of multiplicity $d \geqslant 1$ for some $z \in \mathbb{C}$ (or $z \in \mathbb{R}$), then $\tau^{-1}(z)$ (or $\overline{\tau}(z)$) is a multiplier of multiplicity d. Moreover, each $M(z), z \in \mathbb{C}$, has exactly $2m$ multipliers $\tau_j^{\pm 1}(z), j \in \mathbb{N}_m$ and $\sigma(\mathcal{J}) = \bigcup_{j=1}^m \{z \in \mathbb{C} : |\tau_j(z)| = 1\}$.

iii) If $\tau(z)$ is a simple multiplier for some $z \in \mathbb{C}$ and $|\tau(z)| = 1$, then $\tau'(z) \neq 0$.

It is well known that the spectrum of the scalar ($m = 1$) Jacobi operator $\mathcal{J}$ is absolutely continuous and $\sigma(\mathcal{J}) = \bigcup_1^p [\lambda_{n-1}^+, \lambda_n^-]$, where $\lambda_0^+ < \lambda_1^- \leqslant \lambda_1^+ < \cdots < \lambda_p^- \leqslant \lambda_{p-1}^+ < \lambda_p^-$ and $\lambda_n^\pm$ are $2p$-periodic eigenvalues. The intervals $[\lambda_{n-1}^+, \lambda_n^-], n \in \mathbb{N}_p$ are separated by gaps $\gamma_n = (\lambda_n^-, \lambda_n^+)$ of lengths $|\gamma_n| \geqslant 0$. If a gap γ_n is degenerate, i.e. $|\gamma_n| = 0$, then the corresponding segments σ_n, σ_{n+1} merge. Recall that the Lyapunov function $\Delta(z) = \frac{\varphi_{p+1}(z) + \vartheta_p(z)}{2}$ and the spectrum $\sigma(\mathcal{J}) = \{z \in \mathbb{C} : \ \Delta(z) \in [-1,1]\}$. Note that $(-1)^{p-n}\Delta(\lambda_n^\pm) = 1$ for all $n \in \mathbb{N}_p$. We recall well-known facts (see e.g. [**P**] or [**KKu**]).

THEOREM A. *A real polynomial F is the Lyapunov function for some scalar p-periodic Jacobi operator with numbers $a_n > 0, b_n \in \mathbb{R}, n \in \mathbb{Z}$ iff $F(z) = cz^p + O(z^{p-1})$ as $z \to \infty$ for some $c > 0$ and $F'(z_j) = 0, \quad (-1)^{p-j} F(z_j) \geqslant 1$ for all $j \in \mathbb{N}_{p-1}$ and for some $z_1 < \cdots < z_{p-1}$.*

REMARK. 1) The zeros of $\Delta - t$ for any fixed $t \in [-1,1]$ and a constant $c > 0$ determine the Lyapunov function Δ and the spectrum of the scalar periodic Jacobi operator.

2) $\sigma(\mathcal{J}) = \sigma(\widetilde{\mathcal{J}})$ for some Jacobi operators $\mathcal{J}, \widetilde{\mathcal{J}}$ iff $\Delta = \widetilde{\Delta}$, where $\Delta, \widetilde{\Delta}$ are the corresponding Lyapunov functions.

3) If $a_n = 1, b_n = 0$ for all $n \in \mathbb{Z}$, then $\Delta = \mathcal{C}_p(\frac{z}{2})$, where $\mathcal{C}_p$ is the Chebyshev polynomial.

Before we describe the content of our paper, we briefly comment on the background literature for matrix-valued Jacobi operators. The inverse spectral theory for scalar periodic Jacobi operators is well understood; see a book [**T**] and papers [**BGGK**], [**K**], [**KKu**], [**P**], [**vM**] and the references therein. The corresponding theory for periodic Jacobi matrices with matrix-valued coefficients is still largely a wide open field. Some new particular results were recently obtained by Gesztesy and coauthors [**CG**],[**CGR**],[**GKM**] and see the references therein. Note that for finite Jacobi matrices with matrix-valued coefficients the complete solution of the inverse problem was given recently by Brüning, Chelkak and Korotyaev [**BCK**].

Our motivation for studying matrix-valued Jacobi operators is based on the following fact: the magnetic Schrödinger operator on the armchair nanotube is unitarily equivalent to the periodic matrix-valued Jacobi operator; see [**BKK**].

The eigenvalues of $M(z)$ are the zeros of the equation $D(z,\tau)=0$. This is an algebraic equation in τ of degree $2m$, where the coefficients are polynomials in z. It is well known (see e.g., Chapter 8 in [**Fo**]) that the roots $\tau_j(z), j\in\mathbb{N}_{2m}$ constitute one or several branches of one or several analytic functions that have only algebraic singularities in $\mathbb{C}$. Thus the number of eigenvalues of $M(z)$ is a constant N_e with the exception of some special values of z (see below the definition of a resonance). There is a finite number of such points on the plane. If the functions $\tau_j(z), j\in\mathbb{N}_{2m}$ are all distinct, then $N_e=2m$. If some of them are identical, then $N_e<2m$ and $M(z)$ is permanently degenerate.

If $m=1$, then the Riemann surface for the multipliers has 2 sheets, but the Lyapunov function is a polynomial. Similarly, in the case $m\geqslant 2$ it is more convenient for us to construct the Riemann surface for the Lyapunov functions given by $\Delta_j(z)=\frac{1}{2}(\tau_j(z)+\tau_j^{-1}(z)),\ j\in\mathbb{N}_m$, which has m sheets (see the equation (1.7)). Note that the Lyapunov-Poincaré Theorem gives $\sigma(\mathcal{J})=\bigcup_{j=1}^m\{z:\Delta_j(z)\in[-1,1]\}$. The multiplicity of $z\in\sigma(\mathcal{J})$ is the number of multipliers $\tau_j(z)$, satisfying $|\tau_j(z)|=1$. Let $\#A$ be the number of elements of a set A.

THEOREM 1.1. *There exist analytic functions $\widetilde{\Delta}_s, s=1,\dots,m_0\leqslant m$ on the m_s-sheeted Riemann surface $\mathcal{R}_s, m_s\geqslant 1$ having the following properties:*

i) There exist disjoint subsets $\omega_s\subset\mathbb{N}_m, s\in\mathbb{N}_{m_0}, \bigcup\omega_s=\mathbb{N}_m$ such that all branches of $\widetilde{\Delta}_s, s\in\mathbb{N}_{m_0}$ have the form $\Delta_j(z)=\frac{1}{2}(\tau_j(z)+\tau_j^{-1}(z)),\ j\in\omega_s$ and satisfy

$$(1.7)\qquad \frac{D(z,\tau)}{(2\tau)^m}=\prod_1^{m_0}\Phi_s(z,\nu),\qquad \Phi_s(z,\nu)=\prod_{j\in\omega_s}(\nu-\Delta_j(z)),\qquad \nu=\frac{\tau+\tau^{-1}}{2},$$

for any $z,\tau\in\mathbb{C},\ \tau\neq 0$, where the functions $\Phi_s(z,\nu)$ are some polynomials with respect to $\nu,z\in\mathbb{C}$. If $\Delta_i=\Delta_j$ for some $i\in\omega_k, j\in\omega_s$, then $\Phi_k=\Phi_s$ and $\widetilde{\Delta}_k=\widetilde{\Delta}_s$.

ii) Let some branch $\Delta_j, j\in\mathbb{N}_m$ be real analytic on some interval $Y=(\alpha,\beta)\subset\mathbb{R}$ and $-1<\Delta_j(z)<1$ for any $z\in Y$. Then $\Delta_j'(z)\neq 0$ for each $z\in Y$.

iii) All functions $\rho,\rho_s, s\in\mathbb{N}_{m_0}, \#\omega_s\geqslant 2$ given by (1.8) are polynomials and

$$(1.8)\qquad \rho=\prod_1^{N_0}\rho_s,\quad \rho_s(z)=\prod_{i<j,i,j\in\omega_s}(\Delta_i(z)-\Delta_j(z))^2,\quad z\in\mathbb{C},$$

$$\text{where}\quad \rho_s=1,\quad if\ \#\omega_s=1.$$

iv) Each endpoint of a gap $(\lambda_n^-,\lambda_n^+)$ is a periodic (or anti-periodic) eigenvalue or a real branch point of some $\Delta_j, j\in\mathbb{N}_m$, which is a zero of ρ (below such points are called resonances).

v) Let $\tau_j^0, j\in\mathbb{N}_m$ be eigenvalues of a matrix $A_p=(a_1a_2\dots a_p)^{-1}$. Then the following asymptotics hold:

$$(1.9)\qquad \Delta_j(z)=\frac{z^p}{2}(\tau_j^0+O(z^{-\frac{1}{m}})),\qquad \tau_j(z)=z^p\tau_j^0+O(z^{-\frac{1}{m}})),\qquad j\in\mathbb{N}_m,$$

$$(1.10)\qquad \rho_s(z)=z^{p|\omega_s|(|\omega_s|-1)}(c_s+O(z^{-\frac{1}{m}})),$$

$$c_s = 2^{|\omega_s|(1-|\omega_s|)} \prod_{k<j, k,j\in\omega_s} (\tau_k^0 - \tau_j^0)^2, \quad s \in \mathbb{N}_{m_0}$$

as $|z| \to \infty$. *Moreover, if* $\tau_j^0 \neq \tau_k^0$ *for all* $j,k \in \omega_s, j \neq k$, *then* $c_s \neq 0$ *and* $\rho_s(z) = z^{p|\omega_s|(|\omega_s|-1)}(c_s + O(z^{-1}))$ *as* $|z| \to \infty$.

DEFINITION. The number $z_0 \in \mathbb{C}$ is a **resonance** of $\mathcal{J}$ if z_0 is a zero of ρ given by (1.8).

EXAMPLE. Let $a_n = \operatorname{diag}\{a_{1,n}, a_{2,n}, \dots, a_{m,n}\}$ and $b_n = \operatorname{diag}\{b_{1,n}, b_{2,n}, \dots, b_{m,n}\}$ for all $n \in \mathbb{Z}$ and for some $a_{j,n} > 0, b_{j,n} \in \mathbb{R}, (j,n) \in \mathbb{N}_m \times \mathbb{Z}$. Then the operator $\mathcal{J} = \bigoplus_1^m \mathcal{J}_j$, where $\mathcal{J}_j$ is a scalar Jacobi operator acting in $\ell^2(\mathbb{Z})$, is given by

$$(\mathcal{J}_j y)_n = a_{j,n} y_{n+1} + a_{j,n-1} y_{n-1} + b_{j,n} y_n, \quad n \in \mathbb{Z}, \quad y = (y_n)_{n\in\mathbb{Z}} \in \ell^2(\mathbb{Z}).$$

In this case each Δ_j is the standard Lyapunov function for $\mathcal{J}_j$, and the properties of Δ_j are well known; see Theorem A and [**T**]. Thus $\rho = 1$, since $\omega_j = \{j\}$.

Below we show the following identity (see (3.2))

$$D(z,\tau) = c\tau^m \prod_{j=1}^{mp} (z - \lambda_j(\tau)), \quad z,\tau \in \mathbb{C}, \ \tau \neq 0, \ \ c = (-1)^m \det A_p, \tag{1.11}$$

where $\lambda_j(\tau)$ are zeros of the polynomial $D(z,\tau) = 0$ for fixed τ. Then we obtain $\sigma(\mathcal{J}) = \bigcup_{j=1}^{mp} \sigma_j$, where $\sigma_j = \lambda_j(\mathbb{S}^1)$ is the bounded closed set and $\mathbb{S}^1 = \{\tau \in \mathbb{C} : |\tau| = 1\}$. Let $\mathbb{N}_m^0 = \{0, 1, \dots, m\}$ for $m \geqslant 0$. We describe the minimum spectral data which determine all Lyapunov functions $\Delta_j, j \in \mathbb{N}_m$ and the spectrum of the Jacobi operator.

THEOREM 1.2. *Let* $\varkappa_j \in \mathbb{R}, j \in \mathbb{N}_m^0$ *satisfy* $\cos\varkappa_n \neq \cos\varkappa_k$ *for all* $k \neq n$.

i) Let $\Lambda_0 = \mathbb{N}_{mp}$ *and let* $\Lambda_j \subset \mathbb{N}_{mp}, \#\Lambda_j = (m-j)p + 1, j \in \mathbb{N}_m$. *Then the spectral data* $\mathcal{L}_j = (\lambda_n(e^{i\varkappa_j}))_{n\in\Lambda_j}, j \in \mathbb{N}_m^0$, *for the Jacobi operator* $\mathcal{J}$ *determine the polynomial* $D(\cdot,\cdot)$, *all Lyapunov functions* $\Delta_j, j \in \mathbb{N}_m$ *and the spectrum* $\sigma(\mathcal{J})$ *(counted according to its multiplicity).*

ii) Let $\widetilde{\Lambda}_1 \subset \mathbb{N}_{mp}$ *and let* $\#\widetilde{\Lambda}_1 = \#\Lambda_1 - 1$. *Then there exist infinitely many Jacobi operators, having the same spectral data* $\mathcal{L}_0$, $\widetilde{\mathcal{L}}_1 = (\lambda_n(e^{i\varkappa_1}))_{n\in\widetilde{\Lambda}_1}$, $\mathcal{L}_j$, $j = 2, \dots, m$, *but different determinants and different spectra as sets.*

EXAMPLE. 1) Note that if $e^{ik\varkappa_j} = 1$ for some $k \in \mathbb{N}$, then $\lambda_n(e^{i\varkappa_j}), n \in \mathbb{N}_{mp}$ are k-periodic eigenvalues. Consider the case $m = 2$. Let $\varkappa_0 = 0, \varkappa_1 = \pi$. Then $\mathcal{L}_0$ is a set of all periodic eigenvalues and $\mathcal{L}_1$ is a set of some anti-periodic eigenvalues such that $\#\Lambda_0 = 2p, \#\Lambda_1 = p + 1$. Let the set $\mathcal{L}_2 = \{\widetilde{\lambda}\}$, where $\widetilde{\lambda} \in \sigma(\mathcal{J})$ is some 4-periodic eigenvalue, i.e., $\varkappa_2 = \frac{\pi}{2}$. Due to Theorem 1.2, the spectral data $\mathcal{L}_0, \mathcal{L}_1, \mathcal{L}_2$ determine the polynomial $\det(\mathcal{M}_p(z) - \tau I_m)$, the spectrum $\sigma(\mathcal{J})$ and the Lyapunov functions Δ_1, Δ_2.

2) Let $m \geqslant 1$ be an integer. If $\varkappa_j = \frac{2\pi}{j+1}$, $j \in \mathbb{N}_m^0$, then $\mathcal{L}_j$ consists of $(j+1)$-periodic eigenvalues, which are periodic eigenvalues for the operator $\mathcal{J}$, considered as a $p(j+1)$-periodic operator. The set $\mathcal{L}_1$ consists of periodic eigenvalues, the set $\mathcal{L}_2$ consists of anti-periodic eigenvalues, and so on. Moreover, the spectrum $\sigma(\mathcal{J})$ is recovered by the sets $\mathcal{L}_j$ of $(j+1)$-periodic eigenvalues, where $j \in \mathbb{N}_m^0$.

THEOREM 1.3. *i) Let $p \geqslant 3$. Then each sum $\sum_{j=1}^{mp} \lambda_j^s(\tau), s \in \mathbb{N}_{p-1}$ does not depend on $\tau \in \mathbb{C}$. In particular, the following identities and estimate hold true:*

$$\sum_{n=1}^{mp} \lambda_n(\tau) = \sum_{n=1}^{p} \operatorname{Tr} b_n, \qquad \sum_{n=1}^{mp} \lambda_n^2(\tau) = \sum_{n=1}^{p} \operatorname{Tr}(b_n^2 + 2a_n a_n^\top), \qquad all \quad \tau \in \mathbb{C}, \tag{1.12}$$

$$\sum_{n=1}^{pm} \lambda_n^2(\tau) \geqslant 2pm(\det A_p)^{\frac{2}{pm}}. \tag{1.13}$$

Here the identity (1.13) *holds true iff* $b_n \equiv 0$, $a_n a_n^\top = (\det A_p)^{\frac{2}{pm}} I_m$ *for all* $n \in \mathbb{N}_p$.

REMARK. Theorems 1.2 and 1.3 are used to prove Borg-type uniqueness results for periodic Jacobi operators in [**KKu2**].

We describe a priori estimates

PROPOSITION 1.4. *Let $\|\mathcal{J}\|_\infty = \max_{n,j,k}\{|a_n(j,k)|, |b_n(j,k)|\}$, where the matrices $a_n = \{a_n(j,k)\}$, $b_n = \{b_n(j,k)\}$. Then the following estimates hold:*

$$\|\mathcal{J}\|_\infty \leqslant \max\{|\lambda_0^+|, |\lambda_N^-|\} = \|\mathcal{J}\| \leqslant 3m\|\mathcal{J}\|_\infty, \tag{1.14}$$

$$\|\mathcal{J}\|_\infty \leqslant \lambda_N^- - \lambda_0^+ \leqslant 6m\|\mathcal{J}\|_\infty, \quad if \quad \sum_{j=1}^{p} \operatorname{Tr} b_j = 0. \tag{1.15}$$

REMARK. A priori estimates for scalar Jacobi operators were obtained in [**BGGK**], [**K**], [**KKr**], [**KKu1**], [**Ku**].

The plan of our paper is as follows. In Section 2 we prove Theorem 1.1, where we use [**CK**], [**BBK**] devoted to Schrödinger operators with periodic matrix-valued potentials on the real line. In Section 3 we prove Theorems 1.2, 1.3 and Proposition 1.4. In Section 4 we consider examples for the case $m = p = 2$, where we construct the complex and real resonances. In Section 5 we shortly recall the well-known results concerning the properties of the point spectrum and the absence of a singular continuous spectrum.

2. The Lyapunov functions

We recall some well-known results. For any solution $y = (y_n)_{n\in\mathbb{Z}}$ of the equation $a_n y_{n+1} + b_n y_n + a_{n-1}^\top y_{n-1} = z y_n$ we define

$$f_n = \begin{pmatrix} y_n \\ y_{n+1} \end{pmatrix} \quad \text{and} \quad \mathcal{T}_n = \begin{pmatrix} 0 & I_m \\ -a_n^{-1}a_{n-1}^\top & a_n^{-1}(z-b_n) \end{pmatrix}, \quad n \in \mathbb{Z}. \tag{2.1}$$

Then f_n satisfies $f_n = \mathcal{T}_n f_{n-1}$ and the function $\mathcal{M}_n = \begin{pmatrix} \vartheta_n & \varphi_n \\ \vartheta_{n+1} & \varphi_{n+1} \end{pmatrix}$ satisfies $\mathcal{M}_n = \mathcal{T}_n \mathcal{M}_{n-1}, \mathcal{M}_0 = I_{2m}$. This gives the monodromy matrix $\mathcal{M}_n = \prod_1^n \mathcal{T}_j$, $n \geqslant 1$, where $\prod_1^n X_j = X_n \ldots X_1$ for matrices X_j. We rewrite $\mathcal{T}_n$ in the form

$$\mathcal{T}_n = \begin{pmatrix} I_m & 0 \\ 0 & a_n^{-1} \end{pmatrix} \begin{pmatrix} 0 & I_m \\ -I_m & z-b_n \end{pmatrix} \begin{pmatrix} a_{n-1}^\top & 0 \\ 0 & I_m \end{pmatrix} = P_n^{-1} R_n \mathcal{T}_n^0 P_{n-1}, \tag{2.2}$$

where
(2.3)

$$P_n = a_n^\top \oplus I_m, \quad R_n = P_n \begin{pmatrix} I_m & 0 \\ 0 & a_n^{-1} \end{pmatrix} = a_n^\top \oplus a_n^{-1}, \quad \mathcal{T}_n^0 = \begin{pmatrix} 0 & I_m \\ -I_m & z-b_n \end{pmatrix}.$$

Then using (2.2), we get

$$\mathcal{M}_n = \prod_{j=1}^{n} P_j^{-1} R_j \mathcal{T}_j^0 P_{j-1} = P_n^{-1} M_n P_0, \qquad M_n = \prod_{j=1}^{n} R_j \mathcal{T}_j^0. \tag{2.4}$$

Matrices $R_j, \mathcal{T}_j^0$, $j \geqslant 1$ satisfy

$$R_j^\top J R_j = J, \quad (\mathcal{T}_j^0)^\top J \mathcal{T}_j^0 = J, \quad J = \begin{pmatrix} 0 & I_m \\ -I_m & 0 \end{pmatrix}, \tag{2.5}$$

which yields

$$M_n^\top J M_n = J, \quad M_n^{-1} = -J M_n^\top J, \quad n \in \mathbb{Z}. \tag{2.6}$$

Using (2.4) we obtain

$$M_n = P_n \mathcal{M}_n P_0^{-1} = \begin{pmatrix} a_n^\top \vartheta_n (a_0^\top)^{-1} & a_n^\top \varphi_n \\ \vartheta_{n+1} (a_0^\top)^{-1} & \varphi_{n+1} \end{pmatrix}, \tag{2.7}$$

$$M_n^\top = \begin{pmatrix} a_0^{-1} \vartheta_n^\top a_n & a_0^{-1} \vartheta_{n+1}^\top \\ \varphi_n^\top a_n & \varphi_{n+1}^\top \end{pmatrix}$$

and

$$M_n^{-1} = -J M_n^\top J = \begin{pmatrix} \varphi_{n+1}^\top & -\varphi_n^\top a_n \\ -a_0^{-1} \vartheta_{n+1}^\top & a_0^{-1} \vartheta_n^\top a_n \end{pmatrix}. \tag{2.8}$$

Due to $M_n M_n^{-1} = I_{2m}$ and $M_n^{-1} M_n = I_{2m}$ we deduce that

(2.9)

$$a_n^\top \vartheta_n (a_0^\top)^{-1} \varphi_{n+1}^\top - a_n^\top \varphi_n (a_0^\top)^{-1} \vartheta_{n+1}^\top = I_m, \qquad \vartheta_n (a_0^\top)^{-1} \varphi_n^\top = \varphi_n (a_0^\top)^{-1} \vartheta_n^\top,$$

(2.10)

$$\vartheta_n^\top a_n \varphi_{n+1} - \vartheta_{n+1}^\top a_n \varphi_n = a_0, \ \varphi_{n+1}^\top a_n^\top \varphi_n = \varphi_n^\top a_n^\top \varphi_{n+1}, \ \vartheta_{n+1}^\top a_n^\top \vartheta_n = \vartheta_n^\top a_n^\top \vartheta_{n+1}.$$

Recall the asymptotics of fundamental solutions

$$\varphi_{n+1} = z^n A_n + O(z^{n-1}), \ \vartheta_{n+1} = O(z^{n-1}), \ n \geqslant 1, \quad A_n = (a_1 a_2 \ldots a_n)^{-1} \tag{2.11}$$

as $z \to \infty$. Substituting the last asymptotics into $\mathcal{M}_p$ we obtain

$$\mathcal{M}_p(z) = z^p \begin{pmatrix} 0 & 0 \\ 0 & A_p \end{pmatrix} + O(z^{p-1}). \tag{2.12}$$

It is well known that the determinant $D(z,\tau)$ satisfies

(2.13)

$$D(z,\tau) = \det(M(z) - \tau I) = \tau^{2m} + \sum_{j=1}^{2m} \tau^{2m-j} \xi_j(z) = \prod_{j=1}^{2m} (\tau - \tau_j(z)), \quad z, \tau \in \mathbb{C},$$

$$T_n = \operatorname{Tr} M_p^n(z), \quad \xi_0 = 1, \quad \xi_1 = -T_1, \ \ldots \ , \xi_s = -\frac{1}{s} \sum_0^{s-1} T_{s-j} \xi_j, \ \ldots \ ; \tag{2.14}$$

see pp. 331–333 in [**RS**]. The identity (1.6) gives $\xi_j = \xi_{2m-j}$ for all $j \in \mathbb{N}_m$. Using (2.13) we deduce that

$$\Phi(z,\nu) = \frac{D(z,\tau)}{(2\tau)^m} = \sum_{j=0}^{m} \phi_j(z) \nu^{m-j}, \quad \phi_0 = 1, \tag{2.15}$$

where $\phi_2, \ldots, \phi_m$ are some polynomials. Let $a_n = I_m, b_n = 0$ for all $n \in \mathbb{Z}$ and let $\mathcal{J}^0$ be the corresponding Jacobi operator. If D^0 is the corresponding determinant, then we obtain $D^0(\tau,z) = (\tau^2+1-2\tau\mathcal{C}_p(\frac{z}{2}))^m$ and $\Phi^0 = \frac{D^0(z,\tau)}{(2\tau)^m} = (\nu - \mathcal{C}_p(\frac{z}{2}))^m = \sum_0^m \frac{m!}{(m-j)!j!}(-\mathcal{C}_p(\frac{z}{2}))^j\nu^{m-j}$, where $\mathcal{C}_n, n \geqslant 1$ are the Chebyshev polynomials.

Proof of Theorem 1.1. i) We have the identity $\Phi(z,\nu) = \frac{D(z,\tau)}{(2\tau)^m} = \sum_0^m \phi_j(z)\nu^{m-j}$; see (2.15), where the polynomials ϕ_j are some polynomials. $\Delta_1(z),\ldots,\Delta_m(z)$ are the roots of $\Phi(z,\nu)=0$ for fixed $z \in \mathbb{C}$. Then the statement i) follows from the well-known facts about the zeros of a polynomial $\sum_0^m \nu^{m-j}w_j(z)$, where w_j is a polynomials in z; see [**Fo**].

ii) We have $\Delta_j'(z) = \frac{1}{2}(1-\tau^{-2}(z))\tau'(z) \neq 0, z \in Y$, since by the Lyapunov-Poincaré Theorem, $\tau'(z) \neq 0$ for all $z \in Y$.

iii) The fact that ρ is a polynomial follows from the Viète Theorem. Moreover, using (2.13), (2.15) we can determine ρ explicitly in terms of the discriminant of Φ, similar to the continuous case [**CK**].

iv) Each gap has the form $\gamma_n = (\lambda_n^-, \lambda_n^+) = \bigcap_{j=1}^m \gamma_{n,j}$, where $\gamma_{n,j} = (\lambda_{n,j}^-, \lambda_{n,j}^+) \subset \mathbb{R}$ is some finite interval such that $\Delta_j(z) \notin [-1,1]$ for all $z \in \gamma_{n,j}$. Note that $\Delta_j(\lambda_{n,j}^-) = \pm 1$ or $\Delta_j(\lambda_{n,j}^+) = \pm 1$ or $\lambda_{n,j}^\pm$ is the branch point of $\Delta_j(z)$; otherwise we have a contradiction.

v) Using (2.12) we define $F(t) = t^p\mathcal{M}_p = F_0 + tF_1(t)$, where $F_0 = 0_m \oplus A_p$, $t = z^{-1}$ and F_1 is some matrix polynomial and $F_1(t) = \text{const} + O(t)$ as $t \to 0$. Let $\widetilde{\tau}_j(t), j \in \mathbb{N}_{2m}$ be the eigenvalues of $F(t)$. Recall that $\tau_j^0, j \in \mathbb{N}_m$ are the eigenvalues of A_p. Then, the $\widetilde{\tau}_j(t)$ are given by convergent Puiseux series (the Taylor series in τ^{1/m_j})

$$\widetilde{\tau}_j(t) = \tau_j^0 + \sum_{k=1}^\infty \alpha_{j,k} t^{\frac{k}{m_j}}, \quad |t| < r, \quad \sum_1^k m_j = m, \quad m_j \in \mathbb{N},\ j \in w_s, s \in \mathbb{N}_k$$

for some $r > 0$, see page 4, [**RS**], where $w_1 = \mathbb{N}_{m_1}, w_2 = \mathbb{N}_{m_1+m_2} \setminus w_1, \ldots, w_k = \mathbb{N}_m \setminus \mathbb{N}_{m-m_k}$ and $\tau_j^0 = \tau_{j'}^0$ for all $j, j' \in w_s$. Here $\widetilde{\tau}_j(t), j \in w_s$ are the branches of one or more multivalued analytic functions with at worst algebraic branch points at $t=0$ and satisfy

$$\tau_j(z) = z^p\widetilde{\tau}_j(t) = z^p(\tau_j^0 + O(z^{-\frac{1}{m_s}})) = \tau_j^0 z^p + O(z^{p-\frac{1}{m_s}}), \qquad j \in w_s.$$

If each eigenvalue τ_j^0 is simple, then each $\widetilde{\tau}_j(t), j \in \mathbb{N}_m$ is analytic for small t, i.e., $m_s = 1$. Thus we obtain $\tau_j(z) = \tau_j^0 z^p + O(z^{p-1}), j \in \mathbb{N}_m$, which yields $2\Delta_j = \tau_j(z) + \frac{1}{\tau_j(z)} = \tau_j^0 z^p + O(z^{p-1})$ and then (1.8) gives asymptotics for ρ_s. □

3. Inverse problems

In order to prove Theorem 1.2 we need Lemmas 3.1–3.3.

LEMMA 3.1. *Let $W = \{W_{j,r}\}_{r,j=0}^s$ be an $s \times s$ matrix with components $W_{r,j} = \cos j\varkappa_r$ for some $\varkappa_r \in \mathbb{C}, s \geqslant 1$, where $\cos\varkappa_r \neq \cos\varkappa_{r'}$ for all $r \neq r'$. Then* $\det W \neq 0$.

Proof. Using $\cos jz = \mathcal{C}_j(\cos z)$, where the $\mathcal{C}_j$ are Chebyshev polynomials, we obtain $W_{r,j} = \cos j\varkappa_r = \mathcal{C}_j(\cos\varkappa_r)$. Let $Wy = 0$ for some vector $y = (y_n)_0^s \in \mathbb{C}^{s+1}, y \neq 0$. Then we deduce that a polynomial $P(x) = \sum_{j=0}^s y_j\mathcal{C}_j(x), P \neq 0$ has distinct zeros $\cos\varkappa_j, j \in \mathbb{N}_s^0$, which gives a contradiction, since $\deg P \leqslant \deg \mathcal{C}_s = s$. Thus W is invertible. □

LEMMA 3.2. *Let* $c = (-1)^m \det A_p$. *The polynomials* $\xi_j(z), j \in \mathbb{N}_m$, *given by* (2.13), *satisfy*

$$\xi_j(z) = \xi_{2m-j}(z) = O(z^{pj}), \qquad \xi_m(z) = cz^{pm} + O(z^{pm-1}) \quad as\ z \to \infty, \tag{3.1}$$

$$c\prod_{n=1}^{mp}(z-\lambda_n(\tau)) = \frac{D(z,\tau)}{\tau^m} = (\tau^m+\tau^{-m})+\xi_m(z)+\sum_{j=1}^{m-1}(\tau^{m-j}+\tau^{j-m})\xi_j(z),\ \tau, z \in \mathbb{C}. \tag{3.2}$$

Proof. Using (1.6), (2.13) and Viète's formulas, we get

$$\xi_j(z) = \xi_{2m-j}(z) = (-1)^j \sum_{1\leqslant i_1<\cdots<i_j\leqslant 2m} \prod_{s=1}^{j} \tau_{i_s}(z), \quad 1 \leqslant j \leqslant m. \tag{3.3}$$

The asymptotics (1.9) gives $\tau_j(z) = \tau_j^0 z^p + o(z^p)$, $j \in \mathbb{N}_m$ and the Lyapunov-Poincaré Theorem yields $\tau_{m+j}(z) = \frac{1}{\tau_j(z)} = O(z^{-p})$ as $z \to \infty$, since each $\tau_j^0 \neq 0$. Then, using (3.3), we obtain

$$\xi_j(z){=}O(z^{pj}), \quad \xi_m(z) = (-1)^m \prod_{j=1}^{m}\tau_j^0 z^{pm}+o(z^{pm}) = (-1)^m\det A_p z^{pm}+O(z^{pm-1}).$$

Identity (3.2) follows from the last identity and (1.6), (2.13). □

We need the following lemma about polynomials.

LEMMA 3.3. *Let* $(z_n)_1^k \in \mathbb{C}^k$ *and* $(c_j)_0^s \in \mathbb{C}^{s+1}$ *for some integers* $s, k \geqslant 1$. *Then there exists a unique polynomial* $r(z)$ *satisfying for some polynomial* g *the following relations:*

$$r(z) = z^k\sum_{j=0}^{s} c_j z^j+O(z^{k-1}) \quad \text{as } z \to \infty \qquad \text{and} \quad r = hg, \quad h(z) = \prod_{n=1}^{k}(z-z_n). \tag{3.4}$$

Proof. Introduce the linear space of polynomials $\mathcal{P}_s = \{g : \deg g \leqslant s\}$, $\dim \mathcal{P}_s = s+1$. Note that $g \in \mathcal{P}_s$. Define the linear operator $\mathcal{A} : \mathcal{P}_s \to \mathbb{C}^{s+1}$ by

$$(\mathcal{A}g)_j = \frac{1}{n!}\frac{d^n}{dz^n}\Big(h(z)g(z)\Big)\Big|_{z=0}, \quad n = k+j-1,\ j \in \mathbb{N}_{s+1}, \quad g \in \mathcal{P}_s.$$

We rewrite $\mathcal{A}g$ in the form $(\mathcal{A}g)_j = c_{j-1}$, $j \in \mathbb{N}_{s+1}$, where $h(z)g(z) = z^k\sum_{j=0}^{s} c_j z^j + O(z^{k-1})$ as $z \to \infty$. We deduce that $\mathcal{A} : \mathcal{P}_s \to \mathbb{C}^{s+1}$ is an isomorphism, since $\dim \mathcal{P}_s = \dim \mathbb{C}^{s+1} = s+1$ and if $\mathcal{A}g = 0$, then $g = 0$. If $g = \mathcal{A}^{-1}(c_n)_0^s$ and $r(z) = h(z)g(z)$, then $r(z) = z^k\sum_{j=0}^{s} c_j z^j + O(z^{k-1})$ as $z \to \infty$. Moreover, the polynomial r is unique, since $\mathcal{A}$ is an isomorphism. □

Proof of Theorem 1.2. i) Define the polynomials $\zeta_0 = c^{-1}, \zeta_j(z) = c^{-1}\xi_j, j \in \mathbb{N}_m$, where $c = (-1)^m \det A_p$. Lemma 3.2 yields $\deg \zeta_j \leqslant pj$ and then $\zeta_j(z) = \sum_{n=0}^{pj}\zeta_{j,n}z^n$ for some $\zeta_{j,n} \in \mathbb{R}$. Introduce the sets $K_s = \mathbb{N}_{p(m-s)} \setminus \mathbb{N}_{p(m-s-1)}$, $s \in \mathbb{N}_{m-1}^0$, $K_m = \{0\}$, which satisfy

$$\bigcup_{s=j}^{m} K_s = \mathbb{N}_{p(m-j)}^0, \quad j \in \mathbb{N}_m^0, \quad K_r \cap K_s = \emptyset, for\ \ all\ \ r \neq s. \tag{3.5}$$

Introduce the function $s : \mathbb{N}^0_{pm} \to \mathbb{N}^0_m$: if $n \in K_r$, then $s(n) = r$. Using (3.2) and (3.5) and setting $h_j = \tau^{-j} + \tau^j$, we obtain

(3.6)

$$\begin{aligned} q(z,\tau) &= \frac{D(z,\tau)}{c\tau^m} = \prod_{n=1}^{pm}(z-\lambda_n(\tau)) = \sum_{j=0}^{m} h_j\zeta_{m-j}(z) = \sum_{j=0}^{m} h_j \sum_{n=0}^{p(m-j)} \zeta_{m-j,n}z^n \\ &= \sum_{j=0}^{m}\sum_{s=j}^{m}\sum_{n\in K_s} h_j\zeta_{m-j,n}z^n = \sum_{s=0}^{m}\sum_{j=0}^{s}\sum_{n\in K_s} h_j\zeta_{m-j,n}z^n \\ &= \sum_{s=0}^{m}\sum_{n\in K_s}\sum_{j=0}^{s} h_j\zeta_{m-j,n}z^n = \sum_{s=0}^{m}\sum_{n\in K_s}\eta_n(\tau)z^n = \sum_{n=0}^{pm}\eta_n(\tau)z^n, \end{aligned}$$

where

$$\eta_n(\tau) = \sum_{j=0}^{s} h_j\zeta_{m-j,n}, \quad s = s(n), \quad n \in \mathbb{N}^0_{pm}. \tag{3.7}$$

Substituting $\tau = e^{i\varkappa_r}, r \in \mathbb{N}^0_s$ into the identity (3.7) we get

$$\begin{gathered} \widetilde{\eta}_n = 2W_s y_n, \quad \widetilde{\eta}_n = (\eta_n(e^{i\varkappa_r}))_{r=0}^s, \; y_n = (\zeta_{m-j,n})_{j=0}^s, \\ W_s = (\cos j\varkappa_r)_{r,j=0}^s, \quad s = s(n), \end{gathered} \tag{3.8}$$

for all $n \in \mathbb{N}^0_{pm}$. Due to Lemma 3.1, each matrix $W_s, s \in \mathbb{N}^0_m$ is invertible. Thus (3.8), (3.7) give

$$\eta_n(\tau) = \langle(\tau^j + \tau^{-j})_{j=0}^s, (2W_s)^{-1}\widetilde{\eta}_n\rangle, \quad s = s(n), \quad n \in \mathbb{N}^0_{pm}, \tag{3.9}$$

where $\langle\cdot,\cdot\rangle$ is a scalar product in $\mathbb{R}^s$. Then the functions $\eta_n(\cdot), n \in \mathbb{N}^0_{pm}$ are determined by $\widetilde{\eta}_n = (\eta_n(e^{i\varkappa_r}))_{r=0}^{s(n)}$. We will use this fact to determine all functions $\eta_n(\cdot), n \in \mathbb{N}^0_{pm}$. In order to describe recovering we need the following simple fact.

LEMMA 3.4. *Let $0 \leqslant k \leqslant m-1$. Then the numbers $\eta_n(e^{i\varkappa_j}), (n,j) \in \mathbb{N}^0_{pm} \times \mathbb{N}^0_k$, the functions η_n, $n \in \bigcup_{j=0}^k K_j$, the set Λ_{k+1} and the numbers $\varkappa_j$, $j \in \mathbb{N}^0_m$ determine the numbers $\eta_n(e^{i\varkappa_j}), (n,j) \in \mathbb{N}^0_{pm} \times \mathbb{N}^0_{k+1}$, and the functions η_n, $n \in \bigcup_{j=0}^{k+1} K_j$.*

Proof. Let $S_{k+1} = (\eta_n(e^{i\varkappa_{k+1}}))_{n\in\bigcup_{j=0}^k K_j} = (\eta_n(e^{i\varkappa_{k+1}}))_{n=p(m-k-1)+1}^{pm}$. Using (3.6), we see that the elements of $\mathcal{L}_{k+1} = (\lambda_n(e^{i\varkappa_{k+1}}))_{n\in\Lambda_{k+1}}$ are some zeros of the polynomial $q(z, e^{i\varkappa_{k+1}})$ and recall that $\#\Lambda_{k+1} = p(m-k-1)+1$. Also, using (3.6), we see that the components of the vector S_{k+1} are the coefficients of the polynomial $q(z, e^{i\varkappa_{k+1}})$. Then, due to Lemma 3.3, $\mathcal{L}_{k+1}$ and S_{k+1} uniquely determine the polynomial $q(z, e^{i\varkappa_{k+1}})$ and, using (3.6), we determine $(\eta_n(e^{i\varkappa_{k+1}}))_{n=0}^{pm}$. Substituting the obtained values $(\eta_n(e^{i\varkappa_j}))_{j\in\mathbb{N}^0_{k+1}}$, $n \in K_{k+1}$ into (3.8), we determine $\widetilde{\eta}_n$, $n \in K_{k+1}$. Substituting $\widetilde{\eta}_n$ into (3.9), we determine $\eta_n(\tau)$, $n \in K_{k+1}$. ■

Now we describe the recovering procedure. Recall that $q(z,\tau) = \frac{D(z,\tau)}{c\tau^m}$. Substituting the elements of the set Λ_0 into (3.6), we determine $q(z, e^{i\varkappa_0})$ and $(\eta_n(e^{i\varkappa_0}))_{n=0}^{pm}$. Then (3.8)–(3.9) determine all functions $\eta_n(\tau), n \in K_0$. Step by step, using Lemma 3.4 for $k \in \mathbb{N}^0_{m-1}$, we determine $\eta_n(\tau)$, $n \in \bigcup_{j=0}^m K_j = \mathbb{N}^0_{pm}$. Then (3.6) gives $q(z,\tau)$. Using (2.13) we get $cq(z,\tau) = D(z,\tau) = \tau^{2m} + O(\tau^{2m-1})$ as $\tau \to \infty$. Then the polynomial $q(z,\tau)$ uniquely determines the constant c and $D(z,\tau)$.

ii) Assume that $a_n, b_n, n \in \mathbb{Z}$ are diagonal, and each $a_n > 0, n \in \mathbb{Z}$. Then $\mathcal{J} = \bigoplus_1^m \mathcal{J}_j$, where the $\mathcal{J}_j$ are scalar Jacobi operators. Let $\mathcal{J}_j$ be such that $\sigma(\mathcal{J}_{j_1}) \cap \sigma(\mathcal{J}_{j_2}) = \emptyset$ for all $j_1 \neq j_2$. Let Δ_j be the Lyapunov functions and let $\widetilde{D}_j$ be the corresponding determinant for the operators $\mathcal{J}_j$, $j \in \mathbb{N}_m$. Then we deduce that

$$D(z,\tau) = \prod_{j=1}^m \widetilde{D}_j, \ \ \widetilde{D}_j(z,\tau) = \tau(\tau + \tau^{-1} - 2\Delta_j(z)). \tag{3.10}$$

By Theorem A, the Lyapunov function $\Delta_m(z)$ satisfies $\Delta_m'(z_j^0) = 0$ and

$$(-1)^{p-j}\Delta_m(z_j^0) = |\Delta_m(z_j^0)| \geqslant 1, \quad j \in \mathbb{N}_{p-1}$$

for some $z_1^0 < \cdots < z_p^0$. Then the polynomial $q_s(z) = s(\Delta_m(z) - \cos\varkappa_0) + \cos\varkappa_0$, $s \geqslant 1$ satisfies

$$(-1)^{p-j} q_s(z_j^0) = s\biggl(|\Delta_m(z_j^0)| + (-1)^{p-j}\Bigl(-1 + \frac{1}{s}\Bigr)\cos\varkappa_0\biggr) \geqslant 1, \quad (q_s)'(z_j^0) = 0$$

for any $j \in \mathbb{N}_{p-1}$. Using $\Delta_m = c_m z^p + O(z^{p-1})$, $c_m > 0$ as $z \to \infty$, we get $q_s(z) = c_m s z^{p-1} + O(z^{p-1})$, $c_m > 0$. Due to Theorem A, q_s is a Lyapunov function for some scalar p-periodic Jacobi operator $\mathcal{J}_m^s$. Consider the p-periodic m-dimensional Jacobi operator $\mathcal{J}^s = \mathcal{J}_m^s \oplus \bigoplus_1^{m-1} \mathcal{J}_j$. Then the determinant for $\mathcal{J}^s$ is given by

$$D_s(z,\tau) = \tau^m(\tau + \tau^{-1} - 2q_s(z)) \prod_{j=1}^{m-1} (\tau + \tau^{-1} - 2\Delta_j(z)). \tag{3.11}$$

Using (3.10), (3.11), we see that $\mathcal{L}_0(\mathcal{J}) = \mathcal{L}_0(\mathcal{J}^s)$, where $\mathcal{L}_j(\mathcal{J})$, $j = 0, 1, 2\ldots$ is the set $\mathcal{L}_j$ for $\mathcal{J}$. Each Lyapunov function $\Delta_{j_1}(z)$, $j_1 \in \mathbb{N}_{m-1}$ satisfies

$$\Delta_{j_1}(z) - \cos\varkappa_j = c_{j_1} \prod_{n=1}^p (z - z_{n+pj_1,j}), \qquad c_{j_1} > 0, \ j \in \mathbb{N}_m, \ \text{ for some } \ z_{n+pj_1,j} \in \mathbb{R}.$$

Then using (3.10), (3.11), we deduce that $z_{n+p,j}, n \in \mathbb{N}_{p(m-1)}$ are zeros of each polynomial $D_s(z, e^{i\varkappa_j}), j \in \mathbb{N}_m$, $s \geqslant 1$, where $D_1(z,\tau) = D(z,\tau)$. We take $\mathcal{L}_j(\mathcal{J}^s) = \mathcal{L}_j(\mathcal{J}) = (z_{p+n,j})_{n \in \mathbb{N}_{p(m-j)+1}}$, $j = 2, \ldots, m$ and $\widetilde{\mathcal{L}}_1(\mathcal{J}^s) = \widetilde{\mathcal{L}}_1(\mathcal{J}) = (z_{n+p,1})_{n \in \mathbb{N}_{p(m-1)}}$, where $\widetilde{\mathcal{L}}_1(\mathcal{J})$ is the set $\widetilde{\mathcal{L}}_1$ for $\mathcal{J}$. But we have $D_s \neq D_1 = D$ for any $s > 1$, since $q_1 = \Delta_m \neq q_s$. Using the fact that $q_1 = \Delta_m \neq q_s$ we deduce that $\sigma(\mathcal{J}_m^s) \neq \sigma(\mathcal{J}_m), s > 1$, which implies $\sigma(\mathcal{J}) \neq \sigma(\mathcal{J}^s)$, since $\sigma(\mathcal{J}_{j_1}) \cap \sigma(\mathcal{J}_{j_2}) = \emptyset$ for all $j_1 \neq j_2$ and $\mathcal{J} = \bigoplus_1^m \mathcal{J}_j$, and $\mathcal{J}^s = \mathcal{J}_m^s \oplus \bigoplus_1^{m-1} \mathcal{J}_j$. □

Proof of Theorem 1.3. Let $g_n(\tau) = (-1)^n \sum_{1 \leqslant j_1 < \cdots < j_n \leqslant mp} \prod_{s=1}^n \lambda_{j_s}(\tau)$. Lemma 3.2 implies $\xi_j(z) = O(z^{pj})$. Then using (3.2) for fixed $\tau \neq \tau_1 = 1$, $\tau \neq 0$, we obtain

$$\begin{aligned}
c\sum_{j=1}^{mp} z^{mp-j}(g_j(\tau_1) - g_j(\tau)) &= c\prod_{j=1}^{mp}(z - \lambda_j(\tau_1)) - c\prod_{j=1}^{mp}(z - \lambda_j(\tau)) \\
&= (\tau_1^m + \tau_1^{-m} - \tau^m - \tau^{-m}) + \sum_{n=1}^{m-1}(\tau_1^{m-n} + \tau_1^{n-m} - \tau^{m-n} - \tau^{n-m})\xi_n(z) \\
&= O(z^{(m-1)p}),
\end{aligned}$$

which yields $g_j(\tau_1) = g_j(\tau)$ for all $j \in \mathbb{N}_{p-1}$, since $c \neq 0$. The Newton formulas give

$$u_1 = -g_1, \quad u_n = -ng_n - \sum_{j=1}^{n-1} g_j u_{n-j}, \quad \text{where} \quad u_n(\tau) = \sum_{j=1}^{mp} \lambda_j^n(\tau),$$

which yields $u_n(\tau) = u_n(\tau_1) = u_n(1)$ for all $(n,\tau) \in \mathbb{N}_{p-1} \times (\mathbb{C} \setminus \{0\})$. Due to (5.4), we get

$$\sum_{j=1}^{mp} \lambda_j^s(\tau) = \operatorname{Tr} L(\tau)^s, s \geqslant 1, \qquad \sum_{j=1}^{mp} \lambda_j(\tau) = \operatorname{Tr} L(\tau) = \sum_{j=1}^{N} \operatorname{Tr} b_j$$

$$\sum_{j=1}^{mp} \lambda_j^2(\tau) = \operatorname{Tr} L(\tau)^2 = 2\sum_{j=1}^{p} \|a_j\|^2 + \sum_{j=1}^{p} \|b_j\|^2. \tag{3.12}$$

Let $\varepsilon_{j,n}$, $j \in \mathbb{N}_m$ be eigenvalues of the matrix $a_n a_n^\top, n \in \mathbb{N}_p$. Then (3.12) gives (3.13)

$$\sum_{n=1}^{pm} \lambda_n^2(\tau) \geqslant \sum_{n=1}^{p} \operatorname{Tr} 2a_n a_n^\top = 2\sum_{n=1}^{pm}\sum_{j=1}^{m} \varepsilon_{j,n} \geqslant 2pm\Big(\prod_{j,n} \varepsilon_{j,n}\Big)^{\frac{1}{pm}} = 2pm(\det A_p^2)^{\frac{1}{pm}}.$$

Recall that the identity $\sum_{n=1}^{p}\sum_{j=1}^{m} \varepsilon_{j,n} = pm\Big(\prod_{j,n} \varepsilon_{j,n}\Big)^{\frac{1}{pm}}$ holds true iff each $\varepsilon_{j,n}^{pm} = \det A_p^2$, $(j,n) \in \mathbb{N}_m \times \mathbb{N}_p$. Then the statement ii) holds true. □

Proof of Proposition 1.4. For the selfadjoint operator $\mathcal{J}$ we obtain $\|\mathcal{J}\| = \max\{|\lambda_0^+|, |\lambda_0^-|\}$. Moreover, the operator $\mathcal{J}$ has the form $\mathcal{J} = AS + B + S^{-1}A^*$, where $A = \operatorname{diag}_{n\in\mathbb{Z}}(a_n)$, $B = \operatorname{diag}_{n\in\mathbb{Z}}(b_n)$ and $S(y_n)_{n\in\mathbb{Z}} = (y_{n+1})_{n\in\mathbb{Z}}$, $(y_n)_{n\in\mathbb{Z}} \in \ell^2(\mathbb{Z})^m$ is a shift operator. Then

$$\|\mathcal{J}\| = \|AS + B + S^{-1}A^*\| \leqslant 2\|A\| + \|B\| \leqslant 3m\|\mathcal{J}\|_\infty.$$

If $\sum_1^p \operatorname{Tr} b_j = 0$, then $\sum \lambda_j(\tau) = 0$, which yields $\lambda_0^+ \leqslant 0 \leqslant \lambda_0^-$ and $\lambda_0^- - \lambda_0^+ = |\lambda_0^+| + |\lambda_{N_G}^-|$. Thus all these estimates gives (1.15). □

4. Examples for the case $p = 2$, $m = 2$

We consider the case $p = 2$, $m = 2$, where the real 2×2 matrices a_n, b_n satisfy

$$a_{n+2} = a_n = \begin{pmatrix} 1 & \beta_{2n+1} \\ 0 & 1 \end{pmatrix}, \quad b_{n+2} = b_n = \begin{pmatrix} \alpha_{2n} & \beta_{2n} \\ \beta_{2n} & \alpha_{2n+1} \end{pmatrix},$$

$$\alpha_{n+4} = \alpha_n, \quad \beta_{n+4} = \beta_n.$$

The determinant D, the function ρ and the Lyapunov functions are given by

$$D(\cdot,\tau) = \tau^4 - T_1\tau^3 + T_2\tau^2 - T_1\tau + 1 = (\tau^2 - 2\Delta_1\tau + 1)(\tau^2 - 2\Delta_2\tau + 1) = (2\tau)^2\Phi(\cdot,\nu),$$

$$\rho = \frac{T_1^2 - 4T_2 + 8}{4}, \quad \Delta_1 = \frac{T_1 - \sqrt{4\rho}}{4}, \quad \Delta_2 = \frac{T_1 + \sqrt{4\rho}}{4},$$

$$T_1 = 2z^2 + T_{11}z + T_{10}, \quad T_2 = z^4 + T_{23}z^3 + \cdots + T_{20}, \quad \Phi(\cdot,\nu) = \nu^2 - \frac{T_1}{2}\nu + \frac{T_2}{4} - \frac{1}{2},$$

where $T_{11} = -\sum_0^3 \alpha_n + (\beta_0+\beta_2)\beta_3 + \beta_1(\beta_0+\beta_2)$ and

$$T_{10} = -4 + 2\beta_0\beta_2 + 2\beta_1\beta_3 + \prod_0^3 \beta_n + \alpha_0\alpha_2 + \alpha_1\alpha_3 - \beta_0(\alpha_2\beta_3+\alpha_3\beta_1) - \beta_2(\alpha_0\beta_1+\alpha_1\beta_3),$$

(4.1)

$$T_{23} = -\sum_0^3 \alpha_n, \quad T_{22} = \sum_{0\leqslant j<k\leqslant 3} \alpha_j\alpha_k - \sum_0^3 \beta_n^2 - 4,$$

$$T_{21} = (\alpha_2+\alpha_3)\beta_0^2 + (\alpha_0+\alpha_3)\beta_1^2 + (\alpha_0+\alpha_1)\beta_2^2 + (\alpha_1+\alpha_2)\beta_3^2 - 2\beta_0(\beta_1+\beta_3) - 2\beta_2(\beta_1+\beta_3) + (\alpha_1+\alpha_3)(2-\alpha_0\alpha_2) + (\alpha_0+\alpha_2)(2-\alpha_1\alpha_3),$$

$$T_{20} = 6 + \beta_0^2\beta_2^2 + \beta_1^2\beta_3^2 - 4\beta_1\beta_3 - 4\beta_0\beta_2 + \prod_0^3 \alpha_n - \alpha_0(\alpha_1\beta_2^2+\alpha_3\beta_1^2) - \alpha_2(\alpha_1\beta_3^2+\alpha_3\beta_0^2) - 2\alpha_0\alpha_2 - 2\alpha_1\alpha_3 + 2\beta_0(\alpha_2\beta_3+\alpha_3\beta_1) + 2\beta_2(\alpha_0\beta_1+\alpha_1\beta_3).$$

Below we consider some specific cases.

1. All a_n, b_n are diagonal matrices. In this case all $\beta_n = 0$, the function $\rho = 1$ and

$$\Delta_1 = \frac{1}{2}\left(z - \frac{\alpha_0+\alpha_2}{2}\right)^2 - \frac{1}{2}\left(\frac{\alpha_0-\alpha_2}{2}\right)^2 - 1,$$

$$\Delta_2 = \frac{1}{2}\left(z - \frac{\alpha_1+\alpha_3}{2}\right)^2 - \frac{1}{2}\left(\frac{\alpha_1-\alpha_3}{2}\right)^2 - 1.$$

2. The constant coefficients. Let $\alpha_n = 0, \beta_n = \beta$ for all $n \in \mathbb{Z}$. Then

$$T_1 = 2z^2 + 4\beta^2 z + \beta^4 + 4\beta^2 - 4, \quad \rho = \beta^2(2z+\beta^2)^2(4z+\beta^2+8), \quad \beta \neq 0.$$

The function ρ has 3 zeros z_1, z_2, z_3. If $\beta = 0$, then $\rho = 1, \Delta_1 \equiv \Delta_2 = \frac{z^2}{2}$.

If $\beta \neq 0, \pm\sqrt{8}$, then ρ has the zeros $z_1 = z_2 = -\frac{\beta^2}{2}, z_3 = -\frac{\beta^2+8}{4}$.

If $\beta = \pm\sqrt{8}$, then ρ has the zeros $z_1 = z_2 = z_3 - 4$.

We have the following equations for periodic and anti-periodic eigenvalues:

$$4(\Delta_1-1)(\Delta_2-1) = (z+2)^2((z-2)^2-4\beta^2) = 0, \quad 4(\Delta_1+1)(\Delta_2+1) = (z^2-2\beta^2)^2 = 0.$$

Hence for each $\beta \neq 0$ there exist two double anti-periodic eigenvalues $-\beta\sqrt{2}, \beta\sqrt{2}$, one double periodic eigenvalue -2, and two simple periodic eigenvalues $2(1-b)$, $2(1+b)$. If $\beta = \pm 2$, then there exists the periodic eigenvalue -2 of multiplicity 3.

3. Example. Let $(\alpha_0,\alpha_1,\alpha_2,\alpha_3) = (1,0,-1,0)$, $(\beta_0,\beta_1,\beta_2,\beta_3) = (t,0,0,0)$, $t \in \mathbb{R}\setminus\{0\}$. In this case we have

$$T_1 = 2z^2 - 5, \quad T_2 = z^4 - (t^2+5)z^2 - t^2 z + 8, \quad 4\rho_t = 4t^2z^2 + 4t^2 z + 1, \quad t \neq 1.$$

The Lyapunov functions and resonances $r_t^\pm$ (zeros of ρ) are given by

$$\Delta_{1,t} = \frac{2z^2 - 5 - \sqrt{4\rho_t}}{4}, \quad \Delta_{2,t} = \frac{2z^2 - 5 + \sqrt{4\rho_t}}{4}, \quad r_t^\pm = \frac{-1 \pm \sqrt{1-t^{-2}}}{2}.$$

Each resonance $r_t^\pm, t \in (0,1)$ is complex.

Let $t = 1$. Then the Lyapunov functions are given by $\Delta_{1,1} = \frac{z^2-z-3}{2}, \Delta_{2,1} = \frac{z^2+z-2}{2}$, $\rho_1 = 1$ and the spectral bands $\sigma_t^j = \{z \in \mathbb{R} : \Delta_{j,t}(z) \in [-1,1]\}, j = 1,2$

have the forms

$$\sigma_1^1 = [\lambda_1^1, \lambda_2^1] \cup [\lambda_3^1, \lambda_4^1] = \left[-\frac{\sqrt{21}-1}{2}, -\frac{\sqrt{5}-1}{2}\right] \cup \left[\frac{\sqrt{5}+1}{2}, \frac{\sqrt{21}+1}{2}\right],$$

$$\sigma_1^2 = [\lambda_1^2, \lambda_2^2] \cup [\lambda_3^2, \lambda_4^2] = \left[-\frac{\sqrt{17}+1}{2}, -1\right] \cup \left[0, \frac{\sqrt{17}+1}{2}\right],$$

where $\lambda_k^j, j = 1,2, k = 1,4$ (and $\lambda_k^j, k = 2,3$) are periodic (and anti-periodic) eigenvalues.

Let $t = 1+\varepsilon$ for some small $\varepsilon > 0$. Then the spectral bands have the forms

$$\sigma_t^j = [\lambda_{1,t}^j, \lambda_{2,t}^j] \cup [\lambda_{3,t}^j, \lambda_{4,t}^j], \quad \lambda_{i,t}^j = \lambda_i^j + o(1), \quad (i,j) \in \mathbb{N}_4 \times \mathbb{N}_2, \quad \text{as} \quad t \downarrow 1$$

and the "resonances" $r_t^{\pm} = \frac{-1+o(1)}{2}$, and $\rho_t(z) = (z+\frac{1}{2})^2 + o(1)$. Note that the "resonance" gap $\gamma_t = (r_t^-, r_t^+) \neq \emptyset$, since $r_t^- < r_t^+$ and $\sigma(\mathcal{J}) \cap \overline{\gamma_t} = \emptyset$ as $t \downarrow 1$. Thus the interval $\gamma_t \subset (\lambda_{2,t}^2, \lambda_{3,t}^2)$, where $(\lambda_{2,t}^2, \lambda_{3,t}^2)$ is a gap in the spectrum of $\mathcal{J}$. This gives that the "resonance gap" arises in the gap, and the end points of the spectral bands are periodic or anti-periodic eigenvalues.

If $t > 1$, then the resonances $r_1^- < r_1^-$ are the real branch points. If $t \in (0,1)$, then the resonances $r_1^{\pm} \in \mathbb{C}_{\pm}$ are the complex branch points.

4. Example. Let $(\alpha_0, \alpha_1, \alpha_2, \alpha_3) = (0,1,0,1)$, $(\beta_0, \beta_1, \beta_2, \beta_3) = (t,0,0,0)$, $t \in \mathbb{R}$. Then we obtain

$$T_1 = 2z^2 - 2z - 3, \quad T_2 = z^4 - 2z^3 - (3+t^2)z^2 + (4+t^2)z + 4,$$

$$4\rho_t = (2z-1)^2 + 4t^2(z^2 - z), \quad r_t^{\mp} = \frac{1}{2} \mp \frac{t}{2\sqrt{t^2+1}} \in \mathbb{R}, \quad t \neq 0.$$

Here $r_t^{\pm}$ are the zeros (resonances) of ρ_t and if $t \neq 0$, then $r_t^- < r_t^+$. Thus we have the resonance gap $\gamma_t = (r_t^-, r_t^+) \subset (0,1), t \neq 0$. The Lyapunov functions are given by

$$\Delta_{1,t} = \frac{1}{4}\left(2z^2 - 2z - 3 - \sqrt{4\rho_t(z)}\right), \qquad \Delta_{2,t} = \frac{1}{4}\left(2z^2 - 2z - 3 + \sqrt{4\rho_t(z)}\right).$$

Let $t = 0$. Then the Lyapunov functions have the forms $\Delta_{1,0} = \frac{z^2-2z-1}{2}$, $\Delta_{2,0} = \frac{z^2-2}{2}$ and $\Delta_{1,0}(\frac{1}{2}) = \Delta_{2,0}(\frac{1}{2}) = -\frac{7}{8}$. The spectral bands $\sigma_t^j = \{z \in \mathbb{R} : \Delta_{j,t}(z) \in [-1,1]\}, j = 1,2$ have the forms $\sigma_0^1 = [-1,3]$, $\sigma_0^2 = [-2,2]$. The end points of the spectral bands σ_0^1, σ_0^2 are periodic eigenvalues.

If $t \in (0,\varepsilon)$ for some small ε, then we obtain $\sigma_t^j = [\lambda_1^j(t), r_t^-] \cup [r_t^+, \lambda_2^j(t)]$, where $r_t^- < r_t^+$, $r_t^{\mp} = \frac{1+o(1)}{2}$ and

$$\lambda_{1,t}^1 = -1 + o(1), \quad \lambda_{2,t}^1 = 3 + o(1), \quad \lambda_{1,t}^2 = -2 + o(1), \quad \lambda_{2,t}^2 = 2 + o(1) \quad \text{as} \quad t \to 0.$$

This yields $\sigma(\mathcal{J}) = \sigma_t^1 \cup \sigma_t^2 = [\lambda_{1,t}^2, r_t^-] \cup [r_t^+, \lambda_{4,t}^1]$, where $(r_t^-, r_t^+) \neq \emptyset$ is the resonance gap, since $r_t^- < r_t^+$, and $\lambda_{2,t}^1, \lambda_{2,t}^1$ are periodic eigenvalues.

5. Appendix

Let $H = L^2\left([0,2\pi), \frac{dx}{2\pi}, \mathcal{H}\right) = \int_{[0,2\pi)}^{\oplus} \mathcal{H} \frac{dx}{2\pi}$ be a constant fiber direct integral, where $\mathcal{H} = \mathbb{C}^{pm}$. It is the Hilbert space of square integrable $\mathcal{H}$-valued functions. Let $\mathcal{S}(\mathbb{Z})$ be the set of all compactly supported functions $(f_n)_{n\in\mathbb{Z}} \in l^2(\mathbb{Z})^m$.

LEMMA 5.1. *The operator $U : l^2(\mathbb{Z})^m \to H$, given by*

$$(5.1) \qquad (Uf)_j(x) = \sum_{n=-\infty}^{\infty} e^{-ixn} f_{j+np}, \qquad (x,j) \in [0,2\pi] \times \mathbb{N}_p,$$

is well defined for $\mathcal{S}(\mathbb{Z})$ and uniquely extendable to a unitary operator. Moreover,

$$(5.2) \qquad U\mathcal{J}U^{-1} = \int_{[0,2\pi)}^{\oplus} L(e^{ix}) \frac{dx}{2\pi}, \qquad L(\tau) = \begin{pmatrix} b_1 & a_1 & 0 & \dots & \tau^{-1}a_p^{\top} \\ a_1^{\top} & b_2 & a_2 & \dots & 0 \\ 0 & a_2^{\top} & b_3 & \dots & 0 \\ \dots & \dots & \dots & \dots & \dots \\ \tau a_p & 0 & \dots & a_{p-1}^{\top} & b_p \end{pmatrix},$$

$$(5.3) \qquad \sigma(\mathcal{J}) = \sigma_{ac}(\mathcal{J}) = \bigcup_{x\in[0,2\pi)} \sigma(L(e^{ix})) = \bigcup_{x\in[0,2\pi)} \bigcup_{n=1}^{pm} \{\lambda_n(e^{ix})\}.$$

Proof. The proof repeats standard arguments from [**RS**] (see page 290). □

LEMMA 5.2. *For all $(\tau, z) \in \mathbb{C}^2$, $\tau \neq 0$ the following identities hold true:*

$$(5.4) \qquad \det(L(\tau) - zI_{pm}) = \frac{D(z,\tau)}{c\tau^m} = (-1)^{pm} \prod_{n=1}^{pm} (z - \lambda_n(\tau)).$$

Proof. We fix $\tau \neq 0$. Let $L(\tau)f = \lambda(\tau)f$ for some eigenvalue $\lambda(\tau)$ and some eigenfunction $f = (f_n)_{n=1}^p \in \mathbb{C}^{pm}$. If $f_0 = \tau^{-1} f_p$, then the definition of the matrix $\mathcal{M}_p$ gives $\mathcal{M}_p(\lambda(\tau))(f_0, f_1)^{\top} = \tau(f_0, f_1)^{\top}$. Thus τ is a multiplier of $\mathcal{M}_p(\lambda(\tau))$ and $\lambda(\tau)$ is a zero of $D(\cdot, \tau)$.

First, let all eigenvalues $\lambda_n(\tau), n \in \mathbb{N}_{mp}$ of $L(\tau)$ be distinct for some $\tau \in \mathbb{C}$. Then $\lambda_n(\tau), n \in \mathbb{N}_{mp}$ are zeros of $D(\cdot, \tau)$, which yields (5.4) for all $z \in \mathbb{C}$, since the orders of the polynomials $\det(L(\tau) - zI_{mp})$ and $D(z, \tau)$ coincide.

Second, consider the general case. Let $r_n = n, n \in \mathbb{N}_{mp}$ and let $r_{n+pm} = r_n$ for all $n \in \mathbb{Z}$. Define the Jacobi operator $\mathcal{J}_t = \mathcal{J} + t \operatorname{diag}(r_n)_{n\in\mathbb{Z}}, t \in \mathbb{R}$ and the corresponding matrix $L_t(\tau) = L(\tau) + t \operatorname{diag}(r_n)_{n=1}^{pm}$ (given by (5.2)) and let $D_t(z, \tau)$ be the corresponding determinant. Then all eigenvalues $\lambda_{t,n}(\tau), n \in \mathbb{N}_{mp}$ of $L_t(\tau)$ are distinct for any $t \to \infty$, since $\lambda_{t,n}(\tau) = tr_n + o(t)$ as $t \to \infty$. Thus we obtain

$$(5.5) \qquad \det(L_t(\tau) - zI_{pm}) = \frac{D_t(z,\tau)}{c\tau^m} = (-1)^{pm} \prod_{n=1}^{pm} (z - \lambda_{t,n}(\tau)),$$

for all $z \in \mathbb{C}$ and all large t. The functions in (5.5) are polynomials in t. Then the identities (5.5) hold true for all $t \in \mathbb{R}$, which yields (5.4), at $t = 0$. □

Acknowledgments

Evgeny Korotyaev was partly supported by DFG project BR691/23-1. Parts of this paper were written at the Institute für Mathematik of Humboldt Universität, Berlin; Anton Kutsenko is grateful to the Institute for their hospitality.

References

[AS] Abramowitz, M.; Stegun, A., eds. Handbook of Mathematical Functions. Dover Publications Inc., New York, 1992.

[BBK] Badanin, A.; Brüning, J.; Korotyaev, E. The Lyapunov function for Schrödinger operators with a periodic 2×2 matrix potential, J. Funct. Anal. 234 (2006), 106–126.

[BGGK] Bättig, D.; Grebert, B.; Guillot, J.-C.; Kappeler, T. Fibration of the phase space of the periodic Toda lattice. J. Math. Pures Appl. (9) 72 (1993), no. 6, 553–565.

[BCK] Brüning, J.; Chelkak, D.; Korotyaev, E. Inverse spectral analysis for finite matrix-valued Jacobi operators, preprint 2006.

[BKK] Brüning, J.; Korotyaev, E.; Kutsenko, A. Armchair nanotubes in magnetic field, discrete models, in preparation.

[CK] Chelkak, D.; Korotyaev, E. Spectral estimates for Schrödinger operators with periodic matrix potentials on the real line, Int. Math. Res. Not. 2006 (2006), 1–41.

[CG] Clark, S.; Gesztesy, F. On Weyl-Titchmarsh theory for singular finite difference Hamiltonian systems, J. Comput. Appl. Math., 171 (2004) 151–184.

[CGR] Clark, S.; Gesztesy, F.; Renger, W. Trace formulas and Borg-type theorems for matrix-valued Jacobi and Dirac finite difference operators. J. Diff. Eq. 219 (2005), 144–182.

[DS] Dunford, N.; Schwartz, J. Linear Operators, Part II: Spectral Theory, Interscience, New York, 1988.

[Fo] Forster, O. Lectures on Riemann surfaces. Graduate Texts in Mathematics, 81. Springer-Verlag, New York, 1991.

[GKM] Gesztesy, F.; Kiselev, A.; Makarov, K.,A. Uniqueness results for matrix-valued Schrödinger, Jacobi, and Dirac-type operators, Math. Nachr. 239-240 (2002) 103–145.

[K] Korotyaev, E. Gap-length mapping for periodic Jacobi matrices, Russ. J. Math. Phys. 13 (2006), no.1, 64–69.

[KKr] Korotyaev, E.; Krasovsky, I. Spectral estimates for periodic Jacobi matrices, Commun. Math. Phys. 234 (2003), 517–532.

[KKu] Korotyaev, E.; Kutsenko, A. Inverse problem for the discrete 1D Schrödinger operator with small periodic potentials, Commun. Math. Phys. 261 (2006), 673–692.

[KKu1] Korotyaev, E.; Kutsenko, A. Marchenko-Ostrovski mappings for periodic Jacobi matrices, Russ. J. Math. Phys., 14 (2007), no. 4, 448–452.

[KKu2] Korotyaev, E.; Kutsenko, A. Borg type uniqueness theorems for periodic Jacobi operators with matrix valued coefficients. Preprint 2007.

[Ku] Kutsenko, A. Estimates of parameters for conformal mappings related to a periodic Jacobi matrix. Journal of Mathematical Sciences, 134 (2006), no. 4, 2295–2304.

[MO] Marchenko V.; Ostrovski I. A characterization of the spectrum of the Hill operator. Math. USSR Sb. 26 (1975), 493–554.

[P] Perkolab, L. An inverse problem for a periodic Jacobi matrix. (Russian) Teor. Funktsiĭ Funktsional. Anal. i Prilozhen. 42 (1984), 107–121.

[RS] Reed, M.; Simon, B. Methods of Modern Mathematical Physics, Vol. IV, Analysis of operators, Academic Press, New York, 1978.

[T] Teschl, G. Jacobi Operators and Completely Integrable Nonlinear Lattices, Mathematical Surveys and Monographs, vol. 72, Amer. Math. Soc., Providence, RI, 2000.

[vM] van Moerbeke, P. The spectrum of Jacobi matrices. Invent. Math. 37 (1976), no. 1, 45–81.

Institut für Mathematik, Humboldt Universität zu Berlin, Rudower Chaussee 25, 12489, Berlin, Germany

Department of Mathematics, Saint Petersburg State University, St. Petersburg, Russia

Amer. Math. Soc. Transl.
(2) Vol. **225**, 2008

Hardy Inequalities for Simply Connected Planar Domains

Ari Laptev and Alexander V. Sobolev

To Mikhail Shlëmovich Birman on his 80th birthday

ABSTRACT. In 1986, A. Ancona showed, using the Koebe one-quarter theorem, that for a simply connected planar domain the constant in the Hardy inequality with the distance to the boundary is greater than or equal to 1/16. In this paper we consider classes of domains for which there is a stronger version of the Koebe theorem. This implies better estimates for the constant appearing in the Hardy inequality.

1. Main result and discussion

Let Ω be a domain in $\mathbb{R}^2$ and let $\Omega^c = \mathbb{R}^2 \setminus \Omega$ be its complement. It is known that there exists a number $r = r(\Omega) > 0$ such that for any function $u \in \mathsf{C}_0^1(\Omega)$ we have

$$\int_\Omega |\nabla u|^2 d\mathbf{x} \geq r^2 \int_\Omega \frac{|u|^2}{\delta(\mathbf{x})^2} d\mathbf{x}, \quad \delta(\mathbf{x}) = \inf_{\mathbf{y}\in\Omega^c} |\mathbf{y} - \mathbf{x}|; \tag{1.1}$$

see, e.g., E.B. Davies [**4**], [**5**], [**6**] and V.G. Maz′ya [**10**]. For convex domains, $r = 1/2$ and it is sharp; see e.g. [**4**]. However, the sharp constant for nonconvex domains is unknown, although for arbitrary planar simply connected domains, A. Ancona [**1**] proved (1.1) with $r = 1/4$. Some specific examples of nonconvex domains were considered in [**6**] (see also J. Tidblom [**12**]). For example, it was found that if $\Omega = \mathbb{R}^2 \setminus \mathbb{R}_+$, $\mathbb{R}_+ = [0, \infty)$, then $r^2 = 0.20538...$.

Our objective is to obtain the Hardy inequality for simply connected nonconvex domains $\Omega \subset \mathbb{R}^2$, whose degree of nonconvexity can be “quantified”. We introduce two possible “measures” of nonconvexity.

Let $\Lambda \subset \mathbb{C}$ be a simply connected domain such that $0 \subset \partial\Lambda$. Denote by $\Lambda(w, \phi) = e^{i\phi}\Lambda + w$ the transformation of Λ by rotation by an angle $\phi \in (-\pi, \pi]$ in the positive direction and translation by $w \in \mathbb{C}$:

$$\Lambda(w, \phi) = \{z \in \mathbb{C} : e^{-i\phi}(z - w) \in \Lambda\}. \tag{1.2}$$

Denote by $K_\theta \subset \mathbb{C}$, $\theta \in [0, \pi]$ the sector

$$K_\theta = \{z \in \mathbb{C} : |\arg z| < \theta\}. \tag{1.3}$$

2000 *Mathematics Subject Classification.* Primary 26D15, 30C45.
Key words and phrases. Hardy inequality, Koebe theorem, conformal mappings.

In words, this is an open sector symmetric with respect to the real axis, with the angle 2θ at the vertex. Here and below we always assume that $\arg \zeta \in (-\pi, \pi]$ for all $\zeta \in \mathbb{C}$. Our first assumption on the domain Ω is the following

CONDITION 1.1. *There exists a number $\theta \in [0, \pi]$ such that for each $w \in \Omega^c$ one can find a $\phi = \phi_w \in (-\pi, \pi]$ such that*

$$\Omega \subset K_\theta(w, \phi_w).$$

Very loosely speaking, this means that the domain Ω satisfies the exterior cone condition. The difference is of course that the cone is now supposed to be infinite. Because of this, Condition 1.1 is equivalent to itself if stated for the boundary points $w \in \partial\Omega$ only.

Note also that if Condition 1.1 is satisfied for some θ, then automatically $\theta \geq \pi/2$, and the equality $\theta = \pi/2$ holds for convex domains.

THEOREM 1.2. *Suppose that the domain $\Omega \subset \mathbb{R}^2, \Omega \neq \mathbb{R}^2$ satisfies Condition 1.1 with some $\theta \in [\pi/2, \pi]$. Then for any $u \in \mathsf{C}_0^1(\Omega)$ the Hardy inequality (1.1) holds with*

$$r = \frac{\pi}{4\theta}. \tag{1.4}$$

It is clear that the constant r runs from $1/4$ to $1/2$ when θ varies from π to $\pi/2$. For the domain $\Omega = K_\theta$, Theorem 1.2 does not give the best-known result, found in [**6**], saying that the value of r remains equal to $1/2$ for the range $\theta \in [0, \theta_0]$ where $\theta_0 \approx 2.428$, which is considerably greater than $\pi/2$.

To describe another way to characterize the nonconvexity, for $a > 0$ and $\theta \in [0, \pi)$, introduce the domains

$$\tilde{D}_a = \{z \in \mathbb{C} : |z| > a \ \& \ |\arg z| \neq \pi\}, \quad D_{a,\theta} = \tilde{D}_a(-ae^{i\theta}, 0). \tag{1.5}$$

The domain $\tilde{D}_a$ is the exterior of the disk of radius a centered at the origin with an infinite cut along the negative real semi-axis.

CONDITION 1.3. *There exist numbers $a > 0$ and $\theta_0 \in [0, \pi)$ such that for any $w \in \partial\Omega$ one can find a $\phi = \phi_w \in (-\pi, \pi]$ and $\theta \in [-\theta_0, \theta_0]$ such that*

$$\Omega \subset D_{a,\theta}(w, \phi_w).$$

Note that any domain satisfying Conditions 1.1 or 1.3 is automatically simply connected.

The following theorem applies to the domains with a finite in-radius

$$\delta_{\text{in}} = \sup_{z \in \Omega} \delta(z).$$

THEOREM 1.4. *Let $\Omega \subset \mathbb{R}^2$, $\Omega \neq \mathbb{R}^2$ be a domain such that $\delta_{\text{in}} < \infty$. Suppose that Ω satisfies Condition 1.3 with some $\theta_0 \in [0, \pi)$ and that*

$$2\delta_{\text{in}} \leq R_0(a), \quad R_0(a) = \frac{a}{2(2^{1/2}|\tan(\theta_0/2)| + 1)}. \tag{1.6}$$

Then the Hardy inequality (1.1) holds with

$$r = \frac{1}{2}\left[1 - \frac{2\delta_{\text{in}}}{R_0(a)}\right]. \tag{1.7}$$

A natural example of a domain in which to apply the above theorem is the following horseshoe-shaped domain:

$$\Lambda = \{z \in \mathbb{C} : \rho < |z| < \rho + \delta,\ |\arg z| < \psi,\ \operatorname{Re} z > \rho\cos\psi\}, \quad \psi \in (0, \pi),$$

with $\rho, \delta > 0$. Simple geometric considerations show that this domain satisfies Condition 1.3 with $a = \rho$ and

$$\theta_0 = \begin{cases} 0, & \psi \le \pi/2, \\ 2\psi - \pi, & \psi > \pi/2. \end{cases}$$

Assuming that $\delta\rho^{-1}$ is small, so that $\delta_{\rm in} = \delta$, we deduce from Theorem 1.4 that the Hardy inequality holds with a constant r, which gets close to $1/2$ as $\delta\rho^{-1} \to 0$. On the other hand, if $\delta_{\rm in}\rho^{-1}$ is large, one could apply Theorem 1.2, noticing that Λ satisfies Condition 1.1 with $\theta = (\pi + \psi)/2$, which gives the Hardy inequality with constant

$$r = \frac{1}{2\big(1 + \psi\pi^{-1}\big)},$$

which is obviously independent of $\delta_{\rm in}$ or ρ.

Let us mention briefly some other recent results for convex domains, concerning the Hardy inequality with a remainder term. In the paper [**3**], H. Brezis and M. Marcus showed that if $\Omega \in \mathbb{R}^d$, $d \ge 2$, then the inequality could be improved to include the L^2-norm:

$$\int_\Omega |\nabla u|^2 d\mathbf{x} \ge \frac{1}{4}\int_\Omega \frac{|u|^2}{\delta(\mathbf{x})^2} + C(\Omega)\int_\Omega |u|^2 d\mathbf{x}, \tag{1.8}$$

where the constant $C(\Omega) > 0$ depends on the diameter of Ω. They also conjectured that $C(\Omega)$ should depend on the Lebesgue measure of Ω. This conjecture was justified in [**9**] and later generalised to L^p-type inequalities in [**11**]. Later S. Filippas, V.G. Maz′ya and A. Tertikas [**8**] (see also F.G. Avkhadiev [**2**]) obtained for $C(\Omega)$ an estimate in terms of the in-radius $\delta_{\rm in}$.

2. A version of the Koebe theorem

A. Ancona has pointed out in [**1**] (page 278) that the Hardy inequality for simply connected planar domains can be obtained from the famous Koebe one-quarter theorem. Let f be a conformal mapping (i.e. analytic univalent) defined on the unit disk $\mathbb{D} = \{z \in \mathbb{C} : |z| < 1\}$, normalized by the condition $f(0) = 0$, $f'(0) = 1$. Denote by Ω the image of the disk under the function f, i.e. $\Omega = f(\mathbb{D})$, and set

$$\delta(\zeta) = \operatorname{dist}\{\zeta, \partial\Omega\} = \inf_{w \notin \Omega} |w - \zeta|$$

to be the distance from the point $\zeta \in \Omega$ to the boundary $\partial\Omega$. The classical Koebe one-quarter theorem tells us that

$$\delta(0) \ge r,$$

with $r = 1/4$. On the other hand, if the domain Ω is convex, then it is known that $r = 1/2$; see, e.g., P.L. Duren [**7**], Theorem 2.15. Without the normalization conditions $f(0) = 0$, $f'(0) = 1$ the above estimate can be rewritten as follows:

$$\delta(f(0)) \ge r|f'(0)|. \tag{2.1}$$

For any simply connected domain $\Omega \subset \mathbb{C}$, $\Omega \ne \mathbb{C}$ we denote by $\mathbb{A}(\Omega)$ the class of all conformal maps such that $f(\mathbb{D}) = \Omega$.

Our proof of the main Theorems 1.2 and 1.4 relies on a version of the Koebe theorem, in which the constant r assumes values in the interval $[1/4, 1/2]$. We begin with a general statement which deduces the required Koebe-type result by comparing the domain Ω with some suitable "reference" domain. Let $\Lambda \subset \mathbb{C}, \Lambda \neq \mathbb{C}$ be a simply connected domain such that $0 \subset \partial\Lambda$, and let g be a conformal function which maps Λ onto the complex plane with a cut along the negative semi-axis, i.e. onto

$$\Pi = \mathbb{C} \setminus \{z \in \mathbb{C} : \operatorname{Im} z = 0, \operatorname{Re} z \le 0\},$$

such that $g(0) = 0$. We call Λ a *standard domain* and g a *conformal map associated* with the standard domain Λ.

LEMMA 2.1. *Let $w \in \partial\Omega$ and suppose that for some standard domain Λ the inclusion*

$$\Omega \subset \Lambda(w, \phi) \tag{2.2}$$

holds with some $\phi \in (-\pi, \pi]$. Let g be a conformal map associated with Λ.

(i) *Suppose that there are numbers $M > 0$ and $R_0 > 0$ such that for all $R \in (0, R_0]$,*

$$\left|\frac{g'(z)}{g(z)}\right| \ge \frac{\beta}{|z|} \tag{2.3}$$

for $z : 0 < |z| \le R, z \in \Lambda$ with some $\beta = \beta(R) \in (0, M]$. Then for any $f \in \mathbb{A}(\Omega)$ satisfying the condition $M|f'(0)| \le 4R_0$, the inequality

$$|f(0) - w| \ge \frac{\beta(R_1)}{4}|f'(0)|, \quad R_1 = \frac{M|f'(0)|}{4}, \tag{2.4}$$

holds.

(ii) *Suppose that the bound (2.3) holds for all $z \in \Lambda$ with some positive function $\beta = \beta(z)$. Then the inequality*

$$|f(0) - w| \ge \frac{1}{4} \inf_{z\in\Lambda} \beta(z)|f'(0)| \tag{2.5}$$

holds.

PROOF. Since $\Omega \subset \Lambda(w, \phi)$, the function

$$h(z) = g\big(e^{-i\phi}(f(z) - w)\big)$$

is conformal on $\mathbb{D}$, with values in Π. Since $0 \notin \Pi$, by the classical Koebe theorem,

$$|h(0)| \ge \frac{1}{4}|h'(0)|,$$

so that

$$|g\big(e^{-i\phi}(f(0) - w)\big)| \ge \frac{1}{4}|g'\big(e^{-i\phi}(f(0) - w)\big)||f'(0)|. \tag{2.6}$$

Suppose that the conditions of part (i) are fulfilled. If $|f(0) - w| \ge M|f'(0)|/4$, then there is nothing to prove, so we assume that $|f(0) - w| \le M|f'(0)|/4$. Then by the assumption (2.3) we get

$$\frac{1}{\beta\big(M|f'(0)|/4\big)}|f(0) - w| \ge \frac{1}{4}|f'(0)|,$$

which leads to the required estimate (2.4).

If the conditions of part (ii) are fulfilled, then (2.5) follows directly from (2.6) and (2.3). □

COROLLARY 2.2. *Let $\Omega \subset \mathbb{C}$ be a domain. Suppose that for any $w \in \partial\Omega$ there is a standard domain $\Lambda = \Lambda_w$, such that the inclusion (2.2) holds with some $\phi = \phi_w \in (-\pi, \pi]$. Let $g = g_w$ be the conformal map associated with Λ_w.*

(i) *Suppose that g_w satisfies (2.3) for all $0 < |z| \le R, z \in \Lambda_w$, for all $R \in (0, R_0]$ with some $\beta(R) = \beta_w(R) \in (0, M]$, where $M > 0$ and $R_0 > 0$ are independent of w. Then under the condition $M\delta_{\text{in}} \le R_0$ the estimate (2.1) holds for all $f \in \mathbb{A}(\Omega)$ with*

$$r = \frac{1}{4} \inf_{w\in\partial\Omega} \beta_w(R'), \ R' = M\delta_{\text{in}}.$$

(ii) *Suppose that g_w satisfies (2.3) for all $z \in \Lambda_w$ with some $\beta(z) = \beta_w(z) > 0$. Then the estimate (2.1) holds for all $f \in \mathbb{A}(\Omega)$ with*

$$r = \frac{1}{4} \inf_{w\in\partial\Omega} \inf_{z\in\Lambda_w} \beta_w(z).$$

Observe that under the conditions of this corollary, the domain Ω is automatically simply connected and $\Omega \neq \mathbb{C}$.

PROOF. (i) By the classical Koebe theorem, $|f'(0)| \le 4\delta(f(0)) \le 4\delta_{\text{in}}$, so that under the condition $M\delta_{\text{in}} \le R_0$, Lemma 2.1(i) entails the estimate (2.4) for each $w \in \partial\Omega$. Since $R_1 \le R'$ and $\beta_w(\,\cdot\,)$ is a decreasing function, the required result follows.

Part (ii) immediately follows from Lemma 2.1(ii). □

Now we apply the above results in the cases of standard domains K_θ and $D_{a,\theta}$; see (1.3) and (1.5) for definitions.

THEOREM 2.3. *Suppose that Ω satisfies Condition 1.1 with some $\theta \in [\pi/2, \pi]$. Then for any $f \in \mathbb{A}(\Omega)$ the inequality (2.1) holds with r given by (1.4).*

PROOF. Due to Condition 1.3, for each $w \in \partial\Omega$ we have $\Omega \subset K_\theta(w, \phi)$ with some $\phi \in (-\pi, \pi]$. Clearly, the domain K_θ is standard and the function

$$g(z) = z^\alpha, \ \alpha = \frac{\pi}{\theta}$$

is a conformal map associated with K_θ. One immediately obtains:

$$\frac{g'(z)}{g(z)} = \alpha\frac{z^{\alpha-1}}{z^\alpha} = \frac{\alpha}{z}, \, z \in \Lambda,$$

so that the conditions of Corollary 2.2(ii) hold with the constant $\beta = \alpha$. Now Corollary 2.2(ii) leads to the proclaimed result. □

Note that for convex domains the angle θ is $\pi/2$, and hence we recover the known result $r = 1/2$. Actually, the proof of Lemma 2.1 is modelled on that for convex domains, which is featured in [**7**], Theorem 2.15.

Let us prove a similar result for Condition 1.3.

THEOREM 2.4. *Suppose that Ω satisfies Condition 1.3 with some $a > 0$, $\theta_0 \in [0, \pi)$, that $\delta_{\text{in}} < \infty$, and that the condition (1.6) is fulfilled. Then for any $f \in \mathbb{A}(\Omega)$, the inequality (2.1) holds with r given by (1.7).*

PROOF. Due to Condition 1.3, for each $w \in \partial\Omega$ we have $\Omega \subset D_{a,\theta}(w, \phi)$ with some $\theta \in [0, \theta_0]$ and $\phi \in (-\pi, \pi]$. Clearly, the domain $D_{a,\theta}$ is standard and the function

$$g(z) = h^2(za^{-1}), \quad h(\zeta) = \sqrt{\zeta + e^{i\theta}} - \frac{1}{\sqrt{\zeta + e^{i\theta}}} - 2ib, \quad b = \sin\frac{\theta}{2},$$

is a conformal map associated with $D_{a,\theta}$. We write

$$h(\zeta) = \frac{\zeta + e^{i\theta} - 1 - 2ib\sqrt{\zeta + e^{i\theta}}}{\sqrt{\zeta + e^{i\theta}}} = \frac{\zeta + 2ibe^{i\theta/2} - 2ib\sqrt{\zeta + e^{i\theta}}}{\sqrt{\zeta + e^{i\theta}}} = \frac{\psi(\zeta)}{\sqrt{\zeta + e^{i\theta}}}.$$

A direct calculation shows that

$$\frac{g'(z)}{g(z)} = \frac{2\psi'(z/a)}{a\psi(z/a)} - \frac{1}{z + ae^{i\theta}}.$$

Let us investigate the function ψ in more detail. Assume that $|\zeta| \le 1/2$. Rewrite

$$\begin{aligned}\psi(\zeta) &= \zeta + 2ibe^{i\theta/2} - 2ibe^{i\theta/2}\sqrt{1 + \zeta e^{-i\theta}} \\ &= \zeta + 2ibe^{i\theta/2} - 2ibe^{i\theta/2}\left(1 + \frac{1}{2}\zeta e^{-i\theta} + \zeta\gamma_1\right) \\ &= (1 - ibe^{-i\theta/2})\zeta - 2ibe^{i\theta/2}\zeta\gamma_1 = e^{-i\theta/2}\cos\frac{\theta}{2}\,\zeta - 2ibe^{i\theta/2}\zeta\gamma_1,\end{aligned}$$

where

$$|\gamma_1(\zeta)| \le 2^{-3/2}|\zeta|, \ |\zeta| \le 1/2.$$

Let us look at the derivative:

$$\begin{aligned}\psi'(\zeta) &= 1 - \frac{ib}{\sqrt{\zeta + e^{i\theta}}} = 1 - e^{-i\theta/2}\frac{ib}{\sqrt{1 + \zeta e^{-i\theta}}} \\ &= 1 - ibe^{-i\theta/2}(1 + \gamma_2) = e^{-i\theta/2}\cos\frac{\theta}{2} - ibe^{-i\theta/2}\gamma_2,\end{aligned}$$

where

$$|\gamma_2(\zeta)| \le 2^{1/2}|\zeta|, \ |\zeta| \le 1/2.$$

Therefore

$$\begin{aligned}\frac{g'(z)}{g(z)} &= \frac{2e^{-i\theta/2}\cos\frac{\theta}{2} - 2ibe^{-i\theta/2}\gamma_2(z/a)}{e^{-i\theta/2}\cos\frac{\theta}{2}\ z - 2ibe^{i\theta/2}z\gamma_1(z/a)} - \frac{1}{z + ae^{i\theta}} \\ &= \frac{1}{z}\left[\frac{2 - 2i\tan(\theta/2)\gamma_2(z/a)}{1 - 2i\tan(\theta/2)e^{i\theta}\gamma_1(z/a)} - \frac{z}{a}\frac{1}{z/a + e^{i\theta}}\right].\end{aligned}$$

Denote $\zeta = z/a, \ t = |\tan(\theta/2)|$. Using the above bounds on γ_1, γ_2, we obtain for $|\zeta| \le 1/2$ the inequality

$$\left|\frac{g'(z)}{g(z)}\right| \ge \frac{2}{|z|}\left[\frac{1 - t|\gamma_2(\zeta)|}{1 + 2t|\gamma_1(\zeta)|} - |\zeta|\right] \ge \frac{2}{|z|}\left[\frac{1 - 2^{1/2}t|\zeta|}{1 + 2^{-1/2}t|\zeta|} - |\zeta|\right].$$

Remembering the simple inequality

$$\frac{1-a}{1+b} \ge 1 - a - b,$$

satisfied for all nonnegative a and b, from here we deduce that

$$\left|\frac{g'(z)}{g(z)}\right| \ge \frac{2}{|z|}\left[1 - \left(2^{3/2}|\tan(\theta/2)| + 1\right)\frac{|z|}{a}\right], \ |z|/a \le 1/2.$$

Therefore the condition (2.3) is satisfied for all $0 < |z| \le R$, where

$$R \le \min\left\{\frac{a}{2}, \frac{a}{2^{3/2}|\tan(\theta/2)| + 1}\right\}.$$

In any case, (2.3) is satisfied for

$$R \le R_0(a, \theta) = \frac{a}{2^{3/2}|\tan(\theta/2)| + 2},$$

with

$$\beta(R) = \beta(R, \theta) = 2\left[1 - \frac{R}{R_0(a, \theta)}\right].$$

Note that $\beta \le M$ with $M = 2$ and $\beta(R, \theta) \ge \beta(R, \theta_0)$, so that by Corollary 2.2(i) the estimate (2.1) holds with the r given by (1.7). □

3. Proof of Theorems 1.2, 1.4

As soon as the Koebe theorem (2.1) is established, our proof of the Hardy inequality follows that by A. Ancona [**1**]. Namely, our starting point is the inequality for the half-plane, which is an immediate consequence of the classical Hardy inequality in one dimension. Below we use the usual notation $z = x + iy$, $x, y \in \mathbb{R}$.

PROPOSITION 3.1. *Let* $\mathbb{C}_+ = \{z \in \mathbb{C} : \operatorname{Re} z > 0\}$. *For any* $u \in \mathsf{C}_0^1(\mathbb{C}_+)$ *one has*

$$\int_{\mathbb{C}_+} |\nabla u|^2 dxdy \ge \frac{1}{4}\int_{\mathbb{C}_+} \frac{|u|^2}{x^2} dxdy.$$

Theorems 1.2 and 1.4 immediately follow from Proposition 3.1 with the help of the following conditional result:

THEOREM 3.2. *Let* $\Omega \subset \mathbb{C}$, $\Omega \ne \mathbb{C}$ *be a simply connected domain. Suppose that for all* $g \in \mathbb{A}(\Omega)$ *the inequality* (2.1) *holds with some* $r \in [1/4, 1/2]$. *Then for any conformal mapping* $f : \mathbb{C}_+ \to \Omega$ *the following version of the Koebe theorem holds:*

$$\delta(f(z)) \ge 2rx|f'(z)|. \tag{3.1}$$

For $r = 1/4$ the estimate (3.1) can be found in [**1**]. For the reader's convenience we provide a proof of (3.1).

PROOF. For a conformal mapping $f : \mathbb{C}_+ \to \Omega$ and arbitrary $z \in \mathbb{C}_+$ we define

$$g(w) = g_z(w) = f\big(h(w)\big),\ h(w) = \frac{\overline{z}w + z}{1 - w},$$

where $w \in \mathbb{D}$. It is clear that for each fixed $z \in \mathbb{C}_+$ the function h maps $\mathbb{D}$ onto $\mathbb{C}_+$ and $h(0) = z$, $g(0) = f(z)$. The derivative is

$$g'(w) = \frac{z + \overline{z}}{(1 - w)^2} f'\big(h(w)\big),$$

so that

$$g'(0) = 2xf'(z).$$

Therefore, the Koebe theorem (2.1) implies that

$$\delta\big(f(z)\big) = \delta\big(g(0)\big) \ge r|g'(0)| = 2rx|f'(z)|,$$

as required. □

PROOF OF THEOREMS 1.2, 1.4. According to Theorems 2.3 or 2.4, the Koebe theorem for functions $f \in \mathbb{A}(\Omega)$ holds with the values of r given by (1.4) or (1.7) respectively.

Let $f : \mathbb{C}_+ \to \Omega$ be a conformal map. Remembering that conformal maps preserve the Dirichlet integral, from Proposition 3.1 we get for any $u \in \mathsf{C}_0^1(\Omega)$,

$$\int_\Omega |\nabla u|^2 d\mathbf{x} = \int_{\mathbb{C}_+} |\nabla(u \circ f)|^2 dx dy \geq \frac{1}{4} \int_{\mathbb{C}_+} \frac{|(u \circ f)|^2}{x^2} d\mathbf{x}$$

$$= r^2 \int_{\mathbb{C}_+} \frac{|(u \circ f)|^2}{(2rx)^2 |f'(z)|^2} |f'(z)|^2 d\mathbf{x} \geq r^2 \int_\Omega \frac{|u|^2}{\delta(\mathbf{x})^2} d\mathbf{x}.$$

At the last step we have used (3.1).

Now Theorems 1.2, 1.4 follow. □

Acknowledgements

The authors acknowledge a partial support by the SPECT ESF European programme. A. Laptev would like to thank the University of Birmingham for its hospitality. A. Sobolev was partially supported by the Göran Gustafsson Foundation. Both authors are grateful to M. Levitin for valuable suggestions.

References

1. A. Ancona, *On strong barriers and an inequality of Hardy for domains in* $\mathbb{R}^n$, J. London Math. Soc. (2) **34** (1986), 274–290.
2. F.G. Avkhadiev, *Hardy's inequalities in higher dimensions*, Preprint.
3. H. Brezis and M. Marcus, *Hardy's inequalities revisited*, Dedicated to Ennio De Giorgi. Ann. Scuola Norm. Sup. Pisa Cl. Sci. (4) **25** (1997), no. 1-2 (1998), 217–237.
4. E.B. Davies, *Spectral theory and differential operators*. Cambridge Studies in Advanced Mathematics, **42**. Cambridge University Press, Cambridge, 1995. x+182 pp.
5. E.B. Davies, *A review of Hardy inequalities*, The Maz'ya anniversary collection, Vol. 2, Oper. Theory Adv. Appl. **110** (1999) 55–67, Birkhäuser, Basel.
6. E.B. Davies, *The Hardy constant*, Quart. J. Math. Oxford (2), **46** (1995) 417–431.
7. P.L. Duren, *Univalent functions*, Springer, New York, 1983.
8. S. Filippas, V.G. Maz'ya and A. Tertikas, *On a question of Brezis and Marcus*, Calc. Var. Partial Differential Equations **25** (2006), 491–501.
9. M. Hoffmann-Ostenhof, T. Hoffmann-Ostenhof and A. Laptev, *A geometrical version of Hardy's inequalities*, J. Funct. Anal. **189** (2002), no. 2, 539–548.
10. V.G. Maz'ya, *Sobolev spaces*, Springer-Verlag, 1985.
11. J. Tidblom, *A geometrical version of Hardy's inequality for* $W^{1,p}(\Omega)$, Proc. Amer. Math. Soc. **132** (2004), no. 8, 2265–2271.
12. J. Tidblom, *Improved* L^p *Hardy Inequalities*, Ph.D. thesis, Stockholm University, 2005.

DEPARTMENT OF MATHEMATICS, IMPERIAL COLLEGE LONDON, HUXLEY BUILDING 180, QUEEN'S GATE LONDON SW7 2AZ, UNITED KINGDOM AND DEPARTMENT OF MATHEMATICS, ROYAL INSTITUTE OF TECHNOLOGY, SE-100 44 STOCKHOLM, SWEDEN

E-mail address: a.laptev@imperial.ac.uk, laptev@math.kth.se

DEPARTMENT OF MATHEMATICS, UNIVERSITY COLLEGE LONDON, GOWER STREET, LONDON, WC1E 6BT, UNITED KINGDOM

E-mail address: asobolev@math.ucl.ac.uk

Amer. Math. Soc. Transl.
(2) Vol. **225**, 2008

The Spectral Flow, the Fredholm Index, and the Spectral Shift Function

Alexander Pushnitski

To Mikhail Shlёmovich Birman on his 80th birthday

ABSTRACT. We discuss the well-known "Fredholm index = spectral flow" theorem and show that it can be interpreted as a limited case of an identity involving two spectral shift functions.

1. Introduction

1.1. Background. Let $A(t)$, $t \in \mathbb{R}$, be a family of selfadjoint operators in a separable Hilbert space $\mathcal{H}$ such that the limits

$$A^{\pm} = \lim_{t \to \pm\infty} A(t) \tag{1.1}$$

exist in an appropriate sense. In the Hilbert space $L^2(\mathbb{R}; \mathcal{H})$, consider the operator

$$D_A = \frac{d}{dt} + A(t), \text{ i.e. } (D_A u)(t) = \frac{du(t)}{dt} + A(t)u(t). \tag{1.2}$$

It is well known (see e.g. [**4, 8**] and references to earlier work therein) that, under the appropriate assumptions on $A(t)$, the Fredholm index of the operator D_A equals the spectral flow of the family $\{A(t)\}_{t \in \mathbb{R}}$ through zero. The spectral flow through zero should be understood as the number of eigenvalues of $A(t)$ (counting multiplicities) that cross zero from left to right minus the number of eigenvalues of $A(t)$ that cross zero from right to left as t grows from $-\infty$ to $+\infty$. The "Fredholm index = spectral flow through zero" theorem is one of a large family of index theorems; see e.g. [**4**] for a discussion.

The "Fredholm index = spectral flow through zero" theorem is usually considered under the assumption that the spectra of the operators $A(t)$ are discrete, at least on some interval containing zero. The purpose of this note is to show that this assumption can be lifted at the expense of the trace class assumption

$$\int_{-\infty}^{\infty} \|A'(t)\|_{S_1} \, dt < \infty, \quad A'(t) \equiv \frac{dA(t)}{dt}, \tag{1.3}$$

2000 *Mathematics Subject Classification.* Primary 47A53; Secondary 47A55.

where $\|\cdot\|_{S_1}$ is the trace norm in $\mathcal{H}$. Assumption (1.3) ensures that the difference $A^+ - A^-$ is a trace class operator, which allows one to use the notion of *M. G. Krein's spectral shift function* for the pair A^+, A^-. The point we would like to make is that the "Fredholm index = spectral flow" theorem can be understood as a particular limiting case of a fairly general identity (see (1.11) below) involving two spectral shift functions. This identity might be interesting in its own right.

It is a pleasure to dedicate this note to M. Sh. Birman, from whom I have learned (among many other useful things) to think of the spectral shift function whenever two selfadjoint operators are involved.

1.2. Notation. We denote by S_1 and S_2 the trace class and the Hilbert-Schmidt class, with the norms $\|\cdot\|_{S_1}$ and $\|\cdot\|_{S_2}$. For a selfadjoint operator A and an interval $\delta \subset \mathbb{R}$, we denote by $E_A(\delta)$ the spectral projection of A corresponding to δ. We denote by $N_A(\delta) = \operatorname{Tr} E_A(\delta)$ the total number of eigenvalues (counting multiplicities) of A in the interval δ. For selfadjoint semi-bounded from below operators A and B, the inequality $A \leq B$ is understood in the quadratic form sense, i.e. for all sufficiently large $a > 0$, $\operatorname{Dom}((A+a)^{1/2}) \supset \operatorname{Dom}((B+a)^{1/2})$ and for all $f \in \operatorname{Dom}((B+a)^{1/2})$, $\|(A+a)^{1/2} f\| \leq \|(B+a)^{1/2} f\|$.

1.3. The spectral shift function. Here we recall the necessary facts from the spectral shift function theory. See the original paper [**5**], or the survey [**2**] or the book [**9**], for the details.

Let H and $\widetilde{H}$ be selfadjoint operators in a Hilbert space. The simplest situation in which the spectral shift function can be defined is when the difference $\widetilde{H} - H$ belongs to the trace class S_1. Then there exists a unique function $\xi(\cdot; \widetilde{H}, H) \in L^1(\mathbb{R})$ such that the Lifshits–Kreĭn trace formula

$$\operatorname{Tr}(f(\widetilde{H}) - f(H)) = \int_{-\infty}^{\infty} \xi(\lambda; \widetilde{H}, H) f'(\lambda) d\lambda \tag{1.4}$$

holds true for every $f \in C_0^\infty(\mathbb{R})$. The function $\xi(\cdot; \widetilde{H}, H)$ is called the spectral shift function for the pair $\widetilde{H}$, H. In fact, the class of admissible functions f in (1.4) is much wider than $C_0^\infty(\mathbb{R})$; see [**2, 9**] for the details and references to the literature.

The assumption $\widetilde{H} - H \in S_1$ is very restrictive in applications. Suppose instead that $\widetilde{H}$ and H are nonnegative (in the quadratic form sense) selfadjoint operators such that

$$(\widetilde{H} - z)^{-1} - (H - z)^{-1} \in S_1 \tag{1.5}$$

for some (and hence for all) $z \in \mathbb{C} \setminus [0, \infty)$. Then there exists a unique function $\xi(\cdot; \widetilde{H}, H) \in L^1(\mathbb{R}, (1 + \lambda^2)^{-1} d\lambda)$, $\operatorname{supp} \xi \in [0, \infty)$ such that the trace formula (1.4) holds true for all $f \in C_0^\infty(\mathbb{R})$.

Next, assuming that either $\widetilde{H} - H \in S_1$ or (1.5) holds true, suppose that for some (possibly semi-infinite) open interval $\Delta \subset \mathbb{R}$ we have $\sigma_{ess}(H) \cap \Delta = \emptyset$. By Weyl's theorem on the invariance of the essential spectrum with respect to compact perturbations, we also have $\sigma_{ess}(\widetilde{H}) \cap \Delta = \emptyset$. Then, as is not difficult to see from (1.4), for any $a, b \in \Delta \setminus (\sigma(H) \cup \sigma(\widetilde{H}))$, $a < b$, we have

$$\xi(b; \widetilde{H}, H) - \xi(a, \widetilde{H}, H) = N_H(a, b) - N_{\widetilde{H}}(a, b), \tag{1.6}$$

where $N_H(a, b)$ is the number of eigenvalues (counting multiplicities) of H in (a, b). Formula (1.6) remains true in the case $a = -\infty$ (or $b = \infty$).

1.4. The spectral shift function and the spectral flow. Let $H(\alpha)$, $\alpha \in [0,1]$, be a family of selfadjoint operators such that the operators $H(\alpha) - H(0)$ belong to the trace class and depend continuously on α in the trace norm. Then the function $\xi(\cdot; H(\alpha), H(0))$ is well defined and continuous in α as an element of $L^1(\mathbb{R})$.

As noted above, by Weyl's theorem, $\sigma_{ess}(H(\alpha))$ is independent of $\alpha \in [0,1]$. Suppose that $\sigma_{ess}(H(\alpha)) \cap \Delta = \emptyset$ for some interval $\Delta \subset \mathbb{R}$. Then we claim that for any $\lambda \in \Delta \setminus (\sigma(H(1)) \cup \sigma(H(0)))$, the spectral shift function $\xi(\lambda; H(1), H(0))$ equals the spectral flow of the family $H(\alpha)$ through λ as α grows from 0 to 1:

$$\begin{aligned} &\xi(\lambda; H(1), H(0)) \\ &\quad = \langle \text{the number of eigenvalues of } H(\alpha) \text{ which cross } \lambda \text{ rightwards}\rangle \\ &\quad\quad - \langle \text{the number of eigenvalues of } H(\alpha) \text{ which cross } \lambda \text{ leftwards}\rangle \end{aligned} \tag{1.7}$$

as long as both terms in the r.h.s. are finite.

In order to justify (1.7), first suppose that there exists $\lambda_0 < \lambda$ such that

$$\lambda_0 \in \Delta \setminus \left(\bigcup_{\alpha \in [0,1]} \sigma(H(\alpha)) \right). \tag{1.8}$$

Then it is not difficult to check that $\xi(\lambda_0; H(\alpha), H(0)) = 0$ for all α and so (1.6) (with $a = \lambda_0$, $b = \lambda$) yields

$$\xi(\lambda; H(\alpha), H(0)) = N_{H(0)}(\lambda_0, \lambda) - N_{H(\alpha)}(\lambda_0, \lambda). \tag{1.9}$$

Considering the r.h.s. of (1.9) as a function of α, it is easy to see that

$$\langle \text{r.h.s. of (1.9) with } \alpha = 1 \rangle = \langle \text{r.h.s. of (1.7)} \rangle$$

whenever the r.h.s. of (1.7) is well defined. Thus, (1.7) holds true.

In general, λ_0 as in (1.8) may not exist, but we can always split $[0,1]$ into sufficiently small subintervals δ_i such that for each family $\{H(\alpha) \mid \alpha \in \delta_i\}$, λ_0 can be chosen appropriately. Then formula (1.7) can be obtained by adding up the formulas corresponding to all the subintervals.

1.5. Main result. It will be convenient to write $A(t) = A^- + B(t)$, where A^- is an arbitrary selfadjoint operator in $\mathcal{H}$ and $B(t)$ is a family of selfadjoint trace class operators such that the derivative $B'(t) = \frac{dB(t)}{dt}$ exists in the trace norm and

$$\int_{-\infty}^{\infty} \|B'(t)\|_{S_1} \, dt < \infty. \tag{1.10}$$

This assumption ensures that the limits $B^{\pm} = \lim_{t \to \pm\infty} B(t)$ exist in the trace norm. We assume that $B^- = 0$ (of course, this is merely a normalization condition) and define $A^+ = A^- + B^+$. According to this definition, for all $t \in \mathbb{R}$, the operators $A(t)$ have the same domain $\mathrm{Dom}(A(t)) = \mathrm{Dom}(A^-)$.

Consider the operator D_A (see (1.2)) in the Hilbert space $L^2(\mathbb{R}; \mathcal{H})$ with the domain $\mathrm{Dom}(D_A)$ consisting of all u from the Sobolev space $W_2^1(\mathbb{R}; \mathcal{H})$ such that

$$u(t) \in \mathrm{Dom}(A^-) \text{ for all } t \text{ and } \int_{-\infty}^{\infty} (\|u'(t)\|^2 + \|A^- u(t)\|^2) dt < \infty.$$

The operator D_A is closed. This can be seen as follows. Let us write D_A as a sum $D_A = D_{A^-} + B$. By (1.10) the operator B of multiplication by $B(t)$ is bounded in $L^2(\mathbb{R}; \mathcal{H})$, so the question reduces to the closedness of D_{A^-}. Note that $D_{A^-} = i(\frac{1}{i}\frac{d}{dt}) + A^-$, and the selfadjoint operators $(\frac{1}{i}\frac{d}{dt})$ and A^- in $L^2(\mathbb{R};\mathcal{H})$ commute. Using the spectral representations of $(\frac{1}{i}\frac{d}{dt})$ and A^-, it is easy to see that D_{A^-} is closed on the domain $\mathrm{Dom}(D_{A^-}) = \mathrm{Dom}(D_A)$.

The same argument shows that the adjoint operator D_A^* is defined as a closed operator on the same domain as D_A. Below we consider the selfadjoint operators

$$H = D_A^* D_A \quad \text{and} \quad \widetilde{H} = D_A D_A^*.$$

As the difference $A^+ - A^-$ is a trace class operator, we can consider the spectral shift function $\xi(\lambda; A^+, A^-)$, $\lambda \in \mathbb{R}$. The first part of the theorem below shows that the spectral shift function $\xi(\lambda; \widetilde{H}, H)$ is also well defined. The following theorem relates these two spectral shift functions.

THEOREM 1.1. *Assume* (1.3) *and let* H, $\widetilde{H}$ *be as defined above. For any* $z \in \mathbb{C} \setminus [0,\infty)$, *the difference* $(\widetilde{H} - z)^{-1} - (H - z)^{-1}$ *belongs to the trace class. For a.e.* $\lambda > 0$, *we have the identity*

$$\xi(\lambda; \widetilde{H}, H) = \frac{1}{\pi} \int_{-\sqrt{\lambda}}^{\sqrt{\lambda}} \xi(s; A^+, A^-) \frac{ds}{\sqrt{\lambda - s^2}}, \tag{1.11}$$

where the integral in the r.h.s. converges absolutely.

COROLLARY 1.2. *Suppose that* 0 *is not in the spectrum of* A^+ *or* A^-. *Then the operator* D_A *is Fredholm and*

$$\mathrm{index}\, D_A = \dim \mathrm{Ker}\, D_A - \dim \mathrm{Ker}\, D_A^* = \xi(0; A^+, A^-). \tag{1.12}$$

Generally speaking, the spectral shift function $\xi(\lambda; A^+, A^-)$ is defined as an element of $L^1(\mathbb{R})$, so it does not make sense to speak of its value at a fixed point $\lambda = 0$. However, the assumption of Corollary 1.2 implies that $\xi(\lambda; A^+, A^-)$ is constant near $\lambda = 0$, and so $\xi(0; A^+, A^-)$ is well defined.

According to (1.7), the r.h.s. of (1.12) coincides with the spectral flow of the family $A(t)$ through zero as long as the spectral flow is well defined. Thus, Corollary 1.2 can be interpreted as the generalisation of the "Fredholm index = spectral flow" theorem.

Corollary 1.2, under various sets of conditions on $A(t)$, is well known; see, e.g., [**8, 4, 3**] and references to earlier works therein. The main novelty of this work is formula (1.11).

1.6. The strategy of the proof. The proof of Theorem 1.1 is based on an identity due to [**4**, (3.14)], which we state below as Proposition 1.3. In order to state this identity, let us fix the principal branch of the square root in $\mathbb{C} \setminus (-\infty, 0]$. For any $z \in \mathbb{C} \setminus [0,\infty)$ and any $s \in \mathbb{R}$, we denote

$$g_z(s) = \frac{s}{\sqrt{s^2 - z}}.$$

The formula below involves traces of operators in $\mathcal{H}$ and in $L^2(\mathbb{R};\mathcal{H})$. We denote by tr the trace in $\mathcal{H}$ and by Tr the trace in $L^2(\mathbb{R};\mathcal{H})$.

PROPOSITION 1.3. *Assume* (1.3) *and let* $\widetilde{H}$, H *be as defined above. Then for any* $z \in \mathbb{C} \setminus [0,\infty)$, *the difference* $g_z(A^+) - g_z(A^-)$ *belongs to the trace class in* $\mathcal{H}$ *and*

$$\mathrm{Tr}((\widetilde{H} - z)^{-1} - (H - z)^{-1}) = \frac{1}{2z} \mathrm{tr}(g_z(A^+) - g_z(A^-)). \tag{1.13}$$

The identity (1.13) was proven[1] in [**4**, (3.14)] for the case $\dim \mathcal{H} < \infty$ as a particular case of a more general trace identity. We will give a more streamlined proof of (1.13), based on the ideas from [**3**], where (1.13) was proven in the case $\dim \mathcal{H} = 1$. This plan of the proof is as follows. We first note that the operators H and $\widetilde{H}$ can be represented as

$$H = -\frac{d^2}{dt^2} + Q(t), \qquad \widetilde{H} = -\frac{d^2}{dt^2} + \widetilde{Q}(t), \tag{1.14}$$

where

$$Q(t) = A(t)^2 - A'(t), \qquad \widetilde{Q}(t) = A(t)^2 + A'(t). \tag{1.15}$$

Then, following [**6**], we express the integral kernels of the resolvents $(H - z)^{-1}$ and $(\widetilde{H} - z)^{-1}$ in terms of the solutions to the operator differential equation $-F'' + QF = zF$. Integrating the traces of these resolvent kernels over the diagonal, we obtain the expression on the l.h.s. of (1.13).

Given (1.13), one can derive Theorem 1.1 fairly easily by applying the Lifshits–Kreĭn trace formula (1.4) to both sides of (1.13). This is done in Section 2. Proposition 1.3 in the case $\dim \mathcal{H} < \infty$ and Corollary 1.2 are also proven in Section 2.

In Section 3 (which is of a technical nature) we use an approximation argument to extend Proposition 1.3 from the case $\dim \mathcal{H} < \infty$ to the case $\dim \mathcal{H} = \infty$.

2. Proofs

2.1. Trace class inclusions. Our first task is to show that the operators on both sides of (1.13) belong to the trace class.

LEMMA 2.1. *Under the assumptions of Theorem 1.1, the operators $(H - z)^{-1} - (\widetilde{H} - z)^{-1}$ and $g_z(A^+) - g_z(A^-)$ belong to the trace class for any $z \in \mathbb{C} \setminus [0, \infty)$.*

Note that Lemma 2.1 in particular ensures that the spectral shift function $\xi(\cdot; \widetilde{H}, H)$ is well defined.

Let us introduce some notation. We denote by H_0 the selfadjoint operator $-d^2/dt^2$ in $L^2(\mathbb{R}; \mathcal{H})$ and by $R_0(z) = (H_0 - z)^{-1}$ its resolvent. We also denote $R(z) = (H - z)^{-1}$ and $\widetilde{R}(z) = (\widetilde{H} - z)^{-1}$.

First, we need

LEMMA 2.2. *Let $V(t)$, $t \in \mathbb{R}$, be a family of trace class operators in $\mathcal{H}$ such that*

$$\int_{-\infty}^{\infty} \|V(t)\|_{S_1} dt < \infty, \tag{2.1}$$

and let V be the operator in $L^2(\mathbb{R}; \mathcal{H})$, $(Vu)(t) = V(t)u(t)$. Then for any $z < 0$, one has

(2.2)

$$R_0(z)^{1/2} V R_0(z)^{1/2} \in S_1 \text{ and } \|R_0(z)^{1/2} V R_0(z)^{1/2}\|_{S_1} \leq \frac{1}{2\sqrt{|z|}} \int_{-\infty}^{\infty} \|V(t)\|_{S_1} dt.$$

In particular, V is a relatively form-compact perturbation of H_0.

---

[1]This identity is stated in [**4**] with a wrong sign; compare with [**3**, Example 4.1].

PROOF. Write $V(t) = |V(t)|^{1/2}\operatorname{sign}(V(t))|V(t)|^{1/2}$. Using the Fourier transform, one easily checks that

$$|V|^{1/2}R_0(z)^{1/2} \in S_2$$

and

$$\||V|^{1/2}R_0(z)^{1/2}\|_{S_2}^2 \leq \frac{1}{2\sqrt{|z|}}\int_{-\infty}^{\infty}\||V(t)|^{1/2}\|_{S_2}^2 dt = \frac{1}{2\sqrt{|z|}}\int_{-\infty}^{\infty}\|V(t)\|_{S_1}dt.$$

It remains to write

$$\begin{aligned}\|R_0(z)^{1/2}VR_0(z)^{1/2}\|_{S_1} &= \|R_0(z)^{1/2}|V|^{1/2}\operatorname{sign}(V)|V|^{1/2}R_0(z)^{1/2}\|_{S_1}\\ &\leq \||V|^{1/2}R_0(z)^{1/2}\|_{S_2}^2,\end{aligned}$$

which completes the proof. □

PROOF OF LEMMA 2.1.

1. First we consider the difference of resolvents $R(z) - \widetilde{R}(z)$. By a well-known argument, it suffices to prove that $R(z) - \widetilde{R}(z) \in S_1$ for at least one value of z. Using Lemma 2.2, let us choose $z \in (-\infty, -1)$ with $|z|$ sufficiently large so that

$$\|R_0(z)^{1/2}B'R_0(z)^{1/2}\| \leq 1/2. \tag{2.3}$$

We have

$$\begin{aligned}H - z = H_0 + A^2 - B' - z &\geq H_0 - B' - z\\ &= (H_0 - z)^{1/2}(I - R_0(z)^{1/2}B'R_0(z)^{1/2})(H_0 - z)^{1/2};\end{aligned}$$

the inequality is understood in the sense of quadratic forms. Thus we have

$$R(z) \leq R_0(z)^{1/2}(I - R_0(z)^{1/2}B'R_0(z)^{1/2})^{-1}R_0(z)^{1/2} \leq 2R_0(z).$$

It follows that the resolvent $R(z)$ can be represented as

$$R(z) = M(z)R_0(z)^{1/2} = R_0(z)^{1/2}M(z)^* \quad \text{with} \quad \|M(z)\|^2 \leq 2. \tag{2.4}$$

The same argument works for $\widetilde{R}(z)$, so we have

$$\widetilde{R}(z) = \widetilde{M}(z)R_0^{1/2}(z) = R_0(z)^{1/2}\widetilde{M}(z)^* \quad \text{with} \quad \|\widetilde{M}(z)\|^2 \leq 2. \tag{2.5}$$

Now we can rewrite the resolvent identity as

$$R(z) - \widetilde{R}(z) = 2R(z)B'\widetilde{R}(z) = 2M(z)(R_0(z)^{1/2}B'R_0(z)^{1/2})\widetilde{M}(z)^*,$$

and, by Lemma 2.2, the r.h.s. belongs to the trace class.

2. Consider the difference $g_z(A^+) - g_z(A^-)$. We follow a well-known argument (see e.g. [**2**]). Let $\widetilde{g}_z$ be the Fourier transform of the derivative of g_z; then $\widetilde{g}_z$ is absolutely integrable and we can write

$$g_z(\lambda) = g_z(0) + i\int_{-\infty}^{\infty}\frac{e^{-i\lambda t} - 1}{t}\widetilde{g}_z(t)dt \quad \text{with} \quad \int_{-\infty}^{\infty}|\widetilde{g}_z(t)|dt < \infty. \tag{2.6}$$

Recalling the formula

$$e^{-itA^+} - e^{-itA^-} = -i\int_0^t e^{-i(t-s)A^+}(A^+ - A^-)e^{-isA^-}ds,$$

we obtain the representation

$$g_z(A^+) - g_z(A^-) = \int_{-\infty}^{\infty}dt\ \widetilde{g}_z(t)t^{-1}\int_0^t ds\ e^{-i(t-s)A^+}(A^+ - A^-)e^{-isA^-}. \tag{2.7}$$

Since

$$\|e^{-i(t-s)A^+}(A^+ - A^-)e^{-isA^-}\|_{S_1} \leq \|(A^+ - A^-)\|_{S_1},$$

the integral on the r.h.s. of (2.7) converges in the trace norm. $\square$

2.2. Proof of Theorem 1.1. 1. Denote for brevity $\xi(\lambda) = \xi(\lambda; \widetilde{H}, H)$ and $\eta(\lambda) = \xi(\lambda; A^+, A^-)$. According to the Lifshits–Kreĭn trace formula (1.4), the identity (1.13) can be rewritten as

$$-\int_0^\infty \frac{\xi(t)}{(t-z)^2} dt = \frac{1}{2z} \int_{-\infty}^\infty \eta(t) \left(\frac{\partial}{\partial t} g_z(t) \right) dt,$$

which can be rewritten as

$$\int_0^\infty \xi(t) \left(\frac{\partial}{\partial z} (t-z)^{-1} \right) dt = \int_{-\infty}^\infty \eta(t) \left(\frac{\partial}{\partial z} \frac{1}{\sqrt{t^2 - z}} \right) dt.$$

Integrating over z, we get

$$\int_0^\infty \xi(t) \left(\frac{1}{t-z} - \frac{1}{t+1} \right) dt = \int_{-\infty}^\infty \eta(t) \left(\frac{1}{\sqrt{t^2 - z}} - \frac{1}{\sqrt{t^2+1}} \right) dt. \tag{2.8}$$

2. Now we would like to take the imaginary parts of both sides of (2.8) and pass to the limit as $z \to \lambda + i0$ for a.e. $\lambda > 0$. By the well-known properties of the Cauchy integrals, we have for a.e. $\lambda > 0$,

$$\xi(\lambda) = \lim_{\varepsilon \to +0} \frac{1}{\pi} \int_0^\infty \xi(t) \operatorname{Im} \frac{1}{t - \lambda - i\varepsilon} dt.$$

Now consider the r.h.s. of (2.8). As $\eta \in L^1(\mathbb{R})$, it is easy to see that the integral

$$\int_{-\infty}^\infty \frac{|\eta(t)|}{|t^2 - \lambda|^{1/2}} dt$$

converges for a.e. $\lambda > 0$. By the dominated convergence theorem, this ensures that

$$\operatorname{Im} \int_{-\infty}^\infty \frac{\eta(t)}{\sqrt{t^2 - \lambda - i\varepsilon}} dt \to \int_{-\sqrt{\lambda}}^{\sqrt{\lambda}} \frac{\eta(\lambda)}{\sqrt{\lambda - t^2}} dt, \quad \varepsilon \to +0$$

for a.e. $\lambda > 0$. This allows us to pass to the limit in (2.8), which yields the required result. $\square$

2.3. Proof of Corollary 1.2. 1. In order to check that D_A is Fredholm, we use the following necessary and sufficient condition from [**1**, Theorem A.4]: *A closed operator T is Fredholm if and only if $0 \notin \sigma_{ess}(T^*T)$ and $0 \notin \sigma_{ess}(TT^*)$ and then* $\operatorname{index}(T) = \dim \operatorname{Ker}(T^*T) - \dim \operatorname{Ker}(TT^*)$.

Using our assumptions (1.3) and $0 \in \rho(A^-) \cap \rho(A^+)$, we can find $a > 0$ such that for all sufficiently large $|t|$, one has $A(t)^2 \geq 2aI$. One has

$$H = H_0 + A(t)^2 - B'(t) \geq H_0 + aI + V(t),$$

where

$$V(t) = -B'(t) + (A(t)^2 - aI)E_{A(t)^2}([0, a]).$$

The operator $(A(t)^2 - aI)E_{A(t)^2}([0, a])$ is of a finite rank for all $t \in \mathbb{R}$ and vanishes for all sufficiently large $|t|$. Thus, V satisfies (2.1) and therefore, by Lemma 2.2, V is a relatively form-compact perturbation of H_0. It follows that $\sigma_{ess}(H_0 + aI + V) = \sigma_{ess}(H_0 + aI) = [a, \infty)$. Thus, $\inf \sigma_{ess}(H) \geq a > 0$. The same argument applies to $\widetilde{H}$. By the necessary and sufficient condition quoted above, D_A is a Fredholm operator and

$$\text{index}\, D_A = \dim \operatorname{Ker} H - \dim \operatorname{Ker} \widetilde{H}. \tag{2.9}$$

2. By the previous step, the spectra of H and $\widetilde{H}$ are discrete on $[0, a)$. It is well known (see, e.g., [**1**, Lemma A.3]) that, since $H = D_A^* D_A$ and $\widetilde{H} = D_A D_A^*$, one has $\sigma(H) \setminus \{0\} = \sigma(\widetilde{H}) \setminus \{0\}$ and the multiplicities of the eigenvalues coincide. Thus, from (1.6) we get

$$\xi(\lambda; \widetilde{H}, H) = \dim \operatorname{Ker} H - \dim \operatorname{Ker} \widetilde{H}, \qquad \lambda \in (0, a). \tag{2.10}$$

This identity and its relation to the index of D_A is due to [**3**].

3. Combining Theorem 1.1 with (2.9) and (2.10), we obtain

$$\text{index}\, D_A = \frac{1}{\pi} \int_{-\sqrt{\lambda}}^{\sqrt{\lambda}} \frac{\xi(t; A^+, A^-)}{\sqrt{\lambda - t^2}} dt, \quad \lambda \in (0, a).$$

Since $0 \in \rho(A^-) \cap \rho(A^+)$, the function $\xi(t; A^+, A^-)$ is constant near $t = 0$. Taking into account the identity

$$\frac{1}{\pi} \int_{-\sqrt{\lambda}}^{\sqrt{\lambda}} \frac{1}{\sqrt{\lambda - t^2}} dt = 1,$$

we arrive at (1.12). □

2.4. Proof of Proposition 1.3 for "nice" $A(t)$. Our proof of Proposition 1.3 consists of two steps: we first prove the identity (1.13) for very "nice" functions $A(t)$ and then use an approximation argument. Here we present the first step; the approximation argument, which has a more technical nature, is given in section 3.

LEMMA 2.3. *Suppose that* $\dim \mathcal{H} < \infty$. *Assume also that* $A(t) = A^\pm$ *for all sufficiently large* $\pm t > 0$. *Then Proposition* 1.3 *holds true.*

PROOF. 1. Of course, in the finite-dimensional case, all operators belong to the trace class, so we only need to prove the identity (1.13). In the case $\dim \mathcal{H} = 1$, this formula has been proven in [**3**, Example 4.1]. Below is a direct generalisation of the argument of [**3**] to the matrix-valued case. We call bounded operators on $\mathcal{H}$ "matrices".

By analyticity in z it suffices to consider the case of real negative z. In what follows, we assume $z \in (-\infty, 0)$ and denote

$$\varkappa_\pm = \sqrt{(A^\pm)^2 - z}.$$

The matrices $\varkappa_\pm$ in $\mathcal{H}$ are positive definite.

2. We will compute the trace on the l.h.s. of (1.13) by constructing the integral kernels of the resolvents of H and $\widetilde{H}$ and evaluating integrals of these kernels over the diagonal. Our first aim is to construct the resolvent of H in terms of the solutions to the matrix Schrödinger equation. Here we follow [**6**].

Consider the matrix-valued solutions $F_\pm(t)$, $t \in \mathbb{R}$, to the equation (see (1.14), (1.15))

$$-F_\pm'' + QF_\pm = zF_\pm, \tag{2.11}$$

satisfying the asymptotic conditions

$$F_\pm(t) = e^{\mp \varkappa_\pm t}, \quad \pm t > 0 \text{ large}. \tag{2.12}$$

The existence of solutions $F_\pm$ can be proven in the usual way by converting (2.11), (2.12) into Volterra-type integral equations.

We have the relations

$$F_+(t) = e^{-\varkappa_- t}a + e^{\varkappa_- t}b, \quad -t > 0 \text{ large},$$
$$F_-(t) = e^{\varkappa_+ t}c + e^{-\varkappa_+ t}d, \quad t > 0 \text{ large}, \tag{2.13}$$

for some matrices a, b, c, d (which, of course, depend on $z < 0$). From here it is straightforward to see that if either a or c has a nontrivial kernel, then z is an eigenvalue of H. But, by definition, H cannot have negative eigenvalues, so we have

$$\operatorname{Ker} a = \operatorname{Ker} c = \{0\} \tag{2.14}$$

for all $z < 0$.

3. For any two solutions F, G to the equation (2.11), let us define the Wronskian

$$W(F, G) = F(t)^* G'(t) - F'(t)^* G(t).$$

By a direct calculation, the Wronskian does not depend on t. In particular, using the limiting forms (2.13) of the solutions $F_\pm$, we obtain

$$W(F_+, F_-) = 2\varkappa_+ c = 2a^* \varkappa_-. \tag{2.15}$$

By (2.14), it follows that $W(F_+, F_-)$ is invertible (it is here that we need $\dim \mathcal{H} < \infty$). Now we can construct the integral kernel $R(t, s)$ of the resolvent $(H - z)^{-1}$ as in [**6**]. We have

$$R(t, s) = \begin{cases} F_+(t)(W(F_+, F_-)^*)^{-1} F_-^*(s), & t \geq s, \\ F_-(t)(W(F_+, F_-))^{-1} F_+^*(s), & t < s. \end{cases} \tag{2.16}$$

4. As above, we can construct the integral kernel $\widetilde{R}(x, y)$ of the resolvent $(\widetilde{H} - z)^{-1}$ in terms of the solutions $\widetilde{F}_\pm(t)$ to

$$-\widetilde{F}_\pm'' + \widetilde{Q}\widetilde{F}_\pm = z\widetilde{F}_\pm, \text{ and } \widetilde{F}_\pm(t) = e^{\mp\varkappa_\pm t}, \quad \pm t > 0 \text{ large}. \tag{2.17}$$

By a direct calculation, the function

$$\widetilde{F}_\pm(t) = (F_\pm'(t) + A(t)F_\pm(t))(\mp\varkappa_\pm + A^\pm)^{-1}$$

satisfies (2.17). Also by a direct calculation,

$$W(\widetilde{F}_+, \widetilde{F}_-) = z(-\varkappa_+ + A^+)^{-1} W(F_+, F_-)(\varkappa_- + A^-)^{-1}.$$

This allows us to compute the kernel $\widetilde{R}(t, s)$ in terms of the solutions $F_\pm$:

$$\widetilde{R}(t, s) = \begin{cases} \frac{1}{z}(F_+'(t) + A(t)F_+(t))(W(F_+, F_-)^*)^{-1}(F_-'(s) + A(s)F_-(s))^*, & t \geq s, \\ \frac{1}{z}(F_-'(t) + A(t)F_-(t))(W(F_+, F_-))^{-1}(F_+'(s) + A(s)F_+(s))^*, & t < s. \end{cases} \tag{2.18}$$

5. Now we are ready to compute the trace on the l.h.s. of (1.13):

$$\operatorname{Tr}((\widetilde{H} - z)^{-1} - (H - z)^{-1}) = \lim_{R \to \infty} \int_{-R}^{R} \operatorname{tr}(\widetilde{R}(t, t) - R(t, t))dt.$$

Using our formulas (2.16), (2.18) for the kernels $R(t, t)$ and $\widetilde{R}(t, t)$, we obtain

$$\operatorname{tr}(\widetilde{R}(t, t) - R(t, t)) = \operatorname{tr}(W(F_+, F_-)^*)^{-1}(\frac{1}{z}((F_-')^* + F_-^* A)(F_+' + AF_+) - F_-^* F_+).$$

Integrating by parts, after a little algebra we get

$$\int_{-R}^{R} \operatorname{tr}(\widetilde{R}(t,t) - R(t,t))dt = \frac{1}{z} \operatorname{tr}((W(F_+,F_-)^*)^{-1} F_-^*(F_+' + AF_+)) \,|_{-R}^{R} . \tag{2.19}$$

Now we can calculate the r.h.s. of (2.19) for large $R > 0$, using formula (2.15) and the asymptotic forms (2.12), (2.13) of $F_\pm$:

$$\operatorname{tr}(W(F_+,F_-)^{-1}F_-^*F_+') \,|_{-R}^{R} = -\frac{1}{2}\operatorname{tr}((c^*)^{-1}d^*e^{-2\varkappa_+ R}) - \frac{1}{2}\operatorname{tr}(ba^{-1}e^{-2\varkappa_- R}),$$

$$\operatorname{tr}(W(F_+,F_-)^{-1}F_-^*AF_+) \,|_{-R}^{R} = \frac{1}{2}\operatorname{tr}(\varkappa_+^{-1}A^+ - \varkappa_-^{-1}A^-)$$
$$+\frac{1}{2}\operatorname{tr}(\varkappa_+^{-1}(c^*)^{-1}d^*A^+e^{-2\varkappa_+ R}) - \frac{1}{2}\operatorname{tr}(ba^{-1}\varkappa_-^{-1}A^-e^{-2\varkappa_- R}).$$

As $\varkappa_\pm$ are positive definite matrices, we have $\|e^{-2\varkappa_\pm R}\| \to 0$ as $R \to \infty$. Thus,

$$\operatorname{Tr}((\widetilde{H} - z)^{-1} - (H - z)^{-1}) = \frac{1}{2z}\operatorname{tr}(\varkappa_+^{-1}A^+ - \varkappa_-^{-1}A^-),$$

as required. □

3. Approximation argument

First we give a general statement about convergence on both sides of the identity (1.13) and then construct the approximating sequence $A_n(t)$.

3.1. Convergence in (1.13). Let $A(t) = A^- + B(t)$ and $A_n(t) = A_n^- + B_n(t)$ be operator families satisfying (1.10). As above, we assume $B(-\infty) = B_n(-\infty) = 0$ and define $A_n^+ = A_n^- + B_n(+\infty)$. We also define $H_n = D_{A_n}^* D_{A_n}$, $\widetilde{H}_n = D_{A_n} D_{A_n}^*$, and

$$R_n(z) = (H_n - z)^{-1}, \quad \widetilde{R}_n(z) = (\widetilde{H}_n - z)^{-1}.$$

LEMMA 3.1. *Assume that* $\operatorname{Dom} A^- \subset \operatorname{Dom} A_n^-$ *for all* n *and* $A_n^- f \to A^- f$ *for all* $f \in \operatorname{Dom}(A^-)$. *Next, assume that*

$$\int_{-\infty}^{\infty} \|B_n'(t) - B'(t)\|_{S_1} dt \to 0 \qquad \text{as } n \to \infty. \tag{3.1}$$

Then, for all $z \in (-\infty, -1)$ *with sufficiently large* $|z|$, *one has*

$$\|(g_z(A^+) - g_z(A^-)) - (g_z(A_n^+) - g_z(A_n^-))\|_{S_1} \to 0, \tag{3.2}$$

$$\|(R(z) - \widetilde{R}(z)) - (R_n(z) - \widetilde{R}_n(z))\|_{S_1} \to 0. \tag{3.3}$$

REMARK 3.2. 1. Assumption (3.1) implies that

$$\|(A_n^+ - A_n^-) - (A^+ - A^-)\|_{S_1} \to 0. \tag{3.4}$$

2. Our assumptions on A^-, A_n^- imply that $A_n^- \to A^-$ in the strong resolvent sense; see [**7**, Theorem VIII.25(a)] and its proof. Combining this with (3.4), we see that also $A_n^+ \to A^+$ in the strong resolvent sense.

We will repeatedly make use of the following well-known fact, which holds true for any Schatten-von Neumann class S_p, $p \geq 1$, although we will only need it for the case $p = 1, 2$:

PROPOSITION 3.3. *Let* T_n *be a sequence of bounded operators in a Hilbert space which converges strongly to zero and let* $M \in S_p$; *then* $\|T_n M\|_{S_p} \to 0$. *If* T_n^* *also converges strongly to zero, then* $\|MT_n\|_{S_p} \to 0$.

The first part of this proposition can be found, for example, in [**9**, Lemma 6.1.3], and the second part follows immediately by conjugation, since $\|MT_n\|_{S_p} = \|T_n^* M^*\|_{S_p}$.

Proof of (3.2). Writing the representation (2.7), we get

$$(3.5)\quad (g_z(A^+) - g_z(A^-)) - (g_z(A_n^+) - g_z(A_n^-)) = \int_{-\infty}^{\infty} dt\ \widetilde{g}_z(t)t^{-1} \int_0^t ds\ K_n(t,s),$$
$$K_n(t,s) = e^{-i(t-s)A^+}(A^+ - A^-)e^{-isA^-} - e^{-i(t-s)A_n^+}(A_n^+ - A_n^-)e^{-isA_n^-}.$$

We would like to use the dominated convergence theorem in order to prove that the r.h.s. of (3.5) converges to zero in the trace norm. First note that

$$\|K_n(t,s)\|_{S_1} \le \|A^+ - A^-\|_{S_1} + \|A_n^+ - A_n^-\|_{S_1}$$

and, by (3.4), the r.h.s. is bounded uniformly in n by some constant C. This gives an integrable bound for the integrand in the r.h.s. of (3.5).

Next, we claim that $\|K_n(t,s)\|_{S_1} \to 0$ for all t, s. Indeed, we can write

$$(3.6)\quad \begin{aligned} K_n(t,s) = &(e^{-i(t-s)A^+} - e^{-i(t-s)A_n^+})(A^+ - A^-)e^{-isA^-} \\ &+ e^{-i(t-s)A_n^+}(A^+ - A^-)(e^{-isA^-} - e^{-isA_n^-}) \\ &+ e^{-i(t-s)A_n^+}(A^+ - A^- - A_n^+ + A_n^-)e^{-isA_n^-}. \end{aligned}$$

The last term on the r.h.s. converges to zero by (3.4). Next, since $A_n^\pm \to A^\pm$ in strong resolvent sense (see Remark 3.2), by [**7**, Theorem VIII.21] we have strong convergence $e^{itA_n^\pm} \to e^{itA^\pm}$ as $n \to \infty$. Thus, by Proposition 3.3, the first two terms on the r.h.s. of (3.6) converge to zero in the trace norm. By dominated convergence, this proves (3.2). □

The proof of (3.3) requires a little more work. First we need an abstract lemma which ensures the strong convergence of resolvents. I am indebted to Nikolai Filonov for providing the proof of this lemma.

Let D be a closed densely defined operator in a Hilbert space such that D^* is also densely defined. For each n, let D_n be a closed densely defined operator such that D_n^* is also densely defined and $\operatorname{Dom} D \subset \operatorname{Dom} D_n$ and $\operatorname{Dom} D^* \subset \operatorname{Dom} D_n^*$.

Lemma 3.4. *Assume the above conditions and assume that for all* $f \in \operatorname{Dom} D$ *one has* $\|D_n f - Df\| \to 0$ *and for all* $f \in \operatorname{Dom} D^*$ *one has* $\|D_n^* f - D^* f\| \to 0$ *as* $n \to \infty$. *Then for all* $z \in \mathbb{C} \setminus [0,\infty)$, *one has the strong convergence*

$$(3.7)\quad (D_n^* D_n - z)^{-1} \to (D^* D - z)^{-1}, \quad n \to \infty.$$

Proof. 1. By [**7**, Chapter VIII, Problem 20], it suffices to prove the weak convergence in (3.7).

2. Fix $z \in \mathbb{C} \setminus [0,\infty)$, $f \in \mathcal{H}$ and denote $\varphi_n = (D_n^* D_n - z)^{-1} f$. We need to prove that $\varphi_n \to \varphi$ weakly, where $\varphi = (D^* D - z)^{-1} f$. First note that

$$(3.8)\quad \|\varphi_n\| \le \frac{\|f\|}{\operatorname{dist}(z, [0,\infty))}.$$

Next, we have

$$\|D_n \varphi_n\|^2 - z\|\varphi_n\|^2 = (f, \varphi_n) = (f, (D_n^* D_n - z)^{-1} f),$$

and so

$$\|D_n\varphi_n\|^2 \le |z|\|\varphi_n\|^2 + |(f,(D_n^* D_n - z)^{-1}f)| \le C(z)\|f\|^2. \tag{3.9}$$

By (3.8), (3.9) and the weak compactness of the unit ball in a Hilbert space, from the sequence φ_n one can choose a subsequence φ_{n_k} such that $\varphi_{n_k} \to \tilde\varphi$ and $D_{n_k}\varphi_{n_k} \to \psi$ weakly for some elements $\tilde\varphi$, ψ in $\mathcal{H}$.

3. Let us prove that $\tilde\varphi \in \operatorname{Dom} D$ and $\psi = D\tilde\varphi$. For any $\chi \in \operatorname{Dom} D^*$, we have

$$(D_{n_k}\varphi_{n_k}, \chi) \to (\psi, \chi),$$

and so

$$(\varphi_{n_k}, D_{n_k}^*\chi) \to (\psi, \chi).$$

Since $\|D_{n_k}^*\chi - D^*\chi\| \to 0$ by our assumptions, we get

$$(\tilde\varphi, D^*\chi) = (\psi, \chi) \tag{3.10}$$

for all $\chi \in \operatorname{Dom} D^*$; it follows that $\tilde\varphi \in \operatorname{Dom} D$ and $\psi = D\tilde\varphi$.

4. Next, we have for any $\chi \in \operatorname{Dom} D$:

$$(D_n\varphi_n, D_n\chi) - z(\varphi_n, \chi) = (f, \chi). \tag{3.11}$$

By the previous step, $\varphi_{n_k} \to \tilde\varphi$ and $D_{n_k}\varphi_{n_k} \to \psi$ weakly, and by the hypothesis, $D_n\chi \to D\chi$ strongly. Passing to the limit in (3.11) over the subsequence n_k, we get

$$(\psi, D\chi) - z(\tilde\varphi, \chi) = (f, \chi)$$

for all $\chi \in \operatorname{Dom} D$. It follows that $\psi \in \operatorname{Dom} D^*$ and $D^*\psi - z\tilde\varphi = f$. Recalling that $\psi = D\tilde\varphi$, we get that $\tilde\varphi \in \operatorname{Dom}(D^*D)$ and $(D^*D - z)\tilde\varphi = f$. Thus, we have proven that $\tilde\varphi = \varphi$.

5. We have proven weak convergence $\varphi_n \to \varphi$ over a subsequence n_k. But we could have started from an arbitrary subsequence of φ_n and proven that it has a subsubsequence which weakly converges to φ. This proves that actually the whole sequence φ_n weakly converges to φ. □

PROOF OF (3.3). 1. By Lemma 2.2 and the uniform boundedness of the integrals $\int\|B_n'(t)\|_{S_1}dt$, one can choose $a < -1$ such that for all $z \in (-\infty, a)$, the estimates

$$\|R_0(z)^{1/2}B'R_0(z)^{1/2}\| \le 1/2, \quad \sup_n\|R_0(z)^{1/2}B_n'R_0(z)^{1/2}\| \le 1/2$$

hold true. Then, as in the proof of Lemma 2.1, we get (2.4), (2.5), and also

$$R_n(z) = M_n(z)R_0(z)^{1/2} = R_0(z)^{1/2}M_n(z)^*,$$
$$\widetilde{R}_n(z) = \widetilde{M}_n(z)R_0(z)^{1/2} = R_0(z)^{1/2}\widetilde{M}_n(z)^*$$

with $\|M_n(z)\|^2 \le 2$ and $\|\widetilde{M}_n(z)\|^2 \le 2$. In what follows, we fix $z \in (-\infty, a)$ as above and suppress the dependance of z in our notation for brevity.

2. Note that by Lemma 3.4, we have the strong convergence of resolvents $R_n \to R$, $\widetilde{R}_n \to \widetilde{R}$. Moreover, we claim that for the operators M_n, $\widetilde{M}_n$ we have the strong convergence $M_n \to M$, $\widetilde{M}_n \to \widetilde{M}$. Indeed, for all $f \in \operatorname{Dom}(H_0 - z)^{1/2}$ one has

$$M_n f = R_n(H_0 - z)^{1/2}f \to R(H_0 - z)^{1/2}f = Mf.$$

Since the norms of M_n are uniformly bounded, we get the strong convergence $M_n \to M$. In the same way, we obtain the strong convergence $\widetilde{M}_n \to \widetilde{M}$.

3. Using the resolvent identity, we obtain

$$(R_n - \widetilde{R}_n) - (R - \widetilde{R}) = R_n B'_n \widetilde{R}_n - RB'\widetilde{R}$$
$$= R_n(B'_n - B')\widetilde{R}_n + (R_n - R)B'\widetilde{R}_n + RB'(\widetilde{R}_n - \widetilde{R}). \tag{3.12}$$

Let us consider separately each of the three terms on the r.h.s. of (3.12).

4. For the first term, we have

$$R_n(B'_n - B')\widetilde{R}_n = M_n(R_0^{1/2}(B'_n - B')R_0^{1/2})\widetilde{M}_n^*$$

and the r.h.s. converges to zero in the trace norm by Lemma 2.2 and assumption (3.1).

5. Consider the second term on the r.h.s. of (3.12). We have

$$(R_n - R)B'\widetilde{R}_n = (M_n - M)(R_0^{1/2}B'R_0^{1/2})\widetilde{M}_n^*.$$

Since $R_0^{1/2}B'R_0^{1/2}$ is a trace class operator, and $M_n \to M$ strongly as $n \to \infty$, by Proposition 3.3, we obtain that the r.h.s. converges to zero in the trace norm.

6. Finally, the third term on the r.h.s. of (3.12) can be considered similarly to the second one:

$$[RB'(\widetilde{R}_n - \widetilde{R})]^* = (\widetilde{R}_n - \widetilde{R})B'R = (\widetilde{M}_n - \widetilde{M})(R_0^{1/2}B'R_0^{1/2})M^*,$$

and the r.h.s. goes to zero in the trace norm as $n \to \infty$. □

3.2. Constructing the approximating family $A_n(t)$. We will approximate $A(t)$ in two steps. First, we approximate an arbitrary finite rank family $A(t)$ by the ones with compactly supported $A'(t)$. Next, we approximate an arbitrary family by finite rank families.

LEMMA 3.5. *Let* $\dim \mathcal{H} < \infty$. *Then Proposition* 1.3 *holds true.*

PROOF. By analyticity in z, it suffices to prove (1.13) for $z \in (-\infty, -1)$ with sufficiently large $|z|$. Then we can use Lemma 3.1.

For a given family $A(t)$, let us construct a sequence of families $A_n(t)$ such that for each n and all sufficiently large $\pm t > 0$, we have $A_n(t) = A^\pm$. This is not difficult to do. Indeed, let $A_n(t)$ be such that $A_n(t) = A(t)$ for $t \in [-n, n]$, $A_n(t) = A^-$ for $t \le -n-1$, $A_n(t) = A^+$ for $t \ge n+1$, and $A_n(t)$ is obtained by linear interpolation on $[-n-1, -n]$ and $[n, n+1]$. Explicitly,

$$A_n(t) = A^- + (t+n+1)(A(-n) - A^-), \quad t \in [-n-1, -n],$$
$$A_n(t) = (t-n)A^+ + (n+1-t)A(n), \quad t \in [n, n+1].$$

Then we have

$$\int_{-\infty}^{\infty} \|A'_n(t) - A'(t)\|_{S_1} dt \le \int_n^\infty (\|A'_n(t)\|_{S_1} + \|A'(t)\|_{S_1})dt$$
$$+ \int_{-\infty}^{-n} (\|A'_n(t)\|_{S_1} + \|A'(t)\|_{S_1})dt \to 0$$

as $n \to \infty$. By Lemma 2.3, the identity (1.13) holds true for the families $A_n(t)$. By Lemma 3.1, we can pass to the limit as $n \to \infty$ on both sides of (1.13), which yields the required result. □

Next, we approximate an arbitrary family $A(t)$ by finite rank families.

LEMMA 3.6. *There exists a sequence of finite rank orthogonal projections P_n in $\mathcal{H}$ such that:*

(i) $P_n \to I$ *strongly as* $n \to \infty$;

(ii) $\operatorname{Ran} P_n \subset \operatorname{Dom} A^-$ *for all* n;

(iii) for all $f \in \operatorname{Dom}(A^-)$, *one has* $\|P_n A^- P_n f - A^- f\| \to 0$ *as* $n \to \infty$.

PROOF. For any $k \in \mathbb{Z}$, let E_k be the spectral projection of the operator A^- associated with the interval $(k, k+1]$, and let $Q_j^{(k)}$ be a sequence of finite rank projections such that $\operatorname{Ran} Q_j^{(k)} \subset \operatorname{Ran} E_k$ and $Q_j^{(k)} \to E_k$ strongly as $j \to \infty$. Take $P_n = \sum_{k=-n}^{n} Q_n^{(k)}$. Then (i), (ii) are obvious. Let us prove (iii). If $f \in \operatorname{Dom} A^-$, then $f = \sum_{k=-\infty}^{\infty} f_k$, where $f_k = E_k f$ and $\sum_{k=-\infty}^{\infty}(k^2+1)\|f_k\|^2 < \infty$. Given such an f and $\varepsilon > 0$, we can choose N sufficiently large so that $\sum_{|k|\geq N}(k^2+1)\|f_k\|^2 < \varepsilon^2$; denote $g_1 = \sum_{|k|<N} f_k$ and $g_2 = \sum_{|k|\geq N} f_k$. Then $\|P_n A^- P_n g_2\| \leq \varepsilon$ and $\|A^- g_2\| \leq \varepsilon$. On the other hand, it is easy to see that $\|P_n A^- P_n g_1 - A^- g_1\| \to 0$ as $n \to \infty$. This proves (iii). □

PROOF OF PROPOSITION 1.3. 1. Let $A_n(t) = P_n A(t) P_n$, $B_n(t) = P_n B(t) P_n$, where P_n are as constructed in Lemma 3.6. We claim that

$$\int_{-\infty}^{\infty} \|B_n'(t) - B'(t)\|_{S_1} dt \to 0 \quad \text{as } n \to \infty.$$

Indeed, let us apply the dominated convergence theorem. First, we have

$$\|B_n'(t) - B'(t)\|_{S_1} \leq \|B_n'(t)\|_{S_1} + \|B'(t)\|_{S_1} \leq 2\|B'(t)\|_{S_1}.$$

Next, for all $t \in \mathbb{R}$, we have

$$B_n'(t) - B'(t) = (P_n - I)B'(t)P_n + B'(t)(P_n - I),$$

and the right-hand side converges to zero by Proposition 3.3, since $P_n \to I$ and $B'(t) \in S_1$.

2. By Lemma 3.5, the identity (1.13) holds true for the families $A_n(t)$. By Lemma 3.1, we can pass to the limit as $n \to \infty$ on both sides of (1.13) when $z \in (-\infty, -1)$, $|z|$ large. By analyticity, this yields the required result for all z. □

Acknowledgments

This work originated from a question asked by Yuri Tomilov. I am grateful to Yuri Tomilov and Yuri Latushkin for many useful discussions and encouragement, and to Nikolai Filonov for providing the proof to Lemma 3.4. I am also grateful to M. Sh. Birman and to F. Gesztesy for discussing with me the results of the paper and to V. Sloushch for reading the manuscript and making some useful remarks.

References

[1] J. Avron, R. Seiler and B. Simon, *The index of a pair of projections,* J. Funct. Anal. **120** (1994), no. 1, 220–237.

[2] M. Sh. Birman and D. R. Yafaev, *The spectral shift function. The papers of M. G. Kreĭn and their further development. (Russian)* Algebra i Analiz **4** (1992), no. 5, 1–44; translation in St. Petersburg Math. J. **4** (1993), no. 5, 833–870.

[3] D. Bollé, F. Gesztesy, H. Grosse, W. Schweiger, and B. Simon, *Witten index, axial anomaly, and Kreĭn's spectral shift function in supersymmetric quantum mechanics,* J. Math. Phys. **28** (1987), no. 7, 1512–1525

[4] C. Callias, *Axial Anomalies and Index Theorems on Open Spaces,* Commun. Math. Phys. **62** (1978), 213–234.

[5] M. G. Kreĭn, *On the trace formula in perturbation theory (Russian)* Mat. Sb. **33 (75)**, no. 3 (1953), 597–626.

[6] L. Martínez Alonso and E. Olmedilla, *Trace identities in the inverse scattering transform method associated with matrix Schrödinger operators,* J. Math. Phys. **23** (1982), no. 11, 2116–2121

[7] M. Reed and B. Simon, *Methods of modern mathematical physics, I. Functional analysis.* Academic Press, 1980.

[8] J. Robbin and D. Salamon, *The spectral flow and the Maslov index,* Bull. London Math. Soc. **27** (1995), no. 1, 1–33.

[9] D. R. Yafaev, *Mathematical scattering theory. General theory.* American Mathematical Society, Providence, RI, 1992.

Department of Mathematics, King's College, London, Strand, London, WC2R 2LS, United Kingdom

E-mail address: alexander.pushnitski@kcl.ac.uk

Amer. Math. Soc. Transl.
(2) Vol. **225**, 2008

On the Spectrum of a Translationally Invariant Pauli Operator

Georgi Raikov

To Mikhail Shlëmovich Birman on his 80th birthday

ABSTRACT. We consider a class of translationally invariant 3D Pauli operators $\mathbf{G}$ with magnetic fields with circular integral curves. We show that the spectrum of $\mathbf{G}$ is purely absolutely continuous, coincides with $[0, \infty)$, and has an infinite multiplicity.

1. Introduction

In this article we consider the Pauli operator

$$\mathbf{G} := \left(\sum_{j=1,2,3} \hat{\sigma}_j \left(-i\frac{\partial}{\partial x_j} - A_j \right) \right)^2 = (i\nabla + A)^2 I_2 - \sum_{j=1,2,3} B_j \hat{\sigma}_j,$$

selfadjoint in $L^2(\mathbb{R}^3)^2$. Here

$$\hat{\sigma}_1 = \begin{pmatrix} 0 & 1 \\ 1 & 0 \end{pmatrix}, \quad \hat{\sigma}_2 = \begin{pmatrix} 0 & -i \\ i & 0 \end{pmatrix}, \quad \hat{\sigma}_3 = \begin{pmatrix} 1 & 0 \\ 0 & -1 \end{pmatrix}$$

are the Pauli matrices, I_2 is the unit 2×2-matrix, $A = (A_1, A_2, A_3) : \mathbb{R}^3 \to \mathbb{R}^3$ is the magnetic potential, and $B = (B_1, B_2, B_3) := \operatorname{curl} A$ is the magnetic field. The operator $\mathbf{G}$ is the Hamiltonian of a three-dimensional $\frac{1}{2}$-spin nonrelativistic quantum particle subject to the magnetic field B.

Let (r, θ, x_3) be the standard cylindrical coordinates in $\mathbb{R}^3$. We assume that the magnetic potential A has the form

$$A = (0, 0, a), \quad a = a(r). \tag{1.1}$$

Then we have

$$B = a'(r)(\sin\theta, -\cos\theta, 0), \tag{1.2}$$

i.e. $|B| = |a'(r)|$, and the integral curves of the magnetic field are circles in the planes orthogonal to the Ox_3-axis, centered at the origin.

2000 *Mathematics Subject Classification.* Primary 35J10, 47F05.

Applying the partial Fourier transform with respect to x_3, the expansion in the Fourier series with respect to θ, and a unitary transform in the spin space $\mathbb{C}^2$, we find that the operator $\mathbf{G}$ is unitarily equivalent to

$$\bigoplus_{m\in\mathbb{Z}} \int_{\mathbb{R}}^{\oplus} G_m(k)dk, \tag{1.3}$$

where the operator

$$G_m(k) := \begin{pmatrix} H_m(k) - \frac{2m-1}{2r^2} + a' & \frac{2m-1}{2r^2} \\ \frac{2m-1}{2r^2} & H_m(k) - \frac{2m-1}{2r^2} - a' \end{pmatrix}, \quad k \in \mathbb{R}, \quad m \in \mathbb{Z},$$

with

$$H_m(k) := -\frac{1}{r}\frac{d}{dr}r\frac{d}{dr} + \frac{m^2}{r^2} + (a(r) - k)^2,$$

is selfadjoint in $L^2(\mathbb{R}_+; rdr)^2$. Assuming that $a \to \infty$ and $a' = o(a^2)$ as $r \to \infty$, we find that the spectrum of $G_m(k)$ is discrete, and study the asymptotics as $k \to \pm\infty$ of its eigenvalues $\mu_{n,m}(k)$ whose graphics as functions of k are known as dispersion curves. First, we check that $\mu_{n,m}(k) \to \infty$ as $k \to -\infty$ for all $n \in \mathbb{N}$ and $m \in \mathbb{Z}$ (see Proposition 4.1). Further, we show that under mild assumptions $\mu_{1,m}(k) \to 0$ as $k \to \infty$, $m \in \mathbb{Z}$, even if $a'(r)$ grows (super)exponentially fast as $r \to \infty$ (see Proposition 4.2). Consequently, the spectrum $\sigma(\mathbf{G})$ of the operator $\mathbf{G}$ is absolutely continuous,

$$\sigma(\mathbf{G}) = [0, \infty), \tag{1.4}$$

and the multiplicity of $\sigma(\mathbf{G})$ is infinite (see Theorem 2.1).

Investigation of the dispersion curves for various translationally invariant quantum Hamiltonians with absolutely continuous spectrum could be found in [**11**], [**8**], [**5**], [**9**, Chapter 4], [**15**], [**16**]. In the recent article [**16**] the 3D magnetic Schrödinger operator $\mathbf{H} := (i\nabla + A)^2$ with potential of the form (1.1) has been considered. It was shown that $\sigma(\mathbf{H}) = [0, \infty)$ if $a'(r) \to 0$ as $r \to \infty$, but if $a'(r)$ tends to a positive finite limit or to infinity as $r \to \infty$, then $\sigma(\mathbf{H}) = [\mathcal{E}, \infty)$ with $\mathcal{E} > 0$. This phenomenon is related to the diamagnetic inequality valid for Schrödinger operators, while the stability of the zero infimum of $\sigma(\mathbf{G})$ is coherent with the paramagnetic effects typical for Pauli operators (see [**7**] for a detailed and rigorous discussion of the dia- and paramagnetism for nonhomogeneous magnetic fields).

It is known that equality (1.4) holds for a large class of magnetic fields, not necessarily generated by magnetic potentials of the form (1.1). For example, the results of [**10**] imply that (1.4) is valid for magnetic fields satisfying the following condition. For $x \in \mathbb{R}^3$ set

$$\epsilon_0(x) := |B(x)|, \quad \epsilon_s(x) := \frac{\sum_{|\alpha|=s} |D^\alpha B(x)|}{1 + \sum_{|\alpha|<s} |D^\alpha B(x)|}, \quad s \in \mathbb{N}.$$

Suppose that for some $s \geq 0$, in $\mathbb{R}^3$ there exists an infinite sequence of disjoint balls whose radii tend to infinity, such that the restriction of the function $\epsilon_s(x)$ onto the union of these balls tends to zero as $|x| \to \infty$. Then (1.4) holds. More precisely, the authors of [**10**] considered the 3D Dirac operator $\mathbf{D}$, selfadjoint in $L^2(\mathbb{R}^3; \mathbb{C}^4)$, and showed that

$$\sigma(\mathbf{D}) = \mathbb{R} \setminus (-1, 1). \tag{1.5}$$

Since we have

$$\mathbf{D}^2 = \begin{pmatrix} (\mathbf{G}+I)I_2 & 0 \\ 0 & (\mathbf{G}+I)I_2 \end{pmatrix},$$

(1.5) implies (1.4).

Note that the magnetic field B in (1.2) does not satisfy the hypotheses of [**10**] if $a'(r)$ grows exponentially or faster as $r \to \infty$, but nevertheless (1.4) holds by our Theorem 2.1.

The paper is organized as follows. In Section 2 we give a precise formulation of our main result, Theorem 2.1, and describe simple sufficient conditions which guarantee the validity of its assumptions. In Section 3 we define the selfadjoint operator $\mathbf{G}$ and prove that it is unitarily equivalent to the operator defined in (1.3). In Section 4 we prove Theorem 2.1 following, as explained above, the scheme based on the asymptotic analysis of the eigenvalues $\mu_{n,m}(k)$ as $k \to \pm\infty$.

I dedicate this article to my teacher Professor Mikhail Birman on the occasion of his 80th birthday. Among his numerous works which have strongly influenced the present paper, I would like to mention especially [**1**], [**2**], and [**3**], where the absolute continuity of the spectrum of various magnetic quantum Hamiltonians with periodic coefficients was proved.

2. Main result

The main result of the article is the following

THEOREM 2.1. *Let A have the form* (1.1). *Assume*

$$a(r) \to \infty \quad \text{as} \quad r \to \infty, \tag{2.1}$$

$$a' \in L^s_{\rm loc}([0,\infty); rdr), \quad s > 1, \tag{2.2}$$

$$a'(r) = o(a(r)^2), \quad r \to \infty. \tag{2.3}$$

Suppose moreover that for each $R > 0$ we have

$$\lim_{k\to\infty} e^{Rk} I(k)^{-1} = 0, \tag{2.4}$$

where

$$I(k) := \int_0^\infty e^{-2\int_0^r a(s)ds + 2kr} dr, \quad k > 0. \tag{2.5}$$

Then the spectrum $\sigma(\mathbf{G})$ of the operator $\mathbf{G}$ is purely absolutely continuous, (1.4) *holds, and the multiplicity of $\sigma(\mathbf{G})$ is infinite.*

Let us discuss briefly the assumptions of Theorem 2.1. First, (2.2) implies that a is absolutely continuous on $\mathbb{R}_+ := (0,\infty)$, and, hence, $a \in L^\infty_{\rm loc}(\mathbb{R}_+)$. Consequently, (2.3) entails $a' \in L^\infty_{\rm loc}([R,\infty))$ for $R > 0$ large enough. Further, it follows from (2.2) that a satisfies the estimate

$$|a(r)| = \begin{cases} O(r^{(s-2)/s}), & s \in (1,2), \\ O(|\ln r|^{1/2}), & s = 2, \\ O(1), & s > 2, \end{cases} \qquad r \downarrow 0.$$

Therefore,

$$a \in L^1_{\rm loc}([0,\infty)) \cap L^2_{\rm loc}([0,\infty); rdr).$$

In particular, the integral $\int_0^r a(s)ds$ occurring in (2.5) is well defined.

Evidently, (2.1) and (2.3) are satisfied by a very large class of functions, including, for example, functions which are asymptotically equivalent to positive powers of r, or exponentials of positive powers of r, or finite iterations of exponential functions.

The following proposition contains a sufficient condition for the validity of (2.4).

We will say that a given nonnegative function $\tau : [0,\infty) \to [0,\infty)$ belongs to the class $\mathcal{K}$ if:

- $\tau' \in L^s_{\rm loc}([0,\infty); rdr)$, $s > 1$;
- τ' is continuous and monotonically increasing on $[R_0,\infty)$ for some $R_0 \geq 0$;
- $\tau'(r) \to \infty$ as $r \to \infty$;
- for each $R > 0$ there exists $r_0 > 0$ such that

$$(R - 2\rho_k)k + 2\int_0^{\rho_k} \tau(s)ds + \frac{1}{2}\ln \tau'(\rho_k + r_0) \to -\infty, \quad k \to \infty, \tag{2.6}$$

  where $\rho_k := \tau^{-1}(k)$ for $k > 0$ large enough.

For example, all nonnegative functions τ with $\tau' \in L^s_{\rm loc}([0,\infty); rdr)$, $s > 1$, which for large $r > 0$ are proportional to e^{r^γ}, $\gamma > 0$, or to finite iterations of exponential functions $\exp(\exp(\dots(\exp r)))$, are in the class $\mathcal{K}$.

PROPOSITION 2.1. *Suppose that there exists a function $\tau \in \mathcal{K}$ such that*

$$a(r) \leq \tau(r), \quad r \in [0,\infty). \tag{2.7}$$

Then (2.4) *holds.*

The proof of Proposition 2.1 is contained in Subsection 4.4.

3. Representation of Hamiltonians as direct integrals of operators with discrete spectra

Assume $A \in L^2_{\rm loc}(\mathbb{R}^3)^3$, $B = \operatorname{curl} A \in L^1_{\rm loc}(\mathbb{R}^3)^3$. Let $\mathbf{G}$ be the selfadjoint operator generated in $L^2(\mathbb{R}^3)^2$ by the closure of the quadratic form

$$\begin{aligned} g[\mathbf{u}] :&= \int_{\mathbb{R}^3} \left| \sum_{j=1,2,3} \hat{\sigma}_j \left(i\frac{\partial}{\partial x_j} + A_j \right) \mathbf{u} \right|^2 dx \\ &= \sum_{l=1,2} \int_{\mathbb{R}^3} |(i\nabla + A)u_l|^2\, dx - \sum_{j=1,2,3} \int_{\mathbb{R}^3} B_j \langle \hat{\sigma}_j \mathbf{u}, \mathbf{u}\rangle dx, \\ \mathbf{u} &= \begin{pmatrix} u_1 \\ u_2 \end{pmatrix} \in C_0^\infty(\mathbb{R}^3)^2, \end{aligned}$$

where $\langle\cdot,\cdot\rangle$ denotes the scalar product in $\mathbb{C}^2$. Assume now that $A = (0,0,a)$ with $a = a(x_1,x_2)$ satisfying $a \in L^2_{\rm loc}(\mathbb{R}^2)$, $\nabla a \in L^1_{\rm loc}(\mathbb{R}^2)^2$. Evidently, in this case, $C_0^\infty(\mathbb{R}^2_{x_1,x_2}; \mathcal{S}(\mathbb{R}_{x_3}))^2$, where $\mathcal{S}(\mathbb{R})$ denotes the Schwartz class, is also a form core for the operator $\mathbf{G}$.

Further, for $k \in \mathbb{R}$ introduce the quadratic form

$$\begin{aligned} g_k[\mathbf{u}] :&= \int_{\mathbb{R}^2} \left| i \sum_{j=1,2} \hat{\sigma}_j \frac{\partial \mathbf{u}}{\partial x_j} + \hat{\sigma}_3 (a-k)\mathbf{u} \right|^2 dx \\ &= \int_{\mathbb{R}^2} \left(\left| 2i\frac{\partial u_2}{\partial z} + (a-k)u_1 \right|^2 + \left| 2i\frac{\partial u_1}{\partial \overline{z}} - (a-k)u_2 \right|^2 \right) dx \\ &= \int_{\mathbb{R}^2} \left(\sum_{l=1,2} \left(|\nabla u_l|^2 + (a-k)^2 |u_l|^2 \right) - 4\mathrm{Re}\, i \frac{\partial a}{\partial z} u_2 \overline{u_1} \right) dx, \\ \mathbf{u} &= \begin{pmatrix} u_1 \\ u_2 \end{pmatrix} \in C_0^\infty(\mathbb{R}^2)^2, \end{aligned}$$

where $z = x_1 + ix_2$, $\overline{z} = x_1 - ix_2$. Denote by $G(k)$ the selfadjoint operator generated in $L^2(\mathbb{R}^2)$ by the closure of g_k, $k \in \mathbb{R}$. Note that $C_0^\infty(\mathbb{R}^2 \setminus \{0\})^2$ is also a form core for the operator $G(k)$.

Let $\mathcal{F}$ be the partial Fourier transform with respect to x_3, i.e.

$$(\mathcal{F}u)(x_1, x_2, k) = \frac{1}{\sqrt{2\pi}} \int_{\mathbb{R}} e^{-ix_3 k} u(x_1, x_2, x_3) dx_3.$$

Since $C_0^\infty(\mathbb{R}^2_{x_1,x_2}; \mathcal{S}(\mathbb{R}_{x_3}))$ is invariant with respect to $\mathcal{F}$, we have

$$\mathcal{F}\mathbf{G}\mathcal{F}^* = \int_{\mathbb{R}}^{\oplus} G(k) dk. \tag{3.1}$$

Further, assume that a has the form (1.1), and satisfies (2.2). For $m \in \mathbb{Z}$ define the matrix

$$M_m := \frac{1}{2} \begin{pmatrix} m^2 + (m-1)^2 & m^2 - (m-1)^2 \\ m^2 - (m-1)^2 & m^2 + (m-1)^2 \end{pmatrix}.$$

Note that M_m has eigenvalues m^2 and $(m-1)^2$ with corresponding eigenvectors $\begin{pmatrix} 1 \\ 1 \end{pmatrix}$ and $\begin{pmatrix} 1 \\ -1 \end{pmatrix}$. For $k \in \mathbb{R}$ and $m \in \mathbb{Z}$ introduce the quadratic form

$$g_{k,m}[\mathbf{w}] := \sum_{l=1,2} \int_0^\infty \left(|w_l'|^2 + (a-k)^2 |w_l|^2 - (-1)^l a' |w_l|^2 \right) r dr + \int_0^\infty \langle M_m \mathbf{w}, \mathbf{w} \rangle r^{-1} dr$$

defined originally on $C_0^\infty(\mathbb{R}_+)^2$, and then closed in $L_2(\mathbb{R}_+; rdr)^2$. Let $G_m(k)$ be the selfadjoint operator generated in $L^2(\mathbb{R}_+; rdr)^2$ by $g_{k,m}$.

PROPOSITION 3.1. *Let a satisfy (2.2). Then the operator $G(k)$, $k \in \mathbb{R}$, is unitarily equivalent to the orthogonal sum*

$$\bigoplus_{m \in \mathbb{Z}} G_m(k). \tag{3.2}$$

PROOF. Pick $\mathbf{u} \in C_0^\infty(\mathbb{R}^2 \setminus \{0\})^2$. Passing to polar coordinates (r, θ), and setting $\mathbf{v}(r, \theta) := \mathbf{u}(r\cos\theta, r\sin\theta)$, we find that

$$g_k[\mathbf{u}] = \tilde{g}_k[\mathbf{v}], \quad \|\mathbf{u}\|^2_{L^2(\mathbb{R}_+; rdr)^2} = \sum_{l=1,2} \int_0^\infty \int_0^{2\pi} |v_l(r,\theta)|^2 d\theta r dr, \tag{3.3}$$

where

$$\tilde{g}_k[\mathbf{v}] := \sum_{l=1,2} \int_0^\infty \int_0^{2\pi} \left(\left| \frac{\partial v_l}{\partial r} \right|^2 + \frac{1}{r^2} \left| \frac{\partial v_l}{\partial \theta} \right|^2 + (a-k)^2 |v_l|^2 \right) d\theta r dr$$
$$+ \int_0^\infty \int_0^{2\pi} a'(r) \langle \mathcal{M}(\theta)\mathbf{v}, \mathbf{v} \rangle d\theta r dr,$$

and

$$\mathcal{M}(\theta) = \begin{pmatrix} 0 & -ie^{-i\theta} \\ ie^{i\theta} & 0 \end{pmatrix}.$$

The eigenvalues of the matrix $\mathcal{M}(\theta)$ are equal to 1 and -1, and the corresponding orthonormal eigenvectors can be chosen as

$$\frac{1}{\sqrt{2}} \begin{pmatrix} -ie^{-i\theta} \\ 1 \end{pmatrix}, \quad \frac{1}{\sqrt{2}} \begin{pmatrix} ie^{-i\theta} \\ 1 \end{pmatrix}.$$

Set

$$\mathcal{U} := \frac{1}{\sqrt{2}} \begin{pmatrix} -ie^{-i\theta} & ie^{-i\theta} \\ 1 & 1 \end{pmatrix}.$$

Then

$$\mathcal{U}^* = \mathcal{U}^{-1} := \frac{1}{\sqrt{2}} \begin{pmatrix} ie^{i\theta} & 1 \\ -ie^{i\theta} & 1 \end{pmatrix}, \quad \mathcal{U}^* \mathcal{M}(\theta) \mathcal{U} = \begin{pmatrix} 1 & 0 \\ 0 & -1 \end{pmatrix}.$$

Set $\mathbf{w} := \mathcal{U}^* \mathbf{v}$. We have

$$\tilde{g}_{k,m}[\mathbf{v}] = \sum_{l=1,2} \int_0^\infty \int_0^{2\pi} \left(\left| \frac{\partial w_l}{\partial r} \right|^2 + (a-k)^2 |w_l|^2 - (-1)^l a' |w_l|^2 \right) d\theta r dr$$
$$+ \frac{1}{2} \int_0^\infty \int_0^{2\pi} \left(\left| i \frac{\partial w_1}{\partial \theta} + w_1 - i \frac{\partial w_2}{\partial \theta} - w_2 \right|^2 + \left| \frac{\partial w_1}{\partial \theta} + \frac{\partial w_2}{\partial \theta} \right|^2 \right) d\theta r^{-1} dr. \tag{3.4}$$

Decomposing $\mathbf{w}$ into a Fourier series

$$\mathbf{w}(r, \theta) = \frac{1}{\sqrt{2\pi}} \sum_{m \in \mathbb{Z}} \mathbf{w}_m(r) e^{im\theta},$$

with $\mathbf{w}_m(r) = \begin{pmatrix} w_{m,1}(r) \\ w_{m,2}(r) \end{pmatrix}$, $m \in \mathbb{Z}$, and bearing in mind (3.3)–(3.4), we find that

$$g_k[\mathbf{u}] = \sum_{m \in \mathbb{Z}} g_{k,m}[\mathbf{w}_m], \quad \|\mathbf{u}\|^2_{L^2(\mathbb{R}^2)^2} = \sum_{m \in \mathbb{Z}} \|\mathbf{w}_m\|^2_{L^2(\mathbb{R}_+; rdr)^2}, \quad k \in \mathbb{R},$$

which implies the unitary equivalence of $G(k)$ and the orthogonal sum in (3.2). □

Combining (3.1) with Proposition 3.1, we find that $\mathbf{G}$ is unitarily equivalent to the operator defined in (1.3).

Suppose now that (2.1)–(2.3) hold. Then for each $m \in \mathbb{Z}$ and $k \in \mathbb{R}$, the spectrum of the operator $G_m(k)$ is discrete. Moreover, since the quadratic form $\int_0^\infty a|u|^2 r dr$ is relatively compact with respect to the closed quadratic form $g_{0,m}[u]$, we find easily that $\{G_m(k)\}_{k \in \mathbb{R}}$ is a Kato analytic family of type (B) (see [**12**, Chapter VII, Theorem 4.8]). Let $\{\mu_{n,m}(k)\}_{n \in \mathbb{N}}$ be the nonincreasing sequence of the eigenvalues of the operator $G_m(k)$, $m \in \mathbb{Z}$, $k \in \mathbb{R}$. Evidently, for each $n \in \mathbb{N}$ and $m \in \mathbb{Z}$ we have $\mu_{n,m} \in C(\mathbb{R})$.

4. Proof of the main result

4.1. In this subsection we investigate the asymptotic behaviour as $k \to -\infty$ of the eigenvalues $\mu_{n,m}(k)$, $n \in \mathbb{N}$, $m \in \mathbb{Z}$.

Let $\alpha : \mathbb{R}^2 \to \mathbb{R}$, $\beta : \mathbb{R}^2 \to \mathbb{R}$, satisfy

(4.1)
$$\alpha_+ \in L^2_{\rm loc}(\mathbb{R}^2),\ \alpha_- \in L^2(\mathbb{R}^2),\ (\alpha^2-\beta)_+ \in L^1_{\rm loc}(\mathbb{R}^2),\ (\alpha^2-\beta)_- \in L^s(\mathbb{R}^2),\ s>1.$$

Then for each $p \geq 0$ the quadratic form

$$\int_{\mathbb{R}^2} \left(\left(|\nabla u|^2 + \left((p+\alpha)^2 - \beta\right)\right)|u|^2\right) dx, \quad u \in C_0^\infty(\mathbb{R}^2), \tag{4.2}$$

is lower bounded in $L^2(\mathbb{R}^2)$. Denote by $\mathcal{L}(p)$ the selfadjoint operator generated in $L^2(\mathbb{R}^2)$ by the closure of (4.1).

LEMMA 4.1. *Assume that α and β satisfy* (4.1). *Then we have*

$$\liminf_{p\to\infty} p^{-2} \inf \sigma(\mathcal{L}(p)) \geq 1. \tag{4.3}$$

PROOF. If α is essentially lower-bounded, then (4.3) is trivial. Assume that α is not essentially lower-bounded, and pick an arbitrary $t > 0$. Then we have

$$(p+\alpha)^2 - \beta \geq p^2 - 2pt - 2p(\alpha+t)_- - (\alpha^2-\beta)_-.$$

Hence,

(4.4)
$$\begin{aligned}&\int_{\mathbb{R}^2} \left(|\nabla u|^2 + \left((p+\alpha)^2 - \beta\right)|u|^2\right) dx \\ &\quad \geq \int_{\mathbb{R}^2} \left(|\nabla u|^2 + \left((p^2-2pt) - 2p(\alpha+t)_- - (\alpha^2-\beta)_-\right)|u|^2\right) dx, \quad u \in C_0^\infty(\mathbb{R}^2).\end{aligned}$$

Denote by $\mathcal{L}_1 = \mathcal{L}_1(p,t)$ the selfadjoint operator generated in $L^2(\mathbb{R}^2)$ by the closure of the quadratic form

$$\int_{\mathbb{R}^2} \left(\frac{1}{2}|\nabla u|^2 - 2p(\alpha+t)_-|u|^2\right) dx, \quad u \in C_0^\infty(\mathbb{R}^2),$$

and by $\mathcal{L}_2$ the selfadjoint operator generated in $L^2(\mathbb{R}^2)$ by the closure of the quadratic form

$$\int_{\mathbb{R}^2} \left(\frac{1}{2}|\nabla u|^2 - (\alpha^2-\beta)_-|u|^2\right) dx, \quad u \in C_0^\infty(\mathbb{R}^2).$$

The essential spectra of $\mathcal{L}_j$, $j = 1,2$, coincide with $[0,\infty)$, while their discrete spectra are finite. Moreover, for every fixed $t > 0$ and sufficiently large $p > 0$, the negative spectrum of $\mathcal{L}_1(p,t)$ is not empty. Let $\{-\lambda_j(p,t)\}_{j=1}^N$ be the negative eigenvalues of $\mathcal{L}_1(p,t)$ counted with their multiplicities and enumerated in nondecreasing order. Then (4.4) implies

$$\inf \sigma(\mathcal{L}(p)) \geq p^2 - 2pt - \lambda_1(p,t) + \inf \sigma(\mathcal{L}_2). \tag{4.5}$$

On the other hand, the well-known Lieb–Thirring inequality (see [**13**]) yields

$$\lambda_1(p,t) \leq \sum_{j=1}^N \lambda_j(p,t) \leq Cp^2 \int_{\mathbb{R}^2} (\alpha+t)_-^2\, dx \tag{4.6}$$

with $C > 0$ independent of p, α, and t. Combining (4.5) and (4.6), we get

$$\liminf_{p\to\infty} p^{-2} \inf \sigma(\mathcal{L}(p)) \geq 1 - C \int_{\mathbb{R}^2} (\alpha + t)_-^2 \, dx. \tag{4.7}$$

Since $\alpha_- \in L^2(\mathbb{R}^2)$, we have $\lim_{t\to\infty} \int_{\mathbb{R}^2} (\alpha + t)_-^2 \, dx = 0$, and (4.3) follows from (4.7). □

Remark: We formulate Lemma 4.1 in a form suitable for our purposes. A related result is contained in [**16**, Proposition 3.2].

PROPOSITION 4.1. *Let* $n \in \mathbb{N}$, $m \in \mathbb{Z}$. *Assume that* (2.1)–(2.3) *hold. Then we have*

$$\mu_{n,m}(k) = k^2(1 + o(1)), \quad k \to -\infty. \tag{4.8}$$

PROOF. Evidently, $\mu_{n,m}(k) \geq \inf \sigma(G(k))$, $k \in \mathbb{R}$. On the other hand, for $k \leq 0$ we have $G(k) \geq \tilde{H}(k) I_2$, where $\tilde{H}(k) := \mathcal{L}(-k)$ with $\alpha = a$ and $\beta = |a'|$. Then (4.3) entails

$$\liminf_{k\to-\infty} k^{-2} \mu_{n,m}(k) \geq 1. \tag{4.9}$$

For $\varepsilon > 0$ and $m \in \mathbb{Z}$ denote by $Q_m(\varepsilon)$ the selfadjoint operator generated in $L^2(\mathbb{R}_+; rdr)$ by the closure of the quadratic form

$$\int_0^\infty \left(|u'|^2 + \left(\frac{m^2}{r^2} + (1 + \varepsilon^{-1}) a^2 + |a'| \right) |u|^2 \right) rdr, \quad u \in C_0^\infty(\mathbb{R}_+).$$

The spectrum of $Q_m(\varepsilon)$ is discrete and simple; let $\{\nu_{n,m}(\varepsilon)\}_{n\in\mathbb{N}}$ be the increasing sequence of its eigenvalues. It is easy to check that

$$G_m(k) \leq \left(Q_{m^*}(\varepsilon) + (1 + \varepsilon) k^2 \right) I_2,$$

where $m^* := \max\{|m|, |m-1|\}$. By the minimax principle,

$$\mu_{n,m}(k) \leq \nu_{[\frac{n-1}{2}]+1, m^*}(\varepsilon) + (1 + \varepsilon) k^2,$$

and, hence,

$$\limsup_{k\to-\infty} k^{-2} \mu_{n,m}(k) \leq 1 + \varepsilon, \quad \forall \varepsilon > 0,$$

which, combined with (4.9), yields (4.8). □

4.2. In this subsection we study the behaviour of the eigenvalues $\mu_{1,m}(k)$, $m \in \mathbb{Z}$, as $k \to \infty$.

PROPOSITION 4.2. *Assume the hypotheses of Theorem* 2.1. *Then we have*

$$\lim_{k\to\infty} \mu_{1,m}(k) = 0, \quad m \in \mathbb{Z}. \tag{4.10}$$

PROOF. In order to prove (4.10), it suffices to show that for each $\varepsilon > 0$ and $k \in \mathbb{R}$ there exists $\mathbf{u}_{k,\varepsilon} \in D(g_{k,m})$, the domain of the closed quadratic form $g_{k,m}$, such that

$$\lim_{\varepsilon\downarrow 0} \limsup_{k\to\infty} \frac{g_{k,m}[\mathbf{u}_{k,\varepsilon}]}{\|\mathbf{u}_{k,\varepsilon}\|^2_{L^2(\mathbb{R}_+; rdr)^2}} = 0. \tag{4.11}$$

Set

$$\phi_k(r) := e^{-\int_0^r a(s)ds + kr}, \quad r > 0, \quad k \in \mathbb{R}.$$

By (2.1)–(2.3), the functions $\phi_k(r)$ and $\phi_k'(r) = -(a(r)-k)\phi_k(r)$ decay superexponentially as $r \to \infty$. Hence, $\phi_k \in L^2(\mathbb{R}_+)$ for each $k \in \mathbb{R}$ and we have

$$\int_0^\infty \phi_k(r)^2 dr = I(k), \quad k \in \mathbb{R},$$

the integral $I(k)$ being defined in (2.5). Moreover, ϕ_k satisfies the differential equation

$$-\phi_k'' + ((a(r)-k)^2 - a'(r))\phi_k(r) = 0. \tag{4.12}$$

Further, pick a function $\zeta \in C^\infty([0,\infty))$ such that $\zeta(r) = 0$ if $r \in [0, 1/2]$, $0 \leq \zeta(r) \leq 1$ if $r \in [1/2, 1]$, and $\zeta(r) = 1$ if $r \in [1, \infty)$. Fix $\varepsilon > 0$ and for $r > 0$ set

$$\zeta_\varepsilon(r) := \zeta(\varepsilon^{1/2} r), \quad u_{k,\varepsilon}(r) := r^{-1/2}\zeta_\varepsilon(r)\phi_k(r), \quad \mathbf{u}_{k,\varepsilon}(r) := \begin{pmatrix} 0 \\ u_{k,\varepsilon}(r) \end{pmatrix}.$$

Since $\operatorname{supp} u_{k,\varepsilon} \subseteq [R/2, \infty)$, where $R = R(\varepsilon) := \varepsilon^{-1/2}$, and

$$u_{k,\varepsilon} \in L^2(\mathbb{R}_+; (a^2 + r^{-2})rdr), \quad u_{k,\varepsilon}' \in L^2(\mathbb{R}_+; rdr),$$

we easily find that $\mathbf{u}_{k,\varepsilon} \in D(g_{k,m})$. Integrating by parts, and taking into account (4.12), we get

$$\begin{aligned} g_{k,m}[\mathbf{u}_{k,\varepsilon}] &= \int_0^\infty \phi_k(r)^2 \zeta_\varepsilon'(r)^2 dr + \frac{2m^2 + 2(m-1)^2 - 1}{4} \int_0^\infty \frac{\phi_k(r)^2 \zeta_\varepsilon(r)^2}{r^2} dr \\ &\leq \varepsilon \left(\max_{r \in \mathbb{R}_+} \zeta'(r)^2 \int_0^R \phi_k(r)^2 dr + \left(2m^2 + 2(m-1)^2 - 1\right) \int_{R/2}^\infty \phi_k(r)^2 dr \right) \\ &\leq \text{const.}\ \varepsilon I(k). \end{aligned} \tag{4.13}$$

On the other hand, we have

$$\|\mathbf{u}_{k,\varepsilon}\|^2_{L^2(\mathbb{R}_+; rdr)^2} = \int_0^\infty \phi_k(r)^2 \zeta_\varepsilon(r)^2 dr \geq I(k) - \int_0^R \phi_k(r)^2 dr \geq I(k) - ce^{2kR} \tag{4.14}$$

with $c := \max_{r \in [0,R]} \exp\left(-2\int_0^r a(s)ds\right)$, provided that $k \geq 1/2$. Combining (4.13) with (4.14), we obtain

$$\frac{g_{k,m}[\mathbf{u}_{k,\varepsilon}]}{\|\mathbf{u}_{k,\varepsilon}\|^2_{L^2(\mathbb{R}_+; rdr)^2}} \leq \frac{\text{const.}\ \varepsilon}{1 - ce^{2kR} I(k)^{-1}}. \tag{4.15}$$

Now (4.11) follows from (4.15) and (2.4). □

Remark: The method of the proof of Proposition 4.2 was inspired by the supersymmetric model

$$S = \begin{pmatrix} LL^* & 0 \\ 0 & L^*L \end{pmatrix} := \begin{pmatrix} -\frac{d^2}{dx^2} + q^2 + q' & 0 \\ 0 & -\frac{d^2}{dx^2} + q^2 - q' \end{pmatrix}$$

considered in [**6**] (see also [**4**, Section 6.3]). Here $q : \mathbb{R} \to \mathbb{R}$ is a polynomial, and the operators

$$L := \frac{d}{dx} + q, \quad L^* := -\frac{d}{dx} + q,$$

with appropriate (coinciding) domains are mutually adjoint in $L^2(\mathbb{R})$. Then we have

$$\operatorname{Ker} S = \left\{ \mathbf{u} = \begin{pmatrix} u_1 \\ u_2 \end{pmatrix} \,\middle|\, u_1 \in \operatorname{Ker} L^*,\ u_2 \in \operatorname{Ker} L \right\},$$

and the kernels $\operatorname{Ker} L$ and $\operatorname{Ker} L^*$ admit an explicit and simple description.

4.3. In this subsection we prove Theorem 2.1.

Since $\mathbf{G}$ is unitarily equivalent to the operator defined in (1.3), the absolute continuity of $\sigma(\mathbf{G})$ will follow from the absolute continuity of $\sigma(G_m)$, where $G_m := \int_{\mathbb{R}}^{\oplus} G_m(k)dk$, $m \in \mathbb{Z}$. Recall that $\{G_m(k)\}_{k\in\mathbb{R}}$ is a Kato analytic family of type (B). By an appropriate version of the Rellich theorem (see [**14**, Theorem XII.13], [**12**, Chapter VII, Remark 4.22]), there exists a family of analytic functions $\mathbb{R} \ni k \mapsto \tilde{\mu}_{n,m}(k)$, $n \in \mathbb{N}$, such that the set of the eigenvalues of the operator $G_m(k)$, $k \in \mathbb{R}$, counted with its multiplicities, coincides with $\{\tilde{\mu}_{n,m}(k)\}_{n\in\mathbb{N}}$. By the general spectral theory of direct integrals of selfadjoint operators with discrete spectra (see, e.g., [**14**, Theorem XIII.86]), it suffices to check that among $\{\tilde{\mu}_{n,m}(k)\}_{n\in\mathbb{N}}$ there are no functions which are identically constant. This, however, follows immediately from the inequality $\tilde{\mu}_{n,m}(k) \geq \mu_{1,m}(k)$, $n \in \mathbb{N}$, $k \in \mathbb{R}$, and the relation $\mu_{1,m}(k) \to \infty$ as $k \to -\infty$ implied by Proposition 4.1.

Further, since $\mathbf{G} \geq 0$, we have

$$\overline{\mu_{1,m}(\mathbb{R})} \subseteq \sigma(\mathbf{G}) \subseteq [0,\infty), \quad m \in \mathbb{Z}. \tag{4.16}$$

By Propositions 4.1 and 4.2 and the continuity of $\mu_{1,m}$, we have $\overline{\mu_{1,m}(\mathbb{R})} = [0,\infty)$ for each $m \in \mathbb{Z}$, so that (4.16) entails (1.4) and the infinite multiplicity of the spectrum of the operator $\mathbf{G}$.

4.4. In this subsection we prove Proposition 2.1.

Our argument will be close to the well-known Laplace method for the approximate evaluation of integrals depending on a large parameter.

Fix R and pick $r_0 > 0$ such that (2.6) holds. Also, assume that k is so large that $R_0 \leq \rho_k$, and hence τ' is continuous and monotonic on $[\rho_k, \infty)$. Then we have

$$I(k) \geq \int_0^\infty \exp\left(-2\int_0^r \tau(s)ds + 2kr\right)dr \geq \int_{\rho_k}^{\rho_k+r_0} \exp\left(-2\int_0^r \tau(s)ds + 2kr\right)dr.$$

A second-order Taylor expansion of the function $-2\int_0^r \tau(s)ds + 2kr$ at ρ_k, and the monotonicity of $\tau'(r)$ yield

$$-2\int_0^r \tau(s)ds + 2kr = -2\int_0^{\rho_k} \tau(s)ds + 2k\rho_k - 2\int_{\rho_k}^r \tau'(s)(r-s)ds$$

$$\geq -2\int_0^{\rho_k} \tau(s)ds + 2k\rho_k - \tau'(\rho_k + r_0)(r-\rho_k)^2, \quad r \in [\rho_k, \rho_k + r_0].$$

Therefore,

$$I(k) \geq \exp\left(-2\int_0^{\rho_k} \tau(s)ds + 2k\rho_k\right) \int_{\rho_k}^{\rho_k+r_0} \exp\left(-\tau'(\rho_k+r_0)(r-\rho_k)^2\right)dr.$$

Changing the variable $r = \tau'(\rho_k + r_0)^{-1/2}s + \rho_k$, and taking into account that $\tau'(r) \to \infty$ as $r \to \infty$, we get

$$I(k) \geq \frac{\exp\left(-2\int_0^{\rho_k} \tau(s)ds + 2k\rho_k\right)}{\tau'(\rho_k + r_0)^{1/2}} \left(\frac{\sqrt{\pi}}{2} + o(1)\right), \quad k \to \infty. \tag{4.17}$$

Bearing (2.6) in mind, we find that (4.17) implies (2.4).

Acknowledgements

The author thanks D. R. Yafaev who drew his attention to the spectral properties of a special class of magnetic quantum Hamiltonians. Acknowledgements are due to László Erdős and Xue Ping Wang for useful e-mail exchanges. The financial support of the Chilean Science Foundation *Fondecyt* under Grant 1050716, and of the University of Rennes I during the author's visits in 2005 and 2007 is gratefully acknowledged.

References

[1] M. Sh. Birman, T. A. Suslina, *The two-dimensional periodic magnetic Hamiltonian is absolutely continuous*, Algebra i Analiz **9** (1997), 32–48; English transl., St. Petersburg Math. J. **9** (1998), 21–32.

[2] M. Sh. Birman, T. A. Suslina, *Absolute continuity of a two-dimensional periodic magnetic Hamiltonian with discontinuous vector potential*, Algebra i Analiz **10** (1998), 1–36; English transl., St. Petersburg Math. J. **10** (1999), 579–601.

[3] M. Sh. Birman, T. A. Suslina, *The periodic Dirac operator is absolutely continuous*, Integral Equations Operator Theory **34** (1999), 377–395.

[4] H. Cycon, R. Froese, W. Kirsch, B. Simon, *Schrödinger Operators with Application to Quantum Mechanics and Global Geometry*, Texts and Monographs in Physics, Springer-Verlag, Berlin, Heidelberg, New York, 1987.

[5] S. De Bièvre, J. V. Pulé, *Propagating edge states for a magnetic Hamiltonian*, Math. Phys. Electron. J. **5** (1999), Paper 3, 17 pp.

[6] P. A. Deift, *Applications of a commutation formula*, Duke Math. J. **45** (1978), 267–310.

[7] L. Erdős, *Dia- and paramagnetism for nonhomogeneous magnetic fields*, J. Math. Phys. **38** (1997), 1289–1317.

[8] V. Geĭler, M. Senatorov, *The structure of the spectrum of the Schrödinger operator with a magnetic field in a strip, and finite-gap potentials*, Mat. Sb. **188** (1997), 21–32; English transl., Sb. Math. **188** (1997), 657–669.

[9] B. Helffer, *Introduction to Semi-Classical Methods for the Schrödinger Operator with Magnetic Field. Vienna Version*, Lecture notes of a course given at the ESI, 2006, available at http://www.math.u-psud.fr/~helffer/syrievienne2006.pdf.

[10] B. Helffer, J. Nourrigat, X. P. Wang, *Sur le spectre de l'équation de Dirac (dans $\mathbb{R}^3$ ou $\mathbb{R}^2$) avec champ magnétique*, Ann. Sci. École Norm. Sup. (4) **22** (1989), 515–533.

[11] A. Iwatsuka, *Examples of absolutely continuous Schrödinger operators in magnetic fields*, Publ. Res. Inst. Math. Sci. **21** (1985), 385–401.

[12] T. Kato, *Perturbation Theory for Linear Operators*, 2nd Edition, Die Grundlehren der mathematischen Wissenschaften, **132**, Springer-Verlag, New York, Inc., New York 1980.

[13] E. H. Lieb, W. Thirring, *Inequalities for the moments of the eigenvalues of the Schrödinger Hamiltonian and their relation to Sobolev inequalities*, In: Studies in Mathematical Physics (Essays in Honor of Valentine Bargmann), pp. 269–303. Princeton Univ. Press, Princeton, NJ, 1976.

[14] M. Reed, B. Simon, *Methods of Modern Mathematical Physics, IV. Analysis of Operators*, Academic Press, New York, 1978.

[15] D. Yafaev, *A particle in a magnetic field of an infinite rectilinear current*, Math. Phys. Anal. Geom. **6** (2003), 219–230.

[16] D. Yafaev, *On spectral properties of translationally invariant magnetic Schrödinger operators*, Ann. Henri Poincaré **9** (2008), 181–207.

Departamento de Matemática, Facultad de Matemáticas, Pontificia Universidad Católica de Chile, Av. Vicuña Mackenna 4860, Santiago, Chile

E-mail address: graikov@mat.puc.cl

Amer. Math. Soc. Transl.
(2) Vol. **225**, 2008

Discrete Spectrum Distribution of the Landau Operator Perturbed by an Expanding Electric Potential

Grigori Rozenblum and Alexander V. Sobolev

To Mikhail Shlëmovich Birman on his 80th birthday

ABSTRACT. Under a perturbation by a decaying potential, the Landau Hamiltonian acquires some discrete eigenvalues between the Landau levels. We study the perturbation by an "expanding" electric potential $V(t^{-1}x)$, $t > 0$, and derive a quasi-classical formula for the counting function of the discrete spectrum as $t \to \infty$.

1. Introduction and main result

The two-dimensional Landau Hamiltonian $H_0 = (-i\nabla - \mathbf{a})^2$ describing a charged quantum particle moving in the plane in a constant magnetic field $B = \operatorname{curl} \mathbf{a}$ is one of the earliest explicitly solvable models of Quantum Mechanics. Its spectrum consists of infinitely degenerate eigenvalues (Landau levels) $\Lambda_q = B(2q+1)$, $q = 0, 1, \dots$ (see, e.g., [**10**]); we put $\Lambda_{-1} = -\infty$ for reference convenience. Under a perturbation by an electric or magnetic field decaying at infinity, the Landau levels split forming clusters (generically, infinite) of eigenvalues with Landau levels being their limit points. Various asymptotic properties of these clusters have been extensively studied in the literature. For instance, [**4**], [**11**] [**13**], [**14**] studied the rate of convergence of the eigenvalues to their limit points for rapidly decaying potential perturbations. It was found that for a compactly supported electric potential the eigenvalues converge to Landau levels superexponentially fast. A similar effect was observed for the perturbation by a compactly supported magnetic field [**15**] or impenetrable compact obstacle [**12**]. Another natural problem is to analyze the eigenvalue behavior as the coupling constant at the perturbation becomes large. In this case the eigenvalue asymptotics is described by semi-classical formulas, as shown in [**7**], [**8**]. We refer to the above cites for additional references.

Our objective is to study the discrete spectrum of the Landau Hamiltonian H_0 perturbed by an expanding potential $V^{(t)}(\mathbf{x}) = V(t^{-1}\mathbf{x})$, $t > 0$. Under the condition $V \in \mathsf{L}^1(\mathbb{R}^2) \cap \mathsf{L}^2(\mathbb{R}^2)$ the operator $H = H^{(t)} = H_0 + V^{(t)}$ is properly defined as

2000 *Mathematics Subject Classification.* Primary 35J10, 47F05.
Key words and phrases. Landau Hamiltonian, discrete spectrum, Toeplitz operators, spectral asymptotics.

an operator sum in $\mathsf{L}^2(\mathbb{R}^2)$, and $V^{(t)}$ is H_0-compact. Our goal is to investigate the number $N(\lambda_1, \lambda_2; H^{(t)})$ of eigenvalues of $H^{(t)}$ on the interval $(\lambda_1, \lambda_2) \Subset (\Lambda_\nu, \Lambda_{\nu+1})$ with some $\nu = -1, 0, 1, \dots$, as $t \to \infty$. If $\lambda_1 = -\infty$, we write $N(\lambda_2; H^{(t)})$.

The behavior of $N(\lambda_1, \lambda_2; H^{(t)})$ is determined by the potential V as follows. For any $V \in \mathsf{L}^1(\mathbb{R}^2)$ and $-\infty \le \lambda < \mu \le \infty$ we define

$$A(\lambda, \mu; V) = |\{\mathbf{x} : \lambda < V(\mathbf{x}) < \mu\}|, \quad A^{(0)}(\lambda; V) = |\{\mathbf{x} : V(\mathbf{x}) = \lambda\}|. \tag{1.1}$$

The coefficients $A(\lambda, \mu; V)$ are finite if $\lambda\mu > 0$. Say, for $\mu > \lambda > 0$, we obtain by the Chebyshev inequality

$$A(\lambda, \mu; V) \le A(\lambda, \infty; V) < \frac{1}{\lambda}\|V\|_1 < \infty.$$

The coefficient A is monotone in λ, μ, so that the limits $A(\lambda \pm 0, \mu \pm 0; V)$ are well defined with various combinations of signs $\pm$. We will call the number λ a *generic* value for V if $A^{(0)}(\lambda; V) = 0$; otherwise, this value is called *exceptional*. For a given function V, there are at most countably many exceptional values. For a generic λ, $A(\lambda - 0, \mu; V) = A(\lambda, \mu; V)$; otherwise, $A(\lambda - 0, \mu; V) = A(\lambda, \mu; V) + A^{(0)}(\lambda; V)$, and similarly for μ.

For any real $\lambda_1 < \lambda_2$ such that $[\lambda_1, \lambda_2]$ does not contain any of Λ_q, we define

$$\mathscr{A}(\lambda_1, \lambda_2; V) = \frac{B}{2\pi} \sum_{q=0}^{\infty} A(\lambda_1 - \Lambda_q, \lambda_2 - \Lambda_q; V). \tag{1.2}$$

Since the sets $\{\mathbf{x} : \Lambda_q - \lambda_2 < |V(\mathbf{x})| < \Lambda_q - \lambda_1\}$ are disjoint for different q's, this series converges for $V \in \mathsf{L}^1(\mathbb{R}^2)$, and

$$\mathscr{A}(\lambda_1, \lambda_2; V) \le A\big(\min\{|\Lambda_q - \lambda_2|, |\Lambda_q - \lambda_1|\}, \infty; |V|\big) < \infty.$$

The main result of the paper is contained in the following theorem:

THEOREM 1.1. *Let $(\lambda_1, \lambda_2) \Subset (\Lambda_\nu, \Lambda_{\nu+1})$ with some $\nu = -1, 0, 1, \dots$. Suppose that $V \in \mathsf{L}^1(\mathbb{R}^2) \cap \mathsf{L}^2(\mathbb{R}^2)$. Then*

$$\begin{aligned}\mathscr{A}(\lambda_1, \lambda_2; V) \le & \liminf_{t\to\infty} t^{-2} N(\lambda_1, \lambda_2; H^{(t)}) \\ \le & \limsup_{t\to\infty} t^{-2} N(\lambda_1, \lambda_2; H^{(t)}) \le \mathscr{A}(\lambda_1 - 0, \lambda_2 + 0; V).\end{aligned}$$

If $\lambda_1 - \Lambda_q$ and $\lambda_2 - \Lambda_q$ are generic for V for all $q = 0, 1, \dots$, then

$$\lim_{t\to\infty} t^{-2} N(\lambda_1, \lambda_2; H^{(t)}) = \mathscr{A}(\lambda_1, \lambda_2; V). \tag{1.3}$$

Note that the right hand side of the asymptotic formula (1.3) coincides with the natural quasi-classical expression for the counting function of the magnetic Schrödinger operator; see e.g. [**16**]. On the other hand one might compare this result with the classical Szegö Theorem deriving the canonical distribution for the Toeplitz type operators of Fourier type; see [**6**], Theorem 8.6(c). This comparison is even more appropriate since the proof of Theorem 1.1 relies on the spectral analysis of the Toeplitz type operator $T^{(t)} = P_q V^{(t)} P_q$, where P_q is the projection on the spectral subspace associated with the Landau level Λ_q. Remembering that the nonzero spectra of the operators AB and BA (A, B being both compact) coincide, instead of $T^{(t)}$ it is often more convenient to study the operator $S^{(t)} = W P_q \overline{W}$ with a function W. If W is radially symmetric, then this operator splits into an orthogonal sum of one-dimensional operators, whose eigenvalues (and their asymptotics) are computed using the explicit formula for the integral kernel of P_q (see (3.3)). To handle the general case we apply a method which is based on the approach put forth by M. Birman and M. Solomyak for studying weakly polar integral operators; see [**1**]. Precisely, we partition the plane $\mathbb{R}^2$ into disjoint annular sectors, i.e. domains of the form

$$\Omega_{m,l} = \left\{(\rho, \phi) : (m-1)d < \rho \le md,\ \frac{2\pi}{N}(l-1) < \phi \le \frac{2\pi}{N} l\right\}, m \in \mathbb{N}, l = 1, 2, \dots, N,$$

with a fixed $d > 0$ and natural N. Choosing appropriate d and N, we approximate W by a function which is constant on each $\Omega_{m,l}$. This reduces the operator $S^{(t)}$ to a block-matrix form. The crucial point is that the off-diagonal entries do not contribute to the asymptotics, which implies the "additivity" of the asymptotics in the function W. This property allows one to reduce the problem to the radially symmetric case, for which the eigenvalues are found in closed form.

The above Toeplitz operators are linked to the initial Schrödinger operator using the elementary formula

$$N(\lambda_1, \lambda_2; H^{(t)}) = N(-\infty, b^2; L^{(t)}), \quad L^{(t)} = (H_0 + V^{(t)} - a)^2, \tag{1.4}$$

where $a = (\lambda_1 + \lambda_2)/2$, $b = (\lambda_2 - \lambda_1)/2$, which was used previously in [**9**], [**11**] in similar circumstances. We represent $L^{(t)}$ in block-matrix form with entries of the form $P_q L^{(t)} P_{q'}$. The off-diagonal terms do not affect the asymptotics, and the diagonal ones are directly expressed via the Toeplitz operators of the form $T^{(t)}$.

The paper is organized as follows. Section 2 is devoted to some abstract operator theory. For families of semi-bounded operators depending on a parameter $t \in (0, \infty)$ we introduce the *asymptotic coefficients* describing the asymptotic distribution as $t \to \infty$ of eigenvalues in a given interval, and establish their general properties which are used throughout the paper in different concrete environments. In Sections 3 and 4 we prove the crucial asymptotic formulas for the Toeplitz operators $T^{(t)}$. The reduction of the initial problem to the Toeplitz operators is implemented in Sections 5 and 6.

2. Asymptotic coefficients

Let L be a selfadjoint operator, semi-bounded from below, with

$$\eta_0 = \eta_0(L) = \inf \sigma_{\mathrm{ess}}(L).$$

Denote by $N(\eta; L)$, $\eta < \eta_0$, the number of discrete eigenvalues of L strictly below η. If K is a compact operator, then $\sigma_{\mathrm{ess}}(K) = \{0\}$ and we use the traditional notation

$$n_-(\lambda; K) = N(-\lambda; K),\ n_+(\lambda; K) = N(-\lambda; -K), \quad \lambda > 0,$$

for the counting functions of the negative and positive eigenvalues respectively. For an arbitrary compact operator K, not necessarily selfadjoint, introduce also the counting function of its singular values:

$$n(\lambda; K) = n_+(\lambda^2; K^* K).$$

The distribution functions for eigenvalues of the sum (and difference) of operators satisfy certain inequalities. For compact operators they are known as Ky Fan inequalities and are presented in numerous sources. However we need these kinds of relations for general semi-bounded operators. Although these inequalities are by no means new, we were unable to locate a convenient reference for the particular form we intend to use. Therefore we present here a short explanation.

First, recall the version of the min-max principle for the counting function, which is usually referred to as Glazman's Lemma:

LEMMA 2.1. *Let L be a semi-bounded operator with $\eta_0 = \inf \sigma_{\mathrm{ess}}(L)$, and let $\mathbf{d}(L)$ be the domain of the quadratic form $l[u]$ of the operator L. Then $\eta < \eta_0$ if and only if there exists a linear set*[1] *$\mathscr{L} \subset \mathbf{d}(L)$ of finite codimension such that*

$$l[u] \geq \eta \|u\|^2 \quad \text{for all} \quad u \in \mathscr{L}. \tag{2.1}$$

Moreover,

$$N(\eta; L) = \min \operatorname{codim} \mathscr{L}, \tag{2.2}$$

where the minimum is taken over all linear sets $\mathscr{L} \subset \mathbf{d}(L)$, satisfying (2.1).

Glazman's Lemma appeared in this form in [**5**] (Ch. 1, Theorems 12, 12bis); equivalent formulations (however in terms of eigenvalues, not the distribution function) are given in many books on spectral theory.

The most general form of the eigenvalue distribution function inequality we need is the following result.

LEMMA 2.2. *Let L_1, L_2 be semi-bounded (from below) selfadjoint operators such that $\mathbf{d}(L_1) \subset \mathbf{d}(L_2)$ and L_2 is an L_1-form bounded with a bound strictly less than 1. Then $\eta_0(L_1 + L_2) \geq \eta_0(L_1) + \eta_0(L_2)$ and*

$$N(\eta_1 + \eta_2; L_1 + L_2) \leq N(\eta_1; L_1) + N(\eta_2; L_2) \tag{2.3}$$

for any $\eta_j < \eta_0(L_j)$, $j = 1, 2$, and

$$N(\eta_1 - \eta_2; L_1 - L_2) \geq N(\eta_1; L_1) - N(\eta_2; L_2), \tag{2.4}$$

for any η_1, η_2 such that $\eta_1 - \eta_2 < \eta_0(L_1 - L_2)$, $\eta_2 < \eta_0(L_2)$.

PROOF. Denote $N_j = N(\eta_j; L_j)$, and let $\mathscr{L}_j \in \mathbf{d}(L_j)$ be a subspace of codimension N_j on which $(L_j u, u) \geq \eta_j \|u\|^2$, $j = 1, 2$. Then the subspace $\mathscr{L} = \mathscr{L}_1 \cap \mathscr{L}_2 \subset \mathbf{d}(L_1 + L_2)$ has codimension not greater than $N_1 + N_2$, and for $u \in \mathscr{L}$,

$$((L_1 + L_2)u, u) \geq (\eta_1 + \eta_2)\|u\|^2,$$

so (2.3) follows from Glazman's Lemma. The inequality (2.4) follows from (2.3) by an obvious change of notation. □

We also need a simpler version of the above inequalities. Let L be a semi-bounded operator, and let K be compact and selfadjoint. Clearly, $\eta_0(L + K) = \eta_0(L)$, and by Lemma 2.2,

$$N(\eta - \lambda; L + K) \leq N(\eta; L) + n_-(\lambda; K) \tag{2.5}$$

for any $\eta < \eta_0$, $\lambda > 0$. It is useful to write a similar inequality for a pair of compact selfadjoint operators K_1, K_2:

$$n_\pm(\lambda_1 + \lambda_2; K_1 + K_2) \leq n_\pm(\lambda_1; K_1) + n_\pm(\lambda_2; K_2), \tag{2.6}$$

for any $\lambda_1, \lambda_2 > 0$. For a pair of compact operators (not necessarily selfadjoint) a similar inequality holds:

$$n(\lambda_1 + \lambda_2; K_1 + K_2) \leq n(\lambda_1; K_1) + n(\lambda_2; K_2) \tag{2.7}$$

for any $\lambda_1, \lambda_2 > 0$;see [**3**], Section 9.2, Theorem 9.

[1]Subspace that is not necessarily closed

For a family of semi-bounded operators $L = L^{(t)}$ depending on a parameter $t > 0$, with the value $\eta_0 = \eta_0(L^{(t)})$ independent of t, we introduce the asymptotic coefficients

$$\mathfrak{B}(\eta; L) = \limsup_{t\to\infty} t^{-2} N(\eta; L^{(t)}), \quad \mathfrak{b}(\eta; L) = \liminf_{t\to\infty} t^{-2} N(\eta; L^{(t)}), \quad \eta < \eta_0.$$

Clearly, one can introduce similar asymptotic coefficients, with t^{-2} replaced by $t^{-\gamma}$ with any $\gamma > 0$. Although such characteristics of operator families may prove to be useful for some other eigenvalue counting problems, the case $\gamma = 2$ is sufficient for our purposes. General properties of such coefficients are the same as for $\gamma = 2$. The asymptotic coefficients, just introduced, are not necessarily continuous in η, but they are monotone. We systematically use naturally defined limits such as $\mathfrak{B}(\eta \pm 0; L)$.

In order to keep in line with the traditional definition of counting functions $n_\pm$ for compact operators, for a compact selfadjoint family $K = K^{(t)}$ we denote

$$\mathfrak{R}^{(\pm)}(\lambda; K) = \mathfrak{B}(-\lambda; \mp K), \; \mathfrak{r}^{(\pm)}(\lambda; K) = \mathfrak{b}(-\lambda; \mp K),$$

and for an arbitrary compact family introduce also

$$\mathfrak{R}(\lambda; K) = \mathfrak{R}^{(+)}(\lambda^2; K^*K), \; \mathfrak{r}(\lambda; K) = \mathfrak{r}^{(+)}(\lambda^2; K^*K).$$

The bounds (2.3), (2.4), (2.5) and (2.6) imply similar bounds for the functionals $\mathfrak{B}, \mathfrak{b}$. In particular, for a semi-bounded L and compact K it follows from (2.5) that

$$\begin{aligned} &\mathfrak{B}(\eta - \lambda; L + K) \le \mathfrak{B}(\eta; L) + \mathfrak{R}^{(-)}(\lambda; K), \\ &\mathfrak{b}(\eta - \lambda; L + K) \le \mathfrak{b}(\eta; L) + \mathfrak{R}^{(-)}(\lambda; K), \quad \eta < \eta_0, \; \lambda > 0. \end{aligned} \tag{2.8}$$

For compact operators, the inequality (2.6) produces the bounds

$$\begin{aligned} &\mathfrak{R}(\lambda_1 + \lambda_2; K_1 + K_2) \le \mathfrak{R}(\lambda_1; K_1) + \mathfrak{R}(\lambda_2; K_2), \\ &\mathfrak{r}(\lambda_1 + \lambda_2; K_1 + K_2) \le \mathfrak{r}(\lambda_1; K_1) + \mathfrak{R}(\lambda_2; K_2), \end{aligned} \tag{2.9}$$

for any $\lambda_1, \lambda_2 > 0$ and similar bounds hold for the functionals $\mathfrak{R}^{(\pm)}$, $\mathfrak{r}^{(\pm)}$.

We systematically use analogues of the Birman–Solomyak asymptotic perturbation lemma for eigenvalues; see [**2**].

LEMMA 2.3. *Let $L = L^{(t)}, t > 0$, be a family of selfadjoint semi-bounded from below operators with a value of $\eta_0 = \eta_0(L^{(t)})$ independent of t. Suppose that for any $\delta > 0$ the family L can be represented as $L = L_\delta + Y'_\delta + Y''_\delta$ with some $L^{(t)}$-form-compact selfadjoint $Y'_\delta = Y'^{(t)}_\delta$ and $Y''_\delta = Y''^{(t)}_\delta$, such that*

$$\lim_{\delta\to 0} \mathfrak{B}(\tau; L + MY''_\delta) = 0 \tag{2.10}$$

for any $\tau < \eta_0$ and any $M \in \mathbb{R}$. Then for any $\eta < \eta_0$,

$$\begin{aligned} &\lim_{\epsilon\downarrow 0} \liminf_{M\uparrow 1} \liminf_{\delta\to 0} \mathfrak{B}(\eta - \epsilon; L_\delta + MY'_\delta) \\ &\qquad \le \mathfrak{B}(\eta; L) \le \lim_{\epsilon\downarrow 0} \limsup_{M\downarrow 1} \limsup_{\delta\to 0} \mathfrak{B}(\eta + \epsilon; L_\delta + MY'_\delta); \\ &\lim_{\epsilon\downarrow 0} \liminf_{M\uparrow 1} \liminf_{\delta\to 0} \mathfrak{b}(\eta - \epsilon; L_\delta + MY'_\delta) \\ &\qquad \le \mathfrak{b}(\eta; L) \le \lim_{\epsilon\downarrow 0} \limsup_{M\downarrow 1} \limsup_{\delta\to 0} \mathfrak{b}(\eta + \epsilon; L_\delta + MY'_\delta). \end{aligned} \tag{2.11}$$

PROOF. It suffices to prove the lemma for $\eta_0 = 0$. Using (2.3) with $L_1 = (1-\mu)L + Y'_\delta$ and $L_2 = \mu L + Y''_\delta$, we obtain

$$\mathfrak{b}(\eta; L) \le \mathfrak{b}(\eta+\epsilon; (1-\mu)L_\delta + Y'_\delta) + \mathfrak{B}(-\epsilon; \mu L_\delta + Y''_\delta),$$

for any $\eta < 0$, $\mu \in (0,1)$ and any $0 < \epsilon < |\eta|$. Using (2.10), and passing to the limit as $\delta \to 0$, we get

$$\mathfrak{b}(\eta; L) \le \limsup_{\delta\to 0} \mathfrak{b}(\eta+\epsilon; L_\delta + (1-\mu)^{-1} Y'_\delta).$$

Passing to the limit as $\mu \downarrow 0$ and $\epsilon \downarrow 0$, we get the proclaimed upper bound for $\mathfrak{b}(\eta; L)$, where $M = (1-\mu)^{-1}$. A similar result holds for $\mathfrak{B}(\eta; L)$.

For the lower bound we use (2.4) with $L_1 = (1+\mu)L + Y'_\delta$ and $L_2 = \mu L - Y''_\delta$, which gives

$$\mathfrak{b}(\eta; L) \ge \mathfrak{b}(\eta-\epsilon; (1+\mu)L_\delta + Y'_\delta) - \mathfrak{B}(-\epsilon; \mu L_\delta - Y''_\delta)$$

for any $\eta < 0$, $\mu \in (0,1)$ and $\epsilon > 0$. Using (2.10) again, and passing to the limit as $\delta \to 0$, we get

$$\mathfrak{b}(\eta; L) \ge \liminf_{\delta\to 0} \mathfrak{b}(\eta-\epsilon; L_\delta + (1+\mu)^{-1} Y'_\delta).$$

Passing to the limit as $\mu \downarrow 0$ and $\epsilon \downarrow 0$, we get the proclaimed lower bound for $\mathfrak{b}(\eta; L)$, where $M = (1+\mu)^{-1}$. A similar result holds for $\mathfrak{B}(\eta; L)$. □

LEMMA 2.4. *Let $L = L^{(t)}, t > 0$, be a family of selfadjoint semi-bounded from below operators with a value of $\eta_0 = \eta_0(L^{(t)})$ independent of t. Suppose that for any $\delta > 0$ the family L can be represented as $L = L_\delta + K'_\delta$ with a compact selfadjoint $K'_\delta = K'^{(t)}_\delta$, such that*

$$\lim_{\delta\to 0} \mathfrak{N}(\epsilon; K'_\delta) = 0 \tag{2.12}$$

for any $\epsilon > 0$. Then for any $\eta < \eta_0$,

$$\begin{aligned}
&\lim_{\epsilon\downarrow 0}\liminf_{\delta\to 0} \mathfrak{B}(\eta-\epsilon; L_\delta) \le \mathfrak{B}(\eta; L) \le \lim_{\epsilon\downarrow 0}\limsup_{\delta\to 0} \mathfrak{B}(\eta+\epsilon; L_\delta);\\
&\lim_{\epsilon\downarrow 0}\liminf_{\delta\to 0} \mathfrak{b}(\eta-\epsilon; L_\delta) \le \mathfrak{b}(\eta; L) \le \lim_{\epsilon\downarrow 0}\limsup_{\delta\to 0} \mathfrak{b}(\eta+\epsilon; L_\delta).
\end{aligned} \tag{2.13}$$

PROOF. For δ fixed, write (2.8) for families L_δ, K'_δ and pass to lim sup as $\delta \to 0$ and then as $\epsilon \to 0$. This proves the upper bounds in (2.13). The lower bounds are proved similarly. □

The next result is a direct consequence of this lemma applied to compact operators:

LEMMA 2.5. *Let $K = K^{(t)}$ be a family of compact operators. Suppose that for any $\delta > 0$ the family K can be represented as a sum $K = K_\delta + K'_\delta$ such that for any $\epsilon > 0$ the condition (2.12) is satisfied. Then for any $\lambda > 0$,*

$$\begin{aligned}
&\lim_{\epsilon\downarrow 0}\liminf_{\delta\to 0} \mathfrak{N}(\lambda+\epsilon; K_\delta) \le \mathfrak{N}(\lambda, K) \le \lim_{\epsilon\downarrow 0}\limsup_{\delta\to 0} \mathfrak{N}(\lambda-\epsilon; K_\delta);\\
&\lim_{\epsilon\downarrow 0}\liminf_{\delta\to 0} \mathfrak{r}(\lambda+\epsilon; K_\delta) \le \mathfrak{r}(\lambda, K) \le \lim_{\epsilon\downarrow 0}\limsup_{\delta\to 0} \mathfrak{r}(\lambda-\epsilon; K_\delta).
\end{aligned} \tag{2.14}$$

If, moreover, the families K, K_δ, K'_δ are selfadjoint, then the relations (2.14) hold with $\mathfrak{N}$, $\mathfrak{r}$ replaced by $\mathfrak{N}^{(\pm)}, \mathfrak{r}^{(\pm)}$ respectively.

3. Eigenvalue bounds for Toeplitz operators

3.1. Eigenvalue bounds for auxiliary integral operators. Here we obtain spectral estimates for integral operators involving the projections P_q on the spectral subspaces (*Landau subspaces*) $\mathcal{L}_q$ associated with the Landau levels Λ_q, $q = 0, 1, 2, \dots$. Choosing the gauge $\mathbf{a} = (-\frac{B}{2}x_2, \frac{B}{2}x_1)$ for the magnetic potential, one can write the orthonormal basis of the subspace $\mathcal{L}_q$ using the generalized Laguerre polynomials

$$L_q^{(\alpha)}(\xi) = \sum_{m=0}^{q} \binom{q+\alpha}{q-m} \frac{(-\xi)^m}{m!}, \ \xi \geq 0,$$

as follows:

$$\psi_{q,\alpha}(\mathbf{x}) = \sqrt{\frac{q!}{(q+\alpha)!}} \left[\sqrt{\frac{B}{2}}(x_1 + ix_2)\right]^{\alpha} L_q^{(\alpha)}\left(\frac{B|\mathbf{x}|^2}{2}\right) \sqrt{\frac{B}{2\pi}} \exp\left(-\frac{B|\mathbf{x}|^2}{4}\right), \tag{3.1}$$

for $\alpha = -q, -q+1, \dots$. The orthonormality follows from the standard relation

$$\int_0^\infty \xi^\alpha e^{-\xi} L_q^{(\alpha)}(\xi) L_{q'}^{(\alpha)}(\xi) d\xi = \frac{\Gamma(\alpha+q+1)}{q!} \delta_{q,q'}. \tag{3.2}$$

The integral kernel of the projection P_q is

$$\mathcal{P}_q(\mathbf{x}, \mathbf{y}) = \frac{B}{2\pi} L_q^{(0)}\left(\frac{B|\mathbf{x}-\mathbf{y}|^2}{2}\right) \exp\left(-\frac{B}{4}\left(|\mathbf{x}-\mathbf{y}|^2 + 2i\mathbf{x} \wedge \mathbf{y}\right)\right). \tag{3.3}$$

The following important estimate for the Laguerre polynomials can be found in [**14**].

LEMMA 3.1. *Let $k \in \mathbb{Z}_+$. Then*

$$|L_k^{(\alpha)}(\xi)| \leq (\alpha+k)^k e^{\frac{\xi}{\alpha+k}} \tag{3.4}$$

for all $\xi \geq 0$ and $\alpha \geq 1-k$.

For $t > 0$ and any function $f = f(\mathbf{x})$ we denote $f^{(t)}(\mathbf{x}) = f(t^{-1}\mathbf{x})$. We consider the operator families of the form

$$S^{(t)}(W_1, W_2) = S_q^{(t)}(W_1, W_2) = W_1^{(t)} P_q \overline{W_2^{(t)}}, \quad t > 0,$$

where W_1, W_2 are some complex-valued functions. Along with $S^{(t)}$ we also consider

$$T_{q,q'}^{(t)}(V) = P_q V^{(t)} P_{q'}, \ T_q^{(t)}(V) = T_{q,q}^{(t)}(V),$$

with some complex-valued function V; these are Toeplitz type operators for $q' = q$ and Hankel type operators for $q' \neq q$. The labels $q, q' = 0, 1, 2, \dots$ are fixed and as a rule, they are not reflected in the notation of the operators. The superscript $^{(t)}$ is sometimes omitted as well. The functions W_j, V will be referred to as weight functions. It is convenient to represent $S^{(t)}(W_1, W_2)$ and $T^{(t)}(V)$ in terms of the operator $Z^{(t)}(W) = W^{(t)} P_q$, so that

$$S^{(t)}(W_1, W_2) = Z^{(t)}(W_1)(Z^{(t)}(W_2))^*,$$

$$T^{(t)}(V) = (Z^{(t)}(V_1))^*(Z^{(t)}(V_2)), \ V_1 = \sqrt{|V|}, \ V_2 = V|V|^{-1/2}.$$

Under mild assumptions on W_1, W_2, V the above operators are compact.

LEMMA 3.2. *If* $W, W_1, W_2 \in \mathsf{L}^2(\mathbb{R}^2)$ *and* $V \in \mathsf{L}^1(\mathbb{R}^2)$, *then* $Z^{(t)}(W) \in \mathfrak{S}_2$, $S^{(t)}(W_1,W_2)$, $T^{(t)}(V) \in \mathfrak{S}_1$, *and*

$$\|Z^{(t)}(W)\|_{\mathfrak{S}_2} = t\sqrt{\frac{B}{2\pi}}\|W\|_2,$$

$$\|S^{(t)}(W_1,W_2)\|_{\mathfrak{S}_1} \le t^2\frac{B}{2\pi}\|W_1\|_2\|W_2\|_2, \quad \|T^{(t)}(V)\|_{\mathfrak{S}_1} \le t^2\frac{B}{2\pi}\|V\|_1.$$

PROOF. It suffices to prove the equality for the Hilbert–Schmidt norm of $Z^{(t)}(W)$. Using (3.3) and (3.2), we find

$$\begin{aligned}\|Z^{(t)}(W)\|_{\mathfrak{S}_2}^2 &= \int_{\mathbb{R}^2}\int_{\mathbb{R}^2} |W(t^{-1}\mathbf{x})|^2|\mathcal{P}_q(\mathbf{x},\mathbf{y})|^2 d\mathbf{x}d\mathbf{y}\\ &= \frac{B^2}{(2\pi)^2}\int_{\mathbb{R}^2}\int_{\mathbb{R}^2} |W(t^{-1}\mathbf{x})|^2\left(L_q^{(0)}\left(\frac{B|\mathbf{y}|^2}{2}\right)\right)^2 \exp\left(-\frac{B}{2}|\mathbf{y}|^2\right)d\mathbf{x}d\mathbf{y}\\ &= t^2\|W\|_2^2\frac{B}{2\pi}\int_0^\infty \left(L_q^{(0)}(s)\right)^2 e^{-s}ds = t^2\|W\|_2^2\frac{B}{2\pi},\end{aligned}$$

as required. □

Using the notation introduced in Section 2, we set

(3.5) $\mathfrak{M}(\lambda; W_1, W_2) = \mathfrak{R}(\lambda; S^{(t)}(W_1,W_2))$; $\mathfrak{m}(\lambda; W_1, W_2) = \mathfrak{r}(\lambda; S^{(t)}(W_1,W_2))$.

When $W_1 = W_2 = W$, we write $\mathfrak{M}(\lambda; W)$ and $\mathfrak{m}(\lambda; W)$. In this case the operator $S(W_1, W_2)$ is selfadjoint, so that we can also define the functionals $\mathfrak{M}^{(\pm)}(W)$. Clearly, $\mathfrak{M}^{(-)}(W) = 0$ and $\mathfrak{M}^{(+)}(W) = \mathfrak{M}(W)$.

For the operator $T^{(t)} = T^{(t)}_{q,q'}$ we introduce the related quantities:

$$\mathfrak{N}(\lambda; V) = \mathfrak{R}\big(\lambda; T^{(t)}(V)\big), \ \mathfrak{n}(\lambda; V) = \mathfrak{r}\big(\lambda; T^{(t)}(V)\big),$$

and in the case when V is real-valued and $q = q'$, we introduce the natural notation $\mathfrak{N}^{(\pm)}(V)$ and $\mathfrak{n}^{(\pm)}(V)$ as well. Since the nonzero eigenvalues of $S^{(t)}_q(W)$ and $T^{(t)}_q(|W|^2)$ coincide, we have

(3.6) $$\mathfrak{N}^{(+)}(\lambda; |W|^2) = \mathfrak{M}(\lambda; W), \quad \mathfrak{n}^{(+)}(\lambda; |W|^2) = \mathfrak{m}(\lambda; W).$$

If necessary, we reflect the dependence on q, q' in the notation of the above asymptotic coefficients: for instance, we may write $\mathfrak{m}_q(W_1, W_2)$ and $\mathfrak{N}^{(\pm)}_q(V)$, $\mathfrak{N}_{q,q'}(V)$.

Now Lemma 3.2 leads to the following result.

LEMMA 3.3. *If* $W, W_1, W_2 \in \mathsf{L}^2(\mathbb{R}^2)$ *and* $V \in \mathsf{L}^1(\mathbb{R}^2)$, *then for any* $\lambda > 0$ *and* $q, q' = 0, 1, 2, \dots$, *we have*

$$\mathfrak{R}(\lambda; Z^{(t)}_q(W)) \le \frac{B}{2\pi\lambda^2}\|W\|_2^2,$$

$$\mathfrak{M}_q(\lambda; W_1, W_2) \le \frac{B}{2\pi\lambda}\|W_1\|_2\|W_2\|_2, \quad \mathfrak{N}_{q,q'}(\lambda; V) \le \frac{B}{2\pi\lambda}\|V\|_1.$$

PROOF. For any operator $T \in \mathfrak{S}_p$ we have $n(\lambda; T) \le \lambda^{-p}\|T\|^p_{\mathfrak{S}_p}$. It remains to use Lemma 3.2. □

3.2. Localization. In our study of the eigenvalue behavior, we systematically represent the operators as block matrices associated with certain orthogonal decompositions. The results of this subsection help to show that off-diagonal terms do not contribute to the asymptotic coefficients.

We consider an auxiliary integral operator $K^{(t)}(W_1, W_2, f)$ with the kernel

$$\mathcal{K}_q^{(t)}(\mathbf{x}, \mathbf{y}; W_1, W_2, f) = W_1^{(t)}(\mathbf{x})\mathcal{P}_q(\mathbf{x}, \mathbf{y})f^{(t)}(\mathbf{x}-\mathbf{y})W_2^{(t)}(\mathbf{y}),$$

with some functions W_1, W_2 and f.

LEMMA 3.4. *Let $W_1 \in \mathsf{L}^p, W_2 \in \mathsf{L}^s, f \in \mathsf{L}^\infty$ with arbitrary $2 \le p, s \le \infty$ such that $\frac{1}{p} + \frac{1}{s} \ge \frac{1}{2}$. Suppose that for some $\delta > 0$,*

$$f(\mathbf{z}) = 0, \quad \text{for} \quad |\mathbf{z}| \le \delta.$$

Then for any $\lambda > 0$,

$$\mathfrak{N}\big(\lambda;\ K^{(t)}(W_1, W_2, f)\big) = 0. \tag{3.7}$$

PROOF. By assumption $\mathcal{K}^{(t)}(\mathbf{x}, \mathbf{y}; W_1, W_2, f) = 0$ if $|\mathbf{x}-\mathbf{y}| \le \delta t$. For $|\mathbf{x}-\mathbf{y}| > \delta t$ we use (3.3):

$$|\mathcal{P}_q(\mathbf{x}, \mathbf{y})| \le \frac{B}{2\pi}\left|L_q^{(0)}\left(\frac{B|\mathbf{x}-\mathbf{y}|^2}{2}\right)\right| e^{-\frac{B}{8}\delta^2 t^2} e^{-\frac{B}{8}|\mathbf{x}-\mathbf{y}|^2}.$$

Let $r \in [2, \infty]$ be defined by $p^{-1} + s^{-1} + r^{-1} = 1$. By the Hölder and Young inequalities,

$$\begin{aligned}
\|K^{(t)}(W_1, W_2, f)\|_{\mathfrak{S}_2}^2 &\le t^{\frac{4}{p}+\frac{4}{s}} \|W_1\|_p^2 \|W_2\|_s^2 \|f\|_\infty^2 \frac{B^2}{(2\pi)^2} e^{-\frac{B}{4}\delta^2 t^2} \\
&\quad \times \left(\int_{\mathbb{R}^2} \left|L_q^{(0)}\left(\frac{B|\mathbf{x}|^2}{2}\right)\right|^r e^{-\frac{Br}{4}|\mathbf{x}|^2} d\mathbf{x}\right)^{\frac{2}{r}} \\
&\le t^{\frac{4}{p}+\frac{4}{s}} \|W_1\|_p^2 \|W_2\|_s^2 \|f\|_\infty^2 \left(\frac{B}{2\pi}\right)^{2-\frac{2}{r}} e^{-\frac{B}{4}\delta^2 t^2} \left(\int_0^\infty |L_q^{(0)}(\xi)|^r e^{-r\xi/2} d\xi\right)^{\frac{2}{r}} \\
&\le C t^{\frac{4}{p}+\frac{4}{s}} e^{-\frac{B}{4}\delta^2 t^2}.
\end{aligned}$$

Due to the presence of the exponentially decaying factor, for sufficiently large t the operator $K^{(t)}(W_1, W_2, f)$ has no singular values above λ, whence (3.7). □

The above lemma has a few useful corollaries.

COROLLARY 3.5. *Let $V \in \mathsf{L}^2(\mathbb{R}^2)$ and $R^{(t)}(V) = R_q^{(t)}(V) = [P_q, V^{(t)}]$. Then*

$$\mathfrak{N}(\lambda, R^{(t)}(V)) = 0. \tag{3.8}$$

Moreover, if $V \in \mathsf{L}^1(\mathbb{R}^2) + \mathsf{L}^2(\mathbb{R}^2)$ and $q \ne q'$, then also

$$\mathfrak{N}_{q,q'}(\lambda; V) = 0, \tag{3.9}$$

for all $\lambda > 0$.

PROOF. Fix a $\delta > 0$ and find $\tilde{V} \in \mathsf{C}_0^\infty(\mathbb{R}^2)$ such that $\|V - \tilde{V}\|_2 < \delta$. We have

$$[P_q, V^{(t)}] = [P_q, \tilde{V}^{(t)}] + P_q(V^{(t)} - \tilde{V}^{(t)}) - (V^{(t)} - \tilde{V}^{(t)})P_q. \tag{3.10}$$

To the last two terms in (3.10) we can apply Lemma 3.3, which gives

$$\mathfrak{N}(\epsilon; P(V^{(t)} - \tilde{V}^{(t)}) - (V^{(t)} - \tilde{V}^{(t)})P) \le C\delta^2/\epsilon^2, \tag{3.11}$$

for any $\epsilon > 0$. Since δ is arbitrarily small, it suffices to prove that $\mathfrak{R}(\lambda; [\tilde{V}^{(t)}, P]) = 0$ for any $\lambda > 0$ and $\tilde{V} \in \mathsf{C}_0^\infty(\mathbb{R}^2)$, and then apply Lemma 2.5. The integral kernel of $[\tilde{V}^{(t)}, P_q]$ is

$$\big(\tilde{V}(t^{-1}\mathbf{x}) - \tilde{V}(t^{-1}\mathbf{y})\big)\mathcal{P}_q(\mathbf{x}, \mathbf{y}).$$

Denoting $f(\mathbf{z}) = \chi(|\mathbf{z}| \le \epsilon)$, we rewrite this kernel as follows:

$$(3.12)\qquad \big(\tilde{V}(t^{-1}\mathbf{x}) - \tilde{V}(t^{-1}\mathbf{y})\big)\ \mathcal{P}_q(\mathbf{x}, \mathbf{y}) f^{(t)}(\mathbf{x} - \mathbf{y}) \\ + \mathcal{K}^{(t)}(\mathbf{x}; \mathbf{y}; \tilde{V}, 1, 1 - f) - \mathcal{K}^{(t)}(\mathbf{x}; \mathbf{y}; 1, \tilde{V}, 1 - f).$$

The operators corresponding to the last two terms satisfy (3.7). For the first term in (3.12), we use that $\tilde{V}$ has a compact support and so

$$\big|\tilde{V}(t^{-1}\mathbf{x}) - \tilde{V}(t^{-1}\mathbf{y})\big| f^{(t)}(\mathbf{x} - \mathbf{y}) \le t^{-1} \max_{\mathbf{z}} |\nabla\tilde{V}(\mathbf{z})|\ |\mathbf{x} - \mathbf{y}| \le C\epsilon.$$

Thus the norm of the operator corresponding to the first term in (3.12) is bounded by

$$C\epsilon \max_{\mathbf{x}} \int_{\mathbb{R}^2} |\mathcal{P}_q(\mathbf{x}, \mathbf{y})| d\mathbf{y} \le C'\epsilon,$$

and hence it can be made arbitrarily small, which proves (3.8).

If $V \in \mathsf{L}^2(\mathbb{R}^2)$, then (3.9) follows immediately from (3.8) in view of the identity $T^{(t)}_{q,q'}(V) = R^{(t)}_q(V) P_{q'}$. If $V \in \mathsf{L}^1(\mathbb{R}^2)$, then for arbitrary $\delta > 0$ we approximate V with a function $\tilde{V} \in \mathsf{C}_0^\infty(\mathbb{R}^2)$ such that $\|V - \tilde{V}\|_1 < \delta$, and use again Lemmas 3.3, 2.5 and formula (3.8). □

COROLLARY 3.6. *Let $W_1 \in \mathsf{L}^2(\mathbb{R}^2)$ and $W_2 \in \mathsf{L}^\infty(\mathbb{R}^2)$ be such that $W_1 W_2 = 0$. Then $\mathfrak{M}(\lambda; W_1, W_2) = 0$ for all $\lambda > 0$.*

PROOF. Rewrite $S^{(t)}(W_1, W_2) = -R^{(t)}(W_1) W_2^{(t)}$. Consequently,

$$\mathfrak{M}(\lambda; W_1, W_2) \le \mathfrak{R}(\lambda \|W_2\|_\infty^{-1}; R^{(t)}(W_1)) = 0,$$

by Corollary 3.5. □

4. Eigenvalues of Toeplitz operators

4.1. Additivity of asymptotic coefficients. Further on, we will approximate weight functions by piecewise constant functions. To describe these approximations we cut the plane $\mathbb{R}^2$ in the following way. For fixed $N \in \mathbb{N}$, $d > 0$ we tile $\mathbb{R}^2$ by disjoint annular sectors

$$\Omega_{m,l} = \left\{\mathbf{x} = (\rho, \theta) : (m-1)d < \rho \le md, \frac{2\pi}{N}(l-1) < \theta \le \frac{2\pi}{N} l\right\}, \\ m \in \mathbb{N}, \quad l = 1, 2, \dots, N.$$

For any set Ω we denote by $\chi(\mathbf{x} \in \Omega)$ its characteristic function. Let $X_{m,l} = \chi(\mathbf{x} \in \Omega_{m,l})$.

LEMMA 4.1. *If $(m, l) \ne (m', l')$, then $\mathfrak{M}(\lambda; X_{m,l}, X_{m',l'}) = 0$ for all $\lambda > 0$.*

PROOF. This follows immediately from Corollary 3.6. □

This result leads to the additivity of the asymptotic coefficients for piecewise constant functions of the form

$$W = \sum_{m,l} w_{m,l} X_{m,l}, \tag{4.1}$$

where $w_{m,l}$ are some complex numbers and the sum is finite.

LEMMA 4.2. *Let W have the form* (4.1). *Then*

$$\sum \mathfrak{m}(\lambda + 0; w_{m,l} X_{m,l}) \le \mathfrak{m}(\lambda; W) \le \mathfrak{M}(\lambda; W) \le \sum \mathfrak{M}(\lambda - 0; w_{m,l} X_{m,l}). \tag{4.2}$$

PROOF. We prove the upper bound only; the lower bound is established in the same way, with obvious changes. Represent the operator $S^{(t)}(W)$ as

$$\begin{aligned} S^{(t)}(W) &= \sum_{m,l} |w_{m,l}|^2 S^{(t)}(X_{m,l}) + \sum_{(m,l)\neq(m',l')} w_{m,l}\overline{w_{m',l'}} S^{(t)}(X_{m,l}, X_{m',l'}) \\ &= S' + S''. \end{aligned} \tag{4.3}$$

The family S'' in (4.3) is a finite sum of operators of the form considered in Lemma 4.1; therefore $\mathfrak{R}(\epsilon, S'') = 0$ for any $\epsilon > 0$. Further, the operator S' is a direct sum of the operators $|w_{m,l}|^2 S^{(t)}(X_{m,l})$. Therefore its spectrum is the union of spectra of summands, so

$$\mathfrak{R}(\lambda, S') \le \sum \mathfrak{R}(\lambda, |w_{m,l}|^2 S^{(t)}(X_{m,l})). \tag{4.4}$$

Now we can apply Lemma 2.5. □

Let us establish a similar additivity property for the operator $T(V)$ with a real-valued function V of the form

$$V = \sum_{m,l} v_{m,l} X_{m,l}, \tag{4.5}$$

where $v_{m,l}$ are real and the sum is finite.

LEMMA 4.3. *Let V be of the form* (4.5). *Then*

$$\begin{aligned} \sum_{\pm v_{m,l}>0} \mathfrak{n}^{(+)}(\lambda + 0; \pm v_{m,l} X_{m,l}) &\le \mathfrak{n}^{(\pm)}(\lambda; V) \\ \le \mathfrak{N}^{(\pm)}(\lambda; V) &\le \sum_{\pm v_{m,l}>0} \mathfrak{N}^{(+)}(\lambda - 0; \pm v_{m,l} X_{m,l}). \end{aligned} \tag{4.6}$$

PROOF. As in the previous lemma, we prove only the upper bound. Let $\Omega = \bigcup \Omega_{m,l}$ be the set where $V(\mathbf{x}) \neq 0$, and let $\Omega_{0,0} = \mathbb{R}^2 \setminus \Omega$. It is convenient to include the set $\Omega_{0,0}$ in the family of $\Omega_{m,l}$'s. Rewrite

$$T^{(t)}(V) = \sum X_{m,l} T^{(t)}(v_{n,s} X_{n,s}) X_{m',l'} = T' + T'',$$

where

$$T' = \sum T_{m,l}, \quad T_{m,l} = X_{m,l} T^{(t)}(v_{m,l} X_{m,l}) X_{m,l},$$

and T'' is the sum in which at least one of the pairs $(m,l), (m',l')$ is distinct from (n,s). Consider for instance the term with $(m,l) \neq (n,s)$, and rewrite it as follows:

$$X_{m,l} T^{(t)}(v_{n,s}) X_{m',l'} = v_{n,s} S^{(t)}(X_{m,l}, X_{n,s}) P X_{m',l'},$$

so that by Corollary 3.6, the value $\mathfrak{R}(\epsilon;\ \cdot\ )$ for this operator equals zero for any $\epsilon > 0$. Consequently, $\mathfrak{R}(\epsilon; T'') = 0$ for any $\epsilon > 0$. Next, T' is an orthogonal sum of operators $T_{m,l}$. Therefore

$$n_\pm(\lambda; T') = \sum_{\pm v_{ml}>0} n_+(\lambda; \pm T_{m,l}),\ \mathfrak{R}^{(\pm)}(\lambda; T') \le \sum_{\pm v_{ml}>0} \mathfrak{R}^{(+)}(\lambda; \pm T_{m,l}).$$

So, by Lemma 2.5,

$$\mathfrak{N}^{(\pm)}(\lambda; V) \le \sum_{\pm v_{ml}>0} \mathfrak{R}^{(+)}(\lambda - 0; \pm T_{m,l}).$$

Now we apply again Corollary 3.6 and Lemma 2.5 to each of the operators $T_{m,l}$, which gives

$$\mathfrak{R}^{(+)}(\lambda - 0; \pm T_{m,l}) \le \mathfrak{N}^{(+)}(\lambda - 0; \pm v_{m,l} X_{m,l}),$$

and this leads to the required upper bound. □

Another kind of additivity holds with respect to the Landau projections. For $J > 1$ we denote by $P^{(J)}$ the projection $P^{(J)} = \sum_{q \le J} P_q$. Consider the Toeplitz family $T^{(t)}_{(J)}(V) = P^{(J)} V^{(t)} P^{(J)}$.

LEMMA 4.4. *For $V \in \mathsf{L}^1(\mathbb{R}^2) + \mathsf{L}^2(\mathbb{R}^2)$, $\lambda > 0$,*

$$\sum_{q \le J} \mathfrak{n}_q^{(\pm)}(\lambda + 0; V) \le \mathfrak{r}^{(\pm)}\big(\lambda; T^{(t)}_{(J)}\big) \le \mathfrak{R}^{(\pm)}\big(\lambda; T^{(t)}_{(J)}\big) \le \sum_{q \le J} \mathfrak{N}_q^{(\pm)}(\lambda - 0; V). \tag{4.7}$$

PROOF. We split the family $T^{(t)}_{(J)}$ as follows:

$$T^{(t)}_{(J)} = \sum_q T^{(t)}_q + \sum_{q \ne q'} T^{(t)}_{q,q'} = T' + T''.$$

The operators in T' act in orthogonal subspaces, so their distribution functions add up. By Corollary 3.5, $\mathfrak{R}(\epsilon; T^{(t)}_{q,q'}) = 0,\ q \ne q'$ for any $\epsilon > 0$, so that $\mathfrak{R}(\epsilon; T'') = 0$, and (4.7) follows by Lemma 2.5. □

4.2. Model integral operators. In order to pass from the conditional results in Subsection 4.1 to actual calculations, we need at least some operators for which the asymptotic coefficients are known. Here we consider such model operators.

LEMMA 4.5. *Let $0 \le d_1 < d_2 < \infty$, and let $W = \chi(d_1 < |\mathbf{x}| < d_2)$. Then for any $\lambda > 0$,*

$$\mathfrak{M}(\lambda; W) = \mathfrak{m}(\lambda; W) = \begin{cases} \dfrac{B}{2}(d_2^2 - d_1^2), & \lambda < 1, \\ 0, & \lambda \ge 1. \end{cases} \tag{4.8}$$

PROOF. Since the function W is radially symmetric, using (3.1) we can immediately find all eigenvalues of the operator $S^{(t)}_q(W)$ explicitly (see [**14**], Lemma 3.4):

$$\begin{aligned} \lambda_j = \lambda_j^{(t)} = \lambda_j^{(t)}(d_1, d_2) &= \int_{d_1 t < |\mathbf{x}| < d_2 t} |\psi_{q,j}(\mathbf{x})|^2 d\mathbf{x} \\ &= \frac{q!}{(j+q)!} \int_{\eta_1}^{\eta_2} \xi^j e^{-\xi} \big(L_q^{(j)}(\xi)\big)^2 d\xi, \end{aligned} \tag{4.9}$$

where we denote $\eta_k = B(d_k t)^2/2, k = 1, 2$. Note that the λ_j's are not necessarily labeled in the usual decreasing order. The functions $\psi_{q,j}$ are normalized, so $\lambda_j \leq 1$ and therefore (4.8) holds for $\lambda \geq 1$.

Now let $\lambda < 1$. We will find the asymptotics of $\lambda_j = \lambda_j^{(t)}(0, d)$ as $t \to \infty$. Denote $\eta = B(td)^2/2$ and fix an $\epsilon > 0$. Suppose first that $j \geq (1+\epsilon)\eta$. Then (3.4) implies

$$\lambda_j \leq (j+q)^{2q} \frac{q!}{(j+q)!} \int_0^\eta \xi^j \exp\Big(-(1 - \frac{2}{j+q})\xi\Big) d\xi$$
$$\leq (j+q)^{2q} \frac{q!}{(j+q)!} e^{\frac{2}{1+\epsilon}} \int_0^\eta \xi^j e^{-\xi} d\xi.$$

The maximum of the integrand is attained at $\xi = j$ and it grows for $\xi < j$. Therefore, we can estimate it from above by $\eta^j e^{-\eta}$, which leads to the bound

$$\lambda_j \leq C(j+q)^{2q} \frac{q!}{(j+q)!} \eta^{j+1} e^{-\eta}.$$

Using the Stirling formula, we get

(4.10)

$$\lambda_j \leq C \frac{(j+q)^{2q} q!}{(j+q)^{j+q+1/2}} \eta^{j+1} e^{j+q-\eta} \leq e^q (j+q)^{q-1/2} q! \eta \tau_j(\eta), \quad \tau_j(\eta) = \Big(\frac{\eta}{j}\Big)^j e^{j-\eta}.$$

To estimate $\tau_j(\eta)$ we rewrite it as

$$\tau_j(\eta) = \exp\Big[-\int_\eta^j \Big(\frac{j}{s} - 1\Big) ds\Big].$$

Now we fix $\epsilon_1 \in (0, \epsilon)$ and obtain:

$$\tau_j(\eta) \leq \exp\Big[-\int_\eta^{j(1+\epsilon_1)^{-1}} \Big(\frac{j}{s} - 1\Big) ds\Big] \leq \exp\Big[-\epsilon_1 \int_\eta^{j(1+\epsilon_1)^{-1}} ds\Big]$$
$$\leq \exp\Big(-j\epsilon_1 \frac{\epsilon - \epsilon_1}{(1+\epsilon)(1+\epsilon_1)}\Big).$$

This shows that λ_j tends to zero very fast as $\eta \to \infty$ and $j \geq (1+\epsilon)\eta$. Assume now that $j \leq (1-\epsilon)\eta$. Since the functions $\psi_{q,j}$ are normalized, we have

$$\lambda_j = 1 - \mu_j, \quad \mu_j = \frac{q!}{(j+q)!} \int_\eta^\infty \xi^j e^{-\xi} \big(L_q^{(j)}(\xi)\big)^2 d\xi > 0.$$

Then, using (3.4) again, we obtain

$$\mu_j \leq (j+q)^{2q} \frac{q!}{(j+q)!} \int_\eta^\infty \xi^j \exp\Big(-(1 - \frac{2}{j+q})\xi\Big) d\xi.$$

For an arbitrary $\epsilon_1 \in (0, 1)$ rewrite the integrand as follows:

$$\Big[\xi^j e^{-(1-\epsilon_1)\xi}\Big] \exp\Big(-(\epsilon_1 - \frac{2}{j+q})\xi\Big).$$

The maximum of the term in brackets is attained at $j(1-\epsilon_1)^{-1}$. If $\epsilon_1 > \epsilon$, so that $j(1-\epsilon_1)^{-1} < \eta$, we conclude that on the interval $[\eta, \infty)$ the integrand does not exceed

$$\Big[\eta^j e^{-(1-\epsilon_1)\eta}\Big] \exp\Big(-(\epsilon_1 - \frac{2}{j+q})\xi\Big).$$

Integrating, we get

$$\mu_j \le (j+q)^{2q} \frac{q!}{(j+q)!} \frac{1}{\epsilon_1 - 2(j+q)^{-1}} \eta^j e^{-(1-2(j+q)^{-1})\eta}.$$

As at the first step of the proof, by the Stirling formula we obtain

$$\mu_j \le C(j+q)^{q-1/2} q e^q \frac{1}{\epsilon_1 - 2(j+q)^{-1}} e^{2(j+q)^{-1}\eta} \tau_j(\eta),$$

with the function $\tau_j(\eta)$ defined in (4.10). To estimate it, we rewrite

$$\tau_j(\eta) = \exp\left[-\int_j^\eta \left(1 - \frac{j}{s}\right) ds\right].$$

Choose $\epsilon_1 \in (0, \epsilon)$ and estimate

$$\tau_j(\eta) \le \exp\left[-\int_{j(1-\epsilon_1)^{-1}}^\eta \left(1 - \frac{j}{s}\right) ds\right] \le \exp\left[-\epsilon_1 \int_{j(1-\epsilon_1)^{-1}}^\eta ds\right] \le \exp\left(-\epsilon_1 \eta \frac{\epsilon - \epsilon_1}{1 - \epsilon_1}\right).$$

This shows that μ_j tends to zero very fast as $\eta \to \infty$ and $j \le (1-\epsilon)\eta$, that is, $\lambda_j \to 1$ as $t \to \infty$, uniformly for $j \le (1-\epsilon)\eta$. Summarizing the above calculations, we see that for sufficiently large t the following inequalities hold:

$$\lambda_j(d_1, d_2) \equiv \lambda_j(0, d_2) - \lambda_j(0, d_1) \begin{cases} < \lambda, \ j < (1-\epsilon)\eta_1, \\ > \lambda, \ (1+\epsilon)\eta_1 < j < (1-\epsilon)\eta_2, \\ < \lambda, \ j > (1+\epsilon)\eta_2. \end{cases}$$

Consequently, $(1-\epsilon)\eta_2 - (1+\epsilon)\eta_1 \le n(\lambda, S^{(t)}(W)) \le (1+\epsilon)\eta_2 - (1-\epsilon)\eta_1$ for sufficiently large t. Passing to the limit, we obtain

$$\mathfrak{m}(\lambda; W) \ge (1-\epsilon)\frac{Bd_2^2}{2} - (1+\epsilon)\frac{Bd_1^2}{2}, \ \mathfrak{M}(\lambda; W) \le (1+\epsilon)\frac{Bd_2^2}{2} - (1-\epsilon)\frac{Bd_1^2}{2}.$$

Since $\epsilon \in (0,1)$ is arbitrary, this entails (4.8). □

Now using the additivity and the calculations for the model operator we can find $\mathfrak{m}(\lambda; w_{m,l}X_{m,l})$ and $\mathfrak{M}(\lambda; w_{m,l}X_{m,l})$, which turn out to be equal.

LEMMA 4.6. *If* $\lambda \ge |w_{m,l}|^2$, *then* $\mathfrak{M}(\lambda, w_{m,l}X_{m,l}) = 0$. *For any* $\lambda \in (0, |w_{m,l}|^2)$,

$$\mathfrak{m}(\lambda; w_{m,l}X_{m,l}) = \mathfrak{M}(\lambda; w_{m,l}X_{m,l}) = \frac{B}{2\pi}|\Omega_{m,l}|.$$

PROOF. Since $n(\lambda, S^{(t)}(w_{m,l}X_{m,l})) = n(\lambda|w_{m,l}|^{-2}, S^{(t)}(X_{m,l}))$, it suffices to consider the case $w_{m,l} = 1$. The norm of the operator $S^{(t)}(X_{m,l})$ is not greater than 1, and this takes care of the case $\lambda \ge 1$. Next consider $X_m = \sum_{l=1}^N X_{m,l}$. It is clear that the asymptotic coefficients are the same for all sectors $X_{m,l}$, $l = 1, \dots, N$. Therefore, by Lemma 4.2,

$$\mathfrak{m}(\lambda + 0; X_m) \le N\mathfrak{m}(\lambda; X_{m,l}) \le N\mathfrak{M}(\lambda; X_{m,l}) \le \mathfrak{M}(\lambda - 0; X_m). \tag{4.11}$$

Now (4.8) produces the required formula. □

Before we proceed to more general functions W, we introduce, similarly to (1.1), the appropriate asymptotic coefficients. For $V \in \mathsf{L}^1(\mathbb{R}^2)$ and $\lambda > 0$ we define *sup*- and *sub*- measures of V:

$$A^{(\pm)}(\lambda; V) = A(\lambda, \infty; \pm V) = |\{\mathbf{x} : \pm V(\mathbf{x}) > \lambda\}|. \tag{4.12}$$

With this notation, the result of Lemma 4.6 reads

$$\mathfrak{m}(\lambda; w_{m,l}X_{m,l}) = \mathfrak{M}(\lambda; w_{m,l}X_{m,l}) = \frac{B}{2\pi}A^{(+)}(\lambda; |w_{m,l}|^2 X_{m,l}), \tag{4.13}$$

for all $\lambda > 0$.

COROLLARY 4.7. *Let $v_{m,l}$ be real for some m, l. Then*

$$\mathfrak{n}^{(\pm)}(\lambda; v_{m,l}X_{m,l}) = \mathfrak{N}^{(\pm)}(\lambda; v_{m,l}X_{m,l}) = \frac{B}{2\pi}A^{(\pm)}(\lambda; v_{m,l}X_{m,l}),$$

for all $\lambda > 0$.

PROOF. Denote $V = v_{m,l}X_{m,l}$. If $\pm v_{m,l} > 0$, then obviously $\mathfrak{N}^{(\mp)}(\lambda; V) = 0$, and $\mathfrak{N}^{(\pm)}(\lambda; V) = \mathfrak{N}^{(+)}(\lambda; \pm V) = \mathfrak{M}(\lambda; \sqrt{\pm V})$. It remains to use formula (4.13). □

To proceed we need to establish the "continuity" of the coefficients $A^{(\pm)}(\lambda; V)$ in the function V.

LEMMA 4.8. *Let $W_\delta = V + V'_\delta$, $\delta \neq 0$, where V, $V'_\delta \in \mathsf{L}^1(\mathbb{R}^2)$, V does not depend on δ and $\|V'_\delta\|_1 \to 0$ as $\delta \to 0$. Then*

$$A^{(\pm)}(\lambda; V) \leq \liminf_{\delta\to 0} A^{(\pm)}(\lambda; W_\delta) \leq \limsup_{\delta\to 0} A^{(\pm)}(\lambda; W_\delta) \leq A^{(\pm)}(\lambda - 0; V). \tag{4.14}$$

PROOF. It suffices to consider the + sign. By definition, for any $\epsilon \in (0, \lambda)$,

$$\begin{aligned} A^{(+)}(\lambda; W_\delta) &\leq |\{\mathbf{x} : W_\delta(\mathbf{x}) > \lambda\} \cap \{\mathbf{x} : |V'_\delta(\mathbf{x})| \leq \epsilon\}| + |\{\mathbf{x} : |V'_\delta(\mathbf{x})| > \epsilon\}| \\ &\leq A^{(+)}(\lambda - \epsilon; V) + \epsilon^{-1}\|V'_\delta\|_1. \end{aligned}$$

For the last estimate we have used the Chebyshev inequality. Passing to the limit as $\delta \to 0$ and $\epsilon \downarrow 0$, we get the upper bound in (4.14). Similarly, write

$$A^{(+)}(\lambda; W_\delta) \geq A^{(+)}(\lambda + \epsilon; V) - \epsilon^{-1}\|V'_\delta\|_1.$$

Passing again to the limit as $\delta \to 0$ and $\epsilon \downarrow 0$, we get the lower bound in (4.14). □

4.3. Eigenvalue asymptotics for Toeplitz operators. Corollary 4.7 enables us to establish the spectral asymptotics for the Toeplitz operator $T^{(t)}(V)$ with a piecewise constant function V.

LEMMA 4.9. *Let $V = \sum v_{ml}X_{ml}$, where the sum is finite and v_{ml} are real-valued. Then for any $q \geq 0$,*

$$\frac{B}{2\pi}A^{(\pm)}(\lambda; V) \leq \mathfrak{n}^{(\pm)}(\lambda; V) \leq \mathfrak{N}^{(\pm)}(\lambda; V) \leq \frac{B}{2\pi}A^{(\pm)}(\lambda - 0; V), \tag{4.15}$$

for all $\lambda > 0$.

PROOF. Use Lemma 4.3 and Corollary 4.7. □

We are now in position to treat the general case.

THEOREM 4.10. *For any $V \in \mathsf{L}^1(\mathbb{R}^2)$,*

$$\frac{B}{2\pi}A^{(\pm)}(\lambda; V) \leq \mathfrak{n}^{(\pm)}(\lambda; V) \leq \mathfrak{N}^{(\pm)}(\lambda; V) \leq \frac{B}{2\pi}A^{(\pm)}(\lambda - 0; V). \tag{4.16}$$

Moreover, if λ is a generic value for $\pm V$, we have the asymptotics

$$\mathfrak{n}^{(\pm)}(\lambda; V) = \mathfrak{N}^{(\pm)}(\lambda; V) = \frac{B}{2\pi}A^{(\pm)}(\lambda; V). \tag{4.17}$$

PROOF. Since $n_-(\lambda; T^{(t)}(V)) = n_+(\lambda; T^{(t)}(-V))$, it suffices to consider the + sign only.

For a positive δ we find a sufficiently fine tiling of the plane by annular sectors $\Omega_{m,l}$ and a piecewise constant function $\tilde{V}_\delta$ represented in the form (4.5) with a finite sum, such that $\|V - \tilde{V}_\delta\|_1 < \delta$. Hence by Lemma 3.3, $\mathfrak{N}(\epsilon; V - \tilde{V}_\delta) \le B\delta(2\pi\epsilon)^{-1}$. Furthermore, by Lemma 4.9,

$$\mathfrak{N}^{(+)}(\lambda; \tilde{V}_\delta) \le \frac{B}{2\pi} A^{(+)}(\lambda - 0; \tilde{V}_\delta). \tag{4.18}$$

Thus by Lemma 2.5,

$$\mathfrak{N}^{(+)}(\lambda; V) \le \frac{B}{2\pi} \limsup_{\epsilon \downarrow 0} \limsup_{\delta \to 0} A^{(+)}(\lambda - \epsilon; \tilde{V}_\delta).$$

By virtue of Lemma 4.8, the right hand side does not exceed $B(2\pi)^{-1}A^{(+)}(\lambda - 0; V)$, as required.

The corresponding lower bound for $\mathfrak{n}^{(+)}(\lambda; V)$ is established similarly. □

5. Reduction to Toeplitz operators

Let $(\lambda_1, \lambda_2) \Subset (\Lambda_\nu, \Lambda_{\nu+1})$ with some $\nu = -1, 0, \dots,$ as in Theorem 1.1. We denote

$$a = (\lambda_1 + \lambda_2)/2, \ \ b = (\lambda_2 - \lambda_1)/2, \ \ H_{0a} = H_0 - a, \ \ H_a = H_a(V^{(t)}) = H + V^{(t)} - a.$$

Our analysis of the counting function $N(\lambda_1, \lambda_2; H)$ is based on the obvious relations (cf. (1.4))

$$N(\lambda_1, \lambda_2; H) = N(\lambda_1 - a, \lambda_2 - a; H - a) = N(-b, b; H - a) = N\big(b^2; (H - a)^2\big), \tag{5.1}$$

where, recall, $N(\eta; L)$ stands for the number of eigenvalues of a semi-bounded operator L below η. For methodological purposes we need to consider an operator having a somewhat more general form. Namely, with a fixed, for real-valued functions $V \in \mathsf{L}^2(\mathbb{R}^2), Z \in \mathsf{L}^1(\mathbb{R}^2)$ we set

$$L^{(t)}(V, Z) = H_{0a}^2 + V^{(t)} H_{0a} + H_{0a} V^{(t)} + Z^{(t)}, \tag{5.2}$$

so that $(H - a)^2 = L^{(t)}(V, V^2)$. Under these conditions the operator is well defined via the quadratic form

$$\|H_{0a} u\|^2 + 2\,\mathrm{Re}(H_{0a} u, V^{(t)} u) + (Z^{(t)} u, u)$$

for $u \in \mathrm{Dom}(H_0)$. Using the diamagnetic inequality, it is easy to check that each term of the perturbation $V^{(t)} H_{0a} + H_{0a} V^{(t)} + Z^{(t)}$ is H_{0a}^2-form-compact. The lowest point of the essential spectrum of $L^{(t)}(V, Z)$ is

$$\eta_0 = \min_{q \ge 0} (\Lambda_q - a)^2 > 0. \tag{5.3}$$

We are going to study the discrete spectrum of the operator $L^{(t)}$, independently of its connection with $H^{(t)}$. The following important lemma establishes asymptotic relations of the spectrum of $L^{(t)}(V, Z)$ below η_0 and the spectra of certain Toeplitz type operators.

LEMMA 5.1. *Suppose that $V, Z \in \mathsf{C}_0^\infty(\mathbb{R}^2)$ and let $\eta < \eta_0$. Then for any $\epsilon > 0$ there exists an $\tilde\epsilon \downarrow 0$ as $\epsilon \downarrow 0$, and an integer $J_0 = J_0(\epsilon)$ such that for all $J \geq J_0$ and the projection*

$$P^{(J)} = \sum_{q \leq J} P_q \tag{5.4}$$

we have

$$\begin{aligned} &\mathfrak{b}\big(\eta; L(V,Z)\big) \geq \mathfrak{b}\big(\eta - \tilde\epsilon; P^{(J)} L((1-\epsilon)V, (1-\epsilon)Z) P^{(J)}\big), \\ &\mathfrak{B}\big(\eta; L(V,Z)\big) \leq \mathfrak{B}\big(\eta + \tilde\epsilon; P^{(J)} L((1+\epsilon)V, (1+\epsilon)Z) P^{(J)}\big). \end{aligned} \tag{5.5}$$

PROOF. We denote $P = P^{(J)}$, $Q = I - P$. Let v be a constant such that

$$\sup_{\mathbf{x}} \big(|V(\mathbf{x})| + |\nabla V(\mathbf{x})| + |\Delta V(\mathbf{x})| + |Z(\mathbf{x})|\big) \leq v.$$

Our first aim is to bound the counting function $N(\eta, L^{(t)})$ from above and from below by similar counting functions for $PL^{(t)}P$ and $QL^{(t)}Q$.

Recalling that $H_0 = \Pi_1^2 + \Pi_2^2$ with $\Pi_k = -i\partial_k - a_k$, $k = 1, 2$, and denoting $\tilde V_k(\mathbf{x}) = -i(\partial_k V)(\mathbf{x})$, we conclude that

$$\begin{aligned} [H_0, V^{(t)}] &= \sum_{k=1}^2 \big(\Pi_k [\Pi_k, V^{(t)}] + [\Pi_k, V^{(t)}]\Pi_k\big) \\ &= t^{-1} \sum_{k=1}^2 \big(\Pi_k \tilde V_k^{(t)} + \tilde V_k^{(t)} \Pi_k\big) = t^{-1} \sum_{k=1}^2 \Pi_k \tilde V_k^{(t)} + t^{-2} (\Delta V)^{(t)}. \end{aligned}$$

Consequently,

$$\begin{aligned} |(P[H_0, V^{(t)}]Qu, u)| &\leq t^{-1} \sum_{k=1}^2 |(\tilde V_k^{(t)} Qu, \Pi_k Pu)| + vt^{-2}\|u\|^2 \\ &\leq (2t^2)^{-1}(H_0 Pu, Pu) + 2^{-1}v^2\|Qu\|^2 + vt^{-2}\|u\|^2 \\ &\leq (4t^2)^{-1}(H_{0a}^2 Pu, Pu) + 2^{-1}v^2\|Qu\|^2 + (v + a + 1)t^{-2}\|u\|^2. \end{aligned}$$

Now it follows from (5.2) that for any $\delta > 0$,

$$\begin{aligned} |(PL^{(t)}Qu, u)| &\leq 2|(V^{(t)}Qu, H_{0a}Pu)| + |(P[H_0, V^{(t)}]Qu, u)| + |(Z^{(t)}Qu, Pu)| \\ &\leq \delta\|H_{0a}Pu\|^2 + \frac{v}{\delta}\|Qu\|^2 + (4t^2)^{-1}(H_{0a}^2 Pu, Pu) \\ &\quad + 2^{-1}v^2\|Qu\|^2 + (v + a + 1)t^{-2}\|u\|^2 + \frac{\delta}{2}\|Pu\|^2 + \frac{v^2}{2\delta}\|Qu\|^2 \\ &\leq \big(\delta + (4t^2)^{-1}\big)(H_{0a}^2 Pu, Pu) + \frac{\delta}{2}\|Pu\|^2 \\ &\quad + 2^{-1}v^2\big(1 + \delta^{-1}\big)\|Qu\|^2 + (v + a + 1)t^{-2}\|u\|^2. \end{aligned}$$

Therefore,

$$(L^{(t)}u,u) \le (L^{(t)}Pu,Pu) + 2\big(\delta + (4t^2)^{-1}\big)(H_{0a}Pu,Pu) + \delta\|Pu\|^2$$
$$+ v^2\big(1+\delta^{-1}\big)\|Qu\|^2 + 2(v+a+1)t^{-2}\|u\|^2 + (QL^{(t)}Qu,u),$$

and

$$(L^{(t)}u,u) \ge (L^{(t)}Pu,Pu) - 2\big(\delta + (4t^2)^{-1}\big)(H_{0a}^2Pu,Pu) - \delta\|Pu\|^2$$
$$- v^2\big(1+\delta^{-1}\big)\|Qu\|^2 - 2(v+a+1)t^{-2}\|u\|^2 + (QL^{(t)}Qu,u).$$

Denote

$$\epsilon_1 = 2\big(\delta + (4t^2)^{-1}\big),\quad \epsilon_2 = \delta + 2(v+a+1)t^{-2},$$
$$M = v^2\big(1+\delta^{-1}\big) + 2(v+a+1)t^{-2},$$

so that

(5.6)
$$(L^{(t)}u,u) \le (L^{(t)}Pu,Pu) + \epsilon_1(H_{0a}^2Pu,Pu) + \epsilon_2\|Pu\|^2 + M\|Qu\|^2 + (QL^{(t)}Qu,u),$$
$$(L^{(t)}u,u) \ge (L^{(t)}Pu,Pu) - \epsilon_1(H_{0a}^2Pu,Pu) - \epsilon_2\|Pu\|^2 - M\|Qu\|^2 + (QL^{(t)}Qu,u).$$

Thus we have bounded $L^{(t)}$ from above and from below by orthogonal sums of operators acting in ranges of P and Q. We show now that for sufficiently large J the operators containing Q do not contribute to the t-asymptotics of the counting function. We have

$$(QL^{(t)}Qu,u) = \|H_{0a}Qu\|^2 + (H_{0a}Qu,V^{(t)}Qu) + (V^{(t)}Qu,H_{0a}Qu) + (Z^{(t)}Qu,Qu)$$
$$\ge \|H_{0a}Qu\|^2 - \frac{1}{2}\|H_{0a}Qu\|^2 - 2\|V^{(t)}Qu\|^2 - v\|Qu\|^2$$
$$\ge \frac{1}{2}\|H_{0a}Qu\|^2 - (2v^2+v)\|Qu\|^2$$
$$\ge \frac{1}{2}\|(\Lambda_J - a)Qu\|^2 - (2v^2+v)\|Qu\|^2.$$

So, for sufficiently large J (depending on M and δ), we have $QL^{(t)}Q - MQ \ge \eta Q$ and therefore the operator $QL^{(t)}Q - MQ$ acting in the range of Q has no spectrum below η.

Finally, to estimate the operator containing P in (5.6), we can write

$$L^{(t)}(V,Z) \pm \epsilon_1 H_{0a}^2 = (1\pm\epsilon_1)L^{(t)}\big(V(1\pm\epsilon_1)^{-1}, Z(1\pm\epsilon_1)^{-1}\big).$$

Therefore, for sufficiently large J it follows from (5.6) that

$$N(\eta;L^{(t)}) \le N\big(\eta+\epsilon_2;(1-\epsilon_1)PL^{(t)}\big(V(1-\epsilon_1)^{-1}, Z(1-\epsilon_1)^{-1}\big)P\big),$$
$$N(\eta;L^{(t)}) \ge N\big(\eta-\epsilon_2;(1+\epsilon_1)PL^{(t)}\big(V(1+\epsilon_1)^{-1}, Z(1+\epsilon_1)^{-1}\big)P\big),$$

and (5.5) follows. □

Next we are going to derive an asymptotic estimate for the operator $PL^{(t)}P$ as $t\to\infty$. We define the asymptotic coefficient as

(5.7) $$\mathscr{B}(\eta;V,Z) = \sum_{q=0}^{\infty} \frac{B}{2\pi} A^{(+)}\big((\Lambda_q - a)^2 - \eta, -2(\Lambda_q - a)V - Z\big).$$

Under the assumptions $V \in \mathsf{L}^2(\mathbb{R}^2) \cap \mathsf{L}^1(\mathbb{R}^2)$, $Z \in \mathsf{L}^1(\mathbb{R}^2)$ the above series is absolutely convergent uniformly in η varying on a compact set. Indeed,

$$\begin{aligned} A^{(+)}\big((\Lambda_q - a)^2 - \eta,\ -2(\Lambda_q - a)V - Z\big) &\le A^{(+)}\big((\Lambda_q - a)^2 - \eta, |\Lambda_q - a|\ |V| + |Z|\big) \\ &\le A^{(+)}\big((\Lambda_q - a)^2 - \eta, \frac{1}{2}(\Lambda_q - a)^2 + 2|V|^2 + |Z|\big) \\ &\le A^{(+)}\big(\frac{1}{2}(\Lambda_q - a)^2 - \eta, 2|V|^2 + |Z|\big). \end{aligned}$$

For sufficiently large q, by the Chebyshev inequality, the last term is bounded from above by

$$\big(\frac{1}{2}(\Lambda_q - a)^2 - \eta\big)^{-1}\big(2\|V\|_2^2 + \|Z\|_1\big).$$

This shows that indeed the series in (5.7) converges. Each term in (5.7) is a left semi-continuous nondecreasing function of η. As a result, $\mathscr{B}(\eta; V, Z)$ is left semi-continuous as well.

Let $P^{(J)}$ be as defined in (5.4).

LEMMA 5.2. *Let $V \in \mathsf{L}^1(\mathbb{R}^2) \cap \mathsf{L}^2(\mathbb{R}^2)$, $Z \in \mathsf{L}^1(\mathbb{R}^2)$. Then for any $\eta < \eta_0$,*

$$\begin{aligned} &\liminf_{J\to\infty} \mathfrak{b}(\eta; P^{(J)}L(V,Z)P^{(J)}) \ge \mathscr{B}(\eta; V, Z), \\ (5.8)\qquad &\limsup_{J\to\infty} \mathfrak{B}(\eta; P^{(J)}L(V,Z)P^{(J)}) \le \mathscr{B}(\eta + 0; V, Z). \end{aligned}$$

In particular, if $(\Lambda_q - a)^2 - \eta$ are generic values for $-2(\Lambda_q - a)V - Z$ for all $q = 0, 1, \dots,$ then

$$\lim_{J\to\infty}\lim_{t\to\infty} t^{-2} N(\eta; P^{(J)}L^{(t)}(V,Z)P^{(J)}) = \mathscr{B}(\eta; V, Z). \tag{5.9}$$

PROOF. Fix J and rewrite $PL^{(t)}P$, $P = P^{(J)}$ as

$$\begin{aligned} PL^{(t)}P = &\sum_{q\le J}\big((\Lambda_q - a)^2 P_q + 2(\Lambda_q - a)T_q^{(t)}(V) + T_q^{(t)}(Z)\big) \\ &+ \sum_{q\le J}\ \sum_{q'\le J, q'\ne q}\big((\Lambda_{q'} - a)T_{q,q'}^{(t)}(V) + T_{q,q'}^{(t)}(Z)\big) =: G^{(t)} + \tilde{G}^{(t)}. \end{aligned}$$

By Corollary 3.5 and (2.9) we have $\mathfrak{R}(\epsilon; \tilde{G}) = 0$ for any $\epsilon > 0$.

The operator $G^{(t)}$ is a finite orthogonal sum of the operators $G_q^{(t)} = (\Lambda_q - a)^2 + T_q^{(t)}(2(\Lambda_q - a)V + Z)$ acting in Landau subspaces $\mathfrak{L}_q$, so it suffices to consider each of them individually. Recall that $\eta < \eta_0 \le (\Lambda_q - a)^2$, so that $\eta - (\Lambda_q - a)^2 < 0$ and

$$N(\eta; G_q^{(t)}) = n_+\big((\Lambda_q - a)^2 - \eta; T_q^{(t)}(-2(\Lambda_q - a)V - Z)\big).$$

Applying Theorem 4.10, we obtain

$$\mathfrak{N}^{(+)}((\Lambda_q - a)^2 - \eta; -2(\Lambda_q - a)V - Z) \le \frac{B}{2\pi} A^{(+)}((\Lambda_q - a)^2 - \eta - 0, -2(\Lambda_q - a)V - Z),$$

and therefore, after summation over q,

$$\mathfrak{B}(\eta; G) \le \mathscr{B}(\eta + 0; V, Z);$$

see definition (5.7). Applying Lemma 2.4 to the sum $G + \tilde{G}$, we obtain the sought upper bound. The lower bound in the lemma is derived in the same way. $\square$

COROLLARY 5.3. *Suppose that V, Z are as in Lemma 5.1 and that $\eta < \eta_0$. Then*

$$\mathscr{B}(\eta; V, Z) \le \mathfrak{b}(\eta; L(V, Z)) \le \mathfrak{B}(\eta; L(V, Z)) \le \mathscr{B}(\eta + 0; V, Z). \tag{5.10}$$

PROOF. According to (5.5) and (5.8),

$$\mathfrak{b}(\eta; L(V, Z)) \ge \mathscr{B}\big(\eta - \tilde{\epsilon}; V(1-\epsilon), Z(1-\epsilon)\big),$$

where $\tilde{\epsilon} \downarrow 0$ as $\epsilon \downarrow 0$. Passing to the limit as $\epsilon \downarrow 0$ with the help of (4.14), we arrive at the required lower bound. The upper bound is obtained in the same way. □

6. Reduction to a smooth potential

6.1. Further estimates for the operator $L^{(t)}(V, Z)$. In this section we continue the study of the operator $L^{(t)}(V, Z)$. Our aim now is to extend Corollary 5.3 to nonsmooth functions V and Z.

Recall again the notation

$$L^{(t)} = L^{(t)}(V, Z) = H_{0a}^2 + V^{(t)} H_{0a} + H_{0a} V^{(t)} + Z^{(t)}$$

for some real-valued functions $V \in \mathsf{L}^1(\mathbb{R}^2) \cap \mathsf{L}^2(\mathbb{R}^2)$, $Z \in \mathsf{L}^1(\mathbb{R}^2)$. We start with an eigenvalue estimate for the operator $L^{(t)}$.

LEMMA 6.1. *Suppose that $V \in \mathsf{L}^2(\mathbb{R}^2)$ and $Z \in \mathsf{L}^1(\mathbb{R}^2)$. Then*

$$\mathfrak{B}(\eta; L(V, Z)) \le \frac{C}{(\eta_0 - \eta)^2}(\|V\|_2^2 + \|Z\|_1), \tag{6.1}$$

for all $\eta < \eta_0$ and some constant C independent of η.

PROOF. Since $|(V^{(t)}u, H_{0a}u)| \le \frac{1}{2z}\|V^{(t)}u\|^2 + \frac{z}{2}\|H_{0a}u\|^2$, we have for $z \in (0,1)$:

$$L^{(t)}(V, Z) \ge (1-z)H_{0a}^2 - z^{-1}(V^{(t)})^2 - |Z^{(t)}|.$$

We chose $z = \frac{\eta_0 - \eta}{2\eta_0}$ so that $(1-z)H_{0a}^2 - \eta > 0$ and $(1-z)H_{0a}^2 - \eta \ge c(\eta_0 - \eta)H_{0,-1}^2$, and denote

$$Y_{a,\eta}^{(t)} = \big((1-z)H_{0a}^2 - \eta\big)^{-\frac{1}{2}} \big(z^{-1}(V^{(t)})^2 + |Z^{(t)}|\big)^{\frac{1}{2}}.$$

By the Birman-Schwinger principle we have

$$N\big(\eta; L^{(t)}(V, Z)\big) \le n\big(1; (Y_{a,\eta}^{(t)})^* Y_{a,\eta}^{(t)}\big).$$

By the diamagnetic inequality we have

$$\begin{aligned}
\|Y_{a,\eta}^{(t)}\|_{\mathfrak{S}_2}^2 &\le C(\eta_0 - \eta)^{-1} \|Y_{-1,0}^{(t)}\|_{\mathfrak{S}_2}^2 \\
&\le C(\eta_0 - \eta)^{-1} \|(-\Delta + I)^{-1}\big(z^{-1}(V^{(t)})^2 + |Z^{(t)}|\big)^{\frac{1}{2}}\|_{\mathfrak{S}_2}^2 \\
&\le C'(\eta_0 - \eta)^{-2} t^2 \big(\|V\|_2^2 + \|Z\|_1\big).
\end{aligned}$$

Therefore $(Y_{a,\eta}^{(t)})^* Y_{a,\eta}^{(t)}$ is a trace class operator and

$$\|(Y_{a,\eta}^{(t)})^* Y_{a,\eta}^{(t)}\|_{\mathfrak{S}_1} \le C(\eta_0 - \eta)^{-2} t^2 \big(\|V\|_2^2 + \|Z\|_1\big).$$

The inequality (6.1) follows now by applying the bound $n(\lambda; K) \le \lambda^{-1}\|K\|_{\mathfrak{S}_1}$. □

THEOREM 6.2. *Let $V \in \mathsf{L}^1(\mathbb{R}^2) \cap \mathsf{L}^2(\mathbb{R}^2)$ and $Z \in \mathsf{L}^1(\mathbb{R}^2)$. Then formula (5.10) holds for any $\eta < \eta_0$.*

PROOF. The idea is to approximate V, Z by smooth functions and then use Lemma 2.3 to show that the error does not contribute.

For a fixed $\delta > 0$ represent V, Z as $V = V'_\delta + V''_\delta$, $Z = Z'_\delta + Z''_\delta$ so that $V'_\delta, Z'_\delta \in \mathsf{C}_0^\infty(\mathbb{R}^2)$ and

$$\|V''_\delta\|_1 + \|V''_\delta\|_2^2 + \|Z''_\delta\|_1 < \delta.$$

We apply Lemma 2.3. Clearly,

$$L(V, Z) = L(0,0) + Y'_\delta + Y''_\delta,$$

$$Y'_\delta = V'_\delta H_{0a} + H_{0a}V'_\delta + Z'_\delta,\ Y''_\delta = V''_\delta H_{0a} + H_{0a}V''_\delta + Z''_\delta.$$

$L(V, Z) = L(0, 0)$. According to (6.1),

$$\mathfrak{B}(\tau; L(0,0) + MY''_\delta) \le C\frac{C(1+M^2)\delta}{(\eta_0 - \tau)^2}$$

for any $\tau < \eta_0$ and any $M \in R$, and thus the condition (2.10) is fulfilled. Now it follows from Corollary 5.3 and Lemma 2.3 that

$$\mathfrak{B}(\eta; L(V,Z)) \le \lim_{\epsilon\downarrow 0} \limsup_{M\downarrow 1} \limsup_{\delta\to 0} \mathscr{B}(\eta+\epsilon; MV'_\delta, MZ'_\delta).$$

By Lemma 4.8, the right hand side does not exceed $\mathscr{B}(\eta + 0; V, Z)$, as claimed. Similarly, one obtains the appropriate lower bound for $\mathfrak{b}(\eta, L(V, Z))$. □

6.2. The general asymptotic estimate: Proof of Theorem 1.1. Recall that the coefficients $\mathscr{A}$ and $\mathscr{B}$ are defined in (1.2) and (5.7) respectively. By (5.1) we have $N(\lambda_1, \lambda_2; H^{(t)}) = N(b^2; L^{(t)}(V, V^2))$. Observe that $\mathscr{A}(\lambda_1, \lambda_2; V) = \mathscr{B}(b^2; V, V^2)$. It remains to apply Theorem 6.2.

Acknowledgements

The asymptotic problem considered in the paper appeared as a result of discussions of the first author (GR) with his colleague physicist R. Shekhter. A decisive part of the work was done by the authors when enjoying the hospitality of the Isaac Newton Institute for Mathematical Sciences (Cambridge, UK) in the framework of the programs "Spectral Theory and Partial Differential Equations" and "Analysis on Graphs". It is our pleasure to thank the Institute and the organizers of the programs for providing this opportunity.

References

1. M. Sh. Birman and M. Z. Solomjak, *Asymptotics of the spectrum of weakly polar integral operators.* (Russian) Izv. Akad. Nauk SSSR Ser. Mat. **34**, no. 5 (1970) 1142–1158.
2. ———, *On the principal term of the spectral asymptotics for "nonsmooth" elliptic problems*, Funktsional. Anal. i Prilozhen. 4, no. 4, (1970), 1-13; English transl. in Functional Anal. Appl. 4 (1970), 265–275.
3. ———, *Spectral theory of selfadjoint operators in Hilbert space*, Reidel, 1987.
4. N. Filonov and A. Pushnitski, *Spectral asymptotics of Pauli operators and orthogonal polynomials in complex domains*, Commun. Math. Phys. **264**, no. 3(2006) 759-772.
5. I. M. Glazman, *Direct methods of qualitative spectral analysis of singular differential operators* (in Russian), Gosudarstv. Izdat. Fiz.-Mat. Lit., Moscow, 1963. English transl. Daniel Davey and Co., New York, 1966.
6. U. Grenander and G. Szegö, *Toeplitz forms and their applications.* 2nd ed., Chelsea, New York, 1984.
7. R. Hempel and A. Besch, *Magnetic barriers of compact support and eigenvalues in spectral gaps*, Electron. J. Differential Equations 2003, No. 48, 25 pp.

8. R. Hempel and S. Levendorsky, *On eigenvalues in gaps for perturbed magnetic Schrödinger operators*, J. Math. Phys. **39** (1998), no. 1, 63–78.
9. A. Iwatsuka and H. Tamura, *Asymptotic distribution of negative eigenvalues for two-dimensional Pauli operators with nonconstant magnetic fields*, Ann. Inst. Fourier (Grenoble) **48**(1998), no. 2, 479–515.
10. L. D. Landau and E. M. Lifshitz, Quantum Mechanics: Non-Relativistic Theory, Course of Theoretical Physics, Vol. 3, Pergamon Press, 1958.
11. M. Melgaard and G. Rozenblum, *Eigenvalue asymptotics for weakly perturbed Dirac and Schrödinger operators with constant magnetic fields of full rank.* Comm. Part. Diff. Equat. **28**, 697–736 (2003).
12. A. Pushnitski and G. Rozenblum, *Eigenvalue clusters of the Landau Hamiltonian in the exterior of a compact domain.* Preprint `http://front.math.ucdavis.edu/0707.4297`, Documenta Mathematica **12** (2007), 569–586.
13. G. Raĭkov, *Eigenvalue asymptotics for the Schrödinger operator with homogeneous magnetic potential and decreasing electric potential. I. Behaviour near the essential spectrum tips*, Comm. Part. Diff. Equat. **15** (1990), 407–434.
14. G. Raĭkov and S. Warzel, *Quasi-classical versus non-classical spectral asymptotics for magnetic Schrödinger operators with decreasing electric potentials*, Rev. Math. Phys. **14** (2002), no. 10, 1051–1072.
15. G. Rozenblum and G Tashchiyan, *On the spectrum of the perturbed Landau Hamiltonian*, Preprint `http://front.math.ucdavis.edu/math-ph/0605038`, to appear in Comm. Part. Diff. Equat.
16. A. V. Sobolev, *Quasi-classical asymptotics for the Pauli operator*, Commun. Math. Phys., **194**(1998), 109–134.

Department Mathematics, Chalmers University of Technology and Department of Mathematics, Gothenburg University, Gothenburg, 41296, Sweden
E-mail address: `grigori@math.chalmers.se`

Department of Mathematics, University College London, Gower Street, London, WC1E 6BT, United Kingdom
E-mail address: `asobolev@math.ucl.ac.uk`

Amer. Math. Soc. Transl.
(2) Vol. **225**, 2008

On the Comparison of the Dirichlet and Neumann Counting Functions

Y. Safarov

To Mikhail Shlëmovich Birman on his 80th birthday

Introduction

Let $N_{\mathrm{N}}(\lambda)$ and $N_{\mathrm{D}}(\lambda)$ be the counting functions of the Dirichlet and Neumann Laplacian on a domain $\Omega \subset \mathbb{R}^n$. If λ is not a Dirichlet or Neumann eigenvalue, then

$$(*) \qquad N_{\mathrm{N}}(\lambda) = N_{\mathrm{D}}(\lambda) + g^-(\lambda),$$

where $g^-(\lambda)$ denotes the number of negative eigenvalues of the Dirichlet-to-Neumann map at $\lambda \in \mathbb{R}$. The equality (*) was proved in [**Fr1**] for domains with sufficiently smooth boundaries. L. Friedlander also noticed that (*) immediately implies Payne's conjecture for the Laplacian on a bounded domain, according to which the $(k+1)$th Neumann eigenvalue does not exceed the kth Dirichlet eigenvalue. Later R. Mazzeo remarked that (*) remains valid for domains with smooth boundaries in any Riemannian symmetric space of noncompact type and gave a geometric explanation of Friedlander's result [**M**].

For irregular boundaries, the Dirichlet-to-Neumann map may not be well-defined and then (*) does not make sense. In 2004, N. Filonov suggested another proof of Payne's conjecture for the Laplacian [**Fi**]. This proof does not use (*) and works for nonsmooth boundaries. The author assumed that the resolvent of the Neumann Laplacian on Ω is compact, but this condition can be removed (see Remark 1.9).

The aim of this note is to show that (*) holds for abstract operators in a Hilbert space H, provided that the Dirichlet-to-Neumann map is understood in a proper sense. Traditionally, one assumes that the Dirichlet-to-Neumann map is a family of operators acting in the same space and depending on the spectral parameter λ (see Subsection 1.3). In our understanding, it is a family of operators $\mathcal{B}_\lambda$ generated by the restrictions of the same sesquilinear form to different subspaces $G_\lambda \subset H^1$. The

2000 *Mathematics Subject Classification.* Primary 47A75, 35P15.
Key words and phrases. Eigenvalues, counting function, boundary value problems.
The research was supported by the EPSRC grant GR/T25552/01.

identity (*) is proved with the use of special isomorphisms between the subspaces G_λ with different values of λ.

This approach is close in spirit to Birman's paper [**B1**] on selfadjoint extensions of symmetric operators. In particular, it removes technical problems related to nonsmooth boundaries and allows one to extend Payne's conjecture to all operators generated by differential quadratic forms with constant coefficients on an arbitrary domain $\Omega \subset \mathbb{R}^n$ with $n \geq 2$ (see Corollary 1.13). Another advantage of our scheme is that, unlike the classical Dirichlet-to-Neumann map, the operators $\mathcal{B}_\lambda$ do not blow up as λ passes through isolated eigenvalues. This enables one to perform a more detailed analysis of the relation between their properties and the spectral characteristics of the Dirichlet and Neumann problems.

The paper is constructed as follows. In Section 1 we introduce some necessary notation and state the main results. Note that the notation is deliberately chosen as if A is a second-order elliptic differential operator acting in the Sobolev spaces on a domain, subject to the Dirichlet or Neumann boundary condition (even though H does not have to be a function space and the ellipticity is irrelevant). In Section 2 we prove some simple auxiliary lemmas on abstract selfadjoint operators. Section 3 is devoted to the proof of the main statements. Finally, Section 4 contains some remarks and by-product results, which are not needed in our proofs but may be of interest in themselves.

1. Basic notation and main results

1.1. Notation. We shall always be assuming that $\lambda, \mu \in \mathbb{R}$ and $z \in \mathbb{C}$.

Let H be an infinite-dimensional separable complex Hilbert space. As usual, $(\cdot,\cdot)$ and $\|\cdot\|$ are the inner product and norm in H, and $\dotplus$ denotes a direct sum in H. Let

- H^1 be a dense subspace of H;
- $\mathbf{a}[\cdot]$ be a closed positive quadratic form on H^1 and $\mathbf{a}[\cdot,\cdot]$ be the corresponding sesquilinear form;
- A_{N} be the selfadjoint operator in H generated by the form $\mathbf{a}[\cdot]$.

We shall consider H^1 as a Hilbert space provided with the inner product $\mathbf{a}[\cdot,\cdot]$. Let

- H_0^1 be a closed subspace of H^1 which is dense in H;
- A_{D} be the selfadjoint operator in H generated by the restriction of the form $\mathbf{a}[\cdot]$ to H_0^1.

Further on we shall write B instead of N or D in the case where the corresponding statement holds or the definition refers to both operators A_{N} and A_{D}. In particular, we shall be using the following notation.

- $\sigma(A_{\mathrm{B}})$ and $\sigma_{\mathrm{ess}}(A_{\mathrm{B}})$ denote the spectrum and the essential spectrum of the operator A_{B}.
- $\lambda_{B,\infty} := \inf \sigma_{\mathrm{ess}}(A_{\mathrm{B}})$.
- $\lambda_{B,1} \leqslant \lambda_{B,2} \leqslant \lambda_{B,3} \ldots$ are the eigenvalues of the operator A_{B} lying in the interval $(-\infty, \lambda_{B,\infty})$ and counted with their multiplicities.
- χ_Λ denotes the characteristic function of the Borel set $\Lambda \subset \mathbb{R}$, so that
- $\chi_\Lambda(A_{\mathrm{B}})$ is the spectral projection of A_{B} corresponding to Λ.

- $E_{\mathrm{B}}(z)$ is the orthogonal projection onto $\ker(A_{\mathrm{B}} - zI)$ and $E'_{\mathrm{B}}(z) := I - E_{\mathrm{B}}(z)$.
- $N_{\mathrm{B}}(\lambda) := \dim \chi_{(-\infty,\lambda)}(A_{\mathrm{B}})\, H$ is the left continuous counting function of the operator A_{B}.

The Rayleigh–Ritz variational formula implies that $N_{\mathrm{D}}(\lambda) \leqslant N_{\mathrm{N}}(\lambda)$ or, in other words, $0 < \lambda_{\mathrm{N},j} \leqslant \lambda_{\mathrm{D},j}$ for all $j = 1, 2, \ldots, \infty$. We have $N_{\mathrm{B}}(\lambda) = \#\{j : \lambda_{\mathrm{B},j} < \lambda\}$ whenever $\lambda \leqslant \lambda_{\mathrm{B},\infty}$ and $N_{\mathrm{B}}(\lambda) = \infty$ otherwise.

Let

- H^1_A be the set of vectors $u \in H^1$ such that the functionals $v \to \mathbf{a}[u, v]$ on H^1_0 are H-continuous;
- A be the operator acting from H^1_A to H such that $(Au, v) = \mathbf{a}[u, v]$ for all $v \in H^1_0$;
- $G_z := \{u \in H^1_A : Au = zu\}$, where $z \in \mathbb{C}$;
- $\mathbf{b}[u, v] := \mathbf{a}[u, v] - (Au, v)$ and $\mathbf{b}[u] := \mathbf{b}[u, u]$, where $u \in H^1_A$ and $v \in H^1$.

Since the operator A is H^1-closed, G_z are closed subspaces of H^1. Denote

- $\mathcal{B}_\lambda := (I - \lambda \Pi'_\lambda A_{\mathrm{N}}^{-1})\big|_{G_\lambda}$, where
- Π'_λ is the H^1-orthogonal projection onto G_λ

(an explicit formula for Π'_λ is given in Subsection 2.3). We shall consider $\mathcal{B}_\lambda$ as an operator in G_λ. Obviously,

$$\mathbf{a}[\mathcal{B}_\lambda u, v] \;=\; \mathbf{a}[u, v] - \lambda\,(u, v) \;=\; \mathbf{b}[u, v], \qquad \forall u, v \in G_\lambda. \tag{1.1}$$

Therefore $\mathcal{B}_\lambda$ is a bounded selfadjoint operator in the Hilbert space G_λ provided with the inner product $\mathbf{a}[\cdot,\cdot]$.

Let

- $\sigma(\mathcal{B}_\lambda)$ and $\sigma_{\mathrm{ess}}(\mathcal{B}_\lambda)$ be the spectrum and essential spectrum of $\mathcal{B}_\lambda$,
- $G^0_\lambda := \ker \mathcal{B}_\lambda$, $G^-_\lambda := \chi_{(-\infty,0)}(\mathcal{B}_\lambda) G_\lambda$ and $G^+_\lambda := \chi_{(0,+\infty)}(\mathcal{B}_\lambda) G_\lambda$,

where $\chi_{(-\infty,0)}(\mathcal{B}_\lambda)$ and $\chi_{(0,+\infty)}(\mathcal{B}_\lambda)$ are the corresponding spectral projections of the operator $\mathcal{B}_\lambda$.

Finally, let

- $\mathcal{H}_0$ be the subspace of H spanned by all common eigenvectors of the operators A_{N} and A_{D};
- $\mathcal{H}$ be the H-orthogonal complement of $\mathcal{H}_0$;
- $n_{\mathrm{B}}(\lambda) := \dim E_{\mathrm{B}}(\lambda)\mathcal{H}$ and $n_{\mathrm{N,D}}(\lambda) := \dim E_{\mathrm{B}}(\lambda)\mathcal{H}_0$, so that $n_{\mathrm{B}}(\lambda) = \dim E_{\mathrm{B}}(\lambda) H - n_{\mathrm{N,D}}(\lambda)$.

Clearly, $\mathcal{H}$ and $\mathcal{H}_0$ are invariant subspaces of the operators A_{N} and A_{D}, whose intersections with H^1 are H^1-orthogonal. Similarly, $G_\lambda \cap \mathcal{H}$ and $G_\lambda \cap \mathcal{H}_0$ are invariant subspaces of $\mathcal{B}_\lambda$. One can easily see that $G_\lambda \cap \mathcal{H}_0 = G^0_\lambda \cap \mathcal{H}_0 = E_{\mathrm{B}}(\lambda)\mathcal{H}_0$ and $\mathcal{B}_\lambda|_{G_\lambda \cap \mathcal{H}_0} = 0$. In particular, $G_\lambda \cap \mathcal{H}_0 = \{0\}$ whenever λ is not an eigenvalue corresponding to a common eigenvector of A_{N} and A_{D}.

1.2. Main results. The following lemma implies that the restriction $\mathcal{B}_\lambda|_{\mathcal{H}}$ analytically depends on λ outside the intersection of the essential spectra $\sigma_{\mathrm{ess}}(A_{\mathrm{N}})$ and $\sigma_{\mathrm{ess}}(A_{\mathrm{D}})$.

LEMMA 1.1. *The H^1-orthogonal projection onto $G_\lambda \cap \mathcal{H}$ is an analytic operator-valued function of λ on the set $\mathbb{R} \setminus (\sigma_{\mathrm{ess}}(A_{\mathrm{N}}) \cap \sigma_{\mathrm{ess}}(A_{\mathrm{D}}))$.*

One can easily show that

$$E_{\mathrm{N}}(\lambda)H + E_{\mathrm{D}}(\lambda)H \subset G^0_\lambda, \qquad \forall\lambda\in\mathbb{R} \tag{1.2}$$

(see Subsection 3.3). The next lemma is less obvious.

LEMMA 1.2. *If* $\lambda\notin\sigma_{\mathrm{ess}}(A_{\mathrm{N}})\cap\sigma_{\mathrm{ess}}(A_{\mathrm{D}})$, *then* $G^0_\lambda = E_{\mathrm{D}}(\lambda)H + E_{\mathrm{N}}(\lambda)H$. *If* $\lambda\notin\sigma_{\mathrm{ess}}(A_{\mathrm{N}})\cup\sigma_{\mathrm{ess}}(A_{\mathrm{D}})$, *then the point* 0 *does not belong to the essential spectrum of the operator* $\mathcal{B}_\lambda$.

Lemmas 1.1 and 1.2 imply

THEOREM 1.3. *Let* $\lambda\notin\sigma_{\mathrm{ess}}(A_{\mathrm{N}})\cup\sigma_{\mathrm{ess}}(A_{\mathrm{D}})$. *Then for each sufficiently small* $\varepsilon>0$ *there exists* $\delta>0$ *such that the intersection* $(-\varepsilon,\varepsilon)\cap\sigma(\mathcal{B}_\mu)$ *consists of*

(1) $n_{\mathrm{N}}(\mu)+n_{\mathrm{D}}(\mu)+n_{\mathrm{N,D}}(\mu)$ *zero eigenvalues if* $\mu=\lambda$,
(2) $n_{\mathrm{D}}(\lambda)$ *negative and* $n_{\mathrm{N}}(\lambda)$ *positive eigenvalues if* $\mu\in(\lambda-\delta,\lambda)$,
(3) $n_{\mathrm{N}}(\lambda)$ *negative and* $n_{\mathrm{D}}(\lambda)$ *positive eigenvalues if* $\mu\in(\lambda,\lambda+\delta)$

(as usual, the eigenvalues are counted according to their multiplicities).

REMARK 1.4. By Lemma 1.2, if $\lambda\notin\sigma_{\mathrm{ess}}(A_{\mathrm{N}})\cup\sigma_{\mathrm{ess}}(A_{\mathrm{D}})$, then we have $[-\varepsilon,\varepsilon]\cap\sigma_{\mathrm{ess}}(\mathcal{B}_\mu)=\varnothing$ for all sufficiently small $\varepsilon,\delta>0$ and all $\mu\in[\lambda-\delta,\lambda+\delta]$. By Lemma 1.1, the eigenvalues $\nu_j(\mu)$ of the restrictions $\mathcal{B}_\mu|_{G_\mu\cap\mathcal{H}}$ lying in $(-\varepsilon,\varepsilon)$ are continuous function of $\mu\in(\lambda-\delta,\lambda+\delta)$. Therefore, if ε and δ are small enough, then $\nu_j(\mu)\in(-\varepsilon,\varepsilon)$ for some $\mu\in(\lambda-\delta,\lambda+\delta)$ if and only if $\nu_j(\lambda)=0$. Theorem 1.3 states that $n_{\mathrm{D}}(\mu)$ eigenvalues $\nu_j(\mu)$ change their sign from minus to plus and $n_{\mathrm{N}}(\mu)$ eigenvalues $\nu_j(\mu)$ change their sign from plus to minus as μ passes through the eigenvalue λ. At the point λ all these eigenvalues are equal to zero and, in addition, there are $n_{\mathrm{N,D}}(\lambda)$ zero eigenvalues of the restriction $\mathcal{B}_\lambda|_{G_\lambda\cap\mathcal{H}_0}$.

REMARK 1.5. A similar result was obtained in [**Fr1**] and [**M**] for differential operators on domains with smooth boundaries under the additional assumption that their spectra are discrete. Theorem 1.3 holds in the abstract setting and remains valid for λ lying in the gaps of the essential spectra.

COROLLARY 1.6. *Let* $a<b$. *If* $[a,b]\cap\sigma_{\mathrm{ess}}(A_{\mathrm{N}})=[a,b]\cap\sigma_{\mathrm{ess}}(A_{\mathrm{D}})=\varnothing$, *then*

$$\dim G^-_b \;=\; \dim G^-_a \;+\; \dim\chi_{[a,b)}(A_{\mathrm{N}})\mathcal{H} \;-\; \dim\chi_{(a,b]}(A_{\mathrm{D}})\mathcal{H}. \tag{1.3}$$

If $a<\inf\sigma(A_{\mathrm{N}})$, then $G^-_a=\{0\}$ and, according to the next theorem, the equality (1.3) remains valid for $b\in[\lambda_{\mathrm{N},\infty},\lambda_{\mathrm{D},\infty})$.

THEOREM 1.7. $N_{\mathrm{N}}(\lambda)=N_{\mathrm{D}}(\lambda)+n_{\mathrm{D}}(\lambda)+\dim G^-_\lambda$ *for all* $\lambda<\lambda_{\mathrm{D},\infty}$.

REMARK 1.8. By Theorem 1.7, $N_{\mathrm{N}}(\lambda)=\dim G^-_\lambda$ for all λ lying below $\sigma(A_{\mathrm{D}})$. In the case where A_{N} and A_{D} are selfadjoint extensions of the same symmetric operator defined on $\mathcal{D}(A_{\mathrm{N}})\cap\mathcal{D}(A_{\mathrm{D}})$, the above identity was obtained by M.S. Birman [**B1**] (see also [**B2**]). Theorem 1.7 extends Birman's result to all $\lambda<\lambda_{\mathrm{D},\infty}$ in a slightly more general setting (see Subsection 4.1).

REMARK 1.9. N. Filonov noticed in [**Fi**] that for the Laplacian on an arbitrary domain $\Omega\subset\mathbb{R}^n$ we have

$$\mathbf{a}[u] \;\leqslant\; \lambda\,\|u\|^2, \qquad \forall u\in\chi_{[0,\lambda]}(A_{\mathrm{D}})H+E_{\mathrm{N}}(\lambda)H+G^0_\lambda+G^-_\lambda. \tag{1.4}$$

Similar arguments show that (1.4) holds for any pair of abstract operators A_{D} and A_{N} (see Subsection 3.2). The estimate (1.4) immediately implies that

$$N_{\mathrm{N}}(\lambda) \;\geqslant\; N_{\mathrm{D}}(\lambda) \;+\; n_{\mathrm{D}}(\lambda) \;+\; \dim G^-_\lambda, \qquad \forall\lambda\in\mathbb{R}. \tag{1.5}$$

The inequality (1.5) is sufficient to prove Payne's conjecture for the Laplacian on a bounded domain (see the proof of Corollary 1.13).

REMARK 1.10. The equality $N_{\mathrm{N}}(\lambda) = N_{\mathrm{D}}(\lambda) + n_{\mathrm{D}}(\lambda) + \dim G_\lambda^-$ remains valid for all $\lambda > \lambda_{\mathrm{D},\infty}$ because $N_{\mathrm{N}}(\lambda) = N_{\mathrm{D}}(\lambda) = \infty$. However, as was pointed out by N. Filonov, it may not be true for $\lambda = \lambda_{\mathrm{D},\infty}$.

REMARK 1.11. Let $\lambda = \lambda_{\mathrm{D},k} < \lambda_{\mathrm{N},\infty}$. Theorem 1.7 implies that the number of eigenvalues $\lambda_{\mathrm{N},j}$ lying below $\lambda_{\mathrm{D},k}$ is equal to $k - 1 + n_{\mathrm{D}}(\lambda_{\mathrm{D},k}) + \dim G^-_{\lambda_{\mathrm{D},k}}$. Therefore

(1) $\lambda_{\mathrm{N},k+q_k+p_k-1} < \lambda_{\mathrm{D},k}$, where $p_k := \dim G^-_{\lambda_{\mathrm{D},k}}$ and $q_k := n_{\mathrm{D}}(\lambda_{\mathrm{D},k})$.

If $n_{\mathrm{D}}(\lambda_{\mathrm{D},k}) = 0$, then $\lambda_{\mathrm{N},k+q_k+p_k} = \lambda_{\mathrm{N},k+p_k} = \lambda_{\mathrm{D},k}$; if $n_{\mathrm{D}}(\lambda_{\mathrm{D},k}) \neq 0$, then $q_k \geqslant 1$. Thus we always have

(2) $\lambda_{\mathrm{N},k+p_k} \leqslant \lambda_{\mathrm{D},k}$.

Note that the estimates (1) and (2) are actually consequences of (1.5). These estimates and Lemma 1.2 imply that

(3) $\lambda_{\mathrm{N},k+1} \leqslant \lambda_{\mathrm{D},k}$ whenever there exists a vector $u \in G_{\lambda_{\mathrm{D},k}}$, such that $\mathbf{b}[u] \leqslant 0$ and $u \notin \mathcal{D}(A_{\mathrm{D}})$;

(4) $\lambda_{\mathrm{N},k+1} < \lambda_{\mathrm{D},k}$ whenever $n_{\mathrm{D}}(\lambda_{\mathrm{D},k}) \geqslant 1$ and there exist two vectors $u_1, u_2 \in G_{\lambda_{\mathrm{D},k}}$, such that $\mathbf{b}[u_1] \leqslant 0$, $\mathbf{b}[u_2] \leqslant 0$ and the linear subspace spanned by u_1 and u_2 does not contain Neumann eigenvectors.

Indeed, if $G^-_{\lambda_{\mathrm{D},k}} \geqslant 1$, then (3) and (4) follow from (2) and (1) respectively. If $G^-_{\lambda_{\mathrm{D},k}} = \{0\}$, then $u \in G^0_{\lambda_{\mathrm{D},k}}$ and $u_1, u_2 \in G^0_{\lambda_{\mathrm{D},k}}$. The inclusion $u \in G^0_{\lambda_{\mathrm{D},k}}$ implies that $\lambda_{\mathrm{D},k}$ is also a Neumann eigenvalue and, consequently, $\lambda_{\mathrm{N},k+1} = \lambda_{\mathrm{D},k}$. The inclusions $u_1, u_2 \in G^0_{\lambda_{\mathrm{D},k}}$ imply that $n_{\mathrm{D}}(\lambda_{\mathrm{D},k}) \geqslant 2$ (otherwise a linear combination of u_1 and u_2 would belong to $E_{\mathrm{N}}(\lambda_{\mathrm{D},k})H$).

Lemma 1.2 and Theorem 1.7 also imply

COROLLARY 1.12. *If $\lambda < \lambda_{\mathrm{D},\infty}$ and $\lambda \notin \sigma(A_{\mathrm{N}}) \cup \sigma(A_{\mathrm{D}})$, then the number of negative eigenvalues of the selfadjoint operator $R'(\lambda) := (A_{\mathrm{N}} - \lambda I)^{-1} - (A_{\mathrm{D}} - \lambda I)^{-1}$ in H coincides with $N_{\mathrm{N}}(\lambda) - N_{\mathrm{D}}(\lambda)$.*

Obviously, the number of negative eigenvalues of the operator $(A_{\mathrm{B}} - \lambda I)^{-1}$ jumps by $n_{\mathrm{B}}(\lambda_0) + n_{\mathrm{N,D}}(\lambda_0)$ as λ passes through an eigenvalue λ_0. Corollary 1.12 shows that the corresponding jump for $R'(\lambda)$ is equal to $n_{\mathrm{N}}(\lambda_0) - n_{\mathrm{D}}(\lambda_0)$, as if $R'(\lambda)$ were the orthogonal sum of the operators $(A_{\mathrm{N}} - \lambda I)^{-1}$ and $-(A_{\mathrm{D}} - \lambda I)^{-1}$.

1.3. The Dirichlet-to-Neumann map. In the theory of boundary value problems, it is often possible to construct a linear isomorphism $W : G_\lambda \to \mathfrak{H}$, where $\mathfrak{H}$ is a Hilbert space of functions defined on the boundary. Then one can consider the operator $W\mathcal{B}_\lambda W^{-1} : \mathfrak{H} \to \mathfrak{H}$ instead of $\mathcal{B}_\lambda$. Clearly, these two operators have the same eigenvalues. If H^1, H^1_0 are the Sobolev spaces and Wv is the restriction of v to the boundary, then $W\mathcal{B}_\lambda W^{-1}$ is usually called the Dirichlet-to-Neumann map. This scheme works under certain smoothness conditions on the boundary and the coefficients, whereas our approach does not rely on the existence of an auxiliary operator W and does not require any additional assumptions.

1.4. Applications to boundary value problems. Let Ω be an arbitrary open subset of $\mathbb{R}^n$ with $n \geqslant 2$. Consider a differential operator L acting from the space of m-vector functions $C^\infty(\Omega, \mathbb{C}^m)$ into the space of l-vector functions $C^\infty(\Omega, \mathbb{C}^l)$ and denote by L^* its formal adjoint. Let us assume that the form $\int_\Omega |Lu(x)|^2 \, dx$ with domain $C^\infty(\Omega, \mathbb{C}^m) \cap L_2(\Omega, \mathbb{C}^m)$ is strictly positive and closable in $H = L_2(\Omega, \mathbb{C}^m)$, and denote its closure by $\mathbf{a}[u]$. If $H^1 := \mathcal{D}(\mathbf{a})$ and H^1_0 is the H^1-closure of $C_0^\infty(\Omega)$, then $A = L^*L$ and A_{B} is the differential operator A with the corresponding boundary condition.

COROLLARY 1.13. *Let L be an operator with constant coefficients. Then we have $\lambda_{\mathrm{N},k+1} \leqslant \lambda_{\mathrm{D},k}$ for all eigenvalues $\lambda_{\mathrm{D},k} \in (0, \lambda_{\mathrm{N},\infty})$. If at least one Dirichlet eigenfunction corresponding to $\lambda_{\mathrm{D},k}$ does not satisfy the Neumann boundary condition, then $\lambda_{\mathrm{N},k+1} < \lambda_{\mathrm{D},k}$.*

REMARK 1.14. Our proof of Corollary 1.13 uses the exponential functions $u_\xi(x) = e^{ix\cdot\xi}$ and is very similar to the proof of the Payne conjecture given in [**Fr1**]. The main difference is that L. Friendlander considered the Dirichlet-to-Neumann map and therefore had to assume that the boundary is smooth enough.

REMARK 1.15. If A is the Laplacian on a convex n-dimensional domain with sufficiently smooth boundary, then $\lambda_{\mathrm{N},k+n} < \lambda_{\mathrm{D},k}$. This estimate was obtained in [**LW**]. Later L. Friedlander found another proof, based on the fact that $G^0_\lambda \cup G^-_\lambda$ contains all first-order derivatives $D_j u$ of the Dirichlet eigenfunctions $u \in E_{\mathrm{D}}(\lambda)H$ (the derivatives obviously belong to G_λ, and the estimate $\mathbf{b}[D_j u] \leqslant 0$ is a consequence of the convexity). The inclusion $D_j u \in G^0_\lambda \cup G^-_\lambda$ also implies that $N_{\mathrm{N}}(\lambda) \geqslant N_{\mathrm{D}}(\lambda) + 2n_{\mathrm{D}}(\lambda)$ (see [**Fr2**] for details).

2. Further notation and auxiliary results

2.1. The inverse A_{N}^{-1} is a bounded selfadjoint operator in H^1 because $(u,v) = \mathbf{a}[A_{\mathrm{N}}^{-1}u, v]$ for all $u, v \in H^1$. Since $\mathbf{a}[u,v] = \lambda\,(u,v)$ for all $v \in E_{\mathrm{N}}(\lambda)H$ and $v \in H^1$, its spectral projections $E_{\mathrm{N}}(\lambda)$ are H^1-orthogonal.

Let

- Π_0 be the orthogonal projection in H^1 onto H^1_0.

From the definition of G_z it clear that $\Pi_0' = I - \Pi_0$ (this well-known result can be found, for example, in [**K**] or [**BS**, Chapter 10, Section 3]). Since $\mathbf{a}[A_{\mathrm{D}}^{-1}u, v] = (u,v) = \mathbf{a}[A_{\mathrm{N}}^{-1}u, v]$ for all $u \in H^1$ and $v \in H^1_0$, we have $A_{\mathrm{D}}^{-1} = \Pi_0 A_{\mathrm{N}}^{-1}$ and $\mathcal{D}(A_{\mathrm{D}}) = \Pi_0 \mathcal{D}(A_{\mathrm{N}})$. The following simple lemma is also well known in the theory of selfadjoint extensions.

LEMMA 2.1. *We have $H^1_A = G_0 \dotplus \mathcal{D}(A_{\mathrm{B}})$. If $w_0 \in G_0$ and $w_{\mathrm{B}} \in \mathcal{D}(A_{\mathrm{B}})$, then $A(w_0 + w_{\mathrm{B}}) = A_{\mathrm{B}} w_{\mathrm{B}}$.*

PROOF. Obviously, $G_0 \dotplus \mathcal{D}(A_{\mathrm{B}}) \subset H^1_A$. On the other hand, if $v \in H^1_A$, then there exists $\tilde v \in H$ such that $(u, \tilde v) = \mathbf{a}[u,v]$ for all $u \in H^1_0$. Since $(u,v) = \mathbf{a}[u, A_{\mathrm{N}}^{-1}v]$, this implies that $\Pi_0 v = \Pi_0 A_{\mathrm{N}}^{-1}\tilde v = A_{\mathrm{D}}^{-1}\tilde v$. Therefore $v = \Pi_0' v + A_{\mathrm{D}}^{-1}\tilde v$ and $v = \Pi_0'\left(v - A_{\mathrm{N}}^{-1}\tilde v\right) + A_{\mathrm{N}}^{-1}\tilde v$. These equalities imply the first statement.

If $w_0 \in G_0$ and $w_{\mathrm{B}} \in \mathcal{D}(A_{\mathrm{B}})$, then $\mathbf{a}[w_0 + w_{\mathrm{B}}, v] = \mathbf{a}[w_{\mathrm{B}}, v] = (A_{\mathrm{B}} w_{\mathrm{B}}, v)$, for all $v \in H^1_0$. This proves the second statement. $\square$

2.2. By Lemma 2.1, H^1_A is dense in H^1. If $u,v\in H^1_A$, then

$$\mathbf{a}[A_{\mathrm{D}}^{-1}Au,v]=\mathbf{a}[A_{\mathrm{D}}^{-1}Au,\Pi_0 v]=(AA_{\mathrm{D}}^{-1}Au,\Pi_0 v)=(Au,\Pi_0 v)=\mathbf{a}[\Pi_0 u,v]\,.$$

This implies that $A_{\mathrm{D}}^{-1}A=\Pi_0|_{H^1_A}$ and, consequently,

$$(2.1)\quad G_z\;:=\;\ker(A-zI)\;=\;\ker A_{\mathrm{D}}^{-1}(A-zI)\;=\;\ker\Pi_0(I-zA_{\mathrm{B}}^{-1})\,,\qquad\forall z\in\mathbb{C}\,.$$

By (2.1), we have $(I-zA_{\mathrm{B}}^{-1})\,G_z\subset G_0\cap(I-zA_{\mathrm{B}}^{-1})H^1$. On the other hand, if $(I-zA_{\mathrm{B}}^{-1})u\in G_0$, then $u\in G_z$ because $(A-zI)u=A(I-zA_{\mathrm{B}}^{-1})u=0$. Therefore

$$(2.2)\qquad (I-zA_{\mathrm{B}}^{-1})\,G_z\;=\;G_0\cap(I-zA_{\mathrm{B}}^{-1})H^1\,,\qquad\forall z\in\mathbb{C}\,.$$

Let

- $R_{\mathrm{B}}(z):=(A_{\mathrm{B}}-zI)^{-1}$ be the resolvent of A_{B}.

For each $z\notin\sigma_{\mathrm{ess}}(A_{\mathrm{B}})$, the operator $R_{\mathrm{B}}(z)E'_{\mathrm{B}}(z)$ is bounded from H to H^1, $\ker\left(R_{\mathrm{B}}(z)E'_{\mathrm{B}}(z)\right)=\ker E'_{\mathrm{B}}(z)=E_{\mathrm{B}}(z)H$, $R_{\mathrm{B}}(z)E'_{\mathrm{B}}(z)H\subset E'_{\mathrm{B}}(z)\mathcal{D}(A_{\mathrm{B}})\subset H^1_A$ and $(A-zI)R_{\mathrm{B}}(z)E'_{\mathrm{B}}(z)=E'_{\mathrm{B}}(z)$. We also have

$$(2.3)\qquad (I-zA_{\mathrm{B}}^{-1})^{-1}\big|_{E'_{\mathrm{B}}(z)H}\;=\;(I+zR_{\mathrm{B}}(z))\,E'_{\mathrm{B}}(z)\,,\qquad\forall z\notin\sigma_{\mathrm{ess}}(A_{\mathrm{B}})\,,$$

where the operators in the right and left hand sides map $E'_{\mathrm{B}}(\lambda)H^1$ onto $E'_{\mathrm{B}}(\lambda)H^1$ and are H^1-bounded. This implies that $(I-zA_{\mathrm{B}}^{-1})H^1=E'_{\mathrm{B}}(z)H^1$ and, by (2.2),

$$(2.4)\qquad (I-zA_{\mathrm{B}}^{-1})\,G_z\;=\;G_0\cap E'_{\mathrm{B}}(z)H^1\,,\qquad\forall z\notin\sigma_{\mathrm{ess}}(A_{\mathrm{B}})\,.$$

2.3. Denote $T_z:=(I-zA_{\mathrm{N}}^{-1})\big|_{H^1_0\cap\mathcal{H}}$ and $T_z^\star:=\Pi_0(I-zA_{\mathrm{N}}^{-1})\big|_{\mathcal{H}}$. Let Σ be the set of $z\in\mathbb{C}$ such that the spectrum of the operator $T_z^\star\,T_z:H^1_0\cap\mathcal{H}\to H^1_0\cap\mathcal{H}$ contains the point 0, and let

- $\Pi(z):=T_z\,(T_z^\star\,T_z)^{-1}\,T_z^\star$ and $\Pi'(z):=I-\Pi(z)$, where $z\in\mathbb{C}\setminus\Sigma$.

By (2.1), we have $G_z\cap\mathcal{H}=\ker T_z^\star$. Since $T_z^\star\Pi'(z)=0$ and $\Pi'(z)u=u$ for all $u\in\ker T_z^\star$, this implies that $\Pi'(z)$ is a projection onto $G_z\cap\mathcal{H}$ in $H^1\cap\mathcal{H}$. Its H^1-adjoint coincides with $\Pi'(\bar z)$; in particular, $\Pi'(\lambda)$ is the H^1-orthogonal projection onto $G_z\cap\mathcal{H}$ in $\mathcal{H}$. Thus we obtain

$$(2.5)\qquad \Pi'_\lambda\;=\;\Pi'(\lambda)\oplus E_{\mathrm{B}}(\lambda)|_{H^1\cap\mathcal{H}_0}\,,\qquad\forall\lambda\in\mathbb{R}\setminus\Sigma\,.$$

2.4. If $z\in\mathbb{C}\setminus\sigma_{\mathrm{ess}}(A_{\mathrm{B}})$, let

- $P_{\mathrm{B}}(z):=R_{\mathrm{B}}(z)E'_{\mathrm{B}}(z)\,(A-zI)$ and $P'_{\mathrm{B}}(z):=I-P_{\mathrm{B}}(z)$;
- $G_{z,B}:=\{v\in H^1_A:(A-zI)v\in E_{\mathrm{B}}(z)H\}$.

The operators $P_{\mathrm{B}}(z)$ and $P'_{\mathrm{B}}(z)$ are projections in H^1_A because

$$\begin{aligned}P_{\mathrm{B}}^2(z)&=R_{\mathrm{B}}(z)E'_{\mathrm{B}}(z)\,(A_{\mathrm{B}}-zI)R_{\mathrm{B}}(z)E'_{\mathrm{B}}(z)\,(A-zI)\\&=E'_{\mathrm{B}}(z)R_{\mathrm{B}}(z)E'_{\mathrm{B}}(z)\,(A-zI)=P_{\mathrm{B}}(z)\,.\end{aligned}$$

One can easily show that $P_{\mathrm{B}}(z)H^1_A=E'_{\mathrm{B}}(z)\mathcal{D}(A_{\mathrm{B}})$ and $P'_{\mathrm{B}}(z)H^1_A=G_{z,B}$. The subspace $G_{z,B}$ is the inverse image of $E_{\mathrm{B}}(z)H$ by the map $A-zI$, whereas G_z is the kernel of $A-zI$. Therefore $G_z\subset G_{z,B}$ and the dimension of the quotient space $G_{z,B}/G_z$ does not exceed $n_{\mathrm{B}}(z)+n_{\mathrm{N,D}}(z)$. This implies that the subspaces $G_{z,B}$ are H^1-closed for all $z\in\mathbb{C}\setminus\sigma_{\mathrm{ess}}(A_{\mathrm{B}})$.

If $z\notin\sigma(A_{\mathrm{B}})$, then $P_{\mathrm{B}}(z)H^1_A=\mathcal{D}(A_{\mathrm{B}})$ and $P'_{\mathrm{B}}(z)H^1_A=G_z=G_{z,B}$. In particular, $P_{\mathrm{D}}(0)|_{H^1_A}=\Pi_0|_{H^1_A}$, $P'_{\mathrm{D}}(0)|_{H^1_A}=\Pi'_0|_{H^1_A}$ and $H^1_A=P'_{\mathrm{B}}(0)H^1_A\dotplus P_{\mathrm{B}}(0)H^1_A$

is the decomposition discussed in Lemma 2.1. By direct calculation, if $u,v\in H^1_A$ and $\lambda,\mu\in\mathbb{R}$, then

$$\begin{aligned}(2.6)\quad &\mathbf{b}[P'_{\mathrm{N}}(\lambda)u, P'_{\mathrm{N}}(\mu)v]\\ &= \mathbf{b}[u,v] - \big(u, E'_{\mathrm{N}}(\mu)(A-\mu I)v\big) + \big((A-\mu I)u, R_{\mathrm{N}}(\mu)E'_{\mathrm{B}}(z)(A-\mu I)v\big),\end{aligned}$$

$$\begin{aligned}(2.7)\quad &\mathbf{b}[P'_{\mathrm{D}}(\lambda)u, P'_{\mathrm{D}}(\mu)v]\\ &= \mathbf{b}[u,v] + \big(E'_{\mathrm{D}}(\lambda)(A-\lambda I)u, v\big) - \big(R_{\mathrm{D}}(\lambda)E'_{\mathrm{B}}(z)(A-\lambda I)u, (A-\lambda I)v\big).\end{aligned}$$

3. Proofs of main results

3.1. Proof of Lemma 1.1. Since the operator-valued function $T_z^\star T_z$ is analytic, the inverse operator $(T_z^\star T_z)^{-1}$ exists and analytically depends on z in a sufficiently small neighbourhood of each point $z_0\notin\Sigma$. Thus it is sufficient to show that $\Sigma\subset(\sigma_{\mathrm{ess}}(A_{\mathrm{N}})\cap\sigma_{\mathrm{ess}}(A_{\mathrm{D}}))$.

Let us fix $\lambda\in\mathbb{R}$, and let $u\in\mathcal{H}$. Since $T_\lambda^\star$ is the H^1-adjoint to T_λ, we have $T_\lambda^\star T_\lambda u=0$ if and only if $T_\lambda u=0$. The latter means that $u\in\mathcal{D}(A_{\mathrm{N}})\cap\mathcal{D}(A_{\mathrm{D}})$ and $A_{\mathrm{N}}u=A_{\mathrm{D}}u=\lambda u$. Since $u\in\mathcal{H}$, it is only possible if $u=0$. This implies that $\ker(T_\lambda^\star T_\lambda)=\{0\}$.

Assume that the essential spectrum of the operator $T_\lambda^\star T_\lambda$ contains 0. Then, for any given finite-dimensional subspace $\mathcal{L}$, there exists a sequence of H^1-orthogonal vectors $u_n\in H_0^1\cap\mathcal{H}$ such that u_n are H-orthogonal to $\mathcal{L}$, $\mathbf{a}[u_n]=1$ and $\mathbf{a}[T_\lambda^\star T_\lambda u_n,u_n]=\mathbf{a}[T_\lambda u_n]\to 0$ as $n\to\infty$. Clearly, $T_\lambda u_n=u_n-\lambda A_{\mathrm{N}}^{-1}u_n\to 0$ and $\Pi_0 T_\lambda u_n=u_n-\lambda A_{\mathrm{D}}^{-1}u_n\to 0$ in H^1. If $\lambda\notin\sigma_{\mathrm{ess}}(A_{\mathrm{B}})$, then, by (2.3), we have $\mathbf{a}\left[u-\lambda A_{\mathrm{B}}^{-1}u\right]\geqslant C\,\mathbf{a}[u]$ with some positive constant C for all vectors $u\in H^1$ which are H-orthogonal to $\mathcal{L}=E_{\mathrm{B}}(\lambda)H$. Therefore, by the above, $\lambda\in\sigma_{\mathrm{ess}}(A_{\mathrm{N}})\cap\sigma_{\mathrm{ess}}(A_{\mathrm{D}})$.

3.2. Proof of the estimate (1.4)**.** We have

$$(3.1)\qquad \mathbf{a}[u,v] = \lambda\,(u,v),\quad \forall u\in E_{\mathrm{N}}(\lambda)H,\ \forall v\in H^1,$$

$$(3.2)\qquad \mathbf{a}[u,v] = \lambda\,(u,v),\quad \forall u\in \mathcal{D}(A_{\mathrm{D}}),\ \forall v\in G_\lambda.$$

Let $u=u_1+u_2+u_3$, where $u_1\in\chi_{[0,\lambda]}(A_{\mathrm{D}})H$, $u_2\in E_{\mathrm{N}}(\lambda)$ and $u_3\in G_\lambda^0+G_\lambda^-$. Then (3.1) and (3.2) imply

$$\begin{aligned}\mathbf{a}[u]-\lambda\,\|u\|^2 &= \mathbf{a}[u_1]-\lambda\,\|u_1\|^2+\mathbf{a}[u_3]-\lambda\,\|u_3\|^2\\ &= ((A_{\mathrm{D}}-\lambda I)u_1,u_1)+\mathbf{a}[\mathcal{B}_\lambda u_3,u_3]\ \leqslant\ 0.\end{aligned}$$

3.3. Proof of Lemma 1.2. The inclusion (1.2) immediately follows from (3.1) and (3.2).

Assume that $\lambda\notin\sigma_{\mathrm{ess}}(A_{\mathrm{N}})\cap\sigma_{\mathrm{ess}}(A_{\mathrm{D}})$ and $u\in G_\lambda^0$. Then $(I-\lambda A_{\mathrm{N}}^{-1})u$ is H^1-orthogonal to G_λ. In view of (2.5), this means that $(I-\lambda A_{\mathrm{N}}^{-1})u=\Pi(\lambda)v$ for some $v\in H^1\cap\mathcal{H}$. Therefore $(I-\lambda A_{\mathrm{N}}^{-1})u=(I-\lambda A_{\mathrm{N}}^{-1})w$ for some $w\in H_0^1$, which is equivalent to the inclusion $u\in G_\lambda\cap\left(E_{\mathrm{N}}(\lambda)H+H_0^1\right)=E_{\mathrm{N}}(\lambda)H+E_{\mathrm{D}}(\lambda)H$.

Assume now that $\lambda\notin\sigma_{\mathrm{ess}}(A_{\mathrm{N}})\cup\sigma_{\mathrm{ess}}(A_{\mathrm{D}})$. If the point 0 belongs to the essential spectrum of the operator $\mathcal{B}_\lambda$, then there exists a sequence of vectors $u_n\in G_\lambda\ominus G_\lambda^0$ such that $\mathbf{a}[u_n]=1$ and $\mathbf{a}[\mathcal{B}_\lambda u_n]\to 0$ as $n\to\infty$. Moreover, since $\dim E_{\mathrm{N}}(\lambda)H<\infty$, we can choose the sequence $\{u_n\}$ in such a way that

$$u_n-(T_\lambda^\star T_\lambda)^{-1}\,\Pi_0(I-\lambda A_{\mathrm{N}}^{-1})^2u_n\ \in\ E'_{\mathrm{N}}(\lambda),\qquad \forall n=1,2,\dots,$$

where T_λ and $T_\lambda^\star$ are the operators defined in Subsection 2.3. Then, by (2.5),

$$\begin{aligned}\mathcal{B}_\lambda u_n &= (I-\lambda A_{\rm N}^{-1})u_n - \Pi_\lambda(I-\lambda A_{\rm N}^{-1})u_n\\ &= (I-\lambda A_{\rm N}^{-1})\left(u_n-(T_\lambda^\star T_\lambda)^{-1}\,\Pi_0(I-\lambda A_{\rm N}^{-1})^2u_n\right)\;\to\;0\end{aligned}$$

in H^1, and (2.3) implies that

$$\begin{aligned}&\mathbf{a}\left[u_n-(T_\lambda^\star T_\lambda)^{-1}\,\Pi_0(I-\lambda A_{\rm N}^{-1})^2u_n\right]\\ &\qquad=\;\mathbf{a}\left[\Pi_0'u_n\right]+\mathbf{a}\left[\Pi_0u_n-(T_\lambda^\star T_\lambda)^{-1}\,\Pi_0(I-\lambda A_{\rm N}^{-1})^2u_n\right]\;\to\;0\,.\end{aligned}$$

Therefore $\mathbf{a}[\Pi_0'u_n]=\mathbf{a}[(I-A_{\rm D}^{-1}A)u_n]=\mathbf{a}[(I-\lambda A_{\rm D}^{-1})u_n]\to 0$ as $n\to\infty$. However, this is not possible because $\lambda\notin\sigma_{\rm ess}(A_{\rm D})$ and the u_n are orthogonal to $E_{\rm D}(\lambda)H\subset G_\lambda^0$. The obtained contradiction proves the second statement of the lemma.

3.4. Proof of Theorem 1.3. If $\lambda\notin\sigma(A_{\rm N})\cup\sigma(A_{\rm D})$, then the theorem is obvious because, in view of Lemmas 1.2 and 1.1, we have $n_{\rm N}(\lambda)=n_{\rm D}(\lambda)=n_{\rm N,D}(\lambda)=0$ and $(-\varepsilon,\varepsilon)\cap\sigma(\mathcal{B}_\mu)=\varnothing$ for all sufficiently small $\varepsilon,\delta>0$ and all $\mu\in(\lambda-\varepsilon,\lambda+\varepsilon)$.

Suppose that λ is an isolated eigenvalue. The first statement of the theorem is an immediate consequence of Lemma 1.2, so we only need to prove (2) and (3). Let us choose ε and δ as explained in Remark 1.4 and assume, in addition, that δ is so small that $\lambda-\delta>0$ and the interval $[\lambda-\delta,\lambda+\delta]$ does not contain any points from $\sigma(A_{\rm N})\cup\sigma(A_{\rm D})$ with the exception of λ.

Let $\mathcal{L}_\mu$ be the subspace of $G_\mu\cap\mathcal{H}$ spanned by the eigenfunction corresponding to the eigenvalues $\nu_j(\mu)$ (see Remark 1.4). By Lemma 1.2, we have $\mathcal{L}_\lambda=E_{\rm N}(\lambda)\mathcal{H}\dot{+}E_{\rm D}(\lambda)\mathcal{H}\subset G_\lambda^0$. Therefore $E_{\rm B}'(\lambda)\mathcal{L}_\lambda\subset\mathcal{L}_\lambda$, $\dim E_{\rm N}'(\lambda)\mathcal{L}_\lambda=n_{\rm D}(\lambda)$ and $\dim E_{\rm D}'(\lambda)\mathcal{L}_\lambda=n_{\rm N}(\lambda)$.

We are going to show that

$$\begin{aligned}(3.3)\quad &\left|\,\mathbf{a}\left[\mathcal{B}_\mu P_{\rm N}'(\mu)u,\chi_{(-\varepsilon,\varepsilon)}(\mathcal{B}_\mu)P_{\rm N}'(\mu)u\right]-(\mu-\lambda)\,\|u\|^2\,\right|\\ &\qquad\qquad\leqslant\;C\,(\lambda-\mu)^2\,\|u\|^2\,,\qquad\forall u\in E_{\rm N}'(\lambda)\mathcal{L}_\lambda\,,\end{aligned}$$

$$\begin{aligned}(3.4)\quad &\left|\,\mathbf{a}\left[\mathcal{B}_\mu P_{\rm D}'(\mu)u,\chi_{(-\varepsilon,\varepsilon)}(\mathcal{B}_\mu)P_{\rm D}'(\mu)u\right]-(\lambda-\mu)\,\|u\|^2\,\right|\\ &\qquad\qquad\leqslant\;C\,(\lambda-\mu)^2\,\|u\|^2\,,\qquad\forall u\in E_{\rm D}'(\lambda)\mathcal{L}_\lambda\,,\end{aligned}$$

where C is a constant independent of u and $\mu\in(\lambda-\varepsilon,\lambda+\varepsilon)$. From (3.3) and (3.4) it follows that

$$(3.5)\qquad (\mu-\lambda)\,\mathbf{a}[\mathcal{B}_\mu w,w]\;\geqslant\;0\,,\qquad\forall w\in\chi_{(-\varepsilon,\varepsilon)}(\mathcal{B}_\mu)P_{\rm N}'(\mu)E_{\rm N}'(\lambda)\mathcal{L}_\lambda\,,$$

$$(3.6)\qquad (\lambda-\mu)\,\mathbf{a}[\mathcal{B}_\mu w,w]\;\geqslant\;0\,,\qquad\forall w\in\chi_{(-\varepsilon,\varepsilon)}(\mathcal{B}_\mu)P_{\rm D}'(\mu)E_{\rm D}'(\lambda)\mathcal{L}_\lambda\,,$$

whenever $|\lambda-\mu|$ is small enough. If $\mu=\lambda$, then $\chi_{(-\varepsilon,\varepsilon)}(\mathcal{B}_\mu)P_{\rm B}'(\mu)E_{\rm B}'(\lambda)u=E_{\rm B}'(\lambda)u$ for all $u\in\mathcal{L}_\lambda$. By continuity, we have $\dim\chi_{(-\varepsilon,\varepsilon)}(\mathcal{B}_\mu)P_{\rm B}'(\mu)E_{\rm B}'(\lambda)\mathcal{L}_\lambda=\dim E_{\rm B}'(\lambda)\mathcal{L}_\lambda=n_{\rm N}(\lambda)+n_{\rm D}(\lambda)-n_{\rm B}(\lambda)$ for all μ sufficiently close to λ. Therefore the estimates (3.5) and (3.6) imply the theorem (with another positive δ).

To prove (3.3) and (3.4), note that $\mathbf{b}[u]=0$ for all $u\in\mathcal{L}_\lambda$ and, in view of (2.6) and (2.7),

$$\begin{aligned}(3.7)\quad \mathbf{a}\left[\mathcal{B}_\mu P_{\rm N}'(\mu)u,P_{\rm N}'(\mu)u\right]\;&=\;\mathbf{b}[P_{\rm N}'(\mu)u]\\ &=\;(\mu-\lambda)\,\|u\|^2+(\lambda-\mu)^2\left(u,R_{\rm N}(\mu)u\right),\qquad\forall u\in\mathcal{L}_\lambda\,,\end{aligned}$$

$$(3.8)\quad \mathbf{a}\left[\mathcal{B}_\mu P'_{\mathrm{D}}(\mu)u, P'_{\mathrm{D}}(\mu)u\right] = \mathbf{b}[P'_{\mathrm{D}}(\mu)u] = (\lambda-\mu)\,\|u\|^2 - (\lambda-\mu)^2\left(u, R_{\mathrm{D}}(\mu)u\right), \qquad \forall u \in \mathcal{L}_\lambda .$$

Therefore, for all $\mu \in (\lambda-\delta, \lambda+\delta)$ we have

$$(3.9)\quad \left|\,\mathbf{a}\left[\mathcal{B}_\mu P'_{\mathrm{N}}(\mu)u, P'_{\mathrm{N}}(\mu)u\right] - (\mu-\lambda)\,\|u\|^2\,\right| \leqslant C_1^{-1}(\lambda-\mu)^2\,\|u\|^2, \qquad \forall u \in E'_{\mathrm{N}}(\lambda)\mathcal{L}_\lambda ,$$

$$(3.10)\quad \left|\,\mathbf{a}\left[\mathcal{B}_\mu P'_{\mathrm{D}}(\mu)u, P'_{\mathrm{D}}(\mu)u\right] - (\lambda-\mu)\,\|u\|^2\,\right| \leqslant C_1^{-1}(\lambda-\mu)^2\,\|u\|^2, \qquad \forall u \in E'_{\mathrm{D}}(\lambda)\mathcal{L}_\lambda ,$$

where C_1 is the distance from $[\lambda-\delta, \lambda+\delta]$ to $(\sigma(A_{\mathrm{N}} \cup \sigma(A_{\mathrm{D}})) \setminus \{\lambda\}$.

Let S_{B} be the projections onto $E_{\mathrm{B}}(\lambda)\mathcal{H}$ in $\mathcal{L}_\lambda$ such that $S_{\mathrm{N}}E_{\mathrm{D}}(\lambda) = 0$ and $S_{\mathrm{D}}E_{\mathrm{N}}(\lambda) = 0$. Then $u = S_{\mathrm{N}}u + S_{\mathrm{D}}u$ for all $u \in \mathcal{L}_\lambda$. Since $\dim \mathcal{L}_\lambda < \infty$ and $E_{\mathrm{N}}(\lambda)\mathcal{H} \cap E_{\mathrm{D}}(\lambda)\mathcal{H} = \{0\}$, the projections S_{B} are well defined and bounded as operators from H to H^1.

If $u \in \mathcal{L}_\lambda$, then $P'_{\mathrm{B}}(\mu)u = P'_{\mathrm{B}}(\mu)(u - S_{\mathrm{B}}u) = (I - (\lambda-\mu)R_{\mathrm{B}}(\mu))(u - S_{\mathrm{B}}u)$ for all $\mu \neq \lambda$ and, by (1.1),

$$\mathbf{a}\left[\mathcal{B}_\mu P'_{\mathrm{N}}(\mu)u, v\right] = \mathbf{a}\left[P'_{\mathrm{N}}(\mu)u, v\right] - \mu\,(P'_{\mathrm{N}}(\mu)u, v) = (\mu-\lambda)\,(S_{\mathrm{D}}u, v), \qquad \forall v \in G_\mu ,$$
$$\mathbf{a}\left[\mathcal{B}_\mu P'_{\mathrm{D}}(\mu)u, v\right] = \mathbf{a}\left[P'_{\mathrm{D}}(\mu)u, v\right] - \mu\,(P'_{\mathrm{D}}(\mu)u, v) = (\lambda-\mu)\,(S_{\mathrm{N}}u, v), \qquad \forall v \in G_\mu ,$$

for all $\mu \in (\lambda-\delta, \lambda+\delta)$. Since $(S_{\mathrm{N}}u, v) = \lambda^{-1}\,\mathbf{a}[S_{\mathrm{N}}u, v]$ and $(S_{\mathrm{D}}u, v) = \mu^{-1}\,\mathbf{a}[S_{\mathrm{D}}u, v]$ whenever $v \in G_\mu$, the above identities imply that

$$(3.11)\quad \mathbf{a}\left[\mathcal{B}_\mu P'_{\mathrm{N}}(\mu)u, v\right] = \mu^{-1}(\mu-\lambda)\,\mathbf{a}\left[S_{\mathrm{D}}u, v\right], \qquad \forall u \in \mathcal{L}_\lambda,\ \forall v \in G_\mu ,$$
$$(3.12)\quad \mathbf{a}\left[\mathcal{B}_\mu P'_{\mathrm{D}}(\mu)u, v\right] = \lambda^{-1}(\lambda-\mu)\,\mathbf{a}\left[S_{\mathrm{N}}u, v\right], \qquad \forall u \in \mathcal{L}_\lambda,\ \forall v \in G_\mu .$$

In view of Lemma 1.1, we have that $\left(I - \chi_{(-\varepsilon,\varepsilon)}(\mathcal{B}_\mu)\right) P'_{\mathrm{B}}(\mu)E'_{\mathrm{B}}(\lambda)\big|_{\mathcal{L}_\lambda}$ is an analytic operator-valued function of $\mu \in (\lambda-\delta, \lambda+\delta)$. Since this operator-valued function vanishes at $\mu = \lambda$, we have

$$(3.13)\quad \mathbf{a}\left[\left(I - \chi_{(-\varepsilon,\varepsilon)}(\mathcal{B}_\mu)\right) P'_{\mathrm{B}}(\mu)E'_{\mathrm{B}}(\lambda)u\right] \leqslant C_2\,(\lambda-\mu)^2\,\mathbf{a}[u] = C_2\,(\lambda-\mu)^2\,\lambda\,\|u\|^2, \qquad \forall u \in \mathcal{L}_\lambda ,$$

with some positive constant C_2 independent of μ and u. Substituting $v = \left(I - \chi_{(-\varepsilon,\varepsilon)}(\mathcal{B}_\mu)\right) P'_{\mathrm{B}}(\mu)u$ into (3.11), (3.12) and applying (3.13), we obtain

$$(3.14)\quad \mathbf{a}\left[\mathcal{B}_\mu P'_{\mathrm{B}}(\mu)u, \left(I - \chi_{(-\varepsilon,\varepsilon)}(\mathcal{B}_\mu)\right) P'_{\mathrm{B}}(\mu)u\right] \leqslant C_3\,(\lambda-\mu)^2\,\|u\|^2, \qquad \forall u \in E'_{\mathrm{B}}(\lambda)\mathcal{L}_\lambda ,$$

with some constant C_3 independent of μ and u. Now (3.3) and (3.4) follow from (3.9), (3.10) and (3.14).

3.5. Proof of Corollary 1.6. Lemma 1.1 and Theorem 1.3 imply that $\dim G_\lambda^-$ is constant on every connected component of the set $\mathbb{R} \setminus (\sigma(A_{\mathrm{N}}) \cup \sigma(A_{\mathrm{D}}))$. If $\lambda \notin \sigma_{\mathrm{ess}}(A_{\mathrm{N}}) \cup \sigma_{\mathrm{ess}}(A_{\mathrm{D}})$ and $\lambda \in \Lambda$ is an eigenvalue, then, by Theorem 1.3,

$$(3.15)\quad \dim G_\mu^- = \begin{cases} \dim G_\lambda^- + n_{\mathrm{D}}(\lambda), & \forall \mu \in (\lambda-\delta, \lambda), \\ \dim G_\lambda^- + n_{\mathrm{N}}(\lambda), & \forall \mu \in (\lambda, \lambda+\delta), \end{cases}$$

provided that $\delta > 0$ is small enough. In other words, the value of $\dim G_\mu^-$ jumps by $n_{\mathrm{N}}(\lambda) - n_{\mathrm{D}}(\lambda)$ as μ passes through the eigenvalue λ. Summing up these jumps over all the eigenvalues lying between a and b and taking into account that $\dim G_\lambda^- = \dim G_{\lambda-0}^- - n_{\mathrm{D}}(\lambda)$, we obtain (1.3).

3.6. Proof of Theorem 1.7. Let $\mathcal{L}$ be the subspace of $\chi_{(-\infty,\lambda)}(A_{\mathrm{N}})H$ spanned by all the vectors $v \in \chi_{(-\infty,\lambda)}(A_{\mathrm{N}})H$ such that

$$\chi_{(-\infty,\lambda]}(A_{\mathrm{D}})\,(A - \lambda I)v \;=\; 0\,. \tag{3.16}$$

The inclusion $\mathcal{L} \subset \mathcal{D}(A_{\mathrm{N}})$ implies that $\dim \mathcal{L} \geqslant N_{\mathrm{N}}(\lambda) - N_{\mathrm{D}}(\lambda) - n_{\mathrm{D}}(\lambda)$ and $\mathbf{b}[v] = 0$ for all $v \in \mathcal{L}$. From the latter identity, (3.16) and (2.7) it follows that $\mathbf{b}[P'_{\mathrm{D}}(\lambda)v] < \mathbf{b}[v]$ for all nonzero $v \in \mathcal{L}$. Since $((A - \lambda I)v, v)) < 0$ for all nonzero $v \in \chi_{(-\infty,\lambda)}(A_{\mathrm{N}})H$ and $((A - \lambda I)v, v) \geqslant 0$ for all $v \in \mathcal{D}(A_{\mathrm{D}})$ satisfying (3.16), we have $\mathcal{L} \cap \mathcal{D}(A_{\mathrm{D}}) = \{0\}$. Therefore $\ker P'_{\mathrm{D}}(\lambda)|_{\mathcal{L}} \subset L \cap \mathcal{D}(A_{\mathrm{D}}) = \{0\}$ and, consequently, $\dim P'_{\mathrm{D}}(\lambda)\mathcal{L} \geqslant N_{\mathrm{N}}(\lambda) - N_{\mathrm{D}}(\lambda) - n_{\mathrm{D}}(\lambda)$. Thus we have $\dim G_\lambda^- \geqslant \dim P'_{\mathrm{D}}(\lambda)\mathcal{L} \geqslant N_{\mathrm{N}}(\lambda) - N_{\mathrm{D}}(\lambda) - n_{\mathrm{D}}(\lambda)$. Now the theorem follows from (1.5).

3.7. Proof of Corollary 1.12. We have $R'(\lambda)H \subset G_\lambda$ and, by (1.1),

$$\begin{aligned} \mathbf{a}[\mathcal{B}_\lambda R'(\lambda)u, v] &= \mathbf{a}[R'(\lambda)u, v] - \lambda(R'(\lambda)u, v) \\ &= (u, v) = \mathbf{a}[A_{\mathrm{N}}^{-1}u, v]\,, \qquad \forall u \in H\,, \quad \forall v \in G_\lambda\,. \end{aligned} \tag{3.17}$$

The above identity implies that $\mathcal{B}_\lambda R'(\lambda)u = \Pi'_\lambda A_{\mathrm{N}}^{-1}u$ for all $u \in H$. In view of Lemma 1.2, the operator $\mathcal{B}_\lambda$ is invertible and, consequently, $R'(\lambda) = \mathcal{B}_\lambda^{-1}\Pi'_\lambda A_{\mathrm{N}}^{-1}$. Since $\mathbf{a}[A_{\mathrm{N}}^{-1}u, v] = (u, v)$, the subspace $\Pi'_\lambda A_{\mathrm{N}}^{-1}H$ is H^1-dense in G_λ. Therefore $R'(\lambda)H$ is an H^1-dense subspace of G_λ. Finally, by (3.17),

$$\mathbf{a}[\mathcal{B}_\lambda R'(\lambda)u, R'(\lambda)v] \;=\; (u, R'(\lambda)v)\,, \qquad \forall u, v \in H\,.$$

Thus we have $\mathbf{a}[\mathcal{B}_\lambda u, u] < 0$ on a k-dimensional subspace of G_λ if and only if $(R'(\lambda)u, u) < 0$ on a k-dimensional subspace of H. Now the corollary follows from Theorem 1.7.

3.8. Proof of Corollary 1.13. Let $a(\xi)$ be the full symbol of L^*L and $\lambda_1(\xi), \dots, \lambda_m(\xi)$ be the eigenvalues of $a(\xi)$. Then $\lambda_{\mathrm{D},k} > \lambda_* := \min_j \inf_\xi \lambda_j(\xi)$ for all k because $\mathbf{a}[u] \geqslant \lambda_* \|u\|^2$ on $C_0^\infty(\Omega)$. On the other hand, since $\lambda_j(\xi)$ are continuous functions of ξ, the equation $\det(a(\xi) - \lambda I) = 0$ has infinitely many ξ-solutions for each fixed $\lambda > \lambda_*$. Therefore G_λ contains an infinite-dimensional set formed by functions of the form $u_\xi = e^{ix\cdot\xi}\,\vec{c}$, where $\vec{c} \in \ker(a(\xi) - \lambda I)$. For each of these functions we have $\mathbf{a}[u_\xi] = \lambda\,\|u_\xi\|^2$. This implies that either $\dim G_\lambda^- \geqslant 1$ or $\dim G_\lambda^0 = \infty$. By Lemma 1.2, the latter is possible only if $\lambda \geqslant \lambda_{\mathrm{N},\infty}$. Therefore, by Remark 1.11(2), we have $\lambda_{\mathrm{N},k+1} \leqslant \lambda_{\mathrm{D},k}$ for all eigenvalues lying below $\lambda_{\mathrm{N},\infty}$. If at least one Dirichlet eigenfunction corresponding to $\lambda_{\mathrm{D},k}$ does not satisfy the Neumann boundary condition, then $n_{\mathrm{D}}(\lambda_{\mathrm{D},k}) \geqslant 1$ and, by Remark 1.11(1), $\lambda_{\mathrm{N},k+1} < \lambda_{\mathrm{D},k}$.

4. Remarks

4.1. A_{D} and A_{N} as selfadjoint extensions. Denote $H_0^2 := H_0^1 \cap \mathcal{D}(A_{\mathrm{N}})$. Since $\mathcal{D}(A_{\mathrm{D}}) = \Pi_0\mathcal{D}(A_{\mathrm{N}})$ (see Subsection 2.1), we have $H_0^2 = \mathcal{D}(A_{\mathrm{D}}) \cap \mathcal{D}(A_{\mathrm{N}})$.

The H-adjoint A^* coincides with the restriction of A to H_0^2. Indeed, if $(u, Av) = (\tilde{u}, v)$ for some $u, \tilde{u} \in H$ and all $v \in H_A^1$, then, taking $v \in \mathcal{D}(A_{\mathrm{N}})$ or $v \in \mathcal{D}(A_{\mathrm{D}})$, we obtain $u = A_{\mathrm{N}}^{-1}\tilde{u} = A_{\mathrm{D}}^{-1}\tilde{u}$. Therefore $\mathcal{D}(A^*) \subset H_0^2$. On the other hand, if

$u \in H_0^2$, then $(u, Av) = \mathbf{a}[u,v]$ because $u \in H_0^1$ and $\mathbf{a}[u,v] = (A_{\mathrm{N}}u, v)$. Thus $\mathcal{D}(A^*) = H_0^2$ and $A^* = A|_{H_0^2}$.

If H_0^2 is not dense in H, then the second adjoint A^{**} does not exist and the operator A is not closable in H (see, for example, [**BS**, Section 3.3]).

If H_0^2 is dense in H, then A_{D} and A_{N} are selfadjoint extensions of A^*, and A^{**} is the closure of A. Note that $\mathcal{D}(A) = H_A^1$ may be strictly smaller than $H^1 \cap \mathcal{D}(A^{**})$. Also, the H^1-closed subspaces G_z may be strictly smaller than $\ker(A^{**} - zI)|_{H^1}$ and may not be closed in H (see the next subsection).

4.2. An example. Let Ω be a bounded domain with smooth boundary, $H = L_2(\Omega)$ and H^s be the Sobolev spaces. If $\mathbf{a}[u] = \|\nabla u\|^2 + \|u\|^2$ and H_0^1 is the H^1-closure of $C_0^\infty(\Omega)$, then $A = -\Delta + I$, $\mathcal{D}(A) = \{u \in H^1 \,:\, Au \in H\}$ and $G_0 = \{u \in H^1 \,:\, Au = 0\}$. The selfadjoint operators A_{D} and A_{N} are obtained by imposing the Dirichlet and Neumann boundary conditions. The H-adjoint A^* coincides with the restriction of A to $H_0^2 := \{u \in H^2 \,:\, u|_{\partial\Omega} = \partial_n u|_{\partial\Omega} = 0\}$, where ∂_n is the normal derivative. The second H-adjoint A^{**} is the extension of A to $\mathcal{D}(A^{**}) = \{u \in H \,:\, Au \in H\}$, and $\mathcal{D}(A) = H^1 \cap \mathcal{D}(A^{**})$.

Let us choose a nonzero function $v_0 \in G_0$, and define $\tilde H_0^1 = H_0^1 \oplus \mathcal{L}_0$, where $\mathcal{L}_0$ is the one-dimensional subspace spanned by v_0 and $\oplus$ denotes the orthogonal sum in H^1. Then the corresponding operator $\tilde A$ is the same differential operator $-\Delta + I$ but $\mathcal{D}(\tilde A) = \{u \in \mathcal{D}(A) \,:\, \langle \partial_n P'_{\mathrm{N}}(0)u, v_0\rangle_{\partial\Omega} = 0\}$, where $\langle\cdot,\cdot\rangle_{\partial\Omega}$ denotes the sesquilinear pairing between $H^{-1/2}(\partial\Omega)$ and $H^{1/2}(\partial\Omega)$. The Neumann operator remains the same, and the domain of the new "Dirichlet" operator is $\mathcal{D}(\tilde A_{\mathrm{D}}) = \mathcal{D}(\tilde A) \cap \tilde H_0^1$. Finally, $\mathcal{D}(\tilde A^*) = \mathcal{D}(A_{\mathrm{N}}) \cap \tilde H_0^1 = \{u \in H^2 \cap \tilde H_0^1 \,:\, \partial_n u|_{\partial\Omega} = 0\}$. Note that $H_0^2 \subset \mathcal{D}(\tilde A^*)$.

Let $v_0 \notin H^2$. Then $v_0|_{\partial\Omega} \notin H^{3/2}(\partial\Omega)$ and, consequently, $u|_{\partial\Omega} \notin H^{3/2}(\partial\Omega)$ for all $u \in \tilde H_0^1 \setminus H_0^1$. This implies that $H^2 \cap \tilde H_0^1 = H^2 \cap H_0^1$ and $\mathcal{D}(\tilde A^*) = \mathcal{D}(A^*) = H_0^2$. Thus we have $\tilde A^{**} = A^{**}$. By the above, in this case, $\mathcal{D}(\tilde A) \neq H^1 \cap \mathcal{D}(\tilde A^{**})$.

The H^1-orthogonal complement $\tilde G_0 := H^1 \ominus \tilde H_0^1 = G_0 \ominus \mathcal{L}_0$ coincides with the kernel of the functional $u \to \mathbf{a}[v_0, u] = \langle \partial_n v_0, u\rangle_{\partial\Omega}$ defined on the space G_0. If $v_0 \notin H^2$, then this functional is not H-continuous and $\tilde G_0$ is not H-closed. Now (2.1) implies that $\tilde G_z := \ker(\tilde A - zI)$ are not H-closed for all $z \in \mathbb{C}$.

4.3. The projections $P_{\mathrm{B}}(\lambda)$. Note that by the spectral theorem, the right hand side of (3.7) is a nondecreasing function of μ and the right hand side of (3.8) is a nonincreasing function of μ. This observation allows one to simplify the proof of Theorem 1.3 in the case where $\dim G_\lambda^- < \infty$ or $\dim G_\lambda^+ < \infty$. The monotonicity is an implicit consequence of the following result.

LEMMA 4.1. *Let Λ be an arbitrary real interval, and let $v \in H_A^1$.*

(1) *If $\Lambda \cap \sigma_{\mathrm{ess}}(A_{\mathrm{N}}) = \varnothing$ and $\chi_\Lambda(A_{\mathrm{N}})P'_{\mathrm{N}}(0)v = 0$, then $P'_{\mathrm{N}}(\lambda)v \in G_\lambda$ for all $\lambda \in \Lambda$ and $\mathbf{b}[P'_{\mathrm{N}}(\lambda)v]$ is a nondecreasing function on Λ.*

(2) *If $\Lambda \cap \sigma_{\mathrm{ess}}(A_{\mathrm{D}}) = \varnothing$ and $\chi_\Lambda(A_{\mathrm{D}})P'_{\mathrm{D}}(0)v = 0$, then $P'_{\mathrm{D}}(\lambda)v \in G_\lambda$ for all $\lambda \in \Lambda$ and $\mathbf{b}[P'_{\mathrm{D}}(\lambda)v]$ is a nonincreasing function on Λ.*

If, in addition, $v \notin \mathcal{D}(A_{\mathrm{B}})$, then the function $\mathbf{b}[P'_{\mathrm{B}}(\lambda)v]$ is strictly monotone.

PROOF. Obviously, $(A - \lambda I)P'_{\mathrm{B}}(\lambda)v = E_{\mathrm{B}}(\lambda)(A - \lambda I)v$. Therefore the condition $\chi_\Lambda(A_{\mathrm{B}})P'_{\mathrm{B}}(0)v = 0$ implies that $(A - \lambda I)P'_{\mathrm{B}}(\lambda)v = -\lambda E_{\mathrm{B}}(\lambda)P'_{\mathrm{B}}(0)v = 0$ and, consequently, $P'_{\mathrm{B}}(\lambda)v \in G_\lambda$ for all $\lambda \in \Lambda$.

If $w_{\rm B}\in\mathcal{D}(A_{\rm B})$, then $P'_{\rm B}(\lambda)w_{\rm B}=E_{\rm B}(\lambda)w_{\rm B}$. Using this identity, one can easily show that

$$\mathbf{b}[P'_{\rm N}(\lambda)(w+w_{\rm N})] = \mathbf{b}[P'_{\rm N}(\lambda)w]-(E_{\rm N}(\lambda)(A-\lambda I)w,w_{\rm N}), \tag{4.1}$$
$$\mathbf{b}[P'_{\rm D}(\lambda)(w+w_{\rm D})] = \mathbf{b}[P'_{\rm D}(\lambda)w]+(w_{\rm D},E_{\rm D}(\lambda)(A-\lambda I)w) \tag{4.2}$$

for all $w\in H^1_A$, $w_{\rm N}\in\mathcal{D}(A_{\rm N})$ and $w_{\rm D}\in\mathcal{D}(A_{\rm D})$. Since $E_{\rm B}(\lambda)P'_{\rm B}(0)v=0$ and $P'_{\rm B}(\lambda)P'_{\rm B}(0)=P'_{\rm B}(\lambda)$ for all $\lambda\in\Lambda$, substituting $w=P'_{\rm B}(0)v$, $w_{\rm B}=P_{\rm B}(0)v$ in (4.1), (4.2) and applying (2.6), (2.7), we obtain

$$\mathbf{b}[P'_{\rm N}(\lambda)v]=\mathbf{b}[P'_{\rm N}(0)v]+\lambda\|P'_{\rm N}(0)v\|^2+\lambda^2(R_{\rm N}(\lambda)P'_{\rm N}(0)v,P'_{\rm N}(0)v), \qquad \forall\lambda\in\Lambda,$$
$$\mathbf{b}[P'_{\rm D}(\lambda)v]=\mathbf{b}[P'_{\rm D}(0)v]-\lambda\|P'_{\rm D}(0)v\|^2-\lambda^2(R_{\rm D}(\lambda)P'_{\rm D}(0)v,P'_{\rm D}(0)v), \qquad \forall\lambda\in\Lambda.$$

Now the required monotonicity results follow from the spectral theorem. □

Note that $P'_{\rm B}(0)=P'_{\rm B}(0)P'_{\rm B}(\mu)=P'_{\rm B}(\mu)-\mu A_{\rm B}^{-1}P'_{\rm B}(\mu)$ whenever $\mu\notin\sigma(A_{\rm B})$. Therefore we have $\chi_\Lambda(A_{\rm B})P'_{\rm B}(0)v=0$ if and only if $\chi_\Lambda(A_{\rm B})P'_{\rm B}(\mu)v=0$ for all $\mu\notin\sigma(A_{\rm B})$.

4.4. Analytic properties of $\Pi(z)$. If the embedding $H^1_0\hookrightarrow H$ is compact, then the operator-valued functions $\Pi(z)$ and $\Pi'(z)$ introduced in Subsection 2.3 are meromorphic in the whole complex plane. Indeed, since $\mathbf{a}[A_{\rm D}^{-1}u]=(A_{\rm D}^{-1}u,u)$, the compactness of the embedding $H^1_0\hookrightarrow H$ implies that $A_{\rm D}^{-1}$ is compact as an operator from H to H^1. Consequently,

$$T_z^\star T_z-I = z^2\,\Pi_0A_{\rm N}^{-2}\big|_{H^1_0}-2z\,\Pi_0A_{\rm N}^{-1}\big|_{H^1_0} = z^2\,A_{\rm D}^{-1}A_{\rm N}^{-1}\big|_{H^1_0}-2z\,A_{\rm D}^{-1}\big|_{H^1_0}$$

are compact operators in H^1_0. Now the required result follows from the analytic Fredholm theorem (see, for example, [**Ya**, Section 1.8]).

Acknowledgements

Preliminary results were reported in the workshop "Spectral analysis of difference and differential operators" (Banach Center, 2005), organized by J. Janas and S. Naboko. I am grateful to the organizers and participants for their encouragement and fruitful discussions. I am also indebted to A. Pushnitski, M. Solomyak and, especially, to N. Filonov for their useful comments.

References

[B1] M. S. Birman, *On the theory of selfadjoint extensions of positive definite operators.* Mat. Sb. N.S. **38(80)**, no. 4 (1956), 431–450 (Russian).

[B2] ______, *Perturbations of the continuous spectrum of a singular elliptic operator by varying the boundary and the boundary conditions.* Vestnik Leningrad Univ. **17**, No. 1 (1962), 22–55 (Russian).

[BS] M. S. Birman and M. Z. Solomyak, *Spectral theory of self-adjoint operators in Hilbert space,* Leningrad University, 1980 (Russian). English transl.: Reidel, 1987.

[Fi] N. Filonov. *On an inequality between Dirichlet and Neumann eigenvalues for the Laplace operator,* Algebra Anal. **16**, No. 2 (2004), 172–176; English translation in St. Petersburg Math. J. **16**, no. 2 (2005), 413-416.

[Fr1] L. Friedlander. *Some inequalities between Dirichlet and Neumann eigenvalues,* Arch. Ration. Mech. Anal. **116** (1991), 153–160.

[Fr2] ______, *Remarks on Dirichlet and Neumann eigenvalues,* Amer. J. Math. **117** (1995), 257–262.

[K] M. G. Kreĭn. *The theory of self-adjoint extensions of semibounded Hermitian operators and its applications* I, Mat. Sb. **20**(62):3 (1947), 431–495 (Russian)

[LW] H. A. Levine and H. F. Weinberger, *Inequalities between Dirichlet and Neumann eigenvalues,* Arch. Rational Mech. Anal. **94**, No. 2 (1986), 193–208.

[M] R. Mazzeo. *Remarks on a paper of Friedlander concerning inequalities between Neumann and Dirichlet eigenvalues,* Int. Math. Res. Not., No. 4 (1991), 41-48.

[Ya] D. Yafaev. *Mathematical scattering theory: General theory,* Amer. Math. Soc., Providence, RI, 1992.

Department of Mathematics, King's College London, Strand, London, United Kingdom

E-mail address: yuri.safarov@kcl.ac.uk

Amer. Math. Soc. Transl.
(2) Vol. **225**, 2008

Absolutely Continuous Spectrum of Multi-dimensional Schrödinger Operators with Slowly Decaying Potentials

O. Safronov

To Mikhail Shlëmovich Birman on his 80th birthday

1. Survey of the results in the field

One of the major challenges in the spectral theory of Schrödinger operators is proving the existence of the absolutely continuous spectrum under mild conditions on the decay of the potential. The purpose of this paper is to give a survey of the recent results devoted to this topic. First, suppose that we have an alloy which has a random distribution of atoms of different types. What is the motion of an electron in such a medium? It would be natural to assume that an electron with a small kinetic energy could be easily reflected by the atoms of the alloy so that it will be trapped in a finite volume of 3-dimensional space. It is reasonable to assume that electrons with a high energy would be able to pass through the medium. The latter would result in the absolutely continuous spectrum.

Proving the existence of an absolutely continuous spectrum in such or similar models has attracted the attention of many mathematicians, and some of them believe that one should start with slowly decaying electric potentials and then pass to the case of random potentials. So, we will start with a description of the results devoted to slowly decaying potentials.

1.1. Slowly decaying potentials. Consider the operator

$$-\Delta + V(x) \tag{1.1}$$

acting in the space $L^2(\mathbb{R}^d)$. This operator describes the motion of an electron in the electric field whose potential is the function $V(x)$. If $V = 0$, then the electron is in free motion. In this case, an electron moves having no obstacles in its way. This results in the fact that the operator (1.1) has a purely absolutely continuous spectrum, which covers the interval $(0, \infty)$. By an absolutely continuous spectrum we mean the set of energies described below (in the case where $H = -\Delta + V$):

2000 *Mathematics Subject Classification.* Primary 35J10.

DEFINITION. Let $E_H(\delta)$ be the spectral projection corresponding to a self-adjoint operator H and the set $\delta \subset \mathbb{R}$. Let $\mathfrak{H}_{ac}$ be the largest invariant subspace of vectors f for which μ_f defined by $\mu_f(\delta) = (E_H(\delta)f, f)$ is absolutely continuous. Then the spectrum of the restriction of H onto $\mathfrak{H}_{ac}$ is called the absolutely continuous spectrum of H.

If we turn to the electric field, then the spectrum of the Schrödinger operator might change, because the electric field is already an obstacle for a particle. Now, an electron can have bound states with negative energies, because the electric field can trap the electron between the potential walls and just the presence of one potential pit would be sufficient to bind the electron.

In 1999, P. Deift and R. Killip (see [**7**]) published a beautiful proof of the so called L^2-conjecture in the one-dimensional case. Namely, they proved that the one-dimensional Schrödinger operator with a real-valued potential $V \in L^2(\mathbb{R})$ has an absolutely continuous component in its spectrum covering the set $(0, \infty)$. Since then, different specialists in the spectral theory of Schrödinger operators have been trying to understand what would be a proper generalization of this result to the multi-dimensional case. In [**24**], B. Simon suggested the following

CONJECTURE 1.1. If the electric potential decays at infinity so that the integral

$$\int_{\mathbb{R}^d} |V(x)|^2 |x|^{1-d} dx < \infty \tag{1.2}$$

converges, then the absolutely continuous spectrum of the Schrödinger operator covers the interval $(0, \infty)$.

The questions considered in [**7**] were also studied before in the papers [**5**], [**6**] and [**16**]. In particular, A. Kiselev and M. Christ proved in [**5**] that the condition $|V(x)| \leq C(1 + |x|)^{-1/2-\varepsilon}$ leads to the presence of the absolutely continuous spectrum on all of $(0, \infty)$. However, the purpose of the present paper is to describe the multi-dimensional case rather than to discuss the (very interesting and significant) results in $d = 1$.

We would like to describe one of the approaches to the study of the absolutely continuous spectrum. We start with the statement that if an electron having a negative energy cannot be trapped by the electric field, then all electrons (even with positive energies) in the system escape the region where this field is different from zero and move as free particles. Without additional assumptions on V this statement is incorrect: one needs not only the information about the bound states of the electron, but similar information about the positron moving in the same field (see the paper of Damanik and Killip [**8**]). In other words, one has to consider both operators

$$-\Delta + V(x) \qquad \text{and} \qquad -\Delta - V(x) \tag{1.3}$$

simultaneously. Then the following question arises: "How many bound states with negative energies are both particles allowed to have in order to guarantee the presence of extended states with positive energies?" Here by extended states we mean states that are similar to the free motion.

The correct answer to this question in the case of finitely many bound states has been given in [**8**] for $d = 1$ and in [**21**] for $d \geq 2$.

THEOREM 1.1. *Let $V \in L^\infty(\mathbb{R}^d)$. Assume that the negative spectra of the operators* (1.3) *consist of finitely many eigenvalues. Then the absolutely continuous spectra of these operators are essentially supported by $\mathbb{R}_+ := (0,\infty)$. This means that the spectral projections of* (1.3) *corresponding to the sets $\delta \subset \mathbb{R}_+$ of positive Lebesgue measure are different from zero.*

The conclusion of Theorem 1.1 in $d = 1$ is stronger than the one mentioned above: the positive spectra of $-d^2/dx^2 + V(x)$ and $-d^2/dx^2 - V(x)$ are purely absolutely continuous (see [**8**]) when their negative spectra are finite.

After that, Damanik and Remling [**9**] obtained a generalization of this result in $d = 1$ to the case where the negative eigenvalues accumulate to zero. The corresponding result for $d \geq 2$ has been obtained in [**20**]. The main theorem of [**9**] and [**20**] is as follows.

THEOREM 1.2. *Let $V \in L^\infty$, and let $E_j(V)$ and $E_j(-V)$, $j = 1, 2, \dots$, be the negative eigenvalues of the operators $-\Delta+V$ and $-\Delta-V$ correspondingly. Suppose that*

$$\sum_j \sqrt{|E_j(V)|} + \sum_j \sqrt{|E_j(-V)|} < \infty. \tag{1.4}$$

Then the absolutely continuous spectra of $-\Delta + V$ and $-\Delta - V$ are essentially supported by the positive real line $[0, \infty)$.

In other words, if the condition (1.4) holds, then there exist extended states with positive energies. Moreover, the energies of these extended states fill the interval $(0, \infty)$. Putting it differently, almost every $E > 0$ is the energy of a certain extended state. Thus, knowing how the possible negative energies in the system accumulate to zero, one can tell what is the positive spectrum of this system.

The reader might guess from the nature of the obtained results that there is a conservation law hiding behind them. The appearance of negative eigenvalues happens at the expense of the absolutely continuous spectrum which becomes "thinner". Therefore, the thickness of the positive spectrum can be estimated by the number of negative energy levels. If the number of these levels is finite, then the positive spectrum is absolutely continuous. All these arguments are valid under one additional condition: besides the spectrum of an electron one has to consider the spectrum of a positron. If $d = 1$, this law is related to the trace formulas discovered by Buslaev and Faddeev [**4**]. The idea of applying it to the study of the absolutely continuous spectrum belongs to Deift and Killip [**7**]. In particular, they found that the derivative of the spectral measure of the one-dimensional Schrödinger operator can be estimated by the so-called scattering coefficient. The mentioned conservation law for a multi-dimensional Schrödinger operator has been found in the paper [**17**]. However, it uses not a trace formula of the "second order", but two trace formulas of the "first order". The conservation law says that the entropy of the spectral measure of the operator $-\Delta + V$ (as well as $-\Delta - V$) can be estimated by the negative eigenvalues of the operators $-\Delta + V$ and $-\Delta - V$. Let us clarify this statement. Let μ be the spectral measure of $-\Delta + V$, constructed for the element f, which means that

$$((-\Delta + V - z)^{-1} f, f) = \int_{-\infty}^{\infty} \frac{d\mu(t)}{t - z}.$$

Let $E_j(\pm V)$ be the negative eigenvalues of the operator $-\Delta \pm V$. Then one can find an element f of the space $L^2(\mathbb{R}^d)$ such that the measure constructed for this element will satisfy the condition

$$\int_a^b \log \mu'(\lambda)\, d\lambda \geq -C(1 + \sum_j |E_j(V)|^{1/2} + \sum_j |E_j(-V)|^{1/2}), \tag{1.5}$$

where $0 < a < b < \infty$ and the constant C depends only on a, b, and $||V||_{L^\infty}$. The proof of this statement can be found in [**17**]. Due to Jensen's inequality, the integral $\int_a^b \log \mu'(\lambda)\, d\lambda$ can diverge only to $-\infty$. But, if it converges, then $\mu'(\lambda) > 0$ almost everywhere on $[a, b]$, which leads to the conclusion that the absolutely continuous spectrum of $-\Delta + V$ covers $[a, b]$.

The relation (1.5) follows if one adds the inequalities

$$\begin{aligned} &\int_a^b \log \mu'_+(\lambda)\, d\lambda \geq -C(1 + \sum_j |E_j(V)|^{1/2} + \int_{\mathbb{R}^d} V(x) \frac{dx}{|x|^{d-1}}), \\ &\int_a^b \log \mu'_-(\lambda)\, d\lambda \geq -C(1 + \sum_j |E_j(-V)|^{1/2} - \int_{\mathbb{R}^d} V(x) \frac{dx}{|x|^{d-1}}) \end{aligned} \tag{1.6}$$

and then uses Jensen's inequality to estimate one of the integrals on the left-hand side. Here $\mu_\pm$ are the spectral measures of the operators $-\Delta \pm V$.

It becomes clear that the estimates of negative eigenvalues also play a very important role in the theory of the absolutely continuous spectrum. For example, the estimate

$$\sum_j \sqrt{|E_j(V)|} \leq C_1 \Big(\int_{|\xi|<\varepsilon} |\hat{V}(\xi))|^2 d\xi \Big)^p + C_2 \Big(\int |V|^{d+1} dx \Big)^q, \qquad d \geq 3, \tag{1.7}$$

allows one to say (this is the main result of the paper [**19**]) that the absolutely continuous spectrum of the operator with the potential $V \in L^{d+1}$ fills the positive half-line under the additional condition that the Fourier transform $\hat{V}$ is square integrable near the origin. The number $\varepsilon > 0$ in (1.7) is arbitrary, but the constants depend on it. This result can be interpreted as the first step towards Simon's conjecture, that is, proving the presence of the absolutely continuous spectrum on all of $(0, \infty)$ under condition (1.2).

Besides the above-mentioned results, we would like to mention the work of S. Denisov [**10**]–[**13**] and G. Perelman [**18**]. Denisov proved that condition (1.2) guarantees the presence of the absolutely continuous spectrum in the case where the support of the potential V is a union of sparse spherical layers embedded in each other. More precisely, the statement can be formulated as follows:

THEOREM 1.3. *Let the support of V be the union of the sets*

$$\{x \in \mathbb{R}^3 : \ R_n < |x| < R_n + 1\}, \qquad n \in \mathbb{N},$$

where

$$R_{n+1} > \exp(\alpha R_n), \quad \alpha > 1.$$

Then the condition (1.2) *implies that the absolutely continuous spectrum of $-\Delta + V$ is essentially supported by $(0, \infty)$.*

The main result of [**18**] says that the operator $-\Delta+V$ in $d=3$ has an absolutely continuous spectrum that is essentially supported by the half-line $[0,\infty)$, provided that

$$|V(x)| + |\nabla_\theta V(x)| \le \frac{C}{(1+|x|)^{1/2+\varepsilon}}, \qquad \varepsilon > 0,\ |x|>1,$$

where $\nabla_\theta = |x|\nabla - x\partial/\partial r$. However, proving the same property of the spectrum under the condition

$$\int_{|x|>1} \Big(|\nabla_\theta V(x)|^2 + |V(x)|^2\Big)|x|^{1-d}\,dx < \infty \tag{1.8}$$

is still an open problem. After the result of [**18**] was published, a hope to understand the difference between the one-dimensional and multi-dimensional cases appeared. Indeed, the condition (1.8) involves the derivatives of the potential with respect to the angular variables. There are no such variables in the one-dimensional case; therefore, (1.8) and (1.2) coincide in $d=1$. That was the reason why several mathematicians tried to look for a counterexample, where V would not have derivatives with respect to θ and would satisfy the condition (1.2), but at the same time the essential support of the absolutely continuous spectrum of the Schrödinger operator would be different from $(0,\infty)$. So far, such a counterexample has not been found. Moreover, we know that the condition (1.2) is the right one for the class of potentials supported by the sparse spherical layers. That is why our conclusion is that we can expect some new results related to Conjecture 1.1.

We would also like to mention a nice idea in the recent paper [**14**] for studying a family of potentials rather than a single one. Namely, if one looks at $H(t) = -\Delta + tV$, where t runs over an arbitrary interval and $|V| < C(1+|x|)^{-5/6-\varepsilon}$, then for almost every t, one will witness that $H(t)$ has absolutely continuous spectrum covering the positive half-line.

1.2. Some results on random potentials. Besides what was mentioned above, Denisov gave a different proof of Bourgain's theorem [**2**], [**3**]. Actually, the difference is that Denisov considered the continuous version of the Schrödinger operator in $d=3$ and Bourgain dealt with the discrete one in $d=2$. Let us formulate the main result of the paper [**11**] for $d=3$.

THEOREM 1.4. *Let $d=3$ and let $\{\omega_n\}_{n\in\mathbb{Z}^d}$ be independent identically distributed random variables. Assume that $\mathbb{E}[\omega_n]=0$ for all n. Then the operator*

$$-\Delta + (1+|x|)^{-s}\sum_{n\in\mathbb{Z}^d}\omega_n\chi(x-n), \qquad s>1/2,$$

almost surely has an absolutely continuous spectrum covering the positive real line $[0,\infty)$. Here χ is the characteristic function of the cube $[0,1)^d$.

The result of [**2**] is formulated in the same way; one only needs to say that $d=2$, x is discrete, and $-\Delta$ is the Laplace operator on the lattice $\mathbb{Z}^2$. The main idea of the proof of Theorem 1.4 is to represent $V=(1+|x|)^{-s}\sum_n\omega_n\chi(x-n)$ in the form $V=\operatorname{div} Q$, where Q admits the same estimate as V, i.e.,

$$|Q| < C(1+|x|)^{-1/2-\varepsilon}.$$

This is possible if one defines Q by

$$Q = \nabla\Delta^{-1}V.$$

Note that due to (1.6), the operator $-\Delta + V + Q^2 = -\Delta + \operatorname{div} Q + Q^2$ already has some absolutely continuous spectrum which is essentially supported by $\mathbb{R}_+$; indeed, this operator does not have negative eigenvalues and the integral on the right-hand side of (1.6) converges if one replaces V by $\operatorname{div} Q + Q^2$. It remained to get rid of Q^2 decaying faster than $|x|^{-1-\varepsilon}$, which turned out to be a very complicated procedure.

The paper [**15**] treats potentials that do not decay at all. It deals with a class of random Schrödinger operators

$$-\Delta + V_\omega, \qquad V_\omega(x) = \sum_{j \in J} \omega_j f_j(x),$$

in $L^2(\mathbb{R}^d)$, $d = 2$ or $d = 3$. Here the ω_j are bounded, independent and identically distributed random variables with zero expectations $\mathbb{E}(\omega_j) = 0$, the f_j are characteristic functions of disjoint sets whose union is the whole space $\mathbb{R}^d$, and J is an index set. In the usual ergodic case, the f_j are translates along the lattice $J = \mathbb{Z}^d$, say, of a fixed bump function. Note that in this case the number of bumps inside the spherical layer $\{x \in \mathbb{R}^d : n - 1 \le |x| < n\}$ increases as n^{d-1} as $n \to \infty$. In [**15**], the authors consider models where this number grows as $n^{s(d-1)}$ for some parameter $s > 1$. Hence, the potentials in [**15**] are nonergodic and the parameter s shows how close they are to the ergodic ones. Note that in [**15**] one does not assume that the potential decays at infinity. We prove that, if $d = 2, 3$ and $s > 1 + 2/(d-1)$, then the wave operators

$$\text{s-lim}_{t \to \pm\infty} \exp(it(-\Delta + V_\omega)) \exp(it\Delta)$$

exist almost surely. However, they are not complete.

Let us give an intuitive picture of what is happening. Divide the plane into spherical layers $\{x \in \mathbb{R}^2 : n - 1 \le |x| < n\}$, $n \in \mathbb{N}$, and divide each layer into (approximately) n^s equal pieces with some fixed $s > 1$. Now paint these pieces randomly in yellow and blue colors. With the growth of n the pieces become finer and finer and, outside of a large disk, the plane will look green!

Something similar happens to an electron in the model considered in [**15**]: the yellow and blue colors correspond to positive and negative bumps of the potential. They average to zero at infinity and allow the electron to propagate.

Let us note, however, that this picture is rather naive and does not represent the difficulty of the problem. For example, in the ergodic case the average of the potential over the sphere $\{|x| = r\}$ decreases as a negative power (presumably as $r^{-(d-1)/2}$), which means that an observer running around such a sphere will see the green colour, but to prove existence of the absolutely continuous spectrum for ergodic potentials seems to be a very hard problem.

The main idea of [**15**] is to use the trace class scattering theory developed by M. Sh. Birman (see, for example, [**1**]), T. Kato, S. T. Kuroda, M. Rosenblum and D. B. Pearson (see [**25**] and the references therein). We employ it in a way which is adjusted to cancellation due to the random nature of the potential. Namely, let $Y(\theta)$ be a normalized real eigenfunction of the Laplace-Beltrami operator on the unit sphere. Define the projection

$$Pu(x) = Y(x/|x|) \int_{\{\theta:\ |\theta|=1\}} Y(\theta) u(|x|\theta)\, d\theta.$$

We show that the operator

$$(-\Delta + V_\omega - i)^{-1} P - P(-\Delta - i)^{-1}$$

belongs to the trace class, which guarantees existence of the limits

$$\text{s-lim}_{t\to\pm\infty} \exp(it(-\Delta + V_\omega)) \exp(it\Delta)P.$$

Now, since Y is an arbitrary eigenfunction, the projection P in the last formula can be omitted.

Ideally, one would like to prove the following

CONJECTURE 1.2. Let $d \geq 3$, and let $\{\omega_n\}_{n\in\mathbb{Z}^d}$ be bounded independent identically distributed random variables. Prove that the operator

$$-\Delta + \sum_{n\in\mathbb{Z}^d} \omega_n \chi(x-n)$$

almost surely has a nonempty absolutely continuous spectrum. Here χ is the characteristic function of the cube $[0,1)^d$.

However, as far as we know, this problem is so difficult to study that no results related to the absolutely continuous spectrum of this operator are published. The paper [**15**] obviously treats a different case. An example of an operator of the form

$$-\text{div}\, a(x)\nabla + \sum_{n\in\mathbb{Z}^d} \omega_n \chi(x-n), \qquad a(x) = \{a_{j,k}(x)\} \geq 1, \ d \geq 5,$$

whose absolutely continuous spectrum almost surely coincides with that of the operator $-\text{div}\, a(x)\nabla$ and equals $[0,\infty)$ is given in the paper [**22**]. Surely, the condition $\mathbb{E}[\omega_n] = 0$ is needed in [**22**] to guarantee that the absolutely continuous spectrum does not shift when one adds the potential.

2. One application of eigenvalue estimates

In the conclusion of this paper, which is more a survey than anything else, we would like to describe an idea that helps in the study of problems that might in a certain sense be called one-dimensional problems. The questions described below were studied by Denisov in [**12**]. However, the paper [**12**] gives a local result on some specific energy interval, but does not prove that the whole half-line $\mathbb{R}_+$ is covered by the absolutely continuous spectrum (as far as we understand). Here we develop a different approach to the problem, which is valid for the case where the mean value of the potential is zero.

Consider the domain

$$\Omega = \{\, (x,y) \in \mathbb{R}^{d+1} : \ x \in \mathbb{R}, \ y \in B \,\},$$

where B is the unit ball in $\mathbb{R}^d$. Let us introduce the operator

$$A = -\Delta + V(x,y), \qquad V = \overline{V} \in L^\infty(\Omega), \tag{2.1}$$

with the Neumann conditions on the boundary of Ω.

THEOREM 2.1. *Assume that* $V \in L^\infty$,

$$\int_B V(x,y)\, dy = 0, \qquad \forall x, \tag{2.2}$$

and

$$\int_\Omega V^2(x,y)\, dx\, dy < \infty. \tag{2.3}$$

Then the absolutely continuous spectrum of the operator (2.1) *is essentially supported by* $[0,\infty)$.

Proof. The proof of this theorem relies on a version of Theorem 1.2 suitable for the domain Ω and the fact that the negative eigenvalues $\beta_j(\mp V)$ of the operator $-\Delta \pm V$ satisfy the condition

$$\sum_j |\beta_j(\mp V)|^{1/2} < \infty. \tag{2.4}$$

The proof of (2.4) is based on the circumstance that the behaviour of the eigenvalues near zero depends only on the structure of the edge of the spectrum of the unperturbed operator. But in the suggested model, this edge has the same structure as that of the one-dimensional Schrödinger operator. Let us introduce the notation P_1 for the spectral projection of the operator $-\Delta$ on Ω corresponding to the interval $[0, \varepsilon)$, where ε is sufficiently small. Actually, ε should be smaller than the second eigenvalue of the Laplace operator on B (with Neumann boundary conditions). Set also $P_2 = I - P_1$. Then

$$V = 2\,\mathrm{Re}\, P_1 V P_2 + P_2 V P_2,$$

due to the condition (2.2) on the mean values of V. Let us note that the operator Γ defined by

$$(\Gamma u)(x, y) = \frac{1}{|B|} \int_B u(x, \tilde{y})\, d\tilde{y}$$

commutes with $-\Delta = A - V$.

Let us introduce $n_+(s, T)$ as the number of eigenvalues of a compact operator T that are greater than s. Then due to the Birman-Schwinger principle,

$$\sum_j |\beta_j(V)|^{1/2} = \int_0^\infty n_+(1, (-\Delta + \lambda)^{-1/2} V (-\Delta + \lambda)^{-1/2}) \frac{d\lambda}{2\sqrt{\lambda}}.$$

In addition to the distribution function n_+, we shall need the distribution function of singular values of a nonselfadjoint operator

$$n(s, T) = n_+(s^2, T^*T), \qquad s > 0$$

(here T is a compact operator). Two of the important properties of this function are

$$n(s, TS) \le n(\frac{s}{||S||}, T)$$

and

$$n(s_1 + s_2, T_1 + T_2) \le n(s_1, T_1) + n(s_2, T_2).$$

Using these properties, we obtain

$$\sum_j |\beta_j(V)|^{1/2} \le \int_0^\infty n(c_1, (-\Delta + \lambda)^{-1/2} P_1 V) \frac{d\lambda}{\sqrt{\lambda}}$$
$$+ \int_0^\infty n_+(c_2, (-\Delta + \lambda)^{-1/2} P_2 V P_2 (-\Delta + \lambda)^{-1/2}) \frac{d\lambda}{2\sqrt{\lambda}}.$$

Let us note that the second term equals zero if the norm $||V||_{L^\infty}$ is sufficiently small. In the general case, this term can be well estimated by the integral (2.3). It remains to consider the first term:

$$\int_0^\infty n(c_1,(-\Delta+\lambda)^{-1/2}P_1V)\frac{d\lambda}{\sqrt{\lambda}}$$
$$=\int_0^\infty n_+(c_1^2,(-\Delta+\lambda)^{-1/2}P_1V^2P_1(-\Delta+\lambda)^{-1/2})\frac{d\lambda}{\sqrt{\lambda}}$$
$$\le\int_0^\infty n_+(c_1^2,(-\Delta+\lambda)^{-1/2}\Gamma V^2\Gamma(-\Delta+\lambda)^{-1/2})\frac{d\lambda}{\sqrt{\lambda}}=2\sum_j|\Lambda_j|^{1/2},$$

where Λ_j are the eigenvalues of the operator $-\Delta\Gamma - c_1^{-2}\Gamma V^2\Gamma$, which, as a matter of fact, is the one-dimensional Schrödinger operator with the potential $-c_1^{-2}\overline{V^2}$, where the bar means averaging:

$$\overline{V^2}(x)=\frac{1}{|B|}\int_B V^2(x,y)\,dy.$$

Therefore, according to the Lieb-Thirring bound for a one-dimensional operator, the sum $\sum_j|\Lambda_j|^{1/2}$ can be estimated by the integral (2.3). □

References

[1] M. Sh. Birman, *A local criterion for the existence of wave operators* (Russian) Izv. Akad. Nauk SSSR Ser. Mat. **32** (1968), no. 4, 914–942. English transl., Math. USSR-Izv. **2** (1968), no. 2, 879–906.

[2] J. Bourgain, *On random Schrödinger operators on $\mathbb{Z}^2$*. Discrete Contin. Dynam. Systems **8** (2002), no. 1, 1–15.

[3] ———, *Random lattice Schrödinger operators with decaying potential: Some multidimensional phenomena*, Geometric Aspects of Functional Analysis: Israel Seminar 2001–2002 (V. D. Milman and G. Schechtman, eds.), Lecture Notes in Mathematics, vol. 1807, Springer, Berlin, 2003, pp. 70–98.

[4] V. Buslaev and L. Faddeev, *Formulas for traces for a singular Sturm-Liouville differential operator*. Dokl. Akad. Nauk SSSR **132**, 13–16 (Russian); translated as Soviet Math. Dokl. **1** (1960), 451–454.

[5] M. Christ and A. Kiselev, *Absolutely continuous spectrum for one-dimensional Schrödinger operators with slowly decaying potentials: Some optimal results*. J. Amer. Math. Soc. **11** (1998), 771–797.

[6] M. Christ, A. Kiselev, and C. Remling, *The absolutely continuous spectrum of one-dimensional Schrödinger operators with decaying potentials*. Math. Res. Lett. **4** (1997), no. 5, 719–723.

[7] P. Deift and R. Killip, *On the absolutely continuous spectrum of one-dimensional Schrödinger operators with square summable potentials*. Comm. Math. Phys. **203** (1999), no. 2, 341–347.

[8] D. Damanik and R. Killip, *Half-line Schrödinger operators with no bound states*. Acta Math. **193** (2004), no. 1, 31–72.

[9] D. Damanik and Ch. Remling, *Schrödinger operators with many bound states*. Duke Math. J. **136** (2007), no. 1, 51–80.

[10] S. Denisov, *On the absolutely continuous spectrum of Dirac operator*. Comm. Partial Differential Equations **29** (2004), no. 9-10, 1403–1428.

[11] ———, *Absolutely continuous spectrum of multidimensional Schrödinger operator*. Int. Math. Res. Not. 2004, no. 74, 3963–3982.

[12] ———, *On the preservation of absolutely continuous spectrum for Schrödinger operators*. J. Funct. Anal. **231** (2006), no. 1, 143–156.

[13] ———, *Wave propagation through sparse potential barriers*, Comm. Pure Appl. Math. **LXI** (2008), 0156-0185.

[14] ———, *Schrödinger operators and associated hyperbolic pencils*, J. Funct. Anal. **254** (2008), no. 8, 2186–2226.

[15] R. Frank and O. Safronov, *Absolutely continuous spectrum of a class of random nonergodic Schrödinger operators*. Int. Math. Res. Not. 2005, no. 42, 2559–2577.

[16] A. Kiselev, Y. Last and B. Simon, *Modified Prüfer and* EFGP *transforms and the spectral analysis of one-dimensional Schrödinger operators*. Commun. Math. Phys. **194** (1998), 1–45.

[17] A. Laptev, S. Naboko and O. Safronov, *Absolutely continuous spectrum of Schrödinger operators with slowly decaying and oscillating potentials*. Comm. Math. Phys. **253** (2005), no. 3, 611–631.

[18] G. Perelman, *Stability of the absolutely continuous spectrum for multidimensional Schrödinger operators*. Int. Math. Res. Not. 2005, no. 37, 2289–2313.

[19] O. Safronov, *On the absolutely continuous spectrum of multi-dimensional Schrödinger operators with slowly decaying potentials*. Comm. Math. Phys. **254** (2005), no. 2, 361–366.

[20] ———, *Multi-dimensional Schrödinger operators with some negative spectrum*. J. Funct. Anal. **238** (2006), no. 1, 327–339.

[21] ———, *Multi-dimensional Schrödinger operators with no negative spectrum*. Ann. Henri Poincaré **7** (2006), no. 4, 781–789.

[22] ———, *Absolutely continuous spectrum of one random elliptic operator*, mp-arc preprint.

[23] O. Safronov and G. Stolz, *Absolutely continuous spectrum of Schrödinger operators with potentials slowly decaying inside a cone*. J. Math. Anal. Appl. **326** (2007), no. 1, 192–208.

[24] B. Simon, *Schrödinger operators in the twenty-first century*. Mathematical Physics 2000, 283–288, Imp. Coll. Press, London, 2000.

[25] D. R. Yafaev, *Mathematical Scattering Theory*. General Theory, Translations of Mathematical Monographs, vol. 105, Amer. Math. Soc., Providence, RI, 1992.

Amer. Math. Soc. Transl.
(2) Vol. **225**, 2008

On the Discrete Spectrum of the Perturbed Periodic Magnetic Schrödinger Operator with Degenerate Lower Edge of the Spectrum

Roman Shterenberg

To Mikhail Shlëmovich Birman on his 80th birthday

ABSTRACT. We consider the periodic magnetic Schrödinger operator with degenerate lower edge of the spectrum perturbed by a negative decaying potential. A detailed analysis of the discrete spectrum which appears below the lower edge of the spectrum of the periodic operator is given. The asymptotics for the number of eigenvalues in the large coupling constant limit is found.

1. Introduction

1.1. Let M be a selfadjoint elliptic periodic second-order operator in $L_2(\mathbb{R}^d)$, $d \geq 2$, given by the expression $M = (\mathbf{D} - \mathbf{A}(\mathbf{x}))^* g(\mathbf{x})(\mathbf{D} - \mathbf{A}(\mathbf{x})) + V(\mathbf{x})$, where $\mathbf{D} := -i\nabla$, and let V_0 be the operator of multiplication by a function $V_0(\mathbf{x}) \geq 0$ decaying at infinity. Suppose that the lower edge of the spectrum of M is $\lambda_0 = 0$. We put $M_\alpha := M - \alpha V_0$, $\alpha > 0$. By $N(\lambda, M_\alpha)$, $\lambda \leq 0$, we denote the number of eigenvalues (counting multiplicities) of the operator M_α lying below λ. We are interested in the asymptotics of $N(\lambda, M_\alpha)$ as $\alpha \to \infty$ (in the large coupling constant limit). These asymptotics may be rather diverse, depending both on M and V_0. They were studied in a number of papers (see, e.g., [**1**], [**2**], [**3**], [**4**], [**9**], [**10**]). Note that in all these papers the nonmagnetic case $\mathbf{A} = 0$ was considered. But some results can be extended to the magnetic case as well.

If $d \geq 3$, $\mathbf{A} = 0$, and $V_0 \in L_{d/2}$, then, under some not very restrictive conditions on g and V, the function $N(\lambda, M_\alpha)$, $\lambda \leq 0$, has the Weyl asymptotics

$$N(\lambda, M_\alpha) \sim (2\pi)^{-d} \omega_d \alpha^{d/2} \int_{\mathbb{R}^d} V_0^{d/2} (\det g)^{-1/2} d\mathbf{x}, \quad \lambda \leq 0, \quad \alpha \to \infty; \tag{1.1}$$

here ω_d is the volume of the unit ball in $\mathbb{R}^d$. Moreover, the asymptotics (1.1) holds if and only if $V_0 \in L_{d/2}$. It is important to notice that the Weyl-type power $\alpha^{d/2}$, as

2000 *Mathematics Subject Classification.* Primary 35J10, 35P15.
Key words and phrases. Periodic operator, magnetic Schrödinger operator, lower edge of the spectrum, degeneration, discrete spectrum.
Partially supported by NSF grant DMS-0314129.

well as the asymptotics (1.1) itself, is due to the order of the differential operator M. At the same time, the condition on V_0 which guarantees the validity of the asymptotics (1.1) for $\lambda = 0$ is determined by the behaviour of the lower edge of the spectrum of M. For the nonmagnetic case $\mathbf{A} = 0$ the lower edge of the spectrum has the so-called "proper" structure (for details see, e.g., [**12**], [**13**]) and provides the condition $V_0 \in L_{d/2}$. The situation may change if the magnetic potential appears.

1.2. In the recent paper [**18**] it was shown that, in general, for the magnetic Schrödinger operator the lower edge of the spectrum does not possess the proper structure, i.e., the first band function can have a degenerate minimum. This leads to the following threshold effect. To provide the Weyl-type asymptotics (1.1), one needs $V_0 \in L_r \cap L_{d/2}$ with r depending on the order of degeneracy and strictly less than $d/2$. We accent that it is a purely magnetic effect which cannot be observed in the nonmagnetic case. We mention as well the papers [**6**], [**16**], where a somewhat similar situation was observed for the three-dimensional Dirac operator. Namely, the Dirac operator is a first-order differential operator, and thus the Weyl type power is α^3. But since the band functions which determine the edges of the spectrum (left and right edges of the internal gap in the spectrum) naturally have a second-order minimum or maximum, the Weyl type asymptotics holds if $V_0 \in L_3(\mathbb{R}^3) \cap L_{3/2}(\mathbb{R}^3)$.

1.3. In the present paper, we give a detailed analysis of the asymptotics for $N(0, M_\alpha)$ in the case where the lower edge of the spectrum of M is degenerate. For the sake of simplicity, we consider only the case $d \geq 3$. In the two-dimensional case more subtle considerations are needed, since even in the proper situation additional difficulties arise (see, e.g., [**7**], [**8**]). We also note that similar questions can be set and results can be obtained not only for the negative spectrum, but for the spectrum in the internal gaps as well. For corresponding results in the proper situation, see, e.g., [**1**], [**2**], [**20**]. It is important to notice that the proper structure of the edges of internal gaps is not known even in the nonmagnetic case.

The paper is organized as follows. In Section 2 we give the definition of the operator M and describe its properties. The main result of the paper (Theorem 3.1) is formulated in Section 3. Finally, in Section 4 we prove two technical lemmas and provide the proof of the main theorem.

Technically, our constructions are similar to the ones presented in [**2**] for the nonmagnetic operator, though in our case some new difficulties arise and additional considerations are needed.

2. The operator M

2.1. Notation. We use the notation $\mathbf{x} = \operatorname{col}(x_1, x_2, \dots, x_d) \in \mathbb{R}^d$, $\nabla = \nabla_{\mathbf{x}} = \operatorname{col}(\partial/\partial x_1, \partial/\partial x_2, \dots, \partial/\partial x_d)$, $\mathbf{D} = \operatorname{col}(D_1, D_2, \dots, D_d) = -i\nabla$. Next, $\langle\cdot,\cdot\rangle$, $|\cdot|$ are the standard inner product and the norm in $\mathbb{C}^d$.

Let Γ be a lattice of periods in $\mathbb{R}^d$. By Ω we denote an elementary cell of the lattice Γ and by $\overline{\Omega}$ the closure of Ω. The Sobolev classes of order $s > 0$ with integrability index $p \geq 1$ are denoted by $W_p^s(\mathbb{R}^d)$, $W_p^s(\Omega)$. The subspace in $W_p^s(\Omega)$ formed by the functions, whose Γ-periodic extension belongs to $W_{p,\mathrm{loc}}^s(\mathbb{R}^d)$, is denoted by $\widetilde{W}_p^s(\Omega)$. In the case $p = 2$ we use the notation $H^s(\mathbb{R}^d)$, $H^s(\Omega)$, $\widetilde{H}^s(\Omega)$.

The set of restrictions to Ω of Γ-periodic functions in $\mathbb{R}^d$ that are Hölder of order δ is denoted by $\widetilde{C}^\delta(\Omega)$, $\delta \in (0,1)$.

Next, $\|\cdot\|_H$ is the norm in a Banach space H. We often omit indices if this does not lead to confusion. The Lebesgue measure of a Borel set Ξ in $\mathbb{R}^d$ is denoted by $\operatorname{meas}\Xi$. By c, C (probably, with indices) we denote various constants in estimates. By $\mathbf{1}$ we denote the $d\times d$ identity matrix.

2.2. We consider the operator

$$M = M(g,\mathbf{A},V) := (\mathbf{D}-\mathbf{A}(\mathbf{x}))^* g(\mathbf{x})(\mathbf{D}-\mathbf{A}(\mathbf{x})) + V(\mathbf{x}), \quad \mathbf{x}\in\mathbb{R}^d, \tag{2.1}$$

in the space $L_2(\mathbb{R}^d)$, $d \geq 3$. All coefficients are assumed to be real-valued and Γ-periodic. Also, we assume that the following conditions are fulfilled:

$$V \in L_{p/2}(\Omega), \quad p > d, \tag{2.2}$$

$$\mathbf{A} \in L_p(\Omega), \quad p > d, \tag{2.3}$$

$$c_0\mathbf{1} \leq g(\mathbf{x}) \leq c_1\mathbf{1}, \quad \text{for a.e. } \mathbf{x}\in\mathbb{R}^d, \quad 0 < c_0 \leq c_1 < \infty. \tag{2.4}$$

Under conditions (2.2)–(2.4), the operator M is defined via the closed lower semi-bounded form

$$\begin{aligned} m[u,u] &= m(g,\mathbf{A},V)[u,u] \\ &:= \int_{\mathbb{R}^d} \langle g(\mathbf{x})(\mathbf{D}-\mathbf{A}(\mathbf{x}))u(\mathbf{x}), (\mathbf{D}-\mathbf{A}(\mathbf{x}))u(\mathbf{x})\rangle\, d\mathbf{x} \\ &\quad + \int_{\mathbb{R}^d} V(\mathbf{x})|u(\mathbf{x})|^2 d\mathbf{x}, \qquad u \in H^1(\mathbb{R}^d). \end{aligned} \tag{2.5}$$

In what follows, we shall also assume that the following gauge condition is satisfied:

$$\operatorname{div}(g\mathbf{A}) = 0, \tag{2.6}$$

which is understood in the sense of distributions.

Remark. Condition (2.6) can always be satisfied by an appropriate choice of the gauge transformation $M \mapsto e^{-i\phi}Me^{i\phi}$ with a periodic function ϕ such that $\int_\Omega \phi(\mathbf{x})d\mathbf{x} = 0$. Namely, for a given $\mathbf{A}$ we choose ϕ to satisfy the equation

$$\operatorname{div}(g\nabla\phi) = \operatorname{div}(g\mathbf{A}).$$

Then the new magnetic potential $\widetilde{\mathbf{A}} := \mathbf{A} - \nabla\phi$ satisfies (2.6). However, we also need to check the condition (2.3) for $\widetilde{\mathbf{A}}$. Unfortunately, we have found only partial results (see Theorem 3.7 from [**21**]) which states that $\phi \in \widetilde{W}_p^1(\Omega)$ provided $\mathbf{A} \in L_p(\Omega)$ and, for some $\delta \in (0,1)$,

$$g \in \widetilde{C}^\delta, \quad \delta \in (0,1). \tag{2.7}$$

Thus, we need to assume either condition (2.6) from the beginning or the additional condition (2.7) on the metric g.

2.3. The periodic operator M can be decomposed in the direct integral of the family of operators $M(\mathbf{k})$, $\mathbf{k} \in \widetilde{\Omega}$. Here $\widetilde{\Omega}$ is a fundamental domain of the dual lattice $\widetilde{\Gamma}$. It is convenient to choose $\widetilde{\Omega}$ to be the Brillouin zone; then $\mathbf{k} = 0$ is an internal point of $\widetilde{\Omega}$. The operator $M(\mathbf{k})$, $\mathbf{k} \in \mathbb{R}^d$, acts in $L_2(\Omega)$ and is defined via the closed lower semi-bounded form

$$\begin{aligned} m(\mathbf{k})[u,u] &= m(\mathbf{k}; g, \mathbf{A}, V)[u,u] \\ &:= \int_\Omega \langle g(\mathbf{x})(\mathbf{D} - \mathbf{A}(\mathbf{x}) + \mathbf{k})u(\mathbf{x}), (\mathbf{D} - \mathbf{A}(\mathbf{x}) + \mathbf{k})u(\mathbf{x})\rangle \, d\mathbf{x} \\ &\quad + \int_\Omega V(\mathbf{x})|u(\mathbf{x})|^2 d\mathbf{x}, \qquad u \in \widetilde{H}^1(\Omega). \end{aligned} \tag{2.8}$$

The spectrum of $M(\mathbf{k})$ is discrete. Let $\{E_j(\mathbf{k})\}$ denote the eigenvalues of the operator $M(\mathbf{k})$ enumerated in nondecreasing order (counting multiplicities):

$$E_1(\mathbf{k}) \le E_2(\mathbf{k}) \le \cdots \le E_n(\mathbf{k}) \le \cdots .$$

Let $\{\varphi_j(\mathbf{x}, \mathbf{k})\}$ be the corresponding (periodic) eigenfunctions orthonormal in $L_2(\Omega)$. In what follows, all functions defined on Ω are assumed to be extended to $\mathbb{R}^d$ Γ-periodically. We shall use the notation $\psi_j(\mathbf{x}, \mathbf{k}) := e^{i\langle \mathbf{k}, \mathbf{x}\rangle} \varphi_j(\mathbf{x}, \mathbf{k})$.

The functions $E_j(\mathbf{k})$ are continuous and $\widetilde{\Gamma}$-periodic. We have

$$\sigma(M) = \bigcup_j \operatorname{Ran} E_j(\cdot). \tag{2.9}$$

Here the symbol $\sigma(M)$ stands for the spectrum of the operator M.

3. Main result

3.1. Preliminary constructions. In what follows we shall assume that the following conditions are fulfilled:

$(i_\circ)$ $\min \sigma(M) = 0$;
$(ii_\circ)$ $E_1(\mathbf{k}) = 0 \Leftrightarrow \mathbf{k} = 0 \ (\operatorname{mod} \widetilde{\Gamma})$;
$(iii_\circ)$ $E_2(0) > 0$.

Remark. Conditions $(i_\circ)$–$(iii_\circ)$ are quite standard (cf. [**1**], [**2**]) though they do not cover all possible situations. Adding an appropriate constant to V, one can always satisfy condition $(i_\circ)$. Condition $(ii_\circ)$ is not very restrictive as well. If $E_1(\mathbf{k}) = 0$ at the unique $(\operatorname{mod} \widetilde{\Gamma})$ point $\mathbf{k}_0$, then, applying an appropriate gauge transformation $M \mapsto e^{-i\langle \mathbf{k}_0, \mathbf{x}\rangle} M e^{i\langle \mathbf{k}_0, \mathbf{x}\rangle}$, one can provide the equality $E_1(0) = 0$. On the other hand, condition $(iii_\circ)$ is crucial for further considerations. It allows us to apply analytic perturbation theory (see below).

It follows from $(iii_\circ)$ that the function $E_1(\mathbf{k})$ is analytic in a neighborhood of zero and the function $\varphi_1(\mathbf{x}, \cdot)$ can also be chosen to be $\widetilde{H}^1(\Omega)$-analytic in a neighborhood of $\mathbf{k} = 0$ (see [**15**]). Next, it is known (see, e.g., Theorem 7.3.1 in [**17**]) that the function $\varphi_1(\cdot, \mathbf{k})$ belongs to the class $\widetilde{C}^s(\Omega)$ for some $s > 0$. (Here, to apply Theorem 7.3.1 in [**17**], we consider the equation for $\varphi_1(\cdot, \mathbf{k})$ as an elliptic system of equations with real-valued coefficients and diagonal principal part.) Then, by using the Cauchy formula, it can be shown (cf. Subsection 4.9 in [**1**]) that the function $\varphi_1(\mathbf{x}, \cdot)$ is $\widetilde{C}^s(\Omega)$-analytic in a neighborhood of $\mathbf{k} = 0$. We fix such a choice of the function φ_1 (normalized in $L_2(\Omega)$ for real-valued $\mathbf{k}$). We also fix $\delta > 0$ to be so small that the ball $\{\mathbf{k} : \ |\mathbf{k}| < \delta\}$ belongs to the corresponding neighborhood of $\mathbf{k} = 0$.

Due to conditions $(i_\circ)$, $(ii_\circ)$ and relation (2.9) the function $E_1(\mathbf{k})$ has a minimum at the point $\mathbf{k} = 0$. Usually it is supposed that this minimum is nondegenerate. The motivation is that for the case $\mathbf{A} = 0$ it is true. Besides, it is traditional to assume that generically the edges of internal gaps have proper structure, though there are no precise results. At the same time, an example of the magnetic Schrödinger operator with degenerate lower edge of the spectrum was presented in [**18**]. More precisely, the flattening with respect to one direction of $\mathbf{k}$ occurs in this example. In [**19**], the problem of homogenization for the operator with degenerate lower edge of the spectrum was considered. In the present paper, we investigate another threshold effect related to the asymptotics of a number of negative eigenvalues which appear when the periodic operator M is perturbed by a negative decaying potential. If the lower edge of the spectrum of M is degenerate, this leads to some effects which are not observed in the case of a proper lower edge of the spectrum. To simplify constructions, we consider only the case of degeneration of the fourth order (with respect to one or several variables). More precisely, we assume that the following condition holds:

$(iv_\circ)$ *The minimum of the function $E_1(\mathbf{k})$ at the point $\mathbf{k} = 0$ is expressed by an anisotropic symbol of the fourth order with respect to $k_1, \dots, k_m$, $1 \le m \le d$, and of the second order with respect to $k_{m+1}, \dots, k_d$; i.e., we have*

$$E_1(\mathbf{k}) = b(\mathbf{k})(1 + O(|\mathbf{k}|)), \quad |\mathbf{k}| < \delta/2, \tag{3.1}$$

where $b(\mathbf{k})$ is a polynomial of the fourth order satisfying the following conditions:

$$\begin{aligned} &b(\beta k_1, \dots, \beta k_m, \beta^2 k_{m+1}, \dots, \beta^2 k_d) = \beta^4 b(k_1, \dots, k_d), \quad \beta > 0, \\ &b(\mathbf{k}) \ge 0, \quad b(\mathbf{k}) = 0 \Leftrightarrow \mathbf{k} = 0, \quad \mathbf{k} \in \mathbb{R}^d. \end{aligned} \tag{3.2}$$

Put $B := b(\mathbf{D})$. We also introduce the operators $\Psi_j : L_2(\mathbb{R}^d) \to L_2(\widetilde{\Omega})$,

$$(\Psi_j u)(\mathbf{k}) := (\operatorname{meas} \widetilde{\Omega})^{-1/2} \int_{\mathbb{R}^d} \overline{\psi_j(\mathbf{x}; \mathbf{k})} u(\mathbf{x}) d\mathbf{x}, \quad j \in \mathbb{N}. \tag{3.3}$$

The operators Ψ_j, $j \in \mathbb{N}$, are partially isometric. Consider the ortho-projectors $P_j := \Psi_j^* \Psi_j$ in $L_2(\mathbb{R}^d)$. The projectors P_j are mutually orthogonal and $\sum_j P_j = I$. We have

$$M = \sum_j \Psi_j^* [E_j(\cdot)] \Psi_j. \tag{3.4}$$

Here by $[E_j(\cdot)]$ we denote the operator (in $L_2(\widetilde{\Omega})$) of multiplication by the function $E_j(\mathbf{k})$.

3.2. Classes of compact operators. The space of all bounded linear operators in a Hilbert space $\mathcal{H}$ is denoted by $\mathcal{B}$. The notation $\mathcal{S}_\infty$ stands for the space of compact operators in $\mathcal{H}$. Let $T \in \mathcal{S}_\infty$, and let $s_k(T)$ be the singular numbers of T, i.e., the consecutive eigenvalues (counted with multiplicities) of the operator $(T^*T)^{1/2}$. We put

$$n(s, T) := \operatorname{card}\{k : s_k(T) > s\}, \quad s > 0.$$

We denote by Σ_q, $0 < q < \infty$, the space (ideal) of compact operators distinguished by the condition

$$\|T\|_{\Sigma_q}^q := \sup_{s>0} s^q n(s, T) < \infty, \quad q > 0.$$

The space Σ_q is complete with respect to the quasi-norm $\|\cdot\|_{\Sigma_q}$ and nonseparable. The condition $\lim_{s\to 0} s^q n(s,T) = 0$ defines the separable subspace Σ_q^0 in Σ_q.

3.3. Setting of the problem. Main result. We consider the operator $M_\alpha := M - \alpha V_0$, where $V_0 \geq 0,\ \alpha > 0$. We assume that the following condition is fulfilled:

$$V_0^{1/2}(M+I)^{-1/2} \in \mathcal{S}_\infty. \tag{3.5}$$

Under condition (3.5) the operator M_α is defined via the closed semi-bounded form

$$m_\alpha[u,u] := m[u,u] - \alpha \int_{\mathbb{R}^d} V_0 |u|^2 d\mathbf{x}, \quad u \in H^1(\mathbb{R}^d). \tag{3.6}$$

Besides, $\sigma_{\mathrm{ess}}(M_\alpha) = \sigma_{\mathrm{ess}}(M)$. Here σ_{ess} stands for the essential spectrum. Let $N(\lambda; M_\alpha)$ be the number of eigenvalues of M_α lying below a point $\lambda \leq 0$. We are interested in the asymptotic behaviour of $N(0; M_\alpha)$ for large α. Denote

$$\Delta_q(0, M_\alpha) := \lim_{\alpha\to\infty} \sup \alpha^{-q} N(0; M_\alpha), \quad \delta_q(0, M_\alpha) := \lim_{\alpha\to\infty} \inf \alpha^{-q} N(0; M_\alpha).$$

We introduce the auxiliary operators $B := b(\mathbf{D})$ and $B_\alpha := B - \alpha V_0$. Let $W := V_0^{1/2}$ and $r_m := (2d-m)/4$. Here m is defined in $(iv_\circ)$. We consider the case $d \geq 3$ and assume that the following conditions hold:
(i) $r_m > 1$;
(ii) $W \in L_d(\mathbb{R}^d),\ d \geq 3$;
(iii) $WB^{-1/2} \in \Sigma_{2q},\ q > r_m$;
(iv) $W(B+I)^{-1/2} \in \Sigma_{2q}^0,\ q > r_m$.

Note that under condition (i), inclusion (iii) is sensible. Moreover, the conditions (ii)–(iv) are satisfied for a wide class of slowly decaying potentials V_0 (see, e.g., [**9**], [**10**]). Examples of such potentials and justification of (ii)–(iv) for them are explicit, but cumbersome. We do not go into details here and refer to [**9**], [**10**]. Condition (i) is technical. It will allow us to apply general estimates for integral operators. Similar but much more technical constructions can be done in the case $r_m \leq 1$. We assume (i) for the sake of simplicity.

Note also that under condition (i), conditions (iii) and (iv) are equivalent to
(iii') $N(0; B_\alpha) = O(\alpha^q)$;
(iv') $N(-1; B_\alpha) = o(\alpha^q)$

correspondingly. Finally, let $\mathcal{J}(g, V_0)$ be the standard Weyl asymptotic coefficient,

$$\mathcal{J}(g, V_0) := (2\pi)^{-d} \omega_d \int_{\mathbb{R}^d} V_0^{d/2} (\det g)^{-1/2} d\mathbf{x}, \quad \omega_d := \mathrm{vol}\{\mathbf{x} \in \mathbb{R}^d : \ |\mathbf{x}| < 1\}.$$

Remark. In all cases studied before, when the non-Weyl character of (iii') was entirely due to the behaviour of V_0 at infinity, it turned out that $N(-1)$ satisfied (iv'). On the contrary, if the non-Weyl character of (iii') is due to local singularities of V_0, then $N(0)$ and $N(-1)$ are of the same order (see [**14**]).

The following theorem is the main result of the present paper.

THEOREM 3.1. *Let conditions* (2.2)–(2.4), (2.6), $(i_\circ)$–$(iv_\circ)$, *and* (i)–(iv) *be fulfilled. Then*
a) for $q < d/2$,

$$\Delta_{d/2}(0; M_\alpha) = \delta_{d/2}(0; M_\alpha) = \mathcal{J}(g, V_0);$$

b) for $q > d/2$,

$$\Delta_q(0; M_\alpha) = \Delta_q(0; B - \alpha V_0(\mathbf{x})|\varphi_1(\mathbf{x}; 0)|^2),$$
$$\delta_q(0; M_\alpha) = \delta_q(0; B - \alpha V_0(\mathbf{x})|\varphi_1(\mathbf{x}; 0)|^2);$$

c) for $q = d/2$,

$$\Delta_{d/2}(0; M_\alpha) = \mathcal{J}(g, V_0) + \Delta_{d/2}(0; B - \alpha V_0(\mathbf{x})|\varphi_1(\mathbf{x}; 0)|^2),$$
$$\delta_{d/2}(0; M_\alpha) = \mathcal{J}(g, V_0) + \delta_{d/2}(0; B - \alpha V_0(\mathbf{x})|\varphi_1(\mathbf{x}; 0)|^2).$$

Remarks. 1) The main result is formulated in terms of the model (auxiliary) operator $B - \alpha V_0(\mathbf{x})|\varphi_1(\mathbf{x}; 0)|^2$. This situation is typical for such problems. We emphasize that the model operator is the differential operator with constant coefficients in the principal part. The order of the operator B is determined by the structure of the lower edge of the spectrum of the initial periodic operator M.

2) As we have mentioned, the condition (i) is technical and allows us to simplify the constructions. In the case $r_m \leq 1$, justification and even formulation of the main result are much more difficult. The model operator becomes more complicated. One can compare, for example, the results for the two-dimensional case (see [**8**], [**20**]) and the higher-dimensional case (see [**1**], [**2**]) for the similar problem for the nonmagnetic Schrödinger operator, where there is no degeneration of the lower edge of the spectrum and $r_m = d/2$.

3) The same question can be asked about the spectrum of M_α in the internal gaps of M. The construction in this situation would be similar to [**2**] and [**20**]. The case of internal gaps is more cumbersome and could become the goal of further investigations.

4) Theorem 3.1 shows the transition between "high-energy" asymptotics and "low-energy" asymptotics (the so-called "threshold effect"). If the potential V_0 is "fast" decaying ($q < d/2$), then we have the Weyl-type asymptotics, which has a high-energy origin. If $q > d/2$, then the asymptotics is determined by the structure of the lower edge of the spectrum of M. Finally, if $q = d/2$, then we have competition between two terms. The asymptotic coefficient is the sum of the Weyl-type term and the "threshold" term. We emphasize that the competition is impossible for the case of the proper lower edge of the spectrum if $d \geq 3$. The competition was observed in the two-dimensional case, but it had a different origin there. Our situation is more similar to the case of the three-dimensional Dirac operator discussed in [**6**].

4. Proof of Theorem 3.1

4.1. Reformulation of the problem. According to the Birman-Schwinger principle we have

$$N(0; M_\alpha) = n(1/\alpha; WM^{-1}W) = n(1/\alpha; M^{-1/2}V_0M^{-1/2}). \tag{4.1}$$

Thus, we consider the operator $Y := M^{-1/2}V_0M^{-1/2}$. Let χ be the characteristic function of the ball $\{\mathbf{k} : |\mathbf{k}| < \delta/2\}$. Put

$$\mathcal{X} := \Psi_1^*\chi\Psi_1, \quad \widetilde{\mathcal{X}} := 1 - \mathcal{X}, \quad \widetilde{\chi} := 1 - \chi;$$
$$Z := WM^{-1/2}, \quad Z_\chi := Z\mathcal{X}, \quad Z_{\widetilde{\chi}} := Z\widetilde{\mathcal{X}};$$
$$Y_\chi := Z_\chi^* Z_\chi, \quad Y_{\widetilde{\chi}} := Z_{\widetilde{\chi}}^* Z_{\widetilde{\chi}}, \quad \widehat{Y} := Z_\chi^* Z_{\widetilde{\chi}}.$$

We have

$$Y = Y_\chi + Y_{\widetilde{\chi}} + \widehat{Y} + \widehat{Y}^*.$$

4.2. Auxiliary statements. Before we proceed to estimating the operator Y, we prove two lemmas. Let $M_0 := M(g, 0, 0)$. We have

$$(4.2)\quad \begin{aligned}(M+I)^{-1} &= (M_0+I)^{-1}\\ &\quad - (M_0+I)^{-1/2}(U_1(M_0+I)^{-1/2})^*(U_2(M+I)^{-1/2})(M+I)^{-1/2},\end{aligned}$$

where

$$(4.3)\qquad U_1 := (-2g\mathbf{A}, g\mathbf{A}, \mathrm{sign}V|V|^{1/2})^T, \quad U_2 := (\mathbf{D}, \mathbf{A}, |V|^{1/2})^T.$$

Here we used condition (2.6). It follows from (2.2)–(2.4) that

$$(4.4)\qquad U_1(M_0+I)^{-1/2} \in \mathcal{S}_\infty, \quad U_2(M+I)^{-1/2} \in \mathcal{B}.$$

Denote

$$(4.5)\qquad K_1 := (U_1(M_0+I)^{-1/2})^*(U_2(M+I)^{-1/2}) \in \mathcal{S}_\infty$$

and

$$(4.6)\qquad K_2 := (M+I)^{-1/2}(M_0+I)^{1/2} \in \mathcal{B}.$$

LEMMA 4.1. *We have*

$$W(M+I)^{-1} \in \Sigma_d^0.$$

PROOF. It is well known (see, e.g., Proposition 3.1 in [**1**]) that under condition (ii) we have

$$(4.7)\qquad W(M_0+I)^{-1} \in \Sigma_d^0, \quad W(M_0+I)^{-1/2} \in \Sigma_d.$$

Then the statement of the lemma follows from (4.2), (4.5) and (4.7). □

LEMMA 4.2. *We have* $W(M+I)^{-1/2} \in \Sigma_d$ *and*

$$(4.8)\qquad \lim_{\alpha\to\infty} \alpha^{-d/2} n(1/\alpha; (M+I)^{-1/2}WW(M+I)^{-1/2}) = \mathcal{J}(g, V_0).$$

PROOF. We have (see (4.2))

$$(4.9)\quad \begin{aligned}&n(1/\alpha; (M+I)^{-1/2}WW(M+I)^{-1/2})\\ &\qquad = n(1/\alpha; W(M+I)^{-1}W) = n(1/\alpha; W(M_0+I)^{-1}W - G),\end{aligned}$$

where

$$G := W(M_0+I)^{-1/2}K_1K_2(M_0+I)^{-1/2}W.$$

It is known (see, e.g., Proposition 3.1 from [**1**]) that

$$\lim_{\alpha\to\infty} \alpha^{-d/2} n(1/\alpha; W(M_0+I)^{-1}W) = \mathcal{J}(g, V_0).$$

It follows from (4.5)–(4.7) that $G \in \Sigma_{d/2}^0$. This completes the proof of the lemma. □

We also need the estimate (see, e.g., Proposition 3.1 from [**1**])

$$\|W(M+I)^{-1/2}\|_{\Sigma_d} \le \|W(M_0+I)^{-1/2}\|_{\Sigma_d}\|K_2\| \le C\|W\|_{L_d(\mathbb{R}^d)}. \tag{4.10}$$

4.3. The operator Y.

1) The operator $Y_{\widetilde{\chi}}$. We have

$$M^{-1/2}\widetilde{\mathcal{X}} = (M+I)^{-1/2}\widetilde{\mathcal{X}} + (M+I)^{-1}T_1,$$

where

$$T_1 := \left(M^{1/2} + (M+I)^{1/2}\right)^{-1}(M+I)^{1/2}M^{-1/2}\widetilde{\mathcal{X}} \in \mathcal{B}.$$

Here we have $T_1 \in \mathcal{B}$, because of the presence of $\widetilde{\mathcal{X}}$. Next,

$$W(M+I)^{-1/2}\widetilde{\mathcal{X}} = W(M+I)^{-1/2} - W(M+I)^{-1}T_2,$$

where $T_2 := (M+I)^{1/2}\mathcal{X} \in \mathcal{B}$. Combining this with Lemma 4.1 and Lemma 4.2, we obtain

$$Y_{\widetilde{\chi}} \in \Sigma_{d/2}, \quad \lim_{\alpha\to\infty} \alpha^{-d/2} n(1/\alpha; Y_{\widetilde{\chi}}) = \mathcal{J}(g, V_0). \tag{4.11}$$

2) The operator Y_χ. We represent the operator Z_χ in the form $Z_\chi = K_\chi\Psi_1$, where K_χ is the integral operator with the kernel

$$K_\chi(\mathbf{x},\mathbf{k}) := (2\pi)^{-d/2}W(\mathbf{x})\chi(\mathbf{k})\varphi_1(\mathbf{x};\mathbf{k})e^{i\langle\mathbf{k},\mathbf{x}\rangle}(E_1(\mathbf{k}))^{-1/2}.$$

Introduce as well the operators $\widetilde{K}_\chi$, $\widehat{K}_\chi$ with the kernels

$$\widetilde{K}_\chi(\mathbf{x},\mathbf{k}) := (2\pi)^{-d/2}W(\mathbf{x})\chi(\mathbf{k})\varphi_1(\mathbf{x};\mathbf{k})e^{i\langle\mathbf{k},\mathbf{x}\rangle}(b(\mathbf{k}))^{-1/2},$$
$$\widehat{K}_\chi(\mathbf{x},\mathbf{k}) := (2\pi)^{-d/2}W(\mathbf{x})\chi(\mathbf{k})\varphi_1(\mathbf{x};0)e^{i\langle\mathbf{k},\mathbf{x}\rangle}(b(\mathbf{k}))^{-1/2}.$$

Now we estimate $\|K_\chi - \widetilde{K}_\chi\|_{\Sigma_{2q}}$. Note that, due to the $\widetilde{C}^s(\Omega)$-analyticity of $\varphi_1(\mathbf{x};\mathbf{k})$ in the ball $\{\mathbf{k} : |\mathbf{k}| < \delta/2\}$, the function $\chi(\mathbf{k})\varphi_1(\mathbf{x};\mathbf{k})$ is a multiplier for integral operators in all classes Σ_s (see, e.g., [**11**]). Thus, since $2q > 2r_m > 1$, it suffices to estimate the operator Ξ with the kernel

$$\begin{aligned}\Xi(\mathbf{x},\mathbf{k}) &:= (2\pi)^{-d/2}W(\mathbf{x})\chi(\mathbf{k})e^{i\langle\mathbf{k},\mathbf{x}\rangle}\left((E_1(\mathbf{k}))^{-1/2} - (b(\mathbf{k}))^{-1/2}\right)\\ &= (2\pi)^{-d/2}W(\mathbf{x})e^{i\langle\mathbf{k},\mathbf{x}\rangle}(b(\mathbf{k}))^{-1/2}\chi(\mathbf{k})\left(\left(\frac{b(\mathbf{k})}{E_1(\mathbf{k})}\right)^{1/2} - 1\right).\end{aligned}$$

By (iii), we have $\|WB^{-1/2}\|_{\Sigma_{2q}} \le C$. Therefore, taking (3.1) into account, we obtain

$$\|\Xi\|_{\Sigma_{2q}} \le C\max_{\mathbf{k}}\chi(\mathbf{k})\left(\left(\frac{b(\mathbf{k})}{E_1(\mathbf{k})}\right)^{1/2} - 1\right) \le \widetilde{C}\delta.$$

Thus,

$$\|K_\chi - \widetilde{K}_\chi\|_{\Sigma_{2q}} \le C\delta. \tag{4.12}$$

Next,

$$\chi(\mathbf{k})\left(\varphi_1(\mathbf{x};\mathbf{k}) - \varphi_1(\mathbf{x};0)\right) =: \sum_{j=1}^{d} k_j\chi(\mathbf{k})\widetilde{\varphi}_j(\mathbf{x};\mathbf{k}).$$

Due to the $\widetilde{C}^s(\Omega)$-analyticity of $\varphi_1(\mathbf{x};\mathbf{k})$ in the ball $\{\mathbf{k}:\ |\mathbf{k}|<\delta/2\}$, the functions $\chi(\mathbf{k})\widetilde{\varphi}_j(\mathbf{x};\mathbf{k})$ are multipliers in Σ_{2q}. Then, by analogy with (4.11), we have

$$\|\widetilde{K}_\chi - \widehat{K}_\chi\|_{\Sigma_{2q}} \le C\delta. \tag{4.13}$$

Let L be the operator with the kernel

$$L(\mathbf{x},\mathbf{k}) := (2\pi)^{-d/2} W(\mathbf{x})\varphi_1(\mathbf{x};0)e^{i\langle\mathbf{k},\mathbf{x}\rangle}(b(\mathbf{k}))^{-1/2}.$$

Then

$$\begin{aligned} L(\mathbf{x},\mathbf{k}) &- \widehat{K}_\chi(\mathbf{x},\mathbf{k}) \\ &= (2\pi)^{-d/2} W(\mathbf{x})\varphi_1(\mathbf{x};0)e^{i\langle\mathbf{k},\mathbf{x}\rangle}(b(\mathbf{k})+1)^{-1/2}\left(\frac{b(\mathbf{k})+1}{b(\mathbf{k})}\right)^{1/2}\widetilde{\chi}(\mathbf{k}). \end{aligned}$$

Note that the functions $\varphi_1(\mathbf{x};0)$ and $\left(\frac{b(\mathbf{k})+1}{b(\mathbf{k})}\right)^{1/2}\widetilde{\chi}(\mathbf{k})$ are bounded. By (iv), we have $W(B+I)^{-1/2}\in\Sigma^0_{2q}$. Thus,

$$L - \widehat{K}_\chi \in \Sigma^0_{2q}. \tag{4.14}$$

Finally, since Ψ_1 is partially isometric, it follows from the definition of the operator Z_χ that *the operators $K^*_\chi K_\chi$ and $\Psi_1^* K_\chi^* K_\chi \Psi_1 (= Y_\chi)$ are unitarily equivalent on the orthogonal complements of their kernels.* Combined with (4.11)–(4.13), this implies that

$$\begin{aligned} \lim_{\delta\to0}\left(\limsup_{\alpha\to\infty}\alpha^{-q}n(1/\alpha;Y_\chi)\right) &= \lim_{\delta\to0}\left(\limsup_{\alpha\to\infty}\alpha^{-q}n(1/\alpha;K^*_\chi K_\chi)\right) \\ &= \limsup_{\alpha\to\infty}\alpha^{-q}n(1/\alpha;L^*L), \end{aligned} \tag{4.15}$$

$$\begin{aligned} \lim_{\delta\to0}\left(\liminf_{\alpha\to\infty}\alpha^{-q}n(1/\alpha;Y_\chi)\right) &= \lim_{\delta\to0}\left(\liminf_{\alpha\to\infty}\alpha^{-q}n(1/\alpha;K^*_\chi K_\chi)\right) \\ &= \liminf_{\alpha\to\infty}\alpha^{-q}n(1/\alpha;L^*L). \end{aligned} \tag{4.16}$$

Finally, note that

$$n(1/\alpha;L^*L) = N(0;B-\alpha V_0(\mathbf{x})|\varphi_1(\mathbf{x};0)|^2). \tag{4.17}$$

3) The operator $\widehat{Y}$. It follows from the constructions in cases 1) and 2) that $\widehat{Y}\in\Sigma^0_{d/2}$ for $q<d/2$ and $\widehat{Y}\in\Sigma^0_q$ for $q>d/2$. Thus, we consider the case $q=d/2$. Then it again follows from 1) and 2) that

$$\widehat{Y} - \Psi_1^* L^* W(M+I)^{-1/2} = \widehat{Y}_1 + \widehat{Y}_2, \quad \text{where } \widehat{Y}_1\in\Sigma^0_{d/2},\ \|\widehat{Y}_2\|_{\Sigma_{d/2}}\le C\delta.$$

We need to show that $L^*W(M+I)^{-1/2}\in\Sigma^0_{d/2}$. Due to the estimate $\|L\|_{\Sigma_d}\le C$ and (4.9) it suffices to check the inclusion $Q := L^*\widetilde{W}(M+I)^{-1/2}\in\Sigma^0_{d/2}$, where $\widetilde{W}\in C_0^\infty(\mathbb{R}^d)$. We rewrite Q in the form $Q=\widetilde{L}^*W(M+I)^{-1/2}$, where $\widetilde{L}$ is the operator with the kernel

$$\widetilde{L}(\mathbf{x},\mathbf{k}) := (2\pi)^{-d/2}\widetilde{W}(\mathbf{x})\varphi_1(\mathbf{x};0)e^{i\langle\mathbf{k},\mathbf{x}\rangle}(b(\mathbf{k}))^{-1/2}.$$

It follows from (i), (3.2) and inclusion (5.2) from [**5**] that $\widetilde{L}\in\Sigma_{2r_m}\subset\Sigma^0_d$. This and (4.9) imply that $Q\in\Sigma^0_{d/2}$.

4.4. Proof of Theorem 3.1.

Proof of a). For $q < d/2$ it follows from 2) and 3) that Y_χ, $\widehat{Y} \in \Sigma^0_{d/2}$. Thus, the statement a) of the theorem follows from (4.1), (4.10).

Proof of b). For $q > d/2$ it follows from 1) and 3) that $Y_{\widetilde{\chi}}$, $\widehat{Y} \in \Sigma^0_q$. Thus, the statement b) of the theorem follows from (4.1), (4.14)–(4.16).

Proof of c). For $q = d/2$ it follows from 3) that

$$Y = Y_\chi + Y_{\widetilde{\chi}} + Y_1 + Y_2, \quad \text{where} \quad Y_1 \in \Sigma^0_{d/2}, \ \|Y_2\|_{\Sigma_{d/2}} \le C\delta.$$

Thus, we have

$$\begin{aligned} &\limsup_{\alpha\to\infty} \alpha^{-d/2} n(1/\alpha; Y) = \lim_{\delta\to 0}\left(\limsup_{\alpha\to\infty} \alpha^{-d/2} n(1/\alpha; Y_\chi + Y_{\widetilde{\chi}})\right), \\ &\liminf_{\alpha\to\infty} \alpha^{-d/2} n(1/\alpha; Y) = \lim_{\delta\to 0}\left(\liminf_{\alpha\to\infty} \alpha^{-d/2} n(1/\alpha; Y_\chi + Y_{\widetilde{\chi}})\right). \end{aligned} \tag{4.18}$$

The operators Y_χ and $Y_{\widetilde{\chi}}$ are mutually orthogonal. Therefore,

$$n(1/\alpha; Y_\chi + Y_{\widetilde{\chi}}) = n(1/\alpha; Y_\chi) + n(1/\alpha; Y_{\widetilde{\chi}}). \tag{4.19}$$

Now, the statement c) of the theorem follows from (4.1), (4.10), and (4.14)–(4.18).

Acknowledgements

The author is sincerely grateful to M. Sh. Birman whose papers stimulated the author's interest in the topics. The author also thanks T. A. Suslina for fruitful discussions during preparation of the present text.

References

[1] M. Sh. Birman, *Discrete spectrum in the gaps of the perturbed periodic Schrödinger operator. I. Regular perturbations*, Boundary value problems, Schrödinger operators, deformation quantization, Math. Top. **8** (1995), Akademie Verlag, Berlin, 334–352.

[2] ———, *Discrete spectrum in the gaps of the perturbed periodic Schrödinger operator. II. Nonregular perturbations*, Algebra i Analiz **9** (1997), no. 6, 62–89; English transl., St. Petersburg Math. J. **9** (1998), no. 6, 1073–1095.

[3] ———, *Discrete spectrum in the gaps of a continuous one for perturbations with large coupling constant*, Estimates and Asymptotics for Discrete Spectra of Integral and Differential Equations, Adv. Soviet Math., vol. 7, Amer. Math. Soc., Providence, RI, 1991, 57–73.

[4] ———, *Discrete spectrum of the periodic Schrödinger operator perturbed by a decreasing potential*, Algebra i Analiz **8** (1996), no. 1, 3–20; English transl., St. Petersburg Math. J. **8** (1997), no. 1, 1–14.

[5] M. Sh. Birman, G. E. Karadzhov, and M. Z. Solomyak, *Boundedness conditions and spectrum estimates for the operators $b(X)a(D)$ and their analogs*, Estimates and Asymptotics for Discrete Spectra of Integral and Differential Equations, Adv. Soviet Math., vol. 7, Amer. Math. Soc., Providence, RI, 1991, 85–106.

[6] M. Sh. Birman and A. Laptev, *Discrete spectrum of the perturbed Dirac operator*, Ark. Math. **32** (1994), no. 1, 13–32.

[7] ———, *The negative discrete spectrum of a two-dimensional Schrödinger operator*, Comm. Pure Appl. Math. **49** (1996), no. 9, 967–997.

[8] M. Sh. Birman, A. Laptev, and T. A. Suslina, *Discrete spectrum of a two-dimensional periodic elliptic second-order operator perturbed by a decreasing potential. I. Semi-infinite gap*, Algebra i Analiz **12** (2000), no. 4, 36–78; English transl., St. Petersburg Math. J. **12** (2001), no. 4, 535–567.

[9] M. Sh. Birman and M. Z. Solomyak, *Asymptotics of the spectra of pseudo-differential operators with anisotropic-homogeneous symbols*, Vestnik Leningrad Univ. **13**, no. 3 (1977), 13–21.

[10] ———, *Estimates for the number of negative eigenvalues of the Schrödinger operator and its generalizations*, Estimates and Asymptotics for Discrete Spectra of Integral and Differential Equations, Adv. Soviet Math., vol. 7, Amer. Math. Soc., Providence, 1991, RI, 1–55.

[11] ———, *Estimates for the singular numbers of integral operators*, Uspekhi Mat. Nauk **32** (1977), no. 1 (193), 17–84.

[12] M. Sh. Birman and T. A. Suslina, *Threshold effects near the lower edge of the spectrum for periodic differential operators of mathematical physics*, Operator Theory: Adv. Appl. **129** (2001), Birkhäuser, 71–107.

[13] ———, *Periodic second-order differential operators. Threshold properties and averaging*, Algebra i Analiz **15** (2003), no. 5, 1–108 (Russian); Engl. transl. in St. Petersburg Math. J. **15** (2004), no. 5, 639–714.

[14] M. Sh. Birman and T. Weidl, *The discrete spectrum in a gap of the continuous one for compact supported perturbations*, Mathematical Results in Quantum Mechanics (Blossin, 1993), Oper. Theory Adv. Appl., vol. 70, Birkhäuser, Basel, 1994, 9–12.

[15] T. Kato, *Perturbation theory for linear operators*, Springer-Verlag, Berlin, 1995.

[16] M. Klaus, *On the point spectrum of Dirac operators*, Helv. Phys. Acta **53** (1980), 453–462.

[17] O. A. Ladyzhenskaya and N. N. Ural′tseva, *Linear and quasilinear equations of elliptic type*, 2nd ed., "Nauka", Moscow, 1973; English transl. of 1st ed., Acad. Press, New York-London, 1968.

[18] R. G. Shterenberg, *An example of a periodic magnetic Schrödinger operator with a degenerate lower edge of the spectrum*, Algebra i Analiz **16** (2004), no. 2, 177–185; English transl., St. Petersburg Math. J. **16** (2005), no. 2, 417–422.

[19] ———, *On the homogenization of a periodic magnetic Schrödinger operator with a degenerate lower edge of the spectrum*, Zapiski Nauchn. Sem. POMI **336** (2006), 264–273.

[20] T. A. Suslina, *Discrete spectrum of a two-dimensional periodic elliptic second order operator perturbed by a decaying potential. II. Internal gaps*, Algebra i Analiz **15** (2003), no. 2, 128–189; English transl., St. Petersburg Math. J. **15** (2004), no. 2, 249–287.

[21] G. M. Troianiello, *Elliptic differential equations and obstacle problems*, University Series in Mathematics, Plenum Press, New York, Vol. 75, 1987.

Department of Mathematics, University of Alabama at Birmingham, Birmingham, Alabama 35294, USA

E-mail address: shterenb@math.uab.edu

Amer. Math. Soc. Transl.
(2) Vol. **225**, 2008

Homogenization of Periodic Second Order Differential Operators Including First Order Terms

T. A. Suslina

To Mikhail Shlëmovich Birman on his 80th birthday

ABSTRACT. In $L_2(\mathbb{R}^d;\mathbb{C}^n)$, we consider matrix periodic elliptic second order differential operators $\mathcal{B}_\varepsilon$ with rapidly oscillating coefficients (depending on $\mathbf{x}/\varepsilon$). The operator $\mathcal{B}_\varepsilon$ is of the form $\mathcal{B}_\varepsilon = \mathcal{A}_\varepsilon + \mathcal{Y}_\varepsilon + \mathcal{Y}_\varepsilon^* + c_0 Q^\varepsilon$, where the second order operator $\mathcal{A}_\varepsilon = \mathcal{X}_\varepsilon^* \mathcal{X}_\varepsilon$ is given in a factorized form, the first order operator $\mathcal{Y}_\varepsilon$ is subordinate to the operator $\mathcal{X}_\varepsilon$, and Q^ε is the operator of multiplication by the positive periodic matrix. The *homogenization problem* in the small period limit is studied. It is shown that the inverse operator $\mathcal{B}_\varepsilon^{-1}$ converges as $\varepsilon \to 0$ to $(\mathcal{B}^0)^{-1}$ in the operator norm in $L_2(\mathbb{R}^d;\mathbb{C}^n)$, where $\mathcal{B}^0$ is the effective operator with constant coefficients. For the norm of the difference $\mathcal{B}_\varepsilon^{-1} - (\mathcal{B}^0)^{-1}$, an estimate of order ε is obtained.

§0. Introduction

0.1. Main result. We study the homogenization problem for matrix differential operators (DO's) $\mathcal{B}_\varepsilon$, $\varepsilon > 0$, acting in the space $L_2(\mathbb{R}^d;\mathbb{C}^n)$. The principal part $\mathcal{A}_\varepsilon$ of the operator $\mathcal{B}_\varepsilon$ is given in a factorized form: $\mathcal{A}_\varepsilon = b(\mathbf{D})^* g(\varepsilon^{-1}\mathbf{x}) b(\mathbf{D})$, where $b(\mathbf{D})$ is a first order homogeneous matrix DO, and $g(\mathbf{x})$ is a bounded and positive periodic matrix-valued function. (For conditions on $b(\mathbf{D})$ and $g(\mathbf{x})$, see §4.) Homogenization problems for the operators $\mathcal{A}_\varepsilon$ have been studied in detail in [**BSu1, BSu2, BSu4, BSu5**]. Now we consider more general operators $\mathcal{B}_\varepsilon$, adding first order and zero order terms to $\mathcal{A}_\varepsilon$. Namely, let $\mathcal{Y}_\varepsilon = \sum_{j=1}^d a_j(\varepsilon^{-1}\mathbf{x}) D_j$, where $a_j(\mathbf{x})$, $j = 1,\dots,d$, are bounded periodic matrix-valued functions, and let $Q^\varepsilon(\mathbf{x}) = Q(\varepsilon^{-1}\mathbf{x})$, where $Q(\mathbf{x})$ is a bounded and positive definite periodic matrix-valued function. The operator

$$\mathcal{B}_\varepsilon = \mathcal{A}_\varepsilon + \mathcal{Y}_\varepsilon + \mathcal{Y}_\varepsilon^* + c_0 Q^\varepsilon \tag{0.1}$$

2000 *Mathematics Subject Classification.* Primary 35B27.

Key words and phrases. Periodic operators, threshold approximations, homogenization, effective operator.

Supported by RFBR (grant no. 08-01–00209-a) and the grant "Scientific schools" (grant no. 816.2008.1).

is *our main object*. The constant c_0 is chosen so that the operator (0.1) is positive definite. The coefficients of the operator (0.1) are rapidly oscillating as $\varepsilon \to 0$. *The homogenization problem for the operator* (0.1) *is to approximate the inverse operator* $\mathcal{B}_\varepsilon^{-1}$ *as* $\varepsilon \to 0$. We show that $\mathcal{B}_\varepsilon^{-1}$ converges to $(\mathcal{B}^0)^{-1}$ in the $L_2(\mathbb{R}^d; \mathbb{C}^n)$-operator norm as $\varepsilon \to 0$. Here $\mathcal{B}^0$ is the *effective* operator with constant coefficients:

$$\mathcal{B}^0 = \mathcal{A}^0 + \mathcal{Y}^0 + (\mathcal{Y}^0)^* + c_0\overline{Q} - W.$$

The operator $\mathcal{A}^0 = b(\mathbf{D})^* g^0 b(\mathbf{D})$ is the effective operator for $\mathcal{A}_\varepsilon$, and $\overline{Q}$ is the mean value of the matrix $Q(\mathbf{x})$ over Ω, where Ω is the cell of the lattice of periods. The first order "homogenized" operator $\mathcal{Y}^0$ with constant coefficients and the matrix W are defined in a nontrivial way (see (6.11), (6.12), (7.2)). For the error we obtain the following sharp-order estimate in the $L_2(\mathbb{R}^d; \mathbb{C}^n)$-operator norm:

$$\|\mathcal{B}_\varepsilon^{-1} - (\mathcal{B}^0)^{-1}\|_{L_2 \to L_2} \leq C\varepsilon, \quad \varepsilon > 0. \tag{0.2}$$

The constant in estimate (0.2) is well controlled in terms of the problem data. *Estimate* (0.2) *is the main result of the paper.*

Note that the operator

$$\widetilde{\mathcal{B}}_\varepsilon = \mathcal{A}_\varepsilon + \varepsilon^{-1} p^\varepsilon + c_0 Q^\varepsilon, \tag{0.3}$$

which contains a "singular" potential $\varepsilon^{-1}p^\varepsilon(\mathbf{x}) = \varepsilon^{-1}p(\varepsilon^{-1}\mathbf{x})$, can also be reduced to the form (0.1). Here $p(\mathbf{x})$ is a Hermitian periodic matrix-valued function with zero mean value. In general, the corresponding effective operator contains nontrivial first order terms (see Subsections 8.4 and 8.5).

0.2. In the absence of first order terms (i.e., if $\mathcal{Y}_\varepsilon = 0$), for the first time, an estimate of the form (0.2) has been established in [**BSu1**], [**BSu2**] by the spectral method. Later, by a different method suggested by V. V. Zhikov, such an estimate was obtained in [**Zh**], [**ZhPas**] for the acoustics operator and the operator of elasticity theory (for $\mathcal{Y}_\varepsilon = 0$, $Q^\varepsilon = 1$).

Homogenization problems for a scalar periodic second order operator including first order terms were studied in the book [**BeLP**, Chapter 1, Section 13]. In [**BeLP**], convergence of solutions (in some weak sense) was obtained, but the error estimates were not studied.

In the presence of first order terms, the homogenization problem for the operator (0.1) has been studied in the recent paper [**Bo**] by D. Borisov. The method of [**Bo**] is close to some technical tricks from the papers of V. V. Zhikov. In [**Bo**], an expression for the effective operator $\mathcal{B}^0$ is found and estimate (0.2) is proved. Moreover, in [**Bo**] it is assumed that coefficients of the operator depend both on rapid and slow variables. However, in [**Bo**] it is assumed that coefficients of the operator are sufficiently smooth. Herewith, the constant C in estimate (0.2) depends on the corresponding "smooth" norms of the coefficients.

In the present paper, the homogenization problem for the operator (0.1) is studied by the method which develops the approach of [**BSu1**], [**BSu2**]. Smoothness assumptions on coefficients are not needed. We prove estimate (0.2) with the constant C, which depends on the norms of the coefficients $g, g^{-1}, \{a_j\}, Q, Q^{-1}$ in L_∞, on the constants α_0, α_1 (see (4.1)), and on the parameters of the lattice.

0.3. Method of investigation. Our method is a further development of the scheme suggested in [**BSu1, BSu2**] (and developed in [**BSu3, BSu4, BSu5**]). Using the scaling transformation, we reduce the problem to the study of the operator

$$\mathcal{B}(\varepsilon) = \mathcal{A} + \varepsilon(\mathcal{Y} + \mathcal{Y}^*) + \varepsilon^2 c_0 Q, \tag{0.4}$$

acting in $L_2(\mathbb{R}^d; \mathbb{C}^n)$. The coefficients of this operator depend on $\mathbf{x}$ (not on $\varepsilon^{-1}\mathbf{x}$). Next, by Floquet–Bloch theory, we decompose the operator (0.4) in the direct integral of the operators $\mathcal{B}(\mathbf{k}, \varepsilon)$ acting in $L_2(\Omega; \mathbb{C}^n)$ and depending on the parameter $\mathbf{k} \in \mathbb{R}^d$ (the quasi-momentum). Following [**BSu1, BSu2**], we distinguish the one-dimensional parameter $t = |\mathbf{k}|$. We study the family of operators $\mathcal{B}(\mathbf{k}, \varepsilon)$ as a quadratic operator pencil with respect to the parameters t and ε. Herewith, a good deal of considerations can be fulfilled in the framework of an abstract operator-theoretic scheme.

In the abstract scheme, we consider the operator pencil of the form

$$B(t, \varepsilon) = A(t) + \varepsilon(Y(t) + Y(t)^*) + \varepsilon^2 c_0 Q, \tag{0.5}$$

acting in a separable Hilbert space $\mathfrak{H}$. We assume that the principal part of the operator (0.5) admits a factorization $A(t) = X(t)^* X(t)$, where $X(t) = X_0 + tX_1$. We assume also that the operator $Y(t) = Y_0 + tY_1$ is "subordinate" to $X(t)$. (See §1.) The lower term Q is a bounded and positive definite operator. *We introduce the one-dimensional parameter* $\tau = (t^2 + \varepsilon^2)^{1/2}$ and study the operator family (0.5) by means of analytic perturbation theory with respect to τ. We distinguish the selfadjoint operator S (the spectral germ of the operator (0.5)), acting in the finite-dimensional space $\mathfrak{N} = \operatorname{Ker} A(0)$. It is possible to approximate the operator $B(t, \varepsilon)^{-1}$ by some finite rank operator given in terms of the spectral germ. The corresponding approximation (Theorem 3.2) is the main result of the abstract part of the paper. Using this abstract theorem, we obtain the required result concerning homogenization of the operator $\mathcal{B}_\varepsilon$.

In [**BSu1, BSu2**], first the operator pencil $A(t)$ was studied by means of analytic perturbation theory with respect to the parameter t, and next the resolvent $(A(t) + \varepsilon^2 I)^{-1}$ or the generalized resolvent $(A(t) + \varepsilon^2 Q)^{-1}$ was studied. Such an approach allowed us to take into account the zero order terms, but not the first order terms. Now we suggest introducing the parameter $\tau = (t^2 + \varepsilon^2)^{1/2}$ and considering the operator family (0.5) as a whole. This new approach might be useful also in the previous situation.

To a large extent, the present paper relies on the ideology and technique of [**BSu2**], and we will refer to [**BSu2**] systematically.

0.4. The structure of the paper. The paper contains two chapters. Chapter 1 (§§1–3) is devoted to the abstract operator-theoretic method. In §1, the main objects are introduced. In §2, estimates for the difference of resolvents on the contour are obtained. In §3, threshold approximations for the operator $B(t, \varepsilon)$ are deduced from these estimates. Chapter 2 (§§4–8) is devoted to the homogenization problem for the periodic operator $\mathcal{B}_\varepsilon$. In §4, periodic operators acting in $L_2(\mathbb{R}^d; \mathbb{C}^n)$ are introduced. The operator family $\mathcal{B}(\varepsilon)$ is defined (see (0.4)). In §5, the operator (0.4) is decomposed into the direct integral. The operators $\mathcal{B}(\mathbf{k}, \varepsilon)$ acting in $L_2(\Omega; \mathbb{C}^n)$ are introduced. In §6, the operator $\mathcal{B}(\mathbf{k}, \varepsilon)$ is included in the framework of the abstract scheme. Applying the abstract Theorem 3.2, we obtain a finite rank approximation for the operator $\mathcal{B}(\mathbf{k}, \varepsilon)^{-1}$ (Theorem 6.1). In §7, by the direct integral decomposition, an approximation for the operator $\mathcal{B}(\varepsilon)^{-1}$ in the $L_2(\mathbb{R}^d;\mathbb{C}^n)$-operator norm is deduced from Theorem 6.1 (see Theorem 7.1). Finally, in §8, using the scaling transformation, we deduce the main result (estimate (0.2)) from Theorem 7.1 (see Theorem 8.1). The case of the operator (0.3) is considered in Subsections 8.4 and 8.5.

0.5. Notation. Let $\mathfrak{H}$ and $\mathfrak{G}$ be two separable Hilbert spaces. The symbols $(\cdot,\cdot)_{\mathfrak{H}}$ and $\|\cdot\|_{\mathfrak{H}}$ denote the inner product and the norm in $\mathfrak{H}$; the symbol $\|\cdot\|_{\mathfrak{H}\to\mathfrak{G}}$ stands for the norm of a bounded operator from $\mathfrak{H}$ to $\mathfrak{G}$. Sometimes, we omit the indices if this does not lead to confusion. Let $I = I_{\mathfrak{H}}$ be the identity operator on $\mathfrak{H}$. If $\mathfrak{N}$ is a subspace in $\mathfrak{H}$, then $\mathfrak{N}^\perp = \mathfrak{H}\ominus\mathfrak{N}$. If P is the orthogonal projection of $\mathfrak{H}$ onto $\mathfrak{N}$, then $P^\perp$ is the orthogonal projection onto $\mathfrak{N}^\perp$. The symbol $\langle\cdot,\cdot\rangle$ stands for the standard inner product in $\mathbb{C}^n$; $|\cdot|$ denotes the norm of a vector in $\mathbb{C}^n$; the unit $(n\times n)$-matrix is denoted by $\mathbf{1}_n$. If a is an $(n\times n)$-matrix, then $|a|$ denotes the norm of the matrix a as an operator in $\mathbb{C}^n$. Next, we use the notation $\mathbf{x} = (x^1,\dots,x^d)\in\mathbb{R}^d$, $iD_j = \partial_j = \partial/\partial x^j$, $j = 1,\dots,d$, $\mathbf{D} = -i\nabla = (D_1,\dots,D_d)$. The L_p-classes of $\mathbb{C}^n$-valued functions in a domain $\mathcal{O}\subset\mathbb{R}^d$ are denoted by $L_p(\mathcal{O};\mathbb{C}^n)$, $1\le p\le\infty$. The Sobolev classes are denoted by $H^s(\mathcal{O};\mathbb{C}^n)$. Sometimes, we use the simplified notation $L_p(\mathcal{O})$, $H^s(\mathcal{O})$ for classes of vector-valued or matrix-valued functions.

Chapter 1. Abstract operator-theoretic scheme

§1. Quadratic two-parameter operator families

1.1. The operators $X(t)$ and $A(t)$. Let $\mathfrak{H}$ and $\mathfrak{H}_*$ be complex separable Hilbert spaces. Suppose that $X_0 : \mathfrak{H}\to\mathfrak{H}_*$ is a densely defined and closed operator and that $X_1 : \mathfrak{H}\to\mathfrak{H}_*$ is a bounded operator. Then the operator $X(t) = X_0 + tX_1 : \mathfrak{H}\to\mathfrak{H}_*$, $t\in\mathbb{R}$, is closed on the domain $\operatorname{Dom} X_0$. Consider the selfadjoint operator $A(t) = X(t)^*X(t)$ in $\mathfrak{H}$. The operator $A(t)$ is generated by the closed quadratic form $\|X(t)u\|^2_{\mathfrak{H}_*}$, $u\in\operatorname{Dom}X_0$. For the operator $A_0 := A(0) = X_0^*X_0$, we put $\mathfrak{N} = \operatorname{Ker} A_0 = \operatorname{Ker} X_0$. We assume that the following condition is fulfilled.

CONDITION 1.1. *The point $\lambda = 0$ is an isolated point in the spectrum of A_0, and $0 < n := \dim\mathfrak{N} < \infty$.*

Let d^0 be the distance from the point $\lambda = 0$ to the rest of the spectrum of A_0. We put $\mathfrak{N}_* = \operatorname{Ker} X_0^*$, $n_* := \dim\mathfrak{N}_*$. *We assume that* $n\le n_*\le\infty$. Let P and P_* be the orthogonal projections of $\mathfrak{H}$ onto $\mathfrak{N}$ and of $\mathfrak{H}_*$ onto $\mathfrak{N}_*$, respectively.

The operator family $A(t)$ has been studied in detail in [**BSu2**, Chapter 1], and further in [**BSu3**] and [**BSu5**, Chapter 1].

1.2. The operators $Y(t)$ and Q. Suppose that $Y_0 : \mathfrak{H}\to\mathfrak{H}$ is a densely defined operator such that $\operatorname{Dom}X_0\subset\operatorname{Dom}Y_0$. Let Y_1 be a bounded operator in $\mathfrak{H}$. We put $Y(t) = Y_0 + tY_1$, $\operatorname{Dom}Y(t) = \operatorname{Dom}Y_0$. We impose the following condition.

CONDITION 1.2. *For some $c_1 > 0$, we have*

$$\|Y(t)u\|_{\mathfrak{H}} \le c_1\|X(t)u\|_{\mathfrak{H}_*},\quad u\in\operatorname{Dom}X_0,\quad t\in\mathbb{R}. \tag{1.1}$$

Estimate (1.1) with $t = 0$ implies that $\operatorname{Ker}X_0\subset\operatorname{Ker}Y_0$, i.e., $Y_0P = 0$.

Finally, let Q be a bounded and positive definite operator in $\mathfrak{H}$.

1.3. The operator $B(t,\varepsilon)$. The two-parameter pencil of selfadjoint operators

$$B(t,\varepsilon) = A(t) + \varepsilon(Y(t) + Y(t)^*) + c_0\varepsilon^2 Q, \quad t,\varepsilon \in \mathbb{R}, \tag{1.2}$$

in $\mathfrak{H}$ is *our main object* in Chapter 1. The constant c_0 is chosen so that

$$c_0 \geq \frac{5}{2}c_1^2\|Q^{-1}\|. \tag{1.3}$$

By definition, the operator $B(t,\varepsilon)$ is generated by the closed quadratic form

$$b(t,\varepsilon)[u,u] = \|X(t)u\|^2_{\mathfrak{H}_*} + 2\varepsilon\mathrm{Re}\,(Y(t)u,u)_{\mathfrak{H}} + \varepsilon^2 c_0(Qu,u)_{\mathfrak{H}}, \quad u \in \mathrm{Dom}\, X_0. \tag{1.4}$$

Using (1.1) and (1.3), it is easy to check that

$$b(t,\varepsilon)[u,u] \geq \frac{1}{2}\|X(t)u\|^2_{\mathfrak{H}_*} + \frac{c_1^2}{2}\varepsilon^2\|u\|^2_{\mathfrak{H}}, \quad u \in \mathrm{Dom}\, X_0. \tag{1.5}$$

Thus, the operator $B(t,\varepsilon)$ is positive.

1.4. Passage to the parameter τ. Obviously, the family (1.2) is an analytic operator family with respect to the parameters t and ε. If $t = 0$ and $\varepsilon = 0$, the operator (1.2) coincides with A_0 and has an isolated eigenvalue $\lambda = 0$ of multiplicity n. We want to use the arguments of analytic perturbation theory. However, if $n > 1$, then analytic perturbation theory does not work, since we deal with the multidimensional parameter and the multiple eigenvalue. In order to avoid this difficulty, we introduce the *one-dimensional parameter* $\tau = (t^2 + \varepsilon^2)^{1/2}$, and also the additional parameters $\vartheta_1 = t\tau^{-1}$, $\vartheta_2 = \varepsilon\tau^{-1}$, $\vartheta = (\vartheta_1, \vartheta_2)$. Then (1.2) can be rewritten as $B(t,\varepsilon) = B(\tau;\vartheta)$,

$$B(\tau;\vartheta) = (X_0^* + \tau\vartheta_1 X_1^*)(X_0 + \tau\vartheta_1 X_1) + \tau\vartheta_2(Y_0 + Y_0^*) + \tau^2\vartheta_1\vartheta_2(Y_1 + Y_1^*) + \tau^2 c_0\vartheta_2^2 Q. \tag{1.6}$$

Now the corresponding form (1.4) will be denoted by $b(\tau;\vartheta)$. We study the operator (1.6) as a quadratic pencil with respect to the one-dimensional parameter τ, by means of analytic perturbation theory. At the same time, we must make our constructions and estimates uniform in the additional parameter ϑ, taking into account that $\vartheta_1^2 + \vartheta_2^2 = 1$. In (1.6) we may assume that $\tau \in \mathbb{R}$.

Let $F(\tau;\vartheta;s)$ denote the spectral projection of the operator $B(\tau;\vartheta)$ for the interval $[0,s]$. We fix a number $\delta \in (0, d^0/14)$, and put

$$\tau_0 = \delta^{1/2}(\|X_1\|^2 + \|Y_1\| + c_0\|Q\|)^{-1/2}. \tag{1.7}$$

PROPOSITION 1.3. *For $|\tau| \leq \tau_0$ we have*

$$F(\tau;\vartheta;\delta) = F(\tau;\vartheta;3\delta), \quad \mathrm{rank}\, F(\tau;\vartheta;\delta) = n. \tag{1.8}$$

Proof. Assume that $\mathrm{rank}\, F(\tau;\vartheta;\delta) < n$. Then there exists an element $f \in \mathfrak{N}$ such that $\|f\| = 1$ and $f \perp F(\tau;\vartheta;\delta)\mathfrak{H}$. Hence, $b(\tau;\vartheta)[f,f] > \delta$. On the other hand,

$$\begin{aligned} b(\tau;\vartheta)[f,f] &= \|X_0 f + \tau\vartheta_1 X_1 f\|^2 + 2\tau\vartheta_2\mathrm{Re}\,(Y_0 f + \tau\vartheta_1 Y_1 f, f) + c_0\tau^2\vartheta_2^2(Qf,f) \\ &= \tau^2\vartheta_1^2\|X_1 f\|^2 + 2\tau^2\vartheta_1\vartheta_2\mathrm{Re}\,(Y_1 f, f) + c_0\tau^2\vartheta_2^2(Qf,f) \\ &\leq \tau^2(\|X_1\|^2 + \|Y_1\| + c_0\|Q\|)\|f\|^2 \leq \delta. \end{aligned}$$

This contradiction shows that

$$\mathrm{rank}\, F(\tau;\vartheta;\delta) \geq n. \tag{1.9}$$

Now, assume that $\operatorname{rank} F(\tau;\vartheta;3\delta) > n$. Then there exists an element $f \in F(\tau;\vartheta;3\delta)\mathfrak{H}$ such that $\|f\| = 1$ and $f \perp \mathfrak{N}$. Therefore, $\|X_0 f\|^2 \geq d^0 > 14\delta$. But $\|X_0 f\| \leq \|X_0 f + \tau\vartheta_1 X_1 f\| + \tau_0\|X_1\|\|f\|$, and, by (1.5) and (1.7), we have

$$\begin{aligned} \|X_0 f\|^2 &\leq 2\|X_0 f + \tau\vartheta_1 X_1 f\|^2 + 2\tau_0^2\|X_1\|^2\|f\|^2 \\ &\leq 4b(\tau;\vartheta)[f,f] + 2\tau_0^2\|X_1\|^2\|f\|^2 \leq 12\delta\|f\|^2 + 2\delta\|f\|^2 = 14\delta. \end{aligned} \tag{1.10}$$

This contradiction proves that

$$\operatorname{rank} F(\tau;\vartheta;3\delta) \leq n. \tag{1.11}$$

Comparing (1.9) and (1.11), we see that $\operatorname{rank} F(\tau;\vartheta;\delta) = \operatorname{rank} F(\tau;\vartheta;3\delta) = n$. This is equivalent to (1.8). □

Below we will often write $F(\tau;\vartheta)$ in place of $F(\tau;\vartheta;\delta)$.

1.5. The operators Z and $\widetilde{Z}$. In Subsections 1.5 and 1.6, we introduce some operators that will be needed in the perturbation theory considerations. We denote $\mathcal{D} := \operatorname{Dom} X_0 \cap \mathfrak{N}^\perp$. Since the point $\lambda = 0$ is isolated in the spectrum of A_0, the form $(X_0\varphi, X_0\zeta)$, $\varphi, \zeta \in \mathcal{D}$, determines an inner product in $\mathcal{D}$, converting $\mathcal{D}$ into a Hilbert space. Let $\omega \in \mathfrak{N}$. Suppose that $\phi \in \mathcal{D}$ satisfies the equation

$$X_0^*(X_0\phi + X_1\omega) = 0, \tag{1.12}$$

which is understood in the weak sense. In other words, we look for an element $\phi \in \mathcal{D}$ satisfying the identity

$$(X_0\phi, X_0\zeta)_{\mathfrak{H}_*} = -(X_1\omega, X_0\zeta)_{\mathfrak{H}_*}, \quad \forall \zeta \in \mathcal{D}. \tag{1.13}$$

Since the right-hand side of (1.13) is an antilinear continuous functional of $\zeta \in \mathcal{D}$, there exists a unique solution. Moreover, this solution satisfies the estimate $\|X_0\phi\|_{\mathfrak{H}_*} \leq \|X_1\omega\|_{\mathfrak{H}_*}$. We introduce the operator $Z : \mathfrak{N} \to \mathfrak{H}$ that takes $\omega \in \mathfrak{N}$ to the solution $\phi \in \mathcal{D}$ of equation (1.12).

Similarly, for a given $\omega \in \mathfrak{N}$, suppose that $\psi \in \mathcal{D}$ satisfies the equation

$$X_0^* X_0 \psi + Y_0^*\omega = 0, \tag{1.14}$$

which is understood in the weak sense, i.e.,

$$(X_0\psi, X_0\zeta)_{\mathfrak{H}_*} = -(\omega, Y_0\zeta)_{\mathfrak{H}}, \quad \forall \zeta \in \mathcal{D}. \tag{1.15}$$

By Condition 1.2, the right-hand side of (1.15) is an antilinear continuous functional of $\zeta \in \mathcal{D}$. Therefore, there exists a unique solution, which satisfies the estimate $\|X_0\psi\|_{\mathfrak{H}_*} \leq c_1\|\omega\|_{\mathfrak{H}}$. We introduce the operator $\widetilde{Z} : \mathfrak{N} \to \mathfrak{H}$ that takes $\omega \in \mathfrak{N}$ to the solution $\psi \in \mathcal{D}$ of equation (1.14).

1.6. The operators R and S. We define the operator $R := X_0 Z + X_1 : \mathfrak{N} \to \mathfrak{N}_*$. As has been shown in [**BSu2**, Chapter 1, (1.11)], the operator R can also be defined by the formula $R = P_* X_1|_{\mathfrak{N}}$. According to [**BSu2**, Chapter 1, Subsection 1.3], the operator $S = R^*R : \mathfrak{N} \to \mathfrak{N}$ is called the *spectral germ of the operator family $A(t)$ at $t = 0$*. The operator S can also be written as $S = PX_1^* P_* X_1|_{\mathfrak{N}}$. The spectral germ S plays an important role in the study of the spectral properties of $A(t)$ near the bottom of the spectrum (threshold properties). In more detail, the properties of the spectral germ are described in [**BSu2**, Chapter 1, Subsections 1.3, 1.4, 1.6]. Below in Subsection 1.7 we will introduce a similar notion for the more general operator family $B(\tau;\vartheta)$.

1.7. The analytic branches of eigenvalues and eigenvectors of the operator $B(\tau;\vartheta)$. The spectral germ of the operator $B(\tau;\vartheta)$. By general analytic perturbation theory (see [**K**]), for $|\tau| \le \tau_0$, there exist real-analytic (in τ) functions $\lambda_l(\tau;\vartheta)$ (branches of eigenvalues) and real-analytic $\mathfrak{H}$-valued functions $\varphi_l(\tau;\vartheta)$ (branches of eigenvectors) such that

$$B(\tau;\vartheta)\varphi_l(\tau;\vartheta) = \lambda_l(\tau;\vartheta)\varphi_l(\tau;\vartheta), \quad |\tau| \le \tau_0, \quad l = 1, \dots, n, \tag{1.16}$$

and the vectors $\varphi_l(\tau;\vartheta)$, $l = 1, \dots, n$, form an orthonormal basis in the eigenspace $F(\tau;\vartheta)\mathfrak{H}$. Herewith, for sufficiently small τ_* $(\tau_* \le \tau_0)$, we have the convergent power series expansions

$$\lambda_l(\tau;\vartheta) = \gamma_l(\vartheta)\tau^2 + \mu_l(\vartheta)\tau^3 + \dots, \quad \gamma_l(\vartheta) \ge 0, \quad l = 1, \dots, n, \quad |\tau| \le \tau_*, \tag{1.17}$$

$$\varphi_l(\tau;\vartheta) = \omega_l(\vartheta) + \tau\varphi_l^{(1)}(\vartheta) + \tau^2\varphi_l^{(2)}(\vartheta) + \dots, \quad l = 1, \dots, n, \quad |\tau| \le \tau_*. \tag{1.18}$$

The elements $\omega_l(\vartheta)$, $l = 1, \dots, n$, form an orthonormal basis in $\mathfrak{N}$, whence

$$P = \sum_{l=1}^{n} (\cdot, \omega_l(\vartheta))_{\mathfrak{H}}\omega_l(\vartheta). \tag{1.19}$$

Relations (1.16) should be understood in the weak sense:

$$b(\tau;\vartheta)[\varphi_l(\tau;\vartheta), \zeta] = \lambda_l(\tau;\vartheta)(\varphi_l(\tau;\vartheta), \zeta)_{\mathfrak{H}}, \quad \zeta \in \operatorname{Dom} X_0. \tag{1.20}$$

The left-hand side of (1.20) is of the form

$$\begin{aligned} b(\tau;\vartheta)[\varphi_l(\tau;\vartheta), \zeta] &= ((X_0 + \tau\vartheta_1 X_1)\varphi_l(\tau;\vartheta), (X_0 + \tau\vartheta_1 X_1)\zeta)_{\mathfrak{H}_*} \\ &+ \tau\vartheta_2((Y_0 + \tau\vartheta_1 Y_1)\varphi_l(\tau;\vartheta), \zeta)_{\mathfrak{H}} + \tau\vartheta_2(\varphi_l(\tau;\vartheta), (Y_0 + \tau\vartheta_1 Y_1)\zeta)_{\mathfrak{H}} \\ &+ c_0\tau^2\vartheta_2^2(Q\varphi_l(\tau;\vartheta), \zeta)_{\mathfrak{H}}. \end{aligned}$$

We substitute (1.17), (1.18) in (1.20) and compare the coefficients of the same powers of τ. For brevity, below we drop the argument ϑ in the coefficients of (1.17), (1.18). Comparing the terms of order τ, we obtain

$$(X_0\varphi_l^{(1)} + \vartheta_1 X_1\omega, X_0\zeta)_{\mathfrak{H}_*} + \vartheta_2(\omega, Y_0\zeta)_{\mathfrak{H}} = 0, \quad \zeta \in \operatorname{Dom} X_0.$$

This means that $\varphi_l^{(1)}$ satisfies the following equation:

$$X_0^*(X_0\varphi_l^{(1)} + \vartheta_1 X_1\omega_l) + \vartheta_2 Y_0^*\omega_l = 0, \tag{1.21}$$

which is understood in the weak sense. In terms of the operators Z and $\widetilde{Z}$ introduced in Subsection 1.5 (cf. (1.12), (1.14)), relation (1.21) means that

$$\varphi_l^{(1)} - (\vartheta_1 Z + \vartheta_2\widetilde{Z})\omega_l \in \mathfrak{N}. \tag{1.22}$$

Comparing the terms of order τ^2, we see that

$$\begin{aligned} &(X_0\varphi_l^{(2)} + \vartheta_1 X_1\varphi_l^{(1)}, X_0\zeta)_{\mathfrak{H}_*} + (X_0\varphi_l^{(1)} + \vartheta_1 X_1\omega_l, \vartheta_1 X_1\zeta)_{\mathfrak{H}_*} + \vartheta_2(Y_0\varphi_l^{(1)}, \zeta)_{\mathfrak{H}} \\ &+ \vartheta_2(\varphi_l^{(1)}, Y_0\zeta)_{\mathfrak{H}} + \vartheta_1\vartheta_2((Y_1 + Y_1^*)\omega_l, \zeta)_{\mathfrak{H}} + c_0\vartheta_2^2(Q\omega_l, \zeta)_{\mathfrak{H}} = \gamma_l(\omega_l, \zeta)_{\mathfrak{H}}, \end{aligned}$$

for $\zeta \in \operatorname{Dom} X_0$. If $\zeta \in \mathfrak{N}$, this implies that

$$\begin{aligned} &(X_0\varphi_l^{(1)} + \vartheta_1 X_1\omega_l, \vartheta_1 X_1\zeta)_{\mathfrak{H}_*} + \vartheta_2(Y_0\varphi_l^{(1)}, \zeta)_{\mathfrak{H}} \\ &+ \vartheta_1\vartheta_2((Y_1 + Y_1^*)\omega_l, \zeta)_{\mathfrak{H}} + c_0\vartheta_2^2(Q\omega_l, \zeta)_{\mathfrak{H}} = \gamma_l(\omega_l, \zeta)_{\mathfrak{H}}, \quad \zeta \in \mathfrak{N}. \end{aligned}$$

By (1.22), we obtain:
(1.23)

$$\vartheta_1^2(X_0Z\omega_l + X_1\omega_l, X_1\zeta)_{\mathfrak{H}_*} + \vartheta_1\vartheta_2(X_0\widetilde{Z}\omega_l, X_1\zeta)_{\mathfrak{H}_*} + \vartheta_1\vartheta_2(Y_0Z\omega_l, \zeta)_{\mathfrak{H}}$$
$$+ \vartheta_2^2(Y_0\widetilde{Z}\omega_l, \zeta)_{\mathfrak{H}} + \vartheta_1\vartheta_2((Y_1+Y_1^*)\omega_l, \zeta)_{\mathfrak{H}} + c_0\vartheta_2^2(Q\omega_l, \zeta)_{\mathfrak{H}} = \gamma_l(\omega_l, \zeta)_{\mathfrak{H}}, \quad \zeta \in \mathfrak{N}.$$

By definition of the operator R (see Subsection 1.6), an equation of the form (1.12) for $Z\omega_l$, and the definition of the operator S, the first term can be rewritten as

$$\begin{aligned}(X_0Z\omega_l + X_1\omega_l, X_1\zeta)_{\mathfrak{H}_*} &= (X_0Z\omega_l + X_1\omega_l, X_0Z\zeta + X_1\zeta)_{\mathfrak{H}_*}\\ &= (R\omega_l, R\zeta)_{\mathfrak{H}_*} = (S\omega_l, \zeta)_{\mathfrak{H}}.\end{aligned}$$

By definition of the operator Z (see (1.13)), we rewrite the term $(X_0\widetilde{Z}\omega_l, X_1\zeta)_{\mathfrak{H}_*}$ as $(X_0\widetilde{Z}\omega_l, X_1\zeta)_{\mathfrak{H}_*} = -(X_0\widetilde{Z}\omega_l, X_0Z\zeta)_{\mathfrak{H}_*}$. Using the definition of the operator $\widetilde{Z}$ (see (1.15)), we have $(Y_0Z\omega_l, \zeta)_{\mathfrak{H}} = -(X_0Z\omega_l, X_0\widetilde{Z}\zeta)_{\mathfrak{H}_*}$ and $(Y_0\widetilde{Z}\omega_l, \zeta)_{\mathfrak{H}} = -(X_0\widetilde{Z}\omega_l, X_0\widetilde{Z}\zeta)_{\mathfrak{H}_*}$. Thus, (1.23) takes the form
(1.24)

$$\vartheta_1^2(S\omega_l, \zeta)_{\mathfrak{H}} - \vartheta_1\vartheta_2(X_0\widetilde{Z}\omega_l, X_0Z\zeta)_{\mathfrak{H}_*} - \vartheta_1\vartheta_2(X_0Z\omega_l, X_0\widetilde{Z}\zeta)_{\mathfrak{H}_*}$$
$$- \vartheta_2^2(X_0\widetilde{Z}\omega_l, X_0\widetilde{Z}\zeta)_{\mathfrak{H}_*} + \vartheta_1\vartheta_2((Y_1+Y_1^*)\omega_l, \zeta)_{\mathfrak{H}} + c_0\vartheta_2^2(Q\omega_l, \zeta)_{\mathfrak{H}} = \gamma_l(\omega_l, \zeta)_{\mathfrak{H}},$$

for any $\zeta \in \mathfrak{N}$. The left-hand side of (1.24) is a sesquilinear form of the selfadjoint operator $\mathcal{S}(\vartheta) : \mathfrak{N} \to \mathfrak{N}$, where

$$\begin{aligned}\mathcal{S}(\vartheta) = \vartheta_1^2 S &- \vartheta_1\vartheta_2(X_0Z)^*(X_0\widetilde{Z}) - \vartheta_1\vartheta_2(X_0\widetilde{Z})^*(X_0Z)\\ &- \vartheta_2^2(X_0\widetilde{Z})^*(X_0\widetilde{Z}) + \vartheta_1\vartheta_2 P(Y_1+Y_1^*)|_{\mathfrak{N}} + c_0\vartheta_2^2 PQ|_{\mathfrak{N}}.\end{aligned} \tag{1.25}$$

DEFINITION 1.4. The operator (1.25) is called the spectral germ of the operator family $B(\tau;\vartheta)$ at $\tau = 0$.

Relation (1.24) means that $(\mathcal{S}(\vartheta)\omega_l(\vartheta), \zeta)_{\mathfrak{H}} = \gamma_l(\vartheta)(\omega_l(\vartheta), \zeta)_{\mathfrak{H}}$, $\forall \zeta \in \mathfrak{N}$. We have proved the following statement.

PROPOSITION 1.5. *The numbers $\gamma_l(\vartheta)$ and the vectors $\omega_l(\vartheta)$ defined by* (1.17), (1.18) *are eigenvalues and eigenvectors of the selfadjoint operator $\mathcal{S}(\vartheta) : \mathfrak{N} \to \mathfrak{N}$:*

$$\mathcal{S}(\vartheta)\omega_l(\vartheta) = \gamma_l(\vartheta)\omega_l(\vartheta), \quad l = 1, \dots, n.$$

Thus,

$$\mathcal{S}(\vartheta)P = \sum_{l=1}^{n} \gamma_l(\vartheta)(\cdot, \omega_l(\vartheta))_{\mathfrak{H}}\omega_l(\vartheta). \tag{1.26}$$

1.8. The operator-valued functions $F(\tau;\vartheta)$ and $B(\tau;\vartheta)F(\tau;\vartheta)$ are analytic in τ for $|\tau| \le \tau_0$. From (1.16) and Proposition 1.3 it follows that $F(\tau;\vartheta) = \sum_{l=1}^{n}(\cdot, \varphi_l(\tau;\vartheta))_{\mathfrak{H}}\varphi_l(\tau;\vartheta)$, $|\tau| \le \tau_0$. By (1.18) and (1.19), for sufficiently small τ_*, we have

$$F(\tau;\vartheta) = P + \tau F_1(\vartheta) + \dots, \quad |\tau| \le \tau_*, \tag{1.27}$$

where $F_1(\vartheta) = \check{F}_1(\vartheta) + \check{F}_1(\vartheta)^*$, $\check{F}_1(\vartheta) = \sum_l(\cdot, \omega_l(\vartheta))\varphi_l^{(1)}(\vartheta)$. Next, by (1.16) and Proposition 1.3,

$$B(\tau;\vartheta)F(\tau;\vartheta) = \sum_{l=1}^{n} \lambda_l(\tau;\vartheta)(\cdot, \varphi_l(\tau;\vartheta))_{\mathfrak{H}}\varphi_l(\tau;\vartheta), \quad |\tau| \le \tau_0.$$

Combining this with (1.17), (1.18), and (1.26), for sufficiently small τ_*, we obtain

$$B(\tau;\vartheta)F(\tau;\vartheta) = \tau^2 \mathcal{S}(\vartheta)P + \dots, \quad |\tau| \le \tau_*. \tag{1.28}$$

However, the power series expansions (1.27) and (1.28) are not quite suitable for our purposes. We only need *estimates for* $F(\tau;\vartheta) - P$ *and* $B(\tau;\vartheta)F(\tau;\vartheta) - \tau^2\mathcal{S}(\vartheta)P$, *but on the wider interval* $|\tau| \le \tau_0$, *and with constants controlled explicitly.* Such estimates can be obtained by integrating the difference of the resolvents $(B(\tau;\vartheta) - zI)^{-1} - (A_0 - zI)^{-1}$ over an appropriate contour $\Gamma_\delta \subset \mathbb{C}$.

§2. Estimates for the difference of the resolvents on the contour

2.1. The resolvent identity. We denote $R_z(0) = (A_0 - zI)^{-1}$, $R_z(\tau) = R_z(\tau;\vartheta) = (B(\tau;\vartheta) - zI)^{-1}$, where z is a regular point both for A_0 and for $B(\tau;\vartheta)$. The ordinary resolvent identity for $R_z(\tau) - R_z(0)$ is not applicable under our assumptions. We use the version of the resolvent identity for operators whose quadratic forms have the same domain (see [**BSu2**, Chapter 1, §2]).

Let $\mathfrak{d}$ be the space $\operatorname{Dom} X_0$ with the metric form

$$\|u\|_{\mathfrak{d}}^2 = \|X_0 u\|_{\mathfrak{H}_*}^2 + \gamma \|u\|_{\mathfrak{H}}^2. \tag{2.1}$$

The number $\gamma > 0$ will be chosen below. Let $T(\tau) = T(\tau;\vartheta)$ be the operator in $\mathfrak{d}$ generated by the quadratic form $b(\tau;\vartheta)[u,u] - \|X_0 u\|^2$, $u \in \mathfrak{d}$. We put

$$\Omega_z(0) = I + (z+\gamma)R_z(0), \quad \Omega_z(\tau) = \Omega_z(\tau;\vartheta) = I + (z+\gamma)R_z(\tau;\vartheta). \tag{2.2}$$

According to [**BSu2**, Chapter 1, (2.13)], the following resolvent identity is true:

$$R_z(\tau) - R_z(0) = -\Omega_z(0)T(\tau)R_z(\tau). \tag{2.3}$$

We put

$$\alpha^2 = \sup_{0 \ne u \in \mathfrak{d}} \frac{\|X_0 u\|^2 + \gamma\|u\|^2}{b(\tau;\vartheta)[u,u] + \gamma\|u\|^2}. \tag{2.4}$$

The following estimates will be needed below (see [**BSu2**, Chapter 1, (2.15)–(2.17)]):

$$\|LR_z(\tau)\|_{\mathfrak{H}\to\mathfrak{H}} \le \alpha\gamma^{-1}\|\Omega_z(\tau)\|\|L\|_{\mathfrak{d}\to\mathfrak{d}}, \tag{2.5}$$

$$\|LR_z(0)\|_{\mathfrak{H}\to\mathfrak{H}} \le \gamma^{-1}\|\Omega_z(0)\|\|L\|_{\mathfrak{d}\to\mathfrak{d}}, \tag{2.6}$$

(2.7)
$$\|L_1\Omega_z(0)L_2R_z(\tau)\|_{\mathfrak{H}\to\mathfrak{H}} \le \alpha\gamma^{-1}\|\Omega_z(\tau)\|(1 + |z+\gamma|\gamma^{-1}\|\Omega_z(0)\|)\|L_1\|_{\mathfrak{d}\to\mathfrak{d}}\|L_2\|_{\mathfrak{d}\to\mathfrak{d}},$$

where L, L_1, L_2 are bounded operators in $\mathfrak{d}$, and $\|\Omega_z(\tau)\| = \|\Omega_z(\tau)\|_{\mathfrak{H}\to\mathfrak{H}}$, $\tau \ge 0$.

2.2. The contour Γ_δ. By Γ_δ we denote the contour that envelops the real interval $[0,\delta]$ equidistantly at the distance δ. We need to integrate the difference of the resolvents $R_z(\tau) - R_z(0)$ over the contour Γ_δ. We assume that $|\tau| \le \tau_0$, where τ_0 is defined by (1.7). By Proposition 1.3, the distance between Γ_δ and the spectrum of $B(\tau;\vartheta)$ is at least δ, whence

$$\|R_z(\tau)\| \le \delta^{-1}, \quad z \in \Gamma_\delta, \quad |\tau| \le \tau_0. \tag{2.8}$$

We estimate the number α defined by (2.4). By (1.5) (cf. (1.10)), we have

(2.9)
$$\begin{aligned}\|X_0 u\|^2 + \gamma\|u\|^2 &\le 4b(\tau;\vartheta)[u,u] + (2\tau_0^2\|X_1\|^2 + \gamma)\|u\|^2\\ &\le 4\left(b(\tau;\vartheta)[u,u] + \frac{2\delta+\gamma}{4}\|u\|^2\right).\end{aligned}$$

We choose the number γ so that $2\delta + \gamma = 4\gamma$, i.e., $\gamma = 2\delta/3$. Then, by (2.9), $\alpha \le 2$.

Next, from (2.8), the identity $\gamma = 2\delta/3$ and the inequality $|z| \le 2\delta$ for $z \in \Gamma_\delta$, it follows that the operators (2.2) satisfy the estimate

(2.10)
$$\|\Omega_z(\tau)\| \le 4, \quad z \in \Gamma_\delta, \quad |\tau| \le \tau_0.$$

The difference of the forms $b(\tau;\vartheta)[u,u] - \|X_0 u\|^2$ can be represented as

$$\begin{aligned}&\tau\left(2\vartheta_1 \mathrm{Re}\,(X_0 u, X_1 u) + 2\vartheta_2 \mathrm{Re}\,(Y_0 u, u)\right)\\ &+\tau^2\left(\vartheta_1^2\|X_1 u\|^2 + 2\vartheta_1\vartheta_2 \mathrm{Re}\,(Y_1 u, u) + c_0\vartheta_2^2(Qu,u)\right).\end{aligned}$$

The operator $T(\tau) = T(\tau;\vartheta)$ generated in $\mathfrak{d}$ by this form can be written as

(2.11)
$$T(\tau;\vartheta) = \tau T_1(\vartheta) + \tau^2 T_2(\vartheta).$$

We estimate the norms of $T_1(\vartheta)$ and $T_2(\vartheta)$ in $\mathfrak{d}$. By Condition 1.2, we have

$$\begin{aligned}|2\vartheta_1 \mathrm{Re}\,(X_0 u, X_1 u)| &\le 2\|X_0 u\|\|X_1 u\| \le \kappa\|X_0 u\|^2 + \kappa^{-1}\|X_1 u\|^2, \quad \forall \kappa > 0,\\ |2\vartheta_2 \mathrm{Re}\,(Y_0 u, u)| &\le 2c_1\|X_0 u\|\|u\| \le \kappa\|X_0 u\|^2 + c_1^2\kappa^{-1}\|u\|^2, \quad \forall\kappa > 0.\end{aligned}$$

Hence,

(2.12)
$$|2\vartheta_1 \mathrm{Re}\,(X_0 u, X_1 u) + 2\vartheta_2 \mathrm{Re}\,(Y_0 u, u)| \le 2\kappa\left(\|X_0 u\|^2 + (c_1^2 + \|X_1\|^2)(2\kappa^2)^{-1}\|u\|^2\right).$$

We choose the number κ in such a way that the condition $(c_1^2 + \|X_1\|^2)(2\kappa^2)^{-1} = \gamma = 2\delta/3$ is fulfilled. Then $\kappa = (\sqrt{3}/2)(c_1^2 + \|X_1\|^2)^{1/2}\delta^{-1/2}$, and (2.12) means that

(2.13)
$$\|T_1(\vartheta)\|_{\mathfrak{d}\to\mathfrak{d}} \le \sqrt{3}(c_1^2 + \|X_1\|^2)^{1/2}\delta^{-1/2}.$$

Now we estimate the norm of the operator $T_2(\vartheta)$. We have

$$\begin{aligned}|\vartheta_1^2\|X_1 u\|^2 + 2\vartheta_1\vartheta_2 \mathrm{Re}\,(Y_1 u, u) + c_0\vartheta_2^2(Qu,u)| &\le (\|X_1\|^2 + \|Y_1\| + c_0\|Q\|)\|u\|^2\\ &\le \gamma^{-1}(\|X_1\|^2 + \|Y_1\| + c_0\|Q\|)(\|X_0 u\|^2 + \gamma\|u\|^2).\end{aligned}$$

This implies that

(2.14)
$$\|T_2(\vartheta)\|_{\mathfrak{d}\to\mathfrak{d}} \le 3(\|X_1\|^2 + \|Y_1\| + c_0\|Q\|)(2\delta)^{-1}.$$

For the norm of the operator (2.11) we obtain

$$\|T(\tau;\vartheta)\|_{\mathfrak{d}\to\mathfrak{d}} \le |\tau|\|T_1(\vartheta)\|_{\mathfrak{d}\to\mathfrak{d}} + \tau^2\|T_2(\vartheta)\|_{\mathfrak{d}\to\mathfrak{d}} \le |\tau|(\|T_1(\vartheta)\|_{\mathfrak{d}\to\mathfrak{d}} + \tau_0\|T_2(\vartheta)\|_{\mathfrak{d}\to\mathfrak{d}}).$$

By (2.13), (2.14), and (1.7), we arrive at

(2.15)
$$\begin{aligned}&\|T(\tau;\vartheta)\|_{\mathfrak{d}\to\mathfrak{d}} \le |\tau|\delta^{-1/2}C_T, \quad |\tau| \le \tau_0,\\ &C_T := \sqrt{3}(c_1^2 + \|X_1\|^2)^{1/2} + \frac{3}{2}(\|X_1\|^2 + \|Y_1\| + c_0\|Q\|)^{1/2}.\end{aligned}$$

2.3. Estimate for the norm of the difference of resolvents. By (2.3), (2.5), and (2.10), we have

$$\|R_z(\tau) - R_z(0)\| \le 4\|T(\tau)R_z(\tau)\| \le 16\alpha\gamma^{-1}\|T(\tau)\|_{\mathfrak{d}\to\mathfrak{d}}.$$

Combining this with (2.15), we arrive at

(2.16)
$$\|R_z(\tau) - R_z(0)\| \le 48\delta^{-3/2}C_T|\tau|, \quad |\tau| \le \tau_0, \quad z \in \Gamma_\delta.$$

2.4. The difference of resolvents. Separation of the main part. Estimate (2.16) does not suffice for our purposes. We separate the terms of the difference of resolvents up to the order of τ^2 and give an estimate of order τ^3 for the remainder term. Using (2.3) and (2.11), we write the difference of resolvents as

$$R_z(\tau) - R_z(0) = -\Omega_z(0)T(\tau)R_z(\tau) = \tau\mathcal{I}_1(\tau) + \tau^2\mathcal{I}_2(\tau), \tag{2.17}$$

$$\mathcal{I}_j(\tau) := -\Omega_z(0)T_jR_z(\tau), \quad j = 1, 2. \tag{2.18}$$

We start with the operator $\mathcal{I}_2(\tau)$. Iterating (2.3), we obtain

$$\mathcal{I}_2(\tau) = -\Omega_z(0)T_2(R_z(0) - \Omega_z(0)T(\tau)R_z(\tau)) = \mathcal{I}_2^0 + \mathcal{I}_2^{(1)}(\tau), \tag{2.19}$$

$$\mathcal{I}_2^0 := -\Omega_z(0)T_2R_z(0). \tag{2.20}$$

The term $\mathcal{I}_2^0$ does not depend on τ. By (2.7) and (2.10), the term $\mathcal{I}_2^{(1)}(\tau) = \Omega_z(0)T_2\Omega_z(0)T(\tau)R_z(\tau)$ satisfies the estimate

$$\|\mathcal{I}_2^{(1)}(\tau)\| \le 4\|T_2\Omega_z(0)T(\tau)R_z(\tau)\| \le 16\alpha\gamma^{-1}(1 + 4|z+\gamma|\gamma^{-1})\|T_2\|_{\mathfrak{d}\to\mathfrak{d}}\|T(\tau)\|_{\mathfrak{d}\to\mathfrak{d}}.$$

Using that $\alpha \le 2$, $\gamma = 2\delta/3$ and $|z+\gamma| \le 8\delta/3$ for $z \in \Gamma_\delta$, and taking (2.14) and (2.15) into account, we have

$$\begin{gathered} \|\mathcal{I}_2^{(1)}(\tau)\| \le \widetilde{C}_T\delta^{-5/2}|\tau|, \quad |\tau| \le \tau_0, \quad z \in \Gamma_\delta, \\ \widetilde{C}_T = 1224C_T(\|X_1\|^2 + \|Y_1\| + c_0\|Q\|). \end{gathered} \tag{2.21}$$

Investigation of the term $\mathcal{I}_1(\tau)$ is somewhat more cumbersome. We need to iterate (2.3) twice. By (2.18) and (2.3), we have

$$\mathcal{I}_1(\tau) = -\Omega_z(0)T_1(R_z(0) - \Omega_z(0)T(\tau)R_z(\tau)) = \mathcal{I}_1^0 + \mathcal{I}_1^{(1)}(\tau), \tag{2.22}$$

$$\mathcal{I}_1^0 := -\Omega_z(0)T_1R_z(0). \tag{2.23}$$

The term $\mathcal{I}_1^0$ does not depend on τ. We consider the operator

$$\begin{aligned} \mathcal{I}_1^{(1)}(\tau) &= \Omega_z(0)T_1\Omega_z(0)T(\tau)R_z(\tau) \\ &= \tau\Omega_z(0)T_1\Omega_z(0)T_1R_z(\tau) + \tau^2\Omega_z(0)T_1\Omega_z(0)T_2R_z(\tau), \end{aligned}$$

and apply (2.3) again:

$$\mathcal{I}_1^{(1)}(\tau) = \tau\Omega_z(0)T_1\Omega_z(0)T_1(R_z(0) - \Omega_z(0)T(\tau)R_z(\tau)) + \tau^2\Omega_z(0)T_1\Omega_z(0)T_2R_z(\tau).$$

As a result, we have

$$\mathcal{I}_1^{(1)}(\tau) = \tau\mathcal{I}_2^{00} + \mathcal{I}_1^{(2)}(\tau), \tag{2.24}$$

$$\mathcal{I}_2^{00} := \Omega_z(0)T_1\Omega_z(0)T_1R_z(0). \tag{2.25}$$

The term $\mathcal{I}_2^{00}$ does not depend on τ. By (2.2), the term $\mathcal{I}_1^{(2)}(\tau)$ can be rewritten as

$$\begin{aligned} \mathcal{I}_1^{(2)}(\tau) &= -\tau\Omega_z(0)T_1\Omega_z(0)T_1\Omega_z(0)T(\tau)R_z(\tau) + \tau^2\Omega_z(0)T_1\Omega_z(0)T_2R_z(\tau) \\ &= -\tau\Omega_z(0)T_1T_1\Omega_z(0)T(\tau)R_z(\tau) - \tau(z+\gamma)\Omega_z(0)T_1R_z(0)T_1\Omega_z(0)T(\tau)R_z(\tau) \\ &\quad + \tau^2\Omega_z(0)T_1\Omega_z(0)T_2R_z(\tau). \end{aligned}$$

Using estimates (2.6), (2.7), and (2.10), we have

$$\begin{aligned} \|\mathcal{I}_1^{(2)}(\tau)\| \le{}& |\tau|16\alpha\gamma^{-1}(1 + 4|z+\gamma|\gamma^{-1})^2\|T_1\|^2_{\mathfrak{d}\to\mathfrak{d}}\|T(\tau)\|_{\mathfrak{d}\to\mathfrak{d}} \\ &+ \tau^2 16\alpha\gamma^{-1}(1 + 4|z+\gamma|\gamma^{-1})\|T_1\|_{\mathfrak{d}\to\mathfrak{d}}\|T_2\|_{\mathfrak{d}\to\mathfrak{d}}. \end{aligned}$$

Combining this with (2.13)–(2.15), we obtain
(2.26)

$$\|\mathcal{I}_1^{(2)}(\tau)\| \le \check{C}_T \delta^{-5/2}\tau^2, \quad |\tau| \le \tau_0, \quad z \in \Gamma_\delta,$$

$$\check{C}_T = 144 \cdot 17^2 C_T (c_1^2 + \|X_1\|^2) + 1224\sqrt{3}(c_1^2 + \|X_1\|^2)^{1/2}(\|X_1\|^2 + \|Y_1\| + c_0\|Q\|).$$

Combining relations (2.17), (2.19), (2.22), and (2.24) with estimates (2.21) and (2.26), we arrive at the following statement.

THEOREM 2.1. *For $z \in \Gamma_\delta$ and $|\tau| \le \tau_0$ we have*

$$R_z(\tau) - R_z(0) = \tau \mathcal{I}_1^0 + \tau^2 \mathcal{I}_2^+ + \Psi^0(\tau), \tag{2.27}$$

where $\mathcal{I}_2^+ = \mathcal{I}_2^0 + \mathcal{I}_2^{00}$. The operators $\mathcal{I}_1^0$, $\mathcal{I}_2^0$ and $\mathcal{I}_2^{00}$ are defined by (2.23), (2.20), *and* (2.25). *The operator $\Psi^0(\tau) = \tau \mathcal{I}_1^{(2)}(\tau) + \tau^2 \mathcal{I}_2^{(1)}(\tau)$ satisfies the estimate*

$$\|\Psi^0(\tau)\| \le C_T^0 \delta^{-5/2}|\tau|^3, \quad C_T^0 = \widetilde{C}_T + \check{C}_T. \tag{2.28}$$

The constants $\widetilde{C}_T$ and $\check{C}_T$ are defined according to (2.15), (2.21), *and* (2.26) *and depend only on c_1, c_0, $\|X_1\|$, $\|Y_1\|$, and $\|Q\|$.*

Note that the operators $\mathcal{I}_1^0$ and $\mathcal{I}_2^+$ depend on z and ϑ: $\mathcal{I}_1^0 = \mathcal{I}_1^0(z,\vartheta)$, $\mathcal{I}_2^+ = \mathcal{I}_2^+(z,\vartheta)$, but do not depend on τ, while $\Psi^0(\tau) = \Psi^0(\tau; z, \vartheta)$.

§3. Threshold approximations

3.1. Approximation for $F(\tau;\vartheta)$. We start with the representation

$$F(\tau;\vartheta) = -\frac{1}{2\pi i}\int_{\Gamma_\delta} R_z(\tau;\vartheta)\,dz. \tag{3.1}$$

If $\tau = 0$, the left-hand side of (3.1) is equal to P, whence

$$F(\tau;\vartheta) - P = -\frac{1}{2\pi i}\int_{\Gamma_\delta} (R_z(\tau;\vartheta) - R_z(0))\,dz. \tag{3.2}$$

Directly from (2.16) and (3.2), it follows that

$$\|F(\tau;\vartheta) - P\| \le 48 C_T (1 + \pi^{-1})\delta^{-1/2}|\tau|, \quad |\tau| \le \tau_0. \tag{3.3}$$

3.2. Approximation for $B(\tau;\vartheta)F(\tau;\vartheta)$. We use the representation

$$B(\tau;\vartheta)F(\tau;\vartheta) = -\frac{1}{2\pi i}\int_{\Gamma_\delta} z R_z(\tau;\vartheta)\,dz.$$

For $\tau = 0$, the left-hand side is equal to $A_0 P = 0$, whence

$$B(\tau;\vartheta)F(\tau;\vartheta) = -\frac{1}{2\pi i}\int_{\Gamma_\delta} z(R_z(\tau;\vartheta) - R_z(0))\,dz. \tag{3.4}$$

By (2.27), from (3.4) we deduce that

$$B(\tau;\vartheta)F(\tau;\vartheta) = \tau I_1(\vartheta) + \tau^2 I_2(\vartheta) + \Psi(\tau;\vartheta), \tag{3.5}$$

$$\begin{aligned} I_1(\vartheta) &:= -\frac{1}{2\pi i}\int_{\Gamma_\delta} z\mathcal{I}_1^0(z,\vartheta)\,dz, \quad I_2(\vartheta) := -\frac{1}{2\pi i}\int_{\Gamma_\delta} z\mathcal{I}_2^+(z,\vartheta)\,dz, \\ \Psi(\tau;\vartheta) &:= -\frac{1}{2\pi i}\int_{\Gamma_\delta} z\Psi^0(\tau;z,\vartheta)\,dz. \end{aligned} \tag{3.6}$$

By (2.28), we have

$$\|\Psi(\tau;\vartheta)\| \le 2(1 + \pi^{-1})\delta^{-1/2} C_T^0 |\tau|^3, \quad |\tau| \le \tau_0. \tag{3.7}$$

The coefficients $I_1(\vartheta)$ and $I_2(\vartheta)$ are defined by (3.6). On the other hand, one can find these coefficients by comparing (3.5) and (1.28). This yields $I_1(\vartheta) = 0$, $I_2(\vartheta) = \mathcal{S}(\vartheta)P$. Relations (3.3), (3.5), and (3.7) directly imply the following theorem. In this theorem we use the notation

$$C_1 := 48(1 + \pi^{-1})\delta^{-1/2}C_T, \quad C_2 := 2(1 + \pi^{-1})\delta^{-1/2}C_T^0. \tag{3.8}$$

THEOREM 3.1. *We have*

$$\|F(\tau;\vartheta) - P\| \leq C_1|\tau|, \quad |\tau| \leq \tau_0, \tag{3.9}$$

$$\|B(\tau;\vartheta)F(\tau;\vartheta) - \tau^2\mathcal{S}(\vartheta)P\| \leq C_2|\tau|^3, \quad |\tau| \leq \tau_0. \tag{3.10}$$

The constants C_1, C_2 depend only on δ, c_1, c_0, $\|X_1\|$, $\|Y_1\|$, and $\|Q\|$.

3.3. Approximation for $B(\tau;\vartheta)^{-1}$. Our goal is to approximate the operator $B(\tau;\vartheta)^{-1}$ in terms of the finite rank operator $(\tau^2\mathcal{S}(\vartheta))^{-1}P$. Both operators are of order τ^{-2}, while for the operator norm of their difference we will obtain the estimate of order τ^{-1}. For this, we need an additional condition on $A(t)$. Namely, we *assume that*

$$A(t) \geq c_* t^2 I, \quad c_* > 0, \quad |t| \leq \tau_0 \tag{3.11}$$

(as in [**BSu2**, Chapter 1, §5]). By (1.5), this implies that

$$b(t,\varepsilon)[u,u] \geq \frac{1}{2}(c_* t^2 + c_1^2\varepsilon^2)\|u\|^2_{\mathfrak{H}}, \quad u \in \operatorname{Dom} X_0, \quad |t| \leq \tau_0.$$

Then

$$B(\tau;\vartheta) \geq \check{c}_*\tau^2 I, \quad |\tau| \leq \tau_0, \quad \check{c}_* := \frac{1}{2}\min\{c_*, c_1^2\}. \tag{3.12}$$

From (3.12) it follows that the eigenvalues $\lambda_l(\tau;\vartheta)$, $l = 1,\dots,n$, of the operator family $B(\tau;\vartheta)$ satisfy the estimates $\lambda_l(\tau;\vartheta) \geq \check{c}_*\tau^2$, $|\tau| \leq \tau_0$. Combining this with (1.17) and Proposition 1.5, we see that $\gamma_l(\vartheta) \geq \check{c}_*$, $l = 1,\dots,n$. Then

$$\mathcal{S}(\vartheta) \geq \check{c}_* I_{\mathfrak{N}}. \tag{3.13}$$

Consider the operator

$$B(\tau;\vartheta)^{-1} - (\tau^2\mathcal{S}(\vartheta))^{-1}P = B(\tau;\vartheta)^{-1}F(\tau;\vartheta)^{\perp} + B(\tau;\vartheta)^{-1}F(\tau;\vartheta) - (\tau^2\mathcal{S}(\vartheta))^{-1}P. \tag{3.14}$$

By Proposition 1.3, we obtain

$$\|B(\tau;\vartheta)^{-1}F(\tau;\vartheta)^{\perp}\| \leq (3\delta)^{-1}, \quad |\tau| \leq \tau_0. \tag{3.15}$$

We have

$$G(\tau;\vartheta) := B(\tau;\vartheta)^{-1}F(\tau;\vartheta) - (\tau^2\mathcal{S}(\vartheta))^{-1}P = G_1(\tau;\vartheta) + G_2(\tau;\vartheta) + G_3(\tau;\vartheta), \tag{3.16}$$

where

$$\begin{aligned}
G_1(\tau;\vartheta) &:= B(\tau;\vartheta)^{-1}F(\tau;\vartheta)(F(\tau;\vartheta) - P),\\
G_2(\tau;\vartheta) &:= (F(\tau;\vartheta) - P)(\tau^2\mathcal{S}(\vartheta))^{-1}P,\\
G_3(\tau;\vartheta) &:= F(\tau;\vartheta)B(\tau;\vartheta)^{-1}P - F(\tau;\vartheta)(\tau^2\mathcal{S}(\vartheta))^{-1}P.
\end{aligned}$$

From (3.12) and (3.9) it follows that

$$\|G_1(\tau;\vartheta)\| \leq C_1\check{c}_*^{-1}|\tau|^{-1}, \quad |\tau| \leq \tau_0. \tag{3.17}$$

Similarly, (3.9) and (3.13) imply that

$$\|G_2(\tau;\vartheta)\| \le C_1\check{c}_*^{-1}|\tau|^{-1}, \quad |\tau|\le\tau_0. \tag{3.18}$$

By (3.12), (3.13), and (3.10), the operator

$$G_3(\tau;\vartheta) = F(\tau;\vartheta)B(\tau;\vartheta)^{-1}(\tau^2\mathcal{S}(\vartheta)P - B(\tau;\vartheta)F(\tau;\vartheta))(\tau^2\mathcal{S}(\vartheta))^{-1}P$$

satisfies the estimate

$$\|G_3(\tau;\vartheta)\| \le C_2\check{c}_*^{-2}|\tau|^{-1}, \quad |\tau|\le\tau_0. \tag{3.19}$$

As a result, relations (3.14)–(3.19) imply that

$$\|B(\tau;\vartheta)^{-1} - (\tau^2\mathcal{S}(\vartheta))^{-1}P\| \le (3\delta)^{-1} + (2C_1\check{c}_*^{-1} + C_2\check{c}_*^{-2})|\tau|^{-1}, \quad |\tau|\le\tau_0.$$

Using (1.25), we rewrite the operator $\tau^2\mathcal{S}(\vartheta)$ in terms of t and ε:

$$\begin{aligned}\tau^2\mathcal{S}(\vartheta) = t^2 S + t\varepsilon\Big(-(X_0Z)^*(X_0\widetilde{Z}) - (X_0\widetilde{Z})^*(X_0Z) + P(Y_1+Y_1^*)|_{\mathfrak{N}}\Big)&\\ +\varepsilon^2\Big(-(X_0\widetilde{Z})^*(X_0\widetilde{Z}) + c_0PQ|_{\mathfrak{N}}\Big) =: L(t,\varepsilon).&\end{aligned} \tag{3.20}$$

Note that, by (3.13), we have

$$L(t,\varepsilon) \ge \check{c}_*(t^2+\varepsilon^2)I_{\mathfrak{N}}. \tag{3.21}$$

We have proved the following theorem.

THEOREM 3.2. *Let $B(t,\varepsilon) = B(\tau;\vartheta)$ be the operator defined in Subsection* 1.3. *Suppose that condition* (3.11) *is satisfied. Then*

$$\|B(t,\varepsilon)^{-1} - L(t,\varepsilon)^{-1}P\| \le C_3|\tau|^{-1}, \quad |\tau|\le\tau_0, \tag{3.22}$$

where the operator $L(t,\varepsilon):\mathfrak{N}\to\mathfrak{N}$ is defined by (3.20). *The constant*

$$C_3 = \tau_0(3\delta)^{-1} + 2C_1\check{c}_*^{-1} + C_2\check{c}_*^{-2} \tag{3.23}$$

depends only on δ, c_1, c_0, $\|X_1\|$, $\|Y_1\|$, $\|Q\|$, and c_*. The constant τ_0 is defined by (1.7) *and depends only on δ, c_0, $\|X_1\|$, $\|Y_1\|$, and $\|Q\|$.*

Chapter 2. Homogenization of periodic differential operators

§4. Periodic operators in $L_2(\mathbb{R}^d;\mathbb{C}^n)$

4.1. The lattices Γ and $\widetilde{\Gamma}$. Let Γ be a lattice in $\mathbb{R}^d$ generated by the basis $\mathbf{a}_1,\dots,\mathbf{a}_d$: $\Gamma = \{\mathbf{a}\in\mathbb{R}^d : \mathbf{a} = \sum_{j=1}^d n^j\mathbf{a}_j,\ n^j\in\mathbb{Z}\}$. By Ω we denote the elementary cell of the lattice Γ: $\Omega = \{\mathbf{x}\in\mathbb{R}^d : \mathbf{x} = \sum_{j=1}^d \xi^j\mathbf{a}_j,\ 0<\xi^j<1\}$. The basis $\mathbf{b}^1,\dots,\mathbf{b}^d$ dual to $\mathbf{a}_1,\dots,\mathbf{a}_d$ is defined by the relations $\langle\mathbf{b}^l,\mathbf{a}_j\rangle = 2\pi\delta_j^l$. This basis generates the *lattice* $\widetilde{\Gamma}$ *dual to* Γ. Let $\widetilde{\Omega}$ denote the *Brillouin zone* of the lattice $\widetilde{\Gamma}$: $\widetilde{\Omega} = \{\mathbf{k}\in\mathbb{R}^d : |\mathbf{k}| < |\mathbf{k}-\mathbf{b}|,\ 0\ne\mathbf{b}\in\widetilde{\Gamma}\}$. The domain $\widetilde{\Omega}$ is a fundamental domain for $\widetilde{\Gamma}$. By r_0 we denote the *radius of the ball inscribed in* $\operatorname{clos}\widetilde{\Omega}$.

4.2. Factorized second order operators. Let $b(\mathbf{D}) : L_2(\mathbb{R}^d;\mathbb{C}^n) \to L_2(\mathbb{R}^d;\mathbb{C}^m)$ be a homogeneous first order DO with constant coefficients. *We assume that* $m \geq n$. The symbol $b(\boldsymbol{\xi})$ of the operator $b(\mathbf{D})$ is an $(m \times n)$-matrix-valued linear homogeneous function of $\boldsymbol{\xi} \in \mathbb{R}^d$. *Suppose that* $\operatorname{rank} b(\boldsymbol{\xi}) = n$, $0 \neq \boldsymbol{\xi} \in \mathbb{R}^d$. Then for some $\alpha_0, \alpha_1 > 0$ we have

$$\alpha_0 \mathbf{1}_n \leq b(\boldsymbol{\theta})^* b(\boldsymbol{\theta}) \leq \alpha_1 \mathbf{1}_n, \quad \boldsymbol{\theta} \in \mathbb{S}^{d-1}, \quad 0 < \alpha_0 \leq \alpha_1 < \infty. \tag{4.1}$$

Let $h(\mathbf{x})$, $\mathbf{x} \in \mathbb{R}^d$, be a Γ-periodic $(m \times m)$-matrix-valued function such that

$$h, h^{-1} \in L_\infty. \tag{4.2}$$

We consider the DO

$$\mathcal{X} = hb(\mathbf{D}) : L_2(\mathbb{R}^d;\mathbb{C}^n) \to L_2(\mathbb{R}^d;\mathbb{C}^m), \quad \operatorname{Dom}\mathcal{X} = H^1(\mathbb{R}^d;\mathbb{C}^n). \tag{4.3}$$

The operator $\mathcal{X}$ is closed. The selfadjoint operator $\mathcal{A} = \mathcal{X}^*\mathcal{X}$ in $L_2(\mathbb{R}^d;\mathbb{C}^n)$ is generated by the quadratic form $a[\mathbf{u},\mathbf{u}] = \|\mathcal{X}\mathbf{u}\|^2_{L_2}$, $\mathbf{u} \in H^1(\mathbb{R}^d;\mathbb{C}^n)$. Formally, $\mathcal{A} = b(\mathbf{D})^* g(\mathbf{x}) b(\mathbf{D})$, where $g(\mathbf{x}) = h(\mathbf{x})^* h(\mathbf{x})$. Using the Fourier transformation and conditions (4.1) and (4.2), it is easy to show that

$$\alpha_0 \|g^{-1}\|^{-1}_{L_\infty} \|\mathbf{D}\mathbf{u}\|^2_{L_2} \leq a[\mathbf{u},\mathbf{u}] \leq \alpha_1 \|g\|_{L_\infty} \|\mathbf{D}\mathbf{u}\|^2_{L_2}, \quad \mathbf{u} \in H^1(\mathbb{R}^d;\mathbb{C}^n).$$

4.3. The operators $\mathcal{Y}$ and Q. Let $a_j(\mathbf{x})$, $j = 1, \dots, d$, be bounded Γ-periodic $(n \times n)$-matrix-valued functions in $\mathbb{R}^d$:

$$a_j \in L_\infty, \quad j = 1, \dots, d. \tag{4.4}$$

We consider the operator

$$\mathcal{Y} = \sum_{j=1}^d a_j(\mathbf{x}) D_j, \quad \operatorname{Dom}\mathcal{Y} = H^1(\mathbb{R}^d;\mathbb{C}^n), \tag{4.5}$$

in $L_2(\mathbb{R}^d;\mathbb{C}^n)$. Using (4.1)–(4.5), it is easy to check that

$$\|\mathcal{Y}\mathbf{u}\|_{L_2} \leq c_1 \|\mathcal{X}\mathbf{u}\|_{L_2}, \quad \mathbf{u} \in H^1(\mathbb{R}^d;\mathbb{C}^n), \quad c_1 = \alpha_0^{-1/2} \|g^{-1}\|^{1/2}_{L_\infty} C_a, \tag{4.6}$$

where

$$C_a^2 := \operatorname*{ess\,sup}_{\mathbf{x} \in \mathbb{R}^d} \sum_{j=1}^d |a_j(\mathbf{x})|^2. \tag{4.7}$$

Let $Q(\mathbf{x})$ be a Γ-periodic positive and bounded $(n \times n)$-matrix-valued function in $\mathbb{R}^d$: $Q(\mathbf{x}) > 0$; $Q, Q^{-1} \in L_\infty$. By Q we denote the operator in $L_2(\mathbb{R}^d;\mathbb{C}^n)$ of multiplication by $Q(\mathbf{x})$.

4.4. The operator pencil $\mathcal{B}(\varepsilon)$. In $L_2(\mathbb{R}^d;\mathbb{C}^n)$, we consider the selfadjoint operator $\mathcal{B}(\varepsilon)$, $\varepsilon > 0$, formally given by

$$\mathcal{B}(\varepsilon) = \mathcal{A} + \varepsilon(\mathcal{Y} + \mathcal{Y}^*) + c_0 \varepsilon^2 Q.$$

The precise definition of the operator $\mathcal{B}(\varepsilon)$ is given in terms of the quadratic form

$$b(\varepsilon)[\mathbf{u},\mathbf{u}] = a[\mathbf{u},\mathbf{u}] + 2\varepsilon \operatorname{Re}(\mathcal{Y}\mathbf{u},\mathbf{u})_{L_2} + c_0 \varepsilon^2 (Q\mathbf{u},\mathbf{u})_{L_2}, \quad \mathbf{u} \in H^1(\mathbb{R}^d;\mathbb{C}^n).$$

The constant c_0 is chosen from the condition

$$c_0 \geq \frac{5}{2} c_1^2 \|Q^{-1}\|_{L_\infty} = \frac{5}{2} \alpha_0^{-1} C_a^2 \|g^{-1}\|_{L_\infty} \|Q^{-1}\|_{L_\infty}. \tag{4.8}$$

This condition together with inequality (4.6) ensures the estimate

$$b(\varepsilon)[\mathbf{u},\mathbf{u}] \geq \frac{1}{2}a[\mathbf{u},\mathbf{u}] + \frac{c_1^2}{2}\varepsilon^2\|\mathbf{u}\|^2_{L_2}, \quad \mathbf{u} \in H^1(\mathbb{R}^d;\mathbb{C}^n). \tag{4.9}$$

Hence, the operator $\mathcal{B}(\varepsilon)$ is positive. *Our nearest goal is to approximate the inverse operator $\mathcal{B}(\varepsilon)^{-1}$ for small ε in the operator norm in $L_2(\mathbb{R}^d;\mathbb{C}^n)$.*

§5. Direct integral decomposition for the operator $\mathcal{B}(\varepsilon)$

5.1. The Gelfand transformation. Initially, the Gelfand transformation $\mathcal{U}$ is defined on the functions of the Schwartz class $\mathbf{v} \in \mathcal{S}(\mathbb{R}^d;\mathbb{C}^n)$ by the formula

$$\widetilde{\mathbf{v}}(\mathbf{k},\mathbf{x}) = (\mathcal{U}\mathbf{v})(\mathbf{k},\mathbf{x}) = |\widetilde{\Omega}|^{-1/2}\sum_{\mathbf{a}\in\Gamma}\exp(-i\langle\mathbf{k},\mathbf{x}+\mathbf{a}\rangle)\mathbf{v}(\mathbf{x}+\mathbf{a}),$$

$\mathbf{x} \in \mathbb{R}^d$, $\mathbf{k} \in \mathbb{R}^d$. The operator $\mathcal{U}$ extends by continuity to a unitary mapping

$$\mathcal{U} : L_2(\mathbb{R}^d;\mathbb{C}^n) \to \int_{\widetilde{\Omega}} \oplus L_2(\Omega;\mathbb{C}^n)\,d\mathbf{k} =: \mathcal{H}. \tag{5.1}$$

Let $\widetilde{H}^1(\Omega;\mathbb{C}^n)$ be the subspace of all functions in $H^1(\Omega;\mathbb{C}^n)$ such that the Γ-periodic extension of them to $\mathbb{R}^d$ belongs to $H^1_{\mathrm{loc}}(\mathbb{R}^d;\mathbb{C}^n)$. The relation $\mathbf{v} \in H^1(\mathbb{R}^d;\mathbb{C}^n)$ is equivalent to the fact that $\widetilde{\mathbf{v}}(\mathbf{k},\cdot) \in \widetilde{H}^1(\Omega;\mathbb{C}^n)$ for almost every $\mathbf{k} \in \widetilde{\Omega}$ and

$$\int_{\widetilde{\Omega}}\int_{\Omega}(|(\mathbf{D}+\mathbf{k})\widetilde{\mathbf{v}}(\mathbf{k},\mathbf{x})|^2 + |\widetilde{\mathbf{v}}(\mathbf{k},\mathbf{x})|^2)\,d\mathbf{x}\,d\mathbf{k} < \infty.$$

Under the transformation $\mathcal{U}$, the operator of multiplication by a bounded periodic matrix-valued function in $L_2(\mathbb{R}^d;\mathbb{C}^n)$ turns into multiplication by the same function on the fibers of the direct integral $\mathcal{H}$. On these fibers, the DO $b(\mathbf{D})$ applied to $\mathbf{v} \in H^1(\mathbb{R}^d;\mathbb{C}^n)$ turns into the operator $b(\mathbf{D}+\mathbf{k})$ applied to $\widetilde{\mathbf{v}}(\mathbf{k},\cdot) \in \widetilde{H}^1(\Omega;\mathbb{C}^n)$.

5.2. The operators $\mathcal{A}(\mathbf{k})$. We put

$$\mathfrak{H} = L_2(\Omega;\mathbb{C}^n), \quad \mathfrak{H}_* = L_2(\Omega;\mathbb{C}^m), \tag{5.2}$$

and consider the closed operator $\mathcal{X}(\mathbf{k}) : \mathfrak{H} \to \mathfrak{H}_*$, defined by the relations

$$\mathcal{X}(\mathbf{k}) = hb(\mathbf{D}+\mathbf{k}), \quad \operatorname{Dom}\mathcal{X}(\mathbf{k}) = \widetilde{H}^1(\Omega;\mathbb{C}^n). \tag{5.3}$$

The selfadjoint operator $\mathcal{A}(\mathbf{k}) = \mathcal{X}(\mathbf{k})^*\mathcal{X}(\mathbf{k}) : \mathfrak{H} \to \mathfrak{H}$ is generated by the quadratic form $a(\mathbf{k})[\mathbf{u},\mathbf{u}] = \|\mathcal{X}(\mathbf{k})\mathbf{u}\|^2_{\mathfrak{H}_*}$, $\mathbf{u} \in \widetilde{H}^1(\Omega;\mathbb{C}^n)$. Using the Fourier series expansion for $\mathbf{u}$ and conditions (4.1) and (4.2), it is easy to see that

$$\begin{gathered}\alpha_0\|g^{-1}\|^{-1}_{L_\infty}\|(\mathbf{D}+\mathbf{k})\mathbf{u}\|^2_{L_2(\Omega)} \leq a(\mathbf{k})[\mathbf{u},\mathbf{u}] \leq \alpha_1\|g\|_{L_\infty}\|(\mathbf{D}+\mathbf{k})\mathbf{u}\|^2_{L_2(\Omega)}, \\ \mathbf{u} \in \widetilde{H}^1(\Omega;\mathbb{C}^n).\end{gathered} \tag{5.4}$$

We put

$$\mathfrak{N} := \operatorname{Ker}\mathcal{A}(0) = \operatorname{Ker}\mathcal{X}(0). \tag{5.5}$$

Relations (5.4) with $\mathbf{k} = 0$ show that the kernel (5.5) consists of the constant vector-valued functions:

$$\mathfrak{N} = \{\mathbf{u} \in L_2(\Omega;\mathbb{C}^n) : \ \mathbf{u} = \mathbf{c} \in \mathbb{C}^n\}, \quad \dim\mathfrak{N} = n. \tag{5.6}$$

Using the Fourier series and the lower estimate (5.4), it is easy to check that

$$\mathcal{A}(\mathbf{k}) \geq c_*|\mathbf{k}|^2 I, \quad \mathbf{k} \in \widetilde{\Omega}, \quad c_* = \alpha_0\|g^{-1}\|^{-1}_{L_\infty} \tag{5.7}$$

(see [**BSu2**, Chapter 2, (2.12)]). According to [**BSu2**, Chapter 2, (3.6)], the distance d^0 from the point $\lambda = 0$ to the rest of the spectrum of $\mathcal{A}(0)$ satisfies the estimate

$$d^0 \geq 4c_* r_0^2. \tag{5.8}$$

5.3. The operators $\mathcal{Y}(\mathbf{k})$ and Q. In $L_2(\Omega;\mathbb{C}^n)$, we consider the operator

$$\mathcal{Y}(\mathbf{k}) = \sum_{j=1}^{d} a_j(\mathbf{x})(D_j + k_j), \quad \operatorname{Dom}\mathcal{Y}(\mathbf{k}) = \widetilde{H}^1(\Omega;\mathbb{C}^n).$$

Using (4.1), (4.2), (4.4), and (4.7), it is easy to check that

$$\|\mathcal{Y}(\mathbf{k})\mathbf{u}\|_{\mathfrak{H}} \leq c_1 \|\mathcal{X}(\mathbf{k})\mathbf{u}\|_{\mathfrak{H}_*}, \quad \mathbf{u} \in \widetilde{H}^1(\Omega;\mathbb{C}^n), \tag{5.9}$$

where the constant c_1 is defined by (4.6).

By Q we denote the operator in $L_2(\Omega;\mathbb{C}^n)$ of multiplication by the periodic matrix-valued function $Q(\mathbf{x})$.

5.4. The operator pencil $\mathcal{B}(\mathbf{k},\varepsilon)$. In $L_2(\Omega;\mathbb{C}^n)$, we consider the selfadjoint operator $\mathcal{B}(\mathbf{k},\varepsilon)$ formally given by

$$\mathcal{B}(\mathbf{k},\varepsilon) = \mathcal{A}(\mathbf{k}) + \varepsilon(\mathcal{Y}(\mathbf{k}) + \mathcal{Y}(\mathbf{k})^*) + c_0\varepsilon^2 Q. \tag{5.10}$$

The precise definition of the operator $\mathcal{B}(\mathbf{k};\varepsilon)$ is given in terms of the quadratic form

$$b(\mathbf{k},\varepsilon)[\mathbf{u},\mathbf{u}] = a(\mathbf{k})[\mathbf{u},\mathbf{u}] + 2\varepsilon \operatorname{Re}(\mathcal{Y}(\mathbf{k})\mathbf{u},\mathbf{u})_{\mathfrak{H}} + c_0\varepsilon^2(Q\mathbf{u},\mathbf{u})_{\mathfrak{H}}, \quad \mathbf{u} \in \widetilde{H}^1(\Omega;\mathbb{C}^n). \tag{5.11}$$

From (5.9) and (4.8) it follows that

$$b(\mathbf{k},\varepsilon)[\mathbf{u},\mathbf{u}] \geq \frac{1}{2} a(\mathbf{k})[\mathbf{u},\mathbf{u}] + \frac{c_1^2}{2}\varepsilon^2 \|\mathbf{u}\|_{\mathfrak{H}}^2, \quad \mathbf{u} \in \widetilde{H}^1(\Omega;\mathbb{C}^n). \tag{5.12}$$

Thus, the operator $\mathcal{B}(\mathbf{k},\varepsilon)$ is positive.

5.5. The direct integral for $\mathcal{B}(\varepsilon)$. Under the Gelfand transformation $\mathcal{U}$, the operator $\mathcal{B}(\varepsilon)$ is decomposed into the direct integral of operators $\mathcal{B}(\mathbf{k},\varepsilon)$:

$$\mathcal{U}\mathcal{B}(\varepsilon)\mathcal{U}^{-1} = \int_{\widetilde{\Omega}} \oplus \mathcal{B}(\mathbf{k},\varepsilon)\, d\mathbf{k}. \tag{5.13}$$

In detail, what was said means the following. For $\mathbf{v} \in H^1(\mathbb{R}^d;\mathbb{C}^n)$, we have

$$\widetilde{\mathbf{v}}(\mathbf{k},\cdot) \in \widetilde{H}^1(\Omega;\mathbb{C}^n) \text{ for a.e. } \mathbf{k} \in \widetilde{\Omega}, \tag{5.14}$$

$$b(\varepsilon)[\mathbf{v},\mathbf{v}] = \int_{\widetilde{\Omega}} b(\mathbf{k},\varepsilon)[\widetilde{\mathbf{v}}(\mathbf{k},\cdot),\widetilde{\mathbf{v}}(\mathbf{k},\cdot)]\, d\mathbf{k}. \tag{5.15}$$

Conversely, if $\widetilde{\mathbf{v}} \in \mathcal{H}$ satisfies (5.14) and the integral in (5.15) is finite, then $\mathbf{v} \in H^1(\mathbb{R}^d;\mathbb{C}^n)$ and (5.15) is valid.

§6. Incorporation of the operators $\mathcal{B}(\mathbf{k},\varepsilon)$ into the abstract scheme

6.1. If $d>1$, the operators $\mathcal{B}(\mathbf{k},\varepsilon)$ depend on the multidimensional parameter $\mathbf{k}$. As in [**BSu2**, Chapter 2], we distinguish the one-dimensional parameter t, putting $\mathbf{k}=t\boldsymbol{\theta}$, $t=|\mathbf{k}|$, $\boldsymbol{\theta}\in\mathbb{S}^{d-1}$. We will apply the method of Chapter 1. Now all the objects will depend on the additional parameter $\boldsymbol{\theta}$, and we must make our constructions and estimates uniform in $\boldsymbol{\theta}$. The spaces $\mathfrak{H}$ and $\mathfrak{H}_*$ are defined by (5.2). We put $X(t)=X(t;\boldsymbol{\theta}):=\mathcal{X}(t\boldsymbol{\theta})$. According to (5.3), $X(t;\boldsymbol{\theta})=X_0+tX_1(\boldsymbol{\theta})$, where $X_0=h(\mathbf{x})b(\mathbf{D})$, $\operatorname{Dom}X_0=\widetilde{H}^1(\Omega;\mathbb{C}^n)$, and $X_1(\boldsymbol{\theta})$ is the bounded operator of multiplication by the matrix $h(\mathbf{x})b(\boldsymbol{\theta})$. Next, we put $A(t)=A(t;\boldsymbol{\theta}):=\mathcal{A}(t\boldsymbol{\theta})$. According to (5.5) and (5.6), the kernel $\mathfrak{N}=\operatorname{Ker}X_0=\operatorname{Ker}\mathcal{A}(0)$ is n-dimensional. Condition 1.1 is fulfilled. The number d^0 satisfies estimate (5.8). As has been shown in [**BSu2**, Chapter 2, §3], the condition $n\le n_*=\dim\operatorname{Ker}X_0^*$ is also satisfied. Moreover, either $n_*=n$ (if $m=n$), or $n_*=\infty$ (if $m>n$). The orthogonal projection P of $\mathfrak{H}=L_2(\Omega;\mathbb{C}^n)$ onto the subspace $\mathfrak{N}=\mathbb{C}^n$ is the operator of averaging over the cell Ω: $P\mathbf{u}=|\Omega|^{-1}\int_\Omega\mathbf{u}(\mathbf{x})\,d\mathbf{x}$.

The role of $Y(t)$ is played by the operator $Y(t;\boldsymbol{\theta}):=\mathcal{Y}(t\boldsymbol{\theta})$. Herewith, $Y(t;\boldsymbol{\theta})=Y_0+tY_1(\boldsymbol{\theta})$, where $Y_0=\sum_{j=1}^d a_j(\mathbf{x})D_j$, $\operatorname{Dom}Y_0=\widetilde{H}^1(\Omega;\mathbb{C}^n)$, and $Y_1(\boldsymbol{\theta})$ is the bounded operator of multiplication by the matrix-valued function $\sum_{j=1}^d a_j(\mathbf{x})\theta_j$. Condition 1.2 is satisfied due to estimate (5.9).

Finally, the role of the operator pencil $B(t,\varepsilon)$ (see (1.2)) is played by the operator family (5.10): $B(t,\varepsilon;\boldsymbol{\theta}):=\mathcal{B}(t\boldsymbol{\theta},\varepsilon)$. Now the corresponding form (5.11) will be denoted by $b(t,\varepsilon;\boldsymbol{\theta})$. Estimate (1.5) corresponds to inequality (5.12).

Thus, now all the assumptions of the abstract scheme are fulfilled.

6.2. According to Subsection 1.4, we should fix a positive number δ such that $\delta<d^0/14$. Taking (5.8) and (5.7) into account, we fix δ as follows:

$$\delta=\frac{1}{4}\alpha_0\|g^{-1}\|^{-1}_{L_\infty}r_0^2. \tag{6.1}$$

Note that

$$\|X_1(\boldsymbol{\theta})\|\le\|g\|^{1/2}_{L_\infty}\alpha_1^{1/2},\quad \|Y_1(\boldsymbol{\theta})\|\le C_a, \tag{6.2}$$

where α_1 is the constant from (4.1), and C_a is defined by (4.7). Instead of the precise value $\tau_0(\boldsymbol{\theta})=\delta^{1/2}(\|X_1(\boldsymbol{\theta})\|^2+\|Y_1(\boldsymbol{\theta})\|+c_0\|Q\|)^{-1/2}$ (see (1.7)), we take

$$\begin{aligned}\tau_0&=\delta^{1/2}(\|g\|_{L_\infty}\alpha_1+C_a+c_0\|Q\|_{L_\infty})^{-1/2}\\&=\frac{1}{2}\alpha_0^{1/2}\|g^{-1}\|^{-1/2}_{L_\infty}r_0(\|g\|_{L_\infty}\alpha_1+C_a+c_0\|Q\|_{L_\infty})^{-1/2}.\end{aligned} \tag{6.3}$$

(The value (6.3) is suitable for all $\boldsymbol{\theta}\in\mathbb{S}^{d-1}$.)

By (5.7), condition (3.11) is valid. Then, by (5.12), the operator $B(t,\varepsilon;\boldsymbol{\theta})$ satisfies a condition of the form (3.12):

$$B(t,\varepsilon;\boldsymbol{\theta})\ge\check{c}_*(t^2+\varepsilon^2)I,\quad \mathbf{k}=t\boldsymbol{\theta}\in\widetilde{\Omega},\quad \varepsilon>0, \tag{6.4}$$

$$\check{c}_*=\frac{1}{2}\min\{c_*,c_1^2\}=\frac{1}{2}\min\{\alpha_0\|g^{-1}\|^{-1}_{L_\infty},\alpha_0^{-1}\|g^{-1}\|_{L_\infty}C_a^2\}. \tag{6.5}$$

6.3. The operators $Z(\boldsymbol{\theta})$ and $\widetilde{Z}$. Now the operator Z introduced in Subsection 1.5 depends on $\boldsymbol{\theta}$. Let $\mathbf{c} \in \mathfrak{N} = \mathbb{C}^n$. Equation (1.12) takes the form

$$b(\mathbf{D})^* g(\mathbf{x})(b(\mathbf{D})\mathbf{v}_{\boldsymbol{\theta}} + b(\boldsymbol{\theta})\mathbf{c}) = 0. \tag{6.6}$$

Let $\mathbf{v}_{\boldsymbol{\theta}} \in \widetilde{H}^1(\Omega;\mathbb{C}^n)$ be the solution of equation (6.6) satisfying the condition $\int_\Omega \mathbf{v}_{\boldsymbol{\theta}}\, d\mathbf{x} = 0$. (Equation (6.6) is understood in the weak sense.) Then $\mathbf{v}_{\boldsymbol{\theta}} = Z(\boldsymbol{\theta})\mathbf{c}$.

Let $\Lambda(\mathbf{x})$ be the periodic $(n\times m)$-matrix-valued function satisfying the equation

$$b(\mathbf{D})^* g(\mathbf{x})(b(\mathbf{D})\Lambda(\mathbf{x}) + \mathbf{1}_m) = 0, \quad \int_\Omega \Lambda(\mathbf{x})\, d\mathbf{x} = 0.$$

Obviously, $Z(\boldsymbol{\theta}) : \mathfrak{N} \to \mathfrak{H}$ is the operator of multiplication by the matrix $\Lambda(\mathbf{x})b(\boldsymbol{\theta})$.

Now we construct the operator $\widetilde{Z}$ (see Subsection 1.5). Let $\mathbf{c} \in \mathfrak{N} = \mathbb{C}^n$. Equation (1.14) takes the form

$$b(\mathbf{D})^* g(\mathbf{x}) b(\mathbf{D})\mathbf{w} + \sum_{j=1}^d D_j(a_j(\mathbf{x})^*\mathbf{c}) = 0. \tag{6.7}$$

Let $\mathbf{w} \in \widetilde{H}^1(\Omega;\mathbb{C}^n)$ be the (weak) periodic solution of the equation (6.7) satisfying the condition $\int_\Omega \mathbf{w}\, d\mathbf{x} = 0$. Then $\mathbf{w} = \widetilde{Z}\mathbf{c}$. Let $\widetilde{\Lambda}(\mathbf{x})$ be the periodic $(n\times n)$-matrix-valued function satisfying the equation

$$b(\mathbf{D})^* g(\mathbf{x}) b(\mathbf{D})\widetilde{\Lambda} + \sum_{j=1}^d D_j a_j(\mathbf{x})^* = 0, \quad \int_\Omega \widetilde{\Lambda}(\mathbf{x})\, d\mathbf{x} = 0. \tag{6.8}$$

Obviously, the operator $\widetilde{Z} : \mathfrak{N} \to \mathfrak{H}$ acts as multiplication by the matrix $\widetilde{\Lambda}$.

6.4. The operator $S(\boldsymbol{\theta})$. Now the operator S defined in Subsection 1.6 depends on $\boldsymbol{\theta}$. According to [**BSu2**, Chapter 3, §1], the operator $S(\boldsymbol{\theta}) : \mathfrak{N} \to \mathfrak{N}$ acts as multiplication by the matrix $S(\boldsymbol{\theta}) = b(\boldsymbol{\theta})^* g^0 b(\boldsymbol{\theta})$, $\boldsymbol{\theta} \in \mathbb{S}^{d-1}$. Here g^0 is a constant positive $(m \times m)$-matrix called the *effective* matrix. The effective matrix g^0 can be defined in terms of the matrix $\Lambda(\mathbf{x})$ introduced above by the relation

$$g^0 = |\Omega|^{-1} \int_\Omega g(\mathbf{x})(b(\mathbf{D})\Lambda(\mathbf{x}) + \mathbf{1}_m)\, d\mathbf{x}. \tag{6.9}$$

6.5. The operator $L(t,\varepsilon)$. Now the operator $L(t,\varepsilon)$ defined by (3.20) depends on $\boldsymbol{\theta}$ and is represented as

$$\begin{aligned} L(t,\varepsilon;\boldsymbol{\theta}) &= t^2 S(\boldsymbol{\theta}) + t\varepsilon\left(-(X_0 Z(\boldsymbol{\theta}))^*(X_0\widetilde{Z}) - (X_0\widetilde{Z})^*(X_0 Z(\boldsymbol{\theta}))\right) \\ &\quad + t\varepsilon\, P(Y_1(\boldsymbol{\theta}) + Y_1(\boldsymbol{\theta})^*)|_{\mathfrak{N}} + \varepsilon^2\left(-(X_0\widetilde{Z})^*(X_0\widetilde{Z}) + c_0 PQ|_{\mathfrak{N}}\right). \end{aligned} \tag{6.10}$$

The operator $X_0 Z(\boldsymbol{\theta}) : \mathfrak{N} \to \mathfrak{H}$ acts on constant vectors $\mathbf{c} \in \mathfrak{N}$ as multiplication by the matrix $h(\mathbf{x})(b(\mathbf{D})\Lambda(\mathbf{x}))b(\boldsymbol{\theta})$, and the operator $X_0\widetilde{Z} : \mathfrak{N} \to \mathfrak{H}$ acts as multiplication by the matrix $h(\mathbf{x})(b(\mathbf{D})\widetilde{\Lambda}(\mathbf{x}))$. Then $(X_0 Z(\boldsymbol{\theta}))^*(X_0\widetilde{Z}) = Pb(\boldsymbol{\theta})^*(b(\mathbf{D})\Lambda)^* g(b(\mathbf{D})\widetilde{\Lambda})|_{\mathfrak{N}}$; i.e., the operator $(X_0 Z(\boldsymbol{\theta}))^*(X_0\widetilde{Z}) : \mathfrak{N} \to \mathfrak{N}$ is multiplication by the matrix $b(\boldsymbol{\theta})^* V$, where

$$V := |\Omega|^{-1} \int_\Omega (b(\mathbf{D})\Lambda(\mathbf{x}))^* g(\mathbf{x})(b(\mathbf{D})\widetilde{\Lambda}(\mathbf{x}))\, d\mathbf{x}. \tag{6.11}$$

The adjoint operator $(X_0\widetilde{Z})^*(X_0 Z(\boldsymbol{\theta})) : \mathfrak{N} \to \mathfrak{N}$ is represented by the matrix $V^* b(\boldsymbol{\theta})$. Finally, $(X_0\widetilde{Z})^*(X_0\widetilde{Z}) : \mathfrak{N} \to \mathfrak{N}$ is the operator of multiplication by the matrix

$$W := |\Omega|^{-1} \int_\Omega (b(\mathbf{D})\widetilde{\Lambda}(\mathbf{x}))^* g(\mathbf{x})(b(\mathbf{D})\widetilde{\Lambda}(\mathbf{x}))\, d\mathbf{x}. \tag{6.12}$$

Obviously, the operator $P(Y_1(\boldsymbol{\theta}) + Y_1(\boldsymbol{\theta})^*)|_{\mathfrak{N}}$ acts as multiplication by the matrix $\sum_{j=1}^d (\overline{a_j + a_j^*})\theta_j$, and the operator $PQ|_{\mathfrak{N}}$ acts as multiplication by the matrix $\overline{Q}$. Here $\overline{Q} = |\Omega|^{-1}\int_\Omega Q(\mathbf{x})\,d\mathbf{x}$, and $\overline{a_j + a_j^*}$ is defined similarly.

As a result, (6.10) takes the form

$$\begin{aligned} L(t,\varepsilon;\boldsymbol{\theta}) = t^2 b(\boldsymbol{\theta})^* g^0 b(\boldsymbol{\theta}) + t\varepsilon \left(-b(\boldsymbol{\theta})^* V - V^* b(\boldsymbol{\theta}) + \sum_{j=1}^d (\overline{a_j + a_j^*})\theta_j \right) \\ + \varepsilon^2 \left(-W + c_0\overline{Q}\right). \end{aligned} \tag{6.13}$$

Since the symbol b is homogeneous of first order, we can rewrite (6.13) in terms of the parameter $\mathbf{k} = t\boldsymbol{\theta}$: $L(t,\varepsilon;\boldsymbol{\theta}) = L(\mathbf{k},\varepsilon)$,

$$L(\mathbf{k},\varepsilon) = b(\mathbf{k})^* g^0 b(\mathbf{k}) + \varepsilon \left(-b(\mathbf{k})^* V - V^* b(\mathbf{k}) + \sum_{j=1}^d (\overline{a_j + a_j^*}) k_j \right) + \varepsilon^2 \left(-W + c_0\overline{Q}\right). \tag{6.14}$$

Due to the presence of the projection P in the operator $L(\mathbf{k},\varepsilon)P$, the vector $\mathbf{k}$ can be replaced by $(\mathbf{D}+\mathbf{k})$ in this operator. We put

$$\begin{aligned} \mathcal{A}^0(\mathbf{k}) = b(\mathbf{D}+\mathbf{k})^* g^0 b(\mathbf{D}+\mathbf{k}), \quad \mathcal{Y}^0(\mathbf{k}) = -b(\mathbf{D}+\mathbf{k})^* V + \sum_{j=1}^d \overline{a_j}(D_j + k_j), \\ \mathcal{B}^0(\mathbf{k},\varepsilon) = \mathcal{A}^0(\mathbf{k}) + \varepsilon(\mathcal{Y}^0(\mathbf{k}) + \mathcal{Y}^0(\mathbf{k})^*) + \varepsilon^2(c_0\overline{Q} - W). \end{aligned} \tag{6.15}$$

Then

$$L(\mathbf{k},\varepsilon)P = \mathcal{B}^0(\mathbf{k},\varepsilon)P. \tag{6.16}$$

6.6. Approximation for the operator $\mathcal{B}(\mathbf{k},\varepsilon)^{-1}$. We rely on Theorem 3.2, which will be applied to the operator $B(t,\varepsilon;\boldsymbol{\theta}) = \mathcal{B}(\mathbf{k},\varepsilon)$. First, we have to specify the constants. Using (6.2), instead of the precise value of the constant C_T (see (2.15)), which depends on $\boldsymbol{\theta}$ now, we take

$$C_T = \sqrt{3}(c_1^2 + \alpha_1\|g\|_{L_\infty})^{1/2} + \frac{3}{2}(\alpha_1\|g\|_{L_\infty} + C_a + c_0\|Q\|_{L_\infty})^{1/2}.$$

Here C_a is defined by (4.7), and c_1 is defined by (4.6). Similarly, instead of the precise values of the constants (2.21) and (2.26), which now depend on $\boldsymbol{\theta}$, we take

$$\begin{aligned} \widetilde{C}_T &= 1224 C_T(\alpha_1\|g\|_{L_\infty} + C_a + c_0\|Q\|_{L_\infty}), \\ \check{C}_T &= 144\cdot 17^2 C_T(c_1^2 + \alpha_1\|g\|_{L_\infty}) \\ &\quad + 1224\sqrt{3}(c_1^2 + \alpha_1\|g\|_{L_\infty})^{1/2}(\alpha_1\|g\|_{L_\infty} + C_a + c_0\|Q\|_{L_\infty}). \end{aligned}$$

By (2.28) and (3.8), $C_1 = 48(1+\pi^{-1})\delta^{-1/2}C_T$, $C_2 = 2(1+\pi^{-1})\delta^{-1/2}(\widetilde{C}_T + \check{C}_T)$. The number δ is defined by (6.1). Finally, according to (3.23), the constant C_3 is defined as follows: $C_3 = \tau_0(3\delta)^{-1} + 2C_1\check{c}_*^{-1} + C_2\check{c}_*^{-2}$, where τ_0 is defined by (6.3), and $\check{c}_*$ is defined by (6.5). We see that the constant C_3 can be expressed in terms of α_0, α_1, $\|g\|_{L_\infty}$, $\|g^{-1}\|_{L_\infty}$, $\|a_j\|_{L_\infty}$, $j = 1, \dots, d$, $\|Q\|_{L_\infty}$, c_0, and r_0. Recall that c_0 is subject to the restriction (4.8), which is determined by α_0, $\|a_j\|_{L_\infty}$, $j = 1, \dots, d$, $\|g^{-1}\|_{L_\infty}$, and $\|Q^{-1}\|_{L_\infty}$.

Applying Theorem 3.2 and taking (6.16) into account, we arrive at the estimate

$$\|\mathcal{B}(\mathbf{k},\varepsilon)^{-1} - \mathcal{B}^0(\mathbf{k},\varepsilon)^{-1}P\|_{\mathfrak{H}\to\mathfrak{H}} \le C_3(|\mathbf{k}|^2+\varepsilon^2)^{-1/2}, \quad |\mathbf{k}|^2+\varepsilon^2 \le \tau_0^2. \tag{6.17}$$

6.7. Estimates in the case where $|\mathbf{k}|^2 + \varepsilon^2 > \tau_0^2$ are trivial. By (6.4), for $|\mathbf{k}|^2 + \varepsilon^2 > \tau_0^2$ we have

$$\|\mathcal{B}(\mathbf{k},\varepsilon)^{-1}\| \le \check{c}_*^{-1}\tau_0^{-1}(|\mathbf{k}|^2+\varepsilon^2)^{-1/2}, \quad \mathbf{k}\in\widetilde{\Omega}, \quad |\mathbf{k}|^2+\varepsilon^2 > \tau_0^2. \tag{6.18}$$

According to (3.21), the operator (6.14) satisfies the estimate

$$L(\mathbf{k},\varepsilon) \ge \check{c}_*(|\mathbf{k}|^2+\varepsilon^2)I_{\mathfrak{N}}, \tag{6.19}$$

whence, by (6.16),

$$\|\mathcal{B}^0(\mathbf{k},\varepsilon)^{-1}P\| \le \check{c}_*^{-1}\tau_0^{-1}(|\mathbf{k}|^2+\varepsilon^2)^{-1/2}, \quad \mathbf{k}\in\widetilde{\Omega}, \quad |\mathbf{k}|^2+\varepsilon^2 > \tau_0^2. \tag{6.20}$$

By (6.18) and (6.20), we obtain

$$\|\mathcal{B}(\mathbf{k},\varepsilon)^{-1} - \mathcal{B}^0(\mathbf{k},\varepsilon)^{-1}P\|_{\mathfrak{H}\to\mathfrak{H}} \le 2\check{c}_*^{-1}\tau_0^{-1}(|\mathbf{k}|^2+\varepsilon^2)^{-1/2}, \quad \mathbf{k}\in\widetilde{\Omega}, \quad |\mathbf{k}|^2+\varepsilon^2 > \tau_0^2. \tag{6.21}$$

Now (6.17) and (6.21) directly imply the following statement.

THEOREM 6.1. *Let $\mathcal{B}(\mathbf{k},\varepsilon)$ be the operator defined in Subsection 5.4, and let $\mathcal{B}^0(\mathbf{k},\varepsilon)$ be the operator defined by (6.15). Let P be the orthogonal projection of $\mathfrak{H} = L_2(\Omega;\mathbb{C}^n)$ onto the subspace $\mathfrak{N} = \mathbb{C}^n$ of constant functions. Then we have*

$$\|\mathcal{B}(\mathbf{k},\varepsilon)^{-1} - \mathcal{B}^0(\mathbf{k},\varepsilon)^{-1}P\|_{\mathfrak{H}\to\mathfrak{H}} \le C_4(|\mathbf{k}|^2+\varepsilon^2)^{-1/2}, \quad \mathbf{k}\in\widetilde{\Omega}, \quad \varepsilon > 0. \tag{6.22}$$

The constant $C_4 = \max\{C_3, 2\check{c}_*^{-1}\tau_0^{-1}\}$ depends only on α_0, α_1, $\|g\|_{L_\infty}$, $\|g^{-1}\|_{L_\infty}$, $\|a_j\|_{L_\infty}$, $j = 1, \dots, d$, $\|Q\|_{L_\infty}$, c_0, and r_0.

§7. Approximation for the operator $\mathcal{B}(\varepsilon)^{-1}$

7.1. We return to the study of the operator $\mathcal{B}(\varepsilon)$ acting in $L_2(\mathbb{R}^d;\mathbb{C}^n)$. It was introduced in Subsection 4.4. According to [**BSu2**, Chapter 3, §1], the operator

$$\mathcal{A}^0 = b(\mathbf{D})^* g^0 b(\mathbf{D}) \tag{7.1}$$

is called the *effective operator* for $\mathcal{A} = b(\mathbf{D})^* g(\mathbf{x}) b(\mathbf{D})$. The effective matrix g^0 is defined by (6.9). Next, we put

$$\mathcal{Y}^0 = -b(\mathbf{D})^* V + \sum_{j=1}^{d} \overline{a_j} D_j, \tag{7.2}$$

where the matrix V is defined by (6.11). Consider the operator

$$\mathcal{B}^0(\varepsilon) = \mathcal{A}^0 + \varepsilon(\mathcal{Y}^0 + (\mathcal{Y}^0)^*) + \varepsilon^2(c_0\overline{Q} - W). \tag{7.3}$$

Here W is defined by (6.12). The operator $\mathcal{B}^0(\varepsilon)$ is a second order DO with constant coefficients. Its symbol $L(\boldsymbol{\xi},\varepsilon)$ is defined by (6.14).

Next, we put $\Pi = \mathcal{U}^{-1}[P]\mathcal{U}$, where $[P]$ is the operator in $\mathcal{H}$ (see (5.1)), which acts on the fibers as the operator P of averaging over the cell. It is easily seen (cf.

[**BSu4**, §6]) that the operator Π is the pseudodifferential operator in $L_2(\mathbb{R}^d;\mathbb{C}^n)$ of zero order with the symbol $\chi_{\widetilde{\Omega}}(\boldsymbol{\xi})$. Here $\chi_{\widetilde{\Omega}}(\boldsymbol{\xi})$ is the indicator of the set $\widetilde{\Omega}$. Thus,

$$(\Pi\mathbf{f})(\mathbf{x}) = (2\pi)^{-d/2}\int_{\widetilde{\Omega}} e^{i\langle\mathbf{x},\boldsymbol{\xi}\rangle}\widehat{\mathbf{f}}(\boldsymbol{\xi})\,d\boldsymbol{\xi},$$

where $\widehat{\mathbf{f}}(\boldsymbol{\xi})$ is the Fourier image of $\mathbf{f}(\mathbf{x})$.

By (5.13) and by a similar decomposition for the operator (7.3), we have

$$\begin{aligned}&\|\mathcal{B}(\varepsilon)^{-1} - \mathcal{B}^0(\varepsilon)^{-1}\Pi\|_{L_2(\mathbb{R}^d;\mathbb{C}^n)\to L_2(\mathbb{R}^d;\mathbb{C}^n)}\\ &\quad= \sup_{\mathbf{k}\in\widetilde{\Omega}}\|\mathcal{B}(\mathbf{k},\varepsilon)^{-1} - \mathcal{B}^0(\mathbf{k},\varepsilon)^{-1}P\|_{L_2(\Omega;\mathbb{C}^n)\to L_2(\Omega;\mathbb{C}^n)}.\end{aligned}$$

Combining this with estimate (6.22), we obtain

$$\|\mathcal{B}(\varepsilon)^{-1} - \mathcal{B}^0(\varepsilon)^{-1}\Pi\|_{L_2(\mathbb{R}^d;\mathbb{C}^n)\to L_2(\mathbb{R}^d;\mathbb{C}^n)} \le C_4\varepsilon^{-1},\quad \varepsilon>0. \tag{7.4}$$

7.2. Elimination of the operator Π. Now we show that the operator Π can be replaced by I in (7.4). Under such a replacement, only the constant in the estimate will change. Consider the operator $\mathcal{B}^0(\varepsilon)^{-1}(I-\Pi)$, which is the pseudodifferential operator with the symbol $L(\boldsymbol{\xi},\varepsilon)^{-1}(1-\chi_{\widetilde{\Omega}}(\boldsymbol{\xi}))$, where the matrix-valued function $L(\boldsymbol{\xi},\varepsilon)$ is defined by (6.14). Obviously, $1-\chi_{\widetilde{\Omega}}(\boldsymbol{\xi})=0$ for $|\boldsymbol{\xi}|<r_0$. By (6.19), we obtain

$$\|\mathcal{B}^0(\varepsilon)^{-1}(I-\Pi)\|_{L_2\to L_2} = \sup_{\boldsymbol{\xi}\in\mathbb{R}^d}|1-\chi_{\widetilde{\Omega}}(\boldsymbol{\xi})||L(\boldsymbol{\xi},\varepsilon)^{-1}| \le (\check{c}_*)^{-1}r_0^{-1}\varepsilon^{-1}. \tag{7.5}$$

Relations (7.4) and (7.5) directly imply the following statement.

THEOREM 7.1. *Let $\mathcal{B}(\varepsilon)$ be the operator defined in Subsection* 4.4, *and let $\mathcal{B}^0(\varepsilon)$ be the operator defined by* (7.3). *Then we have*

$$\|\mathcal{B}(\varepsilon)^{-1} - \mathcal{B}^0(\varepsilon)^{-1}\|_{L_2(\mathbb{R}^d;\mathbb{C}^n)\to L_2(\mathbb{R}^d;\mathbb{C}^n)} \le C_5\varepsilon^{-1},\quad \varepsilon>0. \tag{7.6}$$

The constant $C_5 = C_4 + (\check{c}_*)^{-1}r_0^{-1}$ depends only on α_0, α_1, $\|g\|_{L_\infty}$, $\|g^{-1}\|_{L_\infty}$, $\|a_j\|_{L_\infty}$, $j=1,\dots,d$, $\|Q\|_{L_\infty}$, c_0, and r_0.

§8. Homogenization for the operator $\mathcal{B}_\varepsilon$

8.1. Setting of the problem. For any Γ-periodic function $\phi(\mathbf{x})$, $\mathbf{x}\in\mathbb{R}^d$, we denote $\phi^\varepsilon(\mathbf{x}) = \phi(\varepsilon^{-1}\mathbf{x})$. In $L_2(\mathbb{R}^d;\mathbb{C}^n)$, we consider the operators

$$\mathcal{A}_\varepsilon = b(\mathbf{D})^* g^\varepsilon(\mathbf{x}) b(\mathbf{D}),\quad \mathcal{Y}_\varepsilon = \sum_{j=1}^d a_j^\varepsilon(\mathbf{x})D_j,$$

and also the operator Q^ε of multiplication by the matrix-valued function $Q^\varepsilon(\mathbf{x})$. The conditions on $g(\mathbf{x})$, $b(\mathbf{D})$, $a_j(\mathbf{x})$, and $Q(\mathbf{x})$ are the same as above (see Subsections 4.2 and 4.3).

The operator

$$\mathcal{B}_\varepsilon = \mathcal{A}_\varepsilon + \mathcal{Y}_\varepsilon + \mathcal{Y}_\varepsilon^* + c_0 Q^\varepsilon \tag{8.1}$$

is our main object. The constant c_0 is subject to the condition (4.8). The precise definition of the operator $\mathcal{B}_\varepsilon$ is given in terms of the quadratic form

$$b_\varepsilon[\mathbf{u},\mathbf{u}] = \int_{\mathbb{R}^d}\left(\langle g^\varepsilon(\mathbf{x})b(\mathbf{D})\mathbf{u}, b(\mathbf{D})\mathbf{u}\rangle + 2\mathrm{Re}\sum_{j=1}^d\langle a_j^\varepsilon(\mathbf{x})D_j\mathbf{u},\mathbf{u}\rangle + c_0\langle Q^\varepsilon(\mathbf{x})\mathbf{u},\mathbf{u}\rangle\right)d\mathbf{x},$$

$\mathbf{u} \in H^1(\mathbb{R}^d; \mathbb{C}^n)$. The coefficients of the operator (8.1) are rapidly oscillating as $\varepsilon \to 0$. Similarly to (4.9), we have

$$b_\varepsilon[\mathbf{u}, \mathbf{u}] \geq \frac{1}{2} \int_{\mathbb{R}^d} \langle g^\varepsilon(\mathbf{x}) b(\mathbf{D})\mathbf{u}, b(\mathbf{D})\mathbf{u} \rangle \, d\mathbf{x} + \frac{c_1^2}{2} \|\mathbf{u}\|_{L_2}^2, \quad \mathbf{u} \in H^1(\mathbb{R}^d; \mathbb{C}^n).$$

Thus, the operator (8.1) is positive definite.

The homogenization problem for the operator (8.1) *is to approximate the inverse operator* $\mathcal{B}_\varepsilon^{-1}$ *as* $\varepsilon \to 0$.

8.2. The main result. Let $\mathcal{A}^0$ be the operator defined by (7.1), let $\mathcal{Y}^0$ be defined by (7.2), and let W be the matrix (6.12). The operator

$$\mathcal{B}^0 = \mathcal{A}^0 + \mathcal{Y}^0 + (\mathcal{Y}^0)^* + c_0 \overline{Q} - W \tag{8.2}$$

is called the *effective operator* for $\mathcal{B}_\varepsilon$. The operator (8.2) is a second order DO with constant coefficients.

The following theorem is the main result of the paper.

THEOREM 8.1. *Let* $\mathcal{B}_\varepsilon$ *be the operator defined by* (8.1), *and let* $\mathcal{B}^0$ *be the effective operator* (8.2). *Then we have*

$$\|\mathcal{B}_\varepsilon^{-1} - (\mathcal{B}^0)^{-1}\|_{L_2 \to L_2} \leq C_5 \varepsilon, \quad \varepsilon > 0. \tag{8.3}$$

The constant C_5 *depends only on* α_0, α_1, $\|g\|_{L_\infty}$, $\|g^{-1}\|_{L_\infty}$, $\|a_j\|_{L_\infty}$, $j = 1, \dots, d$, $\|Q\|_{L_\infty}$, c_0, *and* r_0.

REMARK 8.2. 1) Under the additional condition $\sum_{j=1}^d D_j a_j^*(\mathbf{x}) = 0$, we have $\widetilde{\Lambda}(\mathbf{x}) = 0$ and $V = W = 0$ (see (6.8), (6.11), (6.12)). Then the effective operator simplifies: $\mathcal{B}^0 = \mathcal{A}^0 + \sum_{j=1}^d (\overline{a_j + a_j^*}) D_j + c_0 \overline{Q}$, i.e., the lower terms are homogenized in a trivial way. 2) If $\sum_{j=1}^d D_j a_j^*(\mathbf{x})$ is not identically zero, then the periodic solution $\widetilde{\Lambda}(\mathbf{x})$ of the problem (6.8) is nontrivial. In this case, according to (6.12), the matrix W is positive. Hence, the term of zero order in $\mathcal{B}^0$, which is equal to $c_0 \overline{Q} - W$, is less than $c_0 \overline{Q}$.

8.3. Proof of Theorem 8.1. Theorem 8.1 is deduced from Theorem 7.1 with the help of the *scaling transformation.* Let T_ε be the unitary scaling transformation in $L_2(\mathbb{R}^d; \mathbb{C}^n)$ given by $(T_\varepsilon \mathbf{u})(\mathbf{x}) = \varepsilon^{d/2} \mathbf{u}(\varepsilon \mathbf{x})$. Obviously, we have $\mathcal{A}_\varepsilon = \varepsilon^{-2} T_\varepsilon^* \mathcal{A} T_\varepsilon$, $\mathcal{Y}_\varepsilon = \varepsilon^{-1} T_\varepsilon^* \mathcal{Y} T_\varepsilon$, $Q^\varepsilon = T_\varepsilon^* Q T_\varepsilon$. Hence, $\mathcal{B}_\varepsilon = \varepsilon^{-2} T_\varepsilon^* \mathcal{B}(\varepsilon) T_\varepsilon$, where the operator $\mathcal{B}(\varepsilon)$ is defined in Subsection 4.4. The operator (8.2) satisfies a similar relation: $\mathcal{B}^0 = \varepsilon^{-2} T_\varepsilon^* \mathcal{B}^0(\varepsilon) T_\varepsilon$, where the operator $\mathcal{B}^0(\varepsilon)$ is defined by (7.3). Hence,

$$\mathcal{B}_\varepsilon^{-1} - (\mathcal{B}^0)^{-1} = \varepsilon^2 T_\varepsilon^* (\mathcal{B}(\varepsilon)^{-1} - \mathcal{B}^0(\varepsilon)^{-1}) T_\varepsilon. \tag{8.4}$$

Now, inequality (8.3) is a direct consequence of (8.4), (7.6), and the unitary property of the operators T_ε in $L_2(\mathbb{R}^d; \mathbb{C}^n)$. This completes the proof of Theorem 8.1.

8.4. Homogenization problem for the operator with a singular potential.[1] Let $p(\mathbf{x})$ be a Hermitian Γ-periodic $(n \times n)$-matrix-valued function in $\mathbb{R}^d$ such that

$$p \in L_q(\Omega), \quad q > d; \quad \int_\Omega p(\mathbf{x}) \, d\mathbf{x} = 0. \tag{8.5}$$

[1] In essence, the results of Subsection 8.4 were obtained in discussions with M. Sh. Birman.

Consider the homogenization problem for the operator

$$\widetilde{\mathcal{B}}_\varepsilon = \mathcal{A}_\varepsilon + \varepsilon^{-1}p^\varepsilon + c_0 Q^\varepsilon, \tag{8.6}$$

which contains a "singular" potential $\varepsilon^{-1}p(\varepsilon^{-1}\mathbf{x})$. Note that the homogenization problem for the operator (8.6) with $n = 1$, $\mathcal{A}_\varepsilon = -\operatorname{div} g^\varepsilon \nabla$ was considered in the book [**BeLP**, Chapter 1, Section 12]. However, in [**BeLP**] the question of an estimate of the form (8.3) was not studied.

We show that the operator (8.6) is a particular case of (8.1). The matrix-valued function $p(\mathbf{x})$ can be (nonuniquely) represented as

$$p(\mathbf{x}) = -\sum_{j=1}^{d} D_j a_j(\mathbf{x}), \tag{8.7}$$

where $a_j(\mathbf{x})$ are Γ-periodic $(n \times n)$-matrix-valued functions such that

$$a_j \in L_\infty, \quad \overline{a_j} = 0, \quad a_j(\mathbf{x})^* = -a_j(\mathbf{x}), \quad j = 1, \dots, d. \tag{8.8}$$

One possible way to find such a representation is the following. Let $\Phi(\mathbf{x})$ be the Γ-periodic $(n \times n)$-matrix-valued function satisfying the equation

$$\Delta\Phi(\mathbf{x}) = p(\mathbf{x}), \quad \int_\Omega \Phi(\mathbf{x})\, d\mathbf{x} = 0. \tag{8.9}$$

Then the matrix $\Phi(\mathbf{x})$ is Hermitian (together with $p(\mathbf{x})$), and $\Phi \in \widetilde{W}^2_q(\Omega)$. (Here $\widetilde{W}^2_q(\Omega)$ is the "periodic" subspace in the Sobolev space $W^2_q(\Omega)$.) We put $a_j(\mathbf{x}) = D_j\Phi(\mathbf{x})$, $j = 1, \dots, d$. Obviously, the required conditions (8.7) and (8.8) are satisfied. Note also that C_a (see (4.7)) satisfies the estimate

$$C_a \le C_q \|p\|_{L_q(\Omega)}, \tag{8.10}$$

where the constant C_q depends only on d, q, and Γ.

We put $\mathcal{Y}_\varepsilon = \sum_{j=1}^d a_j^\varepsilon D_j$, where the a_j are chosen according to (8.7) and (8.8). Then the operator $\mathcal{Y}_\varepsilon + \mathcal{Y}_\varepsilon^*$ acts as multiplication by $\varepsilon^{-1}p^\varepsilon(\mathbf{x})$:

$$(\mathcal{Y}_\varepsilon + \mathcal{Y}_\varepsilon^*)\mathbf{u} = \sum_{j=1}^d a_j(\varepsilon^{-1}\mathbf{x}) D_j \mathbf{u} - \sum_{j=1}^d D_j(a_j(\varepsilon^{-1}\mathbf{x})\mathbf{u}) = \varepsilon^{-1}p(\varepsilon^{-1}\mathbf{x})\mathbf{u}.$$

Thus, the operator (8.6) takes the form (8.1). Assume that c_0 is such that

$$c_0 \ge \frac{5}{2}\alpha_0^{-1} C_q^2 \|p\|^2_{L_q(\Omega)} \|g^{-1}\|_{L_\infty} \|Q^{-1}\|_{L_\infty}.$$

By (8.10), this estimate assures that condition (4.8) is satisfied if the matrices a_j are chosen as described above.

We construct the effective operator for the operator (8.6). Now equation (6.8) for $\widetilde{\Lambda}(\mathbf{x})$ takes the form

$$b(\mathbf{D})^* g(\mathbf{x}) b(\mathbf{D}) \widetilde{\Lambda}(\mathbf{x}) + p(\mathbf{x}) = 0, \quad \int_\Omega \widetilde{\Lambda}(\mathbf{x})\, d\mathbf{x} = 0. \tag{8.11}$$

The constant matrices V and W are defined according to (6.11) and (6.12). Since $\overline{a_j} = 0$, then, by (7.2), $\mathcal{Y}^0 = -b(\mathbf{D})^*V$. As a result, the effective operator takes the form

$$\widetilde{\mathcal{B}}^0 = \mathcal{A}^0 - b(\mathbf{D})^*V - V^*b(\mathbf{D}) + c_0\overline{Q} - W. \tag{8.12}$$

Applying Theorem 8.1, we arrive at the following result.

THEOREM 8.3. *Let $\widetilde{\mathcal{B}}_\varepsilon$ be the operator defined by* (8.6), *and let $\widetilde{\mathcal{B}}^0$ be the effective operator defined by* (8.12). *Then we have*

$$\|\widetilde{\mathcal{B}}_\varepsilon^{-1} - (\widetilde{\mathcal{B}}^0)^{-1}\|_{L_2\to L_2} \le \widetilde{C}\varepsilon, \quad \varepsilon > 0. \tag{8.13}$$

The constant $\widetilde{C}$ depends only on α_0, α_1, $\|g\|_{L_\infty}$, $\|g^{-1}\|_{L_\infty}$, $\|p\|_{L_q(\Omega)}$ (where $q > d$), $\|Q\|_{L_\infty}$, c_0, and r_0.

REMARK 8.4. The statement of Theorem 8.3 remains true if $p(\mathbf{x})$ is a distribution represented as (8.7), where the Γ-periodic matrix-valued functions $a_j(\mathbf{x})$, $j = 1, \dots, d$, satisfy conditions (8.8). It is assumed that c_0 is subject to condition (4.8). In this case, the constant $\widetilde{C}$ in (8.13) depends on α_0, α_1, $\|g\|_{L_\infty}$, $\|g^{-1}\|_{L_\infty}$, $\|a_j\|_{L_\infty}$, $j = 1, \dots, d$, $\|Q\|_{L_\infty}$, c_0, and r_0.

8.5. As an example, consider the operator

$$\widetilde{\mathcal{B}}_\varepsilon = \mathbf{D}^* g^\varepsilon \mathbf{D} + \varepsilon^{-1} p^\varepsilon + c_0 Q^\varepsilon = -\operatorname{div} g^\varepsilon \nabla + \varepsilon^{-1} p^\varepsilon + c_0 Q^\varepsilon, \tag{8.14}$$

acting in $L_2(\mathbb{R}^d)$. Here $g(\mathbf{x})$ is a positive definite and bounded periodic $(d \times d)$-matrix-valued function *with real entries*, $p(\mathbf{x})$ is a real-valued periodic potential satisfying condition (8.5), and $Q(\mathbf{x})$ is a positive periodic function. It is easily seen that in this case we have $\mathcal{Y}^0 + (\mathcal{Y}^0)^* = 0$, whence

$$\widetilde{\mathcal{B}}^0 = \mathbf{D}^* g^0 \mathbf{D} + c_0 \overline{Q} - W.$$

However, if in the operator (8.14), $g(\mathbf{x})$ is a Hermitian matrix *with complex entries*, then the effective operator, in general, contains nontrivial first order terms. Similarly, for *matrix* operators (8.6) (even with real-valued coefficients) the first order term $\mathcal{Y}^0 + (\mathcal{Y}^0)^*$ in the effective operator, in general, is nontrivial. What was said can be confirmed by particular examples.

Acknowledgements

The author is deeply grateful to M. Sh. Birman for useful discussions.

References

[BeLP] A. Bensoussan, J. L. Lions, and G. Papanicolau, *Asymptotic analysis for periodic structures*, North Holland, Amsterdam, 1978.

[BSu1] M. Sh. Birman and T. A. Suslina, *Threshold effects near the lower edge of the spectrum for periodic differential operators of mathematical physics*, Oper. Theory Adv. Appl. **129** (2001), Birkhäuser, pp. 71–107.

[BSu2] ———, *Second order periodic differential operators. Threshold properties and homogenization*, Algebra i Analiz **15** (2003), no. 5, 1–108; English transl., St. Petersburg Math. J. **15** (2004), no. 5, 639–714.

[BSu3] ———, *Threshold approximations with corrector for the resolvent of a factorized self-adjoint operator family*, Algebra i Analiz **17** (2005), no. 5, 69–90; English transl., St. Petersburg Math. J. **17** (2006), no. 5, 745–762.

[BSu4] ———, *Homogenization with corrector term for periodic elliptic differential operators*, Algebra i Analiz **17** (2005), no. 6, 1–104; English transl., St. Petersburg Math. J. **17** (2006), no. 6, 897–973.

[BSu5] ———, *Homogenization with corrector for periodic differential operators. Approximation of solutions in the Sobolev class $H^1(\mathbb{R}^d)$*, Algebra i Analiz **18** (2006), no. 6, 1–130; English transl., St. Petersburg Math. J. **18** (2007), no. 6, 857–955.

[Bo] D. I. Borisov, *Asymptotics of solutions of elliptic systems with rapidly oscillating coefficients*, Algebra i Analiz **20** (2008), no. 2, 19–42.

[K] T. Kato, *Perturbation theory for linear operators*, Grundlehren Math. Wiss, vol. 132, Springer-Verlag, Berlin–New York, 1972.

[Zh] V. V. Zhikov, *On operator estimates in homogenization theory*, Dokl. Akad. Nauk **403** (2005), no. 3, 305–308; English transl., Dokl. Math. **72** (2005), no. 1, 535–538.

[ZhPas] V. V. Zhikov and S. E. Pastukhova, *On operator estimates for some problems in homogenization theory*, Russian J. Math. Phys. **12** (2005), no. 4, 515–524.

Department of Physics, St. Petersburg State University, Ul'yanovskaya 3, Petrodvorets, St. Petersburg, 198504 Russia

E-mail address: `suslina@list.ru`

Amer. Math. Soc. Transl.
(2) Vol. **225**, 2008

Improved Berezin–Li–Yau Inequalities with a Remainder Term

Timo Weidl

To Mikhail Shlëmovich Birman on his 80th birthday

ABSTRACT. We give an improvement of sharp Berezin-type bounds on the Riesz means $\sum_k (\Lambda - \lambda_k)_+^\sigma$ of the eigenvalues λ_k of the Dirichlet Laplacian in a domain if $\sigma \geq 3/2$. It contains a correction term of the order of the standard second term in the Weyl asymptotics. The result is based on an application of sharp Lieb–Thirring inequalities with operator-valued potential to spectral estimates of the Dirichlet Laplacian in domains.

1. Introduction

Let Ω be an open domain in $\mathbb{R}^d$ and consider the Dirichlet Laplacian $-\Delta_\Omega^D$ defined in the form sense on the form domain $\overset{\circ}{H}{}^1(\Omega)$. If the embedding $\overset{\circ}{H}{}^1(\Omega) \hookrightarrow L^2(\Omega)$ is compact, for example, if the domain Ω is bounded, the spectrum of $-\Delta$ is discrete. It consists of a nondecreasing sequence of positive eigenvalues λ_k accumulating at infinity, which we repeat according to their multiplicities. Let

$$n(\Omega, \Lambda) = \operatorname{card}\{\lambda_k | \lambda_k < \Lambda\}$$

be the corresponding counting function.

In 1912, Hermann Weyl proved the famous asymptotic formula [**17**]

$$n(\Omega, \Lambda) = (1 + o(1))\eta(\Omega, \Lambda) \qquad \text{as } \Lambda \to +\infty\,, \tag{1.1}$$

where $\eta(\Omega, \Lambda)$ denotes the respective classical phase space volume

$$\eta(\Omega, \Lambda) := \int_{x\in\Omega} \int_{\xi\in\mathbb{R}^d : |\xi|^2 < \Lambda} \frac{dx \cdot d\xi}{(2\pi)^d} = L_{0,d}^{cl} \operatorname{vol}(\Omega) \Lambda^{d/2}, \quad L_{0,d}^{cl} = \frac{\omega_d}{(2\pi)^d}\,. \tag{1.2}$$

Here ω_d stands for the volume of the unit ball in $\mathbb{R}^d$.

2000 *Mathematics Subject Classification.* Primary 35G15.

Alongside of the counting function we shall discuss the eigenvalue means ($\sigma > 0$, $\Lambda > 0$)

$$(1.3) \qquad S_{\sigma,d}(\Omega,\Lambda) := \sum_n (\Lambda - \lambda_n)_+^\sigma = \sigma \int_0^\Lambda (\Lambda - \tau)^{\sigma-1} n(\Omega,\tau)\, d\tau\,,$$

$$(1.4) \qquad s_{\sigma,d}(\Omega,N) := \sum_{k=1}^N \lambda_k^\sigma = \sigma \int_0^\infty \tau^{\sigma-1}(N - n(\Omega,\tau))_+ d\tau\,.$$

Inserting (1.1) and (1.2) into the integral (1.3) one obtains the Weyl asymptotics

$$(1.5) \qquad S_{\sigma,d}(\Omega,\Lambda) = (1+o(1)) S^{cl}_{\sigma,d}(\Omega,\Lambda) \quad \text{as } \Lambda \to +\infty\,,$$

where

$$S^{cl}_{\sigma,d}(\Omega,\Lambda) = \int_{x\in\Omega}\int_{\xi\in\mathbb{R}^d} (\Lambda - |\xi|^2)_+^\sigma \frac{dx\cdot d\xi}{(2\pi)^d} = L^{cl}_{\sigma,d}\mathrm{vol}(\Omega)\Lambda^{\sigma+d/2}\,.$$

Here we make use of the well-known Lieb–Thirring constants

$$L^{cl}_{\sigma,d} := \frac{\Gamma(\sigma+1)}{2^d \pi^{d/2}\Gamma(1+\sigma+d/2)} = \sigma B\left(\sigma, 1+\frac{d}{2}\right) L^{cl}_{0,d}\,.$$

A similar computation for the expression (1.4) implies

$$(1.6) \qquad s_{\sigma,d}(\Omega,N) = (1+o(1)) s^{cl}_{\sigma,d}(\Omega,N) \quad \text{as } N \to +\infty\,,$$

with the classical quantity

$$s^{cl}_{\sigma,d}(\Omega,N) = \sigma \int_0^\infty \tau^{\sigma-1}\left(N - \eta(\Omega,\tau)\right)_+ d\tau = c(\sigma,d)\left(\mathrm{vol}(\Omega)\right)^{-\frac{2\sigma}{d}} N^{1+\frac{2\sigma}{d}}\,,$$

where

$$c(\sigma,d) := \frac{2\sigma}{d}\left(L^{cl}_{0,d}\right)^{-\frac{2\sigma}{d}} B\left(\frac{2\sigma}{d}, 2\right) = \frac{d}{2\sigma+d}\left(L^{cl}_{0,d}\right)^{-\frac{2\sigma}{d}}\,.$$

It is an important question whether the semi-classical quantities in the Weyl-type formulae can serve as universal bounds for the corresponding spectral quantities of the Dirichlet Laplacian as well. In particular, one is interested in the validity of the following inequalities ($\Lambda > 0$, $N \in \mathbb{N}$):

$$(1.7) \qquad \sum_k (\Lambda - \lambda_k)_+^\sigma = S_{\sigma,d}(\Omega,\Lambda) \le r(\sigma,d) S^{cl}_{\sigma,d}(\Omega,\Lambda)\,, \quad \sigma \ge 0\,,$$

$$(1.8) \qquad \sum_{k=1}^N \lambda_k^\sigma = s_{\sigma,d}(\Omega,N) \ge \rho(\sigma,d) s^{cl}_{\sigma,d}(\Omega,N)\,, \quad \sigma > 0\,,$$

and in the sharp values of the constants $r(\sigma,d)$ and $\rho(\sigma,d)$ therein. From the Weyl asymptotics (1.5), (1.6) it follows that at the best $r(\sigma,d) \ge 1$ and $\rho(\sigma,d) \le 1$.

Berezin showed in [**2**] that (1.7) holds with $r(\sigma,d) = 1$ for all $d \in N$ and all $\sigma \ge 1$. If one applies the Legendre transformation to (1.7) for the case $\sigma = 1$, one finds that

$$s_{1,d}(\Omega,N) \ge s^{cl}_{1,d}(\Omega,N)\,.$$

The latter result has independently been obtained by Li and Yau [**12**] by other means. By Hölder's inequality we have

$$(1.9) \qquad s^{cl}_{1,d}(\Omega,N) \le s_{1,d}(\Omega,N) \le s^{1/\sigma}_{\sigma,d}(\Omega,N) N^{1/\sigma'}\,, \quad \frac{1}{\sigma} + \frac{1}{\sigma'} = 1\,.$$

From this it follows that (1.8) remains true for $\sigma > 1$ with the estimate

$$\rho(\sigma, d) \geq (c(1,d))^\sigma / c(\sigma, d) = \frac{2\sigma + d}{d}\left(\frac{d}{2+d}\right)^\sigma$$

on the constant. If we pass in (1.9) to the limit $\sigma \to \infty$, we obtain the bound

$$\frac{d}{2+d}(L_{0,d}^{cl}\,\mathrm{vol}(\Omega))^{-\frac{2}{d}} N^{\frac{2}{d}} \leq \lambda_N\,, \quad N \in \mathbb{N}\,.$$

This is equivalent to

$$n(\Omega, \Lambda) \leq r(0,d)\eta(\Omega,\Lambda)\,, \quad \Lambda > 0\,,$$

with the estimate $1 \leq r(0,d) \leq (1 + 2d^{-1})^{d/2}$ on the constant $r(0,d)$, which corresponds to the case $\sigma = 0$ for the inequality (1.7). From here, for example, one can conclude by the Lieb-Aizenman trick [**1**] that $r(\sigma, d) \leq (1+2d^{-1})^{d/2}$ for $0 \leq \sigma < 1$.

Pólya showed in 1961 that for all tiling domains Ω one has in fact $r(0,d) = 1$; see [**14**]. The famous Pólya conjecture suggesting that $r(0,d) = 1$ for general domains remains unresolved so far. For certain improvements in this direction, see [**8, 9**]. Recently it has been shown that Pólya's conjecture fails in the case of a constant magnetic field [**15**].

Along with his formula on the main high energy term for the counting function $n(\Omega, \Lambda)$, Weyl conjectured that for the Dirichlet Laplacian the following two-term formula holds true:

$$(1.10)\quad n(\Omega,\Lambda) = L_{0,d}^{cl}\mathrm{vol}(\Omega)\Lambda^{d/2} - \frac{1}{4}L_{0,d-1}^{cl}|\partial\Omega|\Lambda^{(d-1)/2} + o(\Lambda^{(d-1)/2}) \quad \text{as } \Lambda \to \infty\,.$$

Here $|\partial\Omega|$ denotes the $(d-1)$-dimensional measure of the boundary of Ω. This formula fails in general, but it holds true under certain restrictions on the geodesic flow in the domain [**7, 16**].

If we insert (1.10) into (1.3) and (1.4) we obtain the following two-term formulae for the eigenvalue means:

$$(1.11)\quad S_{\sigma,d}(\Omega,\Lambda) = L_{\sigma,d}^{cl}\mathrm{vol}(\Omega)\Lambda^{\sigma+d/2} - \frac{1}{4}L_{\sigma,d-1}^{cl}|\partial\Omega|\Lambda^{\sigma+(d-1)/2} + o(\Lambda^{\sigma+(d-1)/2}),$$

(1.12)

$$s_{\sigma,d}(\Omega,N) = c(\sigma,d)\,(\mathrm{vol}(\Omega))^{-\frac{2\sigma}{d}}\,N^{1+\frac{2\sigma}{d}} + \frac{L_{\sigma,d-1}^{cl}(L_{\sigma,d}^{cl})^{-1-\frac{2\sigma-1}{d}}}{4(\frac{d-1}{2}+\sigma)} \cdot \frac{\sigma|\partial\Omega|}{(\mathrm{vol}(\Omega))^{1+\frac{2\sigma-1}{d}}} N^{1+\frac{2\sigma-1}{d}} + o(N^{1+\frac{2\sigma-1}{d}})\,.$$

Of course, the signs of the second terms support the sharp inequalities stated above.

In view of these two-term asymptotics one can ask if it is possible to find universal bounds on the spectral quantities, which hold true for *all* values of Λ, contain the sharp first Weyl term *and* reflect the contribution of the second-order term?

Note that a straightforward generalisation of the Berezin bound by just adding the second asymptotic term on the right-hand side

$$S_{\sigma,d}(\Omega,\Lambda) \leq S_{\sigma,d}^{cl}(\Omega,\Lambda) - C\cdot|\partial\Omega|\Lambda^{\sigma+\frac{d-1}{2}}$$

must fail. Indeed, without further restrictions the r.h.s. in this bound can just be negative for fixed values of Λ. Therefore, any improvement of sharp one-term bounds must invoke a suitable replacement for the surface volume $|\partial\Omega|$.

The first step towards this goal has been made by Melas [**13**]. Refining the result by Li and Yau he found that

$$\sum_{k=1}^{N} \lambda_k = s_{1,d}(\Omega, N) \geq s_{1,d}^{cl}(\Omega, N) + M(d)\frac{\mathrm{vol}(\Omega)}{J(\Omega)} N \,. \tag{1.13}$$

Here

$$J(\Omega) = \min_{y \in \mathbb{R}^d} \int_{\Omega} |x-y|^2 dx$$

is the momentum of Ω and the constant $M(d)$ depends only on the dimension d. This bound is remarkable, since it improves the Li-Yau inequality which corresponds via the Legendre transformation to the endpoint $\sigma = 1$ of the scale, where $r(\sigma, d) = 1$ is known to be true. On the other hand, comparing (1.13) with (1.12), we find that the additional term of order N on the r.h.s. of (1.13) does not capture the correct order $N^{1+\frac{1}{d}}$ of the second Weyl term.

In the present note we give with (2.2) in Theorem 2.1 an improvement of a Berezin-type bound (1.7) for $\sigma \geq 3/2$. The result is based on the application of sharp Lieb–Thirring inequalities with operator-valued potentials [**11, 10**] to spectral estimates of the Dirichlet Laplacian in domains; see also [**4, 5**]. In fact, we simply raise the explicit remainder term for the second derivative with Dirichlet boundary condition on one-dimensional intervals to the Dirichlet Laplacian on domains in arbitrary dimensions. Although the proof is straightforward, we find (2.2) noteworthy, since

- it contains a correction term of the order of the standard second term in the Weyl asymptotics,
- it improves even the first term by taking only the volume of a suitable subset $\Omega_\Lambda \subset \Omega$ into account,
- it is applicable even for certain domains Ω with infinite volume,
- it generalises without change of constants to Dirichlet Laplacians with arbitrary magnetic fields.

However, our result does not touch the endpoint $\sigma = 1$ of the scale where the sharp one-term Berezin bound is known. Hence, it does not imply an improvement on the best-known estimates on the counting function. The question whether (2.2) can be generalised in some way to $\sigma = 1$ or whether the Melas bound can be improved with a correction term of the second Weyl term order remains open.

The structure of the paper is as follows. In section 2 we state the main result and comment on the improvements which we gain compared to the one-term bound. The proof of our result will be sketched in section 3.

2. Statement of the result

Notation. First let us prepare some notation which shall be of use below.

Let Ω be an open subset of $\mathbb{R}^d$. We fix a Cartesian coordinate system in $\mathbb{R}^d$ and for $x \in \mathbb{R}^d$ we shall also write $x = (x', t) \in \mathbb{R}^{d-1} \times \mathbb{R}$. Each of the sections $\Omega(x') = \{t \in \mathbb{R} \mid (x', t) \in \Omega\}$ consists of at most countably many open intervals $J_k(x') \subset \mathbb{R}$ of length $l_k(x')$. For given x' and Λ let $\kappa(x', \Lambda) \subset \mathbb{N}$ be the subset of all indices k, where $l_k(x') > l_\Lambda := \pi \Lambda^{-1/2}$.

If $\mathrm{vol}(\Omega_\Lambda)$ is finite, the sets $\kappa(x', \Lambda)$ are finite for a.e. x' and we let $\varkappa(x', \Lambda)$ denote the number of the elements of these sets. We shall assume that this function is measurable in x'.

Put

$$\Omega_\Lambda(x') = \bigcup_{k\in\kappa(x',\Lambda)} J_k(x') \subset \Omega(x') \subset \mathbb{R} \quad \text{and} \quad \Omega_\Lambda = \bigcup_{x'\in\mathbb{R}^{d-1}} \{x'\}\times\Omega_\Lambda(x') \subset \Omega\,.$$

That means Ω_Λ is the subset of Ω where the intervals in the t-direction contained in Ω are longer than l_Λ. The set Ω_Λ is monotonically increasing in Λ.

We shall also make use of the quantity

$$d_\Lambda(\Omega) = \int_{x'\in\mathbb{R}^{d-1}} \varkappa(x',\Lambda)dx'\,,$$

which is an effective measure of the projection of Ω_Λ on the x'-plane counting the number of sufficiently long intervals.

A basic estimate. Consider the function

$$f_\mu(A) = \frac{A}{2}B\left(1+\mu,\frac{1}{2}\right) - \sum_{k=1}^{\infty}\left(1-\frac{k^2}{A^2}\right)_+^\mu, \quad A\geq 1\,,\mu>0\,,$$

which is continuous in A. Since

$$\frac{A}{2}B\left(1+\mu,\frac{1}{2}\right) = \int_0^{+\infty}\left(1-\frac{t^2}{A^2}\right)_+^\mu dt > \sum_{k=1}^{\infty}\left(1-\frac{k^2}{A^2}\right)_+^\mu,$$

the function $f_\mu(A)$ takes positive values for $A\geq 1$. Moreover, it is not difficult to see that

$$\lim_{A\to+\infty} f_\mu(A) = \frac{1}{2}, \quad \mu>0\,.$$

Thus, the minimum

$$\varepsilon_\mu = \min_{A\geq 1} f_\mu(A) > 0$$

exists and

$$\sum_{k=1}^{\infty}\left(1-\frac{k^2}{A^2}\right)_+^\mu \leq \frac{A}{2}B\left(1+\mu,\frac{1}{2}\right) - \varepsilon_\mu\,, \quad A\geq 1\,. \tag{2.1}$$

The cases $A=1$ and $A\to+\infty$ ensure that

$$2\varepsilon_\mu \leq \min\left\{1, B\left(1+\mu,\frac{1}{2}\right)\right\}.$$

Main result.

THEOREM 2.1. *There exists a positive constant $\nu=\nu(\sigma,d)$, such that for any open domain $\Omega\subset\mathbb{R}^d$, $\sigma\geq 3/2$ and any $\Lambda>0$ the bound*

$$\sum_k(\Lambda-\lambda_k)_+^\sigma = S_{\sigma,d}(\Omega,\Lambda) \leq L_{\sigma,d}^{cl}\mathrm{vol}(\Omega_\Lambda)\Lambda^{\sigma+\frac{d}{2}} - \nu(\sigma,d)\frac{L_{\sigma,d-1}^{cl}}{4}d_\Lambda(\Omega)\Lambda^{\sigma+\frac{d-1}{2}} \tag{2.2}$$

holds. The optimal constant $\nu=\nu(\sigma,d)$ satisfies

$$4\varepsilon_{\sigma+\frac{d-1}{2}} \leq \nu(\sigma,d) \leq 2\min\left\{1, B\left(1+\sigma+\frac{d-1}{2},\frac{1}{2}\right)\right\}. \tag{2.3}$$

Comments.

REMARK 2.2. The correction term in (2.2) reflects the correct asymptotical order $O(\Lambda^{\sigma+\frac{d-1}{2}})$ of the standard correction term in the Weyl formula.

REMARK 2.3. The first term on the r.h.s. of (2.2) takes only the volume of the part $\Omega_\Lambda \subset \Omega$ into account, that is, only the part of the original domain where it is sufficiently wide for a Dirichlet bound state below Λ to settle in the t-direction. Therefore, we already have an improvement in the first term of the Berezin bound. In particular, the inequality (2.2) is applicable to domains Ω of infinite volume, as long as for given Λ the volume of Ω_Λ is finite.

REMARK 2.4. Numerical evaluations which can be made rigorous by elementary means show that the lower bound on $\nu(\sigma, d)$ is reasonable. For example, if $d = 2$ and $\sigma = 3/2$, then

$$1.91 < \nu\left(\frac{3}{2}, 2\right) \leq 2\,.$$

The upper bounds. At this point let us discuss the upper bounds on $\nu(\sigma, d)$ more in detail. First of all, consider a long thin rectangle of width w and height $h = \delta w$, which is parallel to the coordinate axes. For sufficiently large Λ we have $\Omega = \Omega_\Lambda$, $d_\Lambda = w$ and $|\partial\Omega| = 2(1+\delta)d_\lambda$. The asymptotical formula (1.11) in comparison with (2.2) yields that $\nu(\sigma, d) \leq 2(1+\delta)$. Since δ can be chosen arbitrarily small, one finds that $\nu(\sigma, d) \leq 2$.

Now take again a rectangle of width w and height $h = (1-\delta)l_\Lambda$ for some small positive δ. Then $d_\Lambda = w$, $\Omega = \Omega_\Lambda$ and $\mathrm{vol}(\Omega_\Lambda) = (1-\delta)\pi\Lambda^{-1}d_\Lambda$. The r.h.s. of (2.2) turns into

$$d_\Lambda(\Omega)\Lambda^{\sigma+\frac{d-1}{2}}\left((1-\delta)\pi L^{cl}_{\sigma,d} - \frac{\nu(\sigma,d)}{4}L^{cl}_{\sigma,d-1}\right).$$

Since this expression has to be nonnegative for the bound (2.2) to make sense, we find that

$$\nu(\sigma, d) \leq 4\pi L^{cl}_{\sigma,d}(L^{cl}_{\sigma,d-1})^{-1} = 2B\left(\frac{1}{2}, 1+\sigma+\frac{d-1}{2}\right).$$

Actually, this condition ensures that the right-hand side of (2.2) is nonnegative for arbitrary Ω. Indeed, since the total width of $\Omega_\Lambda(x')$ exceeds $\varkappa(x',\Lambda)\cdot l_\Lambda \geq \pi\Lambda^{-1/2}\varkappa(x',\Lambda)$, the volume of Ω_Λ can be estimated from below by

$$\mathrm{vol}(\Omega_\Lambda) = \int_{x'\in\mathbb{R}^{d-1}} \varkappa(x',\Lambda)\cdot l_\Lambda dx' \geq \pi\Lambda^{-1/2}d_\Lambda(\Omega).$$

Thus, the expression on the r.h.s. of (2.2) satisfies

$$\begin{aligned} L^{cl}_{\sigma,d}\mathrm{vol}(\Omega_\Lambda)\Lambda^{\sigma+\frac{d}{2}} &- \frac{\nu(\sigma,d)}{4}L^{cl}_{\sigma,d-1}d_\Lambda(\Omega)\Lambda^{\sigma+\frac{d-1}{2}} \\ &\geq d_\Lambda(\Omega)\Lambda^{\sigma+\frac{d-1}{2}}\left(\pi L^{cl}_{\sigma,d} - \frac{\nu(\sigma,d)}{4}L^{cl}_{\sigma,d-1}\right) \geq 0\,. \end{aligned}$$

3. The proof of the main result

The proof proceeds in two steps. We start with a variational argument, which transforms our initial problem into a spectral estimate on the negative eigenvalues of a Schrödinger-type operator with operator-valued potential. Our result follows then from an operator-valued Lieb–Thirring bound [**11, 6, 10**].

Step 1: A variational argument. We use the notation introduced in section 2. Moreover, by ∇' and $-\Delta'$ we denote the gradient and the Laplacian in the first $d-1$ dimensions of $\mathbb{R}^d = \mathbb{R}^{d-1} \times \mathbb{R} \ni (x', t)$.

Consider a function u from the form core $C_0^\infty(\Omega)$ of $-\Delta_\Omega^D$. For the quadratic form of $-\Delta_\Omega^D - \Lambda$ we have the identity

$$\begin{aligned}\|\nabla u\|^2_{L^2(\Omega)} &- \Lambda \|u\|^2_{L^2(\Omega)} \\ &= \|\nabla' u\|^2_{L^2(\Omega)} + \int_{\mathbb{R}^{d-1}} dx' \int_{\Omega(x')} \left(\left| \frac{\partial}{\partial t} u(x',t) \right|^2 - \Lambda |u(x',t)|^2 \right) dt \, .\end{aligned}$$

Note that $\Omega(x') = \bigcup_k J_k(x')$. The functions $u(x', \cdot)$ satisfy Dirichlet boundary conditions at the endpoints of $J_k(x')$. The lowest Dirichlet eigenvalue of $-d^2/dt^2$ on $J_k(x')$ equals $\pi^2 l_k^{-2}(x')$. Hence,

$$\int_{J_k(x')} \left| \frac{\partial}{\partial t} u(x',t) \right|^2 dt \geq \frac{\pi^2}{l_k^2(x')} \|u(x', \cdot)\|^2_{L^2(J_k(x'))} \, .$$

In particular, for all $k \notin \kappa(x', \Lambda)$ we have

$$\int_{J_k(x')} \left(\left| \frac{\partial}{\partial t} u(x',t) \right|^2 - \Lambda |u(x',t)|^2 \right) dt \geq 0 \, .$$

It follows that

$$\begin{aligned}\int_{\Omega(x')} &\left(\left| \frac{\partial}{\partial t} u(x',t) \right|^2 - \Lambda |u(x',t)|^2 \right) dt \\ &\geq \sum_{k \in \kappa(x',\Lambda)} \int_{J_k(x')} \left(\left| \frac{\partial}{\partial t} u(x',t) \right|^2 - \Lambda |u(x',t)|^2 \right) dt \\ &\geq - \sum_{k \in \kappa(x',\Lambda)} \langle V_k u(x', \cdot), u(x', \cdot) \rangle_{L^2(J_k(x'))} \, ,\end{aligned}$$

where the operators $V_k = V_k(x', \Lambda)$ denote the negative parts of the Sturm-Liouville problems $-\frac{d^2}{dt^2} - \Lambda$ with Dirichlet boundary conditions on $L^2(J_k(x'))$. Put

$$W(x', \Lambda) = \left(\bigoplus_{k \in \kappa(x',\Lambda)} V_k(x', \Lambda) \right) \oplus \mathbb{O} \quad \text{on} \quad L^2(\Omega_\Lambda(x')) \oplus L^2(\Omega(x') \setminus \Omega_\Lambda(x')) \, .$$

This operator is bounded on $L^2(\Omega(x'))$ and for any $u \in C_0^\infty(\Omega)$ it follows that

$$\begin{aligned}\|\nabla u\|^2_{L^2(\Omega)} - \Lambda \|u\|^2_{L^2(\Omega)} &\geq \|\nabla' u\|^2_{L^2(\Omega)} \\ &\quad - \int_{\mathbb{R}^{d-1}} \langle W(x', \Lambda) u(x', \cdot), u(x', \cdot) \rangle_{L^2(\Omega(x'))} \, dx' \, .\end{aligned}$$

In a last step we study now the function $f = u + v$ where both functions $u \in C_0^\infty(\Omega)$ and $v \in C_0^\infty(\hat{\Omega})$ are extended by zero to $\mathbb{R}^d$ and where $\hat{\Omega} = \mathbb{R}^d \setminus \overline{\Omega}$. Since $\|\nabla v\|^2_{L^2(\hat{\Omega})} \geq \|\nabla' v\|^2_{L^2(\hat{\Omega})}$ we have

$$\begin{aligned}\|\nabla v\|^2_{L^2(\hat{\Omega})} &+ \|\nabla u\|^2_{L^2(\Omega)} - \Lambda \|u\|^2_{L^2(\Omega)} \\ &\geq \|\nabla' f\|^2_{L^2(\mathbb{R}^d)} - \int_{\mathbb{R}^{d-1}} \langle W f(x', \cdot), f(x', \cdot) \rangle_{L^2(\mathbb{R})} \, dx' \, .\end{aligned} \tag{3.1}$$

Here we extend (in a slight abuse of notation) $W = W(x', \Lambda)$ by an orthogonal sum with $\mathbb{O}|_{L_2(\mathbb{R}\setminus\overline{\Omega(x')})}$ to a bounded operator on $L^2(\mathbb{R})$.

The inequality (3.1) holds true for $f \in C_0^\infty(\mathbb{R}^d \setminus \partial\Omega)$, which is a form core for $(-\Delta_{\hat\Omega}^D)\oplus(-\Delta_\Omega^D-\Lambda)$ corresponding to the expression on the l.h.s. The semi-bounded form on the r.h.s. is closed on the larger domain $H^1(\mathbb{R}^{d-1}, L^2(\mathbb{R}))$, where it corresponds to the Schrödinger-type operator $-\Delta' \otimes \mathbb{I} - W(x', \Lambda)$ on $L^2(\mathbb{R}^{d-1}, L^2(\mathbb{R}))$. Due to the positivity of $-\Delta_{\hat\Omega}^D$ the variational principle implies that

$$\begin{aligned}\sum_k (\Lambda - \lambda_k)_+^\sigma = \operatorname{tr}(-\Delta_\Omega^D - \Lambda)_-^\sigma &= \operatorname{tr}\left((-\Delta_{\hat\Omega}^D) \oplus (-\Delta_\Omega^D - \Lambda)\right)_-^\sigma \\ &\le \operatorname{tr}\left(-\Delta' \otimes \mathbb{I} - W(x', \Lambda)\right)_-^\sigma .\end{aligned} \tag{3.2}$$

Step 2: Sharp Lieb–Thirring bounds. We are now in a position to apply a sharp Lieb–Thirring inequality for operator-valued potentials (Theorem 3.1 in [**11**]) in the dimension $d-1$ with $\sigma \ge 3/2$. In our setting it reads as follows:

$$\operatorname{tr}\left(-\Delta' \otimes \mathbb{I} - W(x', \Lambda)\right)_-^\sigma \le L_{\sigma,d-1}^{cl} \int_{\mathbb{R}^{d-1}} \operatorname{tr} W^{\sigma+\frac{d-1}{2}}(x', \Lambda)\, dx' . \tag{3.3}$$

The eigenvalues of the operator $W(x', \Lambda)$ are known explicitly and the nonzero ones equal $\mu_{j,k} = \Lambda - j^2\pi^2 l_k^{-2}(x')$ for $k \in \kappa(x', \Lambda)$ and $j = 1, 2, \dots, [l_k(x') l_\Lambda^{-1}]$. If we insert this into (3.3) and (3.2) we obtain

$$\begin{aligned}&\sum_k (\Lambda - \lambda_k)_+^\sigma \\ &\le L_{\sigma,d-1}^{cl} \int_{\mathbb{R}^{d-1}} \sum_{k\in\kappa(x',\Lambda)} \sum_{j=1}^{[l_k(x')l_\Lambda^{-1}]} \left(\Lambda - j^2\pi^2 l_k^{-2}(x')\right)^{\sigma+\frac{d-1}{2}} dx' \\ &\le \Lambda^{\sigma+\frac{d-1}{2}} L_{\sigma,d-1}^{cl} \int_{\mathbb{R}^{d-1}} \sum_{k\in\kappa(x',\Lambda)} \sum_{j=1}^{[l_k(x')l_\Lambda^{-1}]} \left(1 - \frac{j^2}{l_k^2(x') l_\Lambda^{-2}}\right)^{\sigma+\frac{d-1}{2}} dx' .\end{aligned} \tag{3.4}$$

Since for $k \in \kappa(x', \Lambda)$ we have $l_k(x') l_\Lambda^{-1} \ge 1$, from (2.1) it follows that

$$\sum_{j=1}^{[l_k(x')l_\Lambda^{-1}]} \left(1 - \frac{j^2}{l_k^2(x') l_\Lambda^{-2}}\right)^{\sigma+\frac{d-1}{2}} \le \frac{l_k(x') l_\Lambda^{-1}}{2} B\left(1 + \sigma + \frac{d-1}{2}, \frac{1}{2}\right) - \varepsilon_{\sigma+\frac{d-1}{2}} ,$$

which together with (3.4) amounts to

$$\begin{aligned}\sum_k (\Lambda - \lambda_k)^\sigma \le{}& \frac{\Lambda^{\sigma+\frac{d}{2}}}{2\pi} B\left(1 + \sigma + \frac{d-1}{2}, \frac{1}{2}\right) L_{\sigma,d-1}^{cl} \int_{\mathbb{R}^{d-1}} \sum_{k\in\kappa(x',\Lambda)} l_k(x')\, dx' \\ &- \varepsilon_{\sigma+\frac{d-1}{2}} \Lambda^{\sigma+\frac{d-1}{2}} L_{\sigma,d-1}^{cl} \int_{\mathbb{R}^{d-1}} \sum_{k\in\kappa(x',\Lambda)} 1\, dx' .\end{aligned} \tag{3.5}$$

Note that $\sum_{k\in\kappa(x',\Lambda)} l_k(x')$ stands for the total width of Ω_Λ in the t-direction and

$$\int_{\mathbb{R}^{d-1}} \sum_{k\in\kappa(x',\Lambda)} l_k(x')\, dx' = \operatorname{vol}(\Omega_\Lambda) . \tag{3.6}$$

Moreover, we have

$$\int_{\mathbb{R}^{d-1}} \sum_{k\in\kappa(x',\Lambda)} 1dx' = \int_{\mathbb{R}^{d-1}} \varkappa(x',\Lambda)dx' = d_\Lambda(\Omega)\,. \tag{3.7}$$

A straightforward computation shows that

$$\frac{1}{2\pi}B\left(1+\sigma+\frac{d-1}{2},\frac{1}{2}\right)L^{cl}_{\sigma,d-1} = L^{cl}_{\sigma,d}\,. \tag{3.8}$$

Inserting (3.6)-(3.8) into (3.5), we complete the proof.

The magnetic case. Here we consider the magnetic Dirichlet Laplacian $(i\nabla+A(x))^2$ defined in the forms sense from a form core $C_0^\infty(\Omega)$ for $A\in L^d_{loc}(\Omega,\mathbb{R}^d)$ for $d\geq 3$ and $A\in L^{1+\varepsilon}_{loc}(\Omega,\mathbb{R}^d)$ for $d=2$ and some $\varepsilon>0$. It turns out that the bound (2.2) holds with the same estimates on $\nu(\sigma,d)$ in the magnetic case as well. The proof uses the idea of induction in dimension in the magnetic case, see **[11]** section 3.2, an argument due to Helffer. We sketch it for the benefit of the reader.

First assume for simplicity that the vector potential A is continuous. For $x=(x_1,\dots,x_d)$ let $x=(x_j',x_j'')$ with $x_j'=(x_1,\dots,x_j)$ and $x_j''=(x_{j+1},\dots,x_d)$. We put

$$\Omega(x_j') = \{x_j''\in\mathbb{R}^{d-j}\mid x=(x_j',x_j'')\}\in\Omega.$$

This is an open set in $\mathbb{R}^{d-j}$, and we let $(i\nabla_j''+A''(x''))^2$ be the magnetic Dirichlet Laplacian on it. In a first step we can choose a gauge $A(x)=(0,A_1''(x_1',x_1''))$, where the first component of $A(x)$ vanishes. Using a variational argument similar to the one above and a sharp Lieb–Thirring bound in dimension one, we find that for $\sigma\geq 3/2$,

$$\begin{aligned}\sum_k(\Lambda-\lambda_k)_+^\sigma &= \operatorname{tr}((i\nabla+A(X))^2-\Lambda)_-^\sigma\\ &\leq L^{cl}_{\sigma,1}\int_{\mathbb{R}}\operatorname{tr}((i\nabla_1''+A_1''(x_1',x_1''))^2-\Lambda)_-^{\sigma+\frac12}dx_1\,.\end{aligned}$$

We now apply the same argument to the operator $(i\nabla_1''+A_1''(x_1',x_1''))^2$ on $\Omega(x_1')$ and so on. Decreasing dimensions one by one we get in the $(d-1)$st step,

$$\sum_k(\Lambda-\lambda_k)_+^\sigma \leq L\int_{\mathbb{R}^{d-1}}\operatorname{tr}\left(\left(i\frac{d}{dx_d}+\tilde{A}_d''(x_{d-1}',x_d)\right)^2-\Lambda\right)_-^{\sigma+\frac{d-1}{2}}dx_1\,,$$

where $L=\prod_{s=0}^{d-2}L^{cl}_{\sigma+\frac{s}{2},1}$. With the notation $\tilde{A}_d''$ we take into account that the explicit expression for the vector potential changes at each step due to the necessary gauge transformation.

We make available a family of one-dimensional Sturm-Liouville problems, where the magnetic field simply can be gauged away. That means we can make use of the same arguments for the operator in the $t=x_{d-1}''$-direction as above and obtain

$$\sum_k(\Lambda-\lambda_k)_+^\sigma \leq L\Lambda^{\sigma+\frac{d-1}{2}}\int_{\mathbb{R}^{d-1}}\sum_{k\in\kappa(x',\Lambda)}\sum_{j=1}^{[l_k(x')l_\Lambda^{-1}]}\left(1-\frac{j^2}{l_k^2(x')l_\Lambda^{-2}}\right)^{\sigma+\frac{d-1}{2}}dx'\,.$$

Since $L = L^{cl}_{\sigma,1} \cdots L^{cl}_{\sigma+\frac{d-2}{2},1} = L^{cl}_{\sigma,d-1}$, we complete the proof in the same way as above and find that

$$\text{(3.9)}\quad \operatorname{tr}((i\nabla + A(x))^2 - \Lambda)^{\sigma}_{-} \le L^{cl}_{\sigma,d}\operatorname{vol}(\Omega_\Lambda)\Lambda^{\sigma+\frac{d}{2}} - \nu(\sigma,d)\frac{L^{cl}_{\sigma,d-1}}{4} d_\Lambda(\Omega)\Lambda^{\sigma+\frac{d-1}{2}} .$$

Since the estimates[1] on the constants $\nu(\sigma, d)$ do not depend on the magnetic vector potential, we can reduce the result to a nonsmooth A by a standard argument.

For this let us just note that in the case of a counterexample, even if Λ is an accumulation point of the spectrum, already a finite partial eigenvalue sum will fail the bound (3.9). Thus, it suffices to study the quadratic form $\int_\Omega |(i\nabla + A)^2 u|^2 dx$ on the finite-dimensional subspace of u spanned by the corresponding eigenfunctions. In fact, one can reduce oneself to a finite-dimensional subspace of the form core. Since A can be approximated by a continuous A in L^2_{loc}, we have a uniform convergence $\int_\Omega |(i\nabla + A_\delta)^2 u|^2 dx \to \int_\Omega |(i\nabla + A)^2 u|^2 dx$ on normalised u from this subspace. Variational arguments give a contradiction to the counterexample.

This line of arguments will always work if we have an approximation of the vector potential by continuous ones in terms of the convergence of the forms on any finite-dimensional subspace of a form core in $C_0^\infty(\Omega)$. In particular, it applies to the minimal extensions of the Aharonov–Bohm type operators for $d = 2$.

In conclusion, we remark that the standard one-term Berezin bound

$$\text{(3.10)}\qquad \operatorname{tr}((i\nabla + A(x))^2 - \Lambda)^{\sigma}_{-} \le L^{cl}_{\sigma,d}\operatorname{vol}(\Omega)\Lambda^{\sigma+\frac{d}{2}}$$

holds with the semi-classical constant for constant magnetic fields if $\sigma \ge 1$ [**3**]. It fails even in this special case for $\sigma < 1$ [**15**]. The question whether (3.10) holds true (or can even be improved by a suitable correction term) for general magnetic fields for $\sigma \in [1, 3/2)$ remains open.

Acknowledgements

The author acknowledges support by the DAAD D/04/26013 and by the DFG We 1964-2/1. He is grateful to R. Benguria, P. Exner and A. Laptev for helpful comments. The material has been discussed and improved during the workshop 'Low eigenvalues of Laplace and Schrödinger operators' held at AIM in May 2006.

References

[1] Michael Aizenman and Elliott H. Lieb. *On semiclassical bounds for eigenvalues of Schrödinger operators.* Phys. Lett. A, **66**(6):427–429, 1978.

[2] F. A. Berezin. *Covariant and contravariant symbols of operators.* Izv. Akad. Nauk SSSR Ser. Mat., **36**:1134–1167, 1972.

[3] L. Erdős, M. Loss, and V. Vougalter. *Diamagnetic behavior of sums of Dirichlet eigenvalues.* Ann. Inst. Fourier (Grenoble), **50**(3):891–907, 2000.

[4] Pavel Exner, Helmut Linde, and Timo Weidl. *Lieb-Thirring inequalities for geometrically induced bound states.* Lett. Math. Phys., *70*(1):83–95, 2004.

[5] Pavel Exner and Timo Weidl. *Lieb-Thirring inequalities on trapped modes in quantum wires.* XIIIth International Congress on Mathematical Physics (London, 2000), pp. 437–443. Int. Press, Boston, MA, 2001.

[6] D. Hundertmark, A. Laptev, and T. Weidl. *New bounds on the Lieb-Thirring constants.* Invent. Math., *140*(3):693–704, 2000.

[1] This does not mean that the true optimal constants $\nu(\sigma, d)$ are necessarily independent of A; our estimate on $\nu(\sigma, d)$, however, is independent of the magnetic field.

[7] Victor Ivrii. *Microlocal analysis and precise spectral asymptotics.* Springer-Verlag, Berlin, 1998.

[8] A. Laptev. *On inequalities for the bound states of Schrödinger operators.* Partial differential operators and mathematical physics (Holzhau, 1994), Oper. Theory Adv. Appl., vol. 78, pp. 221–225, Birkhäuser, Basel, 1995.

[9] ———. *Dirichlet and Neumann eigenvalue problems on domains in Euclidean spaces.* J. Funct. Anal., **151**(2):531–545, 1997.

[10] Ari Laptev and Timo Weidl. *Recent results on Lieb-Thirring inequalities.* Journées "Équations aux Dérivées Partielles" (La Chapelle sur Erdre, 2000), Exp. No. XX, 14. Univ. Nantes, Nantes, 2000.

[11] ———. *Sharp Lieb-Thirring inequalities in high dimensions.* Acta Math., **184**(1):87–111, 2000.

[12] Peter Li and Shing Tung Yau. *On the Schrödinger equation and the eigenvalue problem.* Comm. Math. Phys., **88**(3):309–318, 1983.

[13] Antonios D. Melas. *A lower bound for sums of eigenvalues of the Laplacian.* Proc. Amer. Math. Soc., **131**(2):631–636 (electronic), 2003.

[14] G. Pólya. *On the eigenvalues of vibrating membranes.* Proc. London Math. Soc. (3), **11**:419–433, 1961.

[15] Michael Loss, Rupert Frank, and Timo Weidl. *Pólya's conjecture in the presence of a constant magnetic field.* Preprint.

[16] Yu. Safarov and D. Vassiliev. *The asymptotic distribution of eigenvalues of partial differential operators*, Amer. Math. Soc., Providence, RI, 1997.

[17] Hermann Weyl. *Das asymptotische Verteilungsgesetz der Eigenwerte linearer partieller Differentialgleichungen (mit einer Anwendung auf die Theorie der Hohlraumstrahlung).* Math. Ann., **71**(4):441–479, 1912.

Universität Stuttgart, FB Mathematik, Pfaffenwaldring 57, 70569 Stuttgart, Germany

E-mail address: weidl@mathematik.uni-stuttgart.de

Amer. Math. Soc. Transl.
(2) Vol. **225**, 2008

Spectral and Scattering Theory of Fourth Order Differential Operators

D. R. Yafaev

To Mikhail Shlëmovich Birman on his 80th birthday

ABSTRACT. An ordinary differential operator of the fourth order with coefficients converging at infinity sufficiently rapidly to constant limits is considered. Scattering theory for this operator is developed in terms of special solutions of the corresponding differential equation. In contrast to equations of second order, "scattering" solutions contain exponentially decaying terms. A relation between the scattering matrix and a matrix of coefficients at exponentially decaying modes is found. In the second part of the paper the operator D^4 on the half-axis with different boundary conditions at the point zero is studied. Explicit formulas for basic objects of scattering theory are found. In particular, a classification of different types of zero-energy resonances is given.

1. Introduction

1.1. General scattering theory for differential operators does not depend on the order of the operators (see [**4**] for the trace class approach and [**10**] for the smooth approach). Suppose that the coefficients of a differential operator H converge sufficiently rapidly to constant values at infinity, and let H_0 be the operator with these constant coefficients. Then the wave operators $W_\pm(H, H_0)$ for the pair H_0, H exist and are complete, and the corresponding scattering matrix is a unitary operator. The operator H does not have a singular continuous spectrum, and its point spectrum might accumulate only at thresholds (critical values of the symbol of the operator H_0). Moreover, an expansion theorem in eigenfunctions of the operator H is true.

However the behavior of eigenfunctions of the continuous spectrum at infinity is essentially simpler for differential operators of the second order than for higher order differential operators. This is intimately related to the fact that higher order differential operators might have eigenvalues embedded in the continuous spectrum.

2000 *Mathematics Subject Classification.* Primary 34B25, 35P25, 47A40.
Key words and phrases. One-dimensional differential operators, fourth order, exponentially decaying modes, scattering matrix, spectral shift function, zero-energy resonances.

1.2. In this paper we consider one-dimensional differential operators. For definiteness, we choose operators of the fourth order. Set $H_0 = D^4$ and

$$H = D^4 + Dv_1(x)D + v_0(x), \quad D = -id/dx, \tag{1.1}$$

where the functions v_0 and v_1 are real and satisfy a short-range assumption

$$v_j \in L_1(\mathbb{R}), \quad j = 0, 1. \tag{1.2}$$

Smoothness of the function $v_1(x)$ is not required. We are interested in construction of the wave operators and of the scattering matrix in terms of solutions of the corresponding differential equation

$$(u'''(x) - v_1(x)u'(x))' + v_0(x)u(x) = \lambda u(x), \quad \lambda > 0. \tag{1.3}$$

This construction turns out to be more complicated than for equations of the second order. Indeed, the "free" equation $u^{(4)}(x) = \lambda u(x)$ has solutions e^{ikx}, e^{-ikx}, e^{-kx} and e^{kx}, where $\lambda = k^4$, $k > 0$. If, for example, the coefficients $v_0(x)$ and $v_1(x)$ are compactly supported, then every solution of equation (1.3) is a linear combination of these exponentials for large positive and large negative x. Eigenfunctions $\psi_j(x, \lambda)$, $j = 1$, of the operator H (its continuous spectrum has multiplicity two) are special solutions of equation (1.3). It is natural to expect that they do not contain exponentially increasing terms and that the scattering matrix is determined only by coefficients at oscillating terms e^{ikx} and e^{-ikx}. We justify this conjecture under the general short-range assumption (1.2).

Furthermore, if the coefficients $v_1(x)$ and $v_0(x)$ decay super-exponentially at infinity, then it is possible to distinguish exponentially decreasing terms in the asymptotics of the functions $\psi_j(x, \lambda)$, $j = 1, 2$, as $x \to \pm\infty$. Our main observation in the first part of the paper is that the coefficients of these terms determine the scattering matrix. This might be eventually of interest for a study of the (inverse) problem of a reconstruction of the coefficients $v_0(x)$ and $v_1(x)$ from scattering data. We refer to [**2**] for a comprehensive study of the inverse problem.

Recall that for differential operators of the second order the eigenfunctions of the continuous spectrum can be constructed (see [**14**]) with the help of Volterra integral equations. This procedure seems not to work for operators of higher order. Therefore we use a general scheme of scattering theory (see, e.g., [**11, 16**]) relying on the Lippmann–Schwinger equation.

In Section 2 we present a stationary approach to scattering theory for the operator H. This approach is quite general and, up to some technical details, works for multi-dimensional differential operators of an arbitrary order (see [**10**]). The absence of the singular continuous spectrum is verified in Section 3. Here instead of Agmon's bootstrap arguments [**1**], a method specific for ordinary differential equations is used. After this prerequisite, we study in Section 4 asymptotic behavior of eigenfunctions of the continuous spectrum. The relation between the scattering matrix and the coefficients at exponentially decreasing modes is formulated in Theorem 4.8.

1.3. The second part of the paper (Sections 5, 6 and 7) is devoted to a study of the operator $H = D^4$ in the space $L_2(\mathbb{R}_+)$ with some selfadjoint boundary conditions at the point $x = 0$. This operator can be compared with a well-known Hamiltonian D^2 with a boundary condition

$$u'(0) = \alpha u(0), \quad \alpha = \bar{\alpha}. \tag{1.4}$$

This boundary condition is interpreted as a point interaction at the point $x = 0$. The point interaction is a good approximation to a perturbation by a potential for low (but not for high) energies.

We shall write down explicit (although not very simple) formulas for basic objects of scattering theory for the operator H, such as the resolvent kernel, eigenfunctions, the scattering matrix, the perturbation determinant, the spectral shift function and so forth. This model seems to be of interest because it allows us to analyse a behavior of different objects at zero energy as well as at a positive eigenvalue of the operator H. In particular, we discuss different types of zero-energy resonances.

There exist several definitions of zero-energy resonances which are essentially (but not completely) equivalent. According to a general variational definition (see [**3**]) an operator H has a zero-energy resonance if for a small negative perturbation an additional negative eigenvalue appears. Of course this definition depends not only on H but also on a class of perturbations. Other definitions (see [**15**] or [**18**]) are adapted to differential operators. Thus, the Schrödinger operator $H = -\Delta + v(x)$ has a zero-energy resonance if the kernel of its resolvent $R(z) = (H-z)^{-1}$ has a singularity $\varphi(x)\overline{\varphi(x')}(-z)^{-1/2}$ as $|z| \to 0$. Here $\varphi(x)$ is a solution of the equation $H\varphi = 0$ which is bounded at infinity (in the one-dimensional case). The existence of such solutions gives still another criterium for the appearance of zero-energy resonances. Zero-energy resonances can be considered as a weakened version of bound states (for zero energy) and are often called half-bound states. This point of view is confirmed by the behavior of the spectral shift function $\xi(\lambda)$ at the point $\lambda = 0$. Consider, for example, the operators H_0 and $H(\alpha)$ corresponding to the differential expression D^2 in the space $L_2(\mathbb{R}_+)$ with boundary conditions $u(0) = 0$ and (1.4), respectively. The operator $H(\alpha)$ has a zero-energy resonance if and only if $\alpha = 0$. The spectral shift function $\xi(\lambda)$ for the pair H_0, $H(\alpha)$ is continuous at $\lambda = 0$ if $\alpha \neq 0$ and it has the jump $-1/2$ if $\alpha = 0$. This should be compared with the fact that the jump of $\xi(\lambda)$ equals -1 at an eigenvalue of H.

In Section 7 we analyse in some detail zero-energy resonances for the operator $H = D^4$ with different boundary conditions at the point $x = 0$. It turns out that, compared to second order, for fourth order differential operators their notion acquires some new features although the variational definition remains of course valid. A more detailed classification is given in terms of singularities of the resolvent at the point $z = 0$, of a behavior of solutions (which are of course all polynomials of degree 3) of the equation $u^{(4)}(x) = 0$ satisfying the boundary condition and of a jump of the corresponding spectral shift function at the point $\lambda = 0$. This analysis shows that it is natural to introduce now 1/4- and 3/4-bound states. Here we mention only that the operator H has a 1/4-bound state (a 3/4-bound state) if a nontrivial linear function (a constant) satisfies the boundary conditions at $x = 0$.

Thanks are due to V. Suhanov for a useful discussion.

2. General scattering theory

2.1. Under the assumption

$$\sup_{x\in\mathbb{R}} \int_x^{x+1} (|v_1(y)| + |v_0(y)|)dy < \infty$$

we have that, for all $\varepsilon > 0$ and all functions u from the Sobolev class $\mathsf{H}^2(\mathbb{R})$,

$$\begin{aligned}\int_{-\infty}^{\infty} (|v_1(x)||u'(x)|^2 + |v_0(x)||u(x)|^2)dx \\ \leq \varepsilon \int_{-\infty}^{\infty} |u''(x)|^2 dx + C(\varepsilon)\int_{-\infty}^{\infty} |u(x)|^2 dx.\end{aligned} \tag{2.1}$$

Here and below C denotes different positive constants whose precise values are of no importance. Therefore the quadratic form

$$h[u,u] = \int_{-\infty}^{\infty} (|u''(x)|^2 + v_1(x)|u'(x)|^2 + v_0(x)|u(x)|^2)dx$$

is semi-bounded from below and is closed on $\mathsf{H}^2(\mathbb{R})$. Thus, it defines a selfadjoint operator H in the Hilbert space $\mathcal{H} = L_2(\mathbb{R})$ with domain $\mathcal{D}(H) \subset \mathsf{H}^2(\mathbb{R})$. This operator corresponds to the formal differential expression (1.1).

Note that the operators $H_0 = D^4 : \mathsf{H}^2(\mathbb{R}) \to \mathsf{H}^{-2}(\mathbb{R})$ and, by estimate (2.1),

$$V = Dv_1(x)D + v_0(x) : \mathsf{H}^2(\mathbb{R}) \to \mathsf{H}^{-2}(\mathbb{R}) \tag{2.2}$$

are bounded. It can easily be shown that a function $u \in \mathsf{H}^2(\mathbb{R})$ belongs to $\mathcal{D}(H)$ if and only if $H_0u + Vu \in L_2(\mathbb{R})$; in this case $Hu = H_0u + Vu$. It follows that, for $u \in \mathcal{D}(H)$, the function $u'''(x) - v_1(x)u'(x)$ is absolutely continuous and

$$(Hu)(x) = (u'''(x) - v_1(x)u'(x))' + v_0(x)u(x).$$

Let us discuss the main steps of a construction of scattering theory for the operator H. The resolvent $R_0(z) = (H_0 - z)^{-1}$, $z \in \mathbb{C} \setminus [0,\infty)$, of the operator H_0 can be calculated explicitly.

LEMMA 2.1. *Let $z \in \mathbb{C} \setminus [0,\infty)$ and $\zeta^4 = z$, $\arg\zeta \in (0, \pi/2)$. Then*

$$(R_0(z)f)(x) = \frac{1}{4\zeta^3}\int_{-\infty}^{\infty} (ie^{i\zeta|x-y|} - e^{-\zeta|x-y|})f(y)dy. \tag{2.3}$$

PROOF. Using the Fourier transform, we see that

$$(R_0(z)f)(x) = 2\pi^{-1}\int_{-\infty}^{\infty}\int_{-\infty}^{\infty} e^{ik(x-y)}(k^4 - z)^{-1}f(y)dkdy.$$

Therefore the resolvent kernel equals

$$R_0(x,y;z) = 2\pi^{-1}\int_{-\infty}^{\infty} e^{ik|x-y|}(k^4 - z)^{-1}dk.$$

This integral can be complemented in the upper half-plane by a big half-circle and then calculated by residues at the points ζ and $i\zeta$. □

Let us denote by Π the complex plane cut along $[0,\infty)$ (including upper and lower edges). According to Lemma 2.1 the resolvent kernel $R_0(x,y;z)$ is a continuous function of $z \in \Pi$ with the exception of the point $z = 0$. The following two results are immediate consequences of the explicit formula (2.3).

PROPOSITION 2.2. *If $f \in L_1(\mathbb{R})$, then for all $z \in \Pi$, $z \neq 0$, the function $R_0(z)f \in C^3(\mathbb{R})$ and $(R_0(z)f)'''(x)$ is an absolutely continuous function.*

PROPOSITION 2.3. *Let G_j be the operator of multiplication by a function $g_j \in L_2(\mathbb{R})$, $j = 0, 1$. Then the operator-valued functions*

$$G_0 D^l R_0(z) G_1, \quad l = 0, 1, 2, 3,$$

depend in the Hilbert–Schmidt norm continuously on $z \in \Pi$, $z \neq 0$.

To extend the latter result to the resolvent $R(z) = (H - z)^{-1}$ of the operator H, we proceed from the resolvent identity

$$R(z) = R_0(z) - R_0(z) V R(z) = R_0(z) - R(z) V R_0(z), \quad \operatorname{Im} z \neq 0, \tag{2.4}$$

where V is the operator (2.2). Let G_j and Ω_j be the operators of multiplication by the functions $\sqrt{|v_j(x)|}$ and $\operatorname{sgn} v_j(x)$, respectively. We introduce an auxiliary space $\mathcal{G} = \mathcal{H} \oplus \mathcal{H}$ and define (bounded) operators $\mathbf{G}_0$, $\mathbf{G} : \mathsf{H}^2(\mathbb{R}) \to \mathcal{G}$ by the formulas[1]

$$\mathbf{G}_0 = (\Omega_0 G_0, \Omega_1 G_1 D), \quad \mathbf{G} = (G_0, G_1 D). \tag{2.5}$$

It follows from equality (2.2) that $V = \mathbf{G}^* \mathbf{G}_0 = \mathbf{G}_0^* \mathbf{G}$. The resolvent identity (2.4) implies that

$$(I + \mathbf{G}_0 R_0(z) \mathbf{G}^*)(I - \mathbf{G}_0 R(z) \mathbf{G}^*) = (I - \mathbf{G}_0 R(z) \mathbf{G}^*)(I + \mathbf{G}_0 R_0(z) \mathbf{G}^*) = I.$$

Hence the inverse operator $(I + \mathbf{G}_0 R_0(z) \mathbf{G}^*)^{-1}$ exists and is bounded so that using again (2.4) we obtain the representation

$$R(z) = R_0(z) - R_0(z) \mathbf{G}^* (I + \mathbf{G}_0 R_0(z) \mathbf{G}^*)^{-1} \mathbf{G}_0 R_0(z). \tag{2.6}$$

Thus, the resolvent $R(z)$ for $\operatorname{Im} z \neq 0$ considered as a mapping from $\mathsf{H}^{-2}(\mathbb{R})$ to $\mathsf{H}^2(\mathbb{R})$ is a bounded operator.

It follows from Proposition 2.3 that under assumption (1.2) the operator-valued function $\mathbf{G}_0 R_0(z) \mathbf{G}^*$, analytic for $\operatorname{Im} z \in \mathbb{C} \setminus [0, \infty)$, is continuous in the Hilbert–Schmidt norm for $z \in \Pi$ except at the point $z = 0$. Therefore according to the analytic Fredholm alternative (see, e.g., [**16**]) the set $\mathcal{N} \in \mathbb{R}_+$ where at least one of two homogeneous equations

$$\mathbf{f} + \mathbf{G}_0 R_0(\lambda \pm i0) \mathbf{G}^* \mathbf{f} = 0, \quad \mathbf{f} = (f_0, f_1), \tag{2.7}$$

has a nontrivial solution $\mathbf{f} \in \mathcal{G}$ is closed and has Lebesgue measure zero. The operator-valued function $(I + \mathbf{G}_0 R_0(z) \mathbf{G}^*)^{-1}$ of $z \in \Pi$ is continuous in norm except for points from the set $\mathcal{N} \cup \{0\}$. Therefore equation (2.6) leads to the following result.

THEOREM 2.4. *Let assumption (1.2) hold, and let G_j be the same operators as in Proposition 2.3. Then the operator-valued functions $G_1 R(z) G_2$, $G_1 D R(z) G_2$ and $G_1 D R(z) D G_2$ of $z \in \Pi$ are continuous in the Hilbert–Schmidt norm except for points from the set $\mathcal{N} \cup \{0\}$. The set $\Lambda = \mathbb{R}_+ \setminus \mathcal{N}$ is open and has full Lebesgue measure.*

COROLLARY 2.5. *The spectrum of the operator H on the set Λ is absolutely continuous.*

We denote by $P^{(c)}$ the orthogonal projector on the absolutely continuous subspace $\mathcal{H}^{(c)}$ of the operator H.

[1]Although it is more convenient to write vectors as rows, we regard them as columns as far as matrix multiplication is concerned.

2.2. Given Theorem 2.4, an expansion in eigenfunctions of the operator H, a formula representation of the scattering matrix, etc., are consequences of general results of scattering theory (see [**6**] or [**16**]).

Denote by $E_0(\lambda)$ the spectral family of the operator H_0. In the momentum representation the operator $E_0(\lambda)$ acts as multiplication by the characteristic function $\chi_\lambda(\xi)$ of the interval $(-\sqrt[4]{\lambda}, \sqrt[4]{\lambda})$, that is,

$$(\widehat{E_0(\lambda) f_0})(\xi) = \chi_\lambda(\xi)\hat{f}_0(\xi),$$

where $\hat{f}_0(\xi)$ is the Fourier transform of the function $f_0(x)$. It follows that for $f_0 \in L_1(\mathbb{R}) \cap L_2(\mathbb{R})$ the function $(E_0(\lambda)f_0, f_0)$ belongs to the class $C^1(\mathbb{R}_+)$ and

$$d(E_0(\lambda)f_0, f_0)/d\lambda = 4^{-1}\lambda^{-3/4}(|\hat{f}_0(\sqrt[4]{\lambda})|^2 + |\hat{f}_0(-\sqrt[4]{\lambda})|^2). \tag{2.8}$$

Next we construct the canonical spectral representation of the operator H_0. We set

$$\Gamma_0(\lambda)f_0 = 2^{-1}\lambda^{-3/8}(\hat{f}_0(\lambda^{1/4}), \hat{f}_0(-\lambda^{1/4})), \quad \Gamma_0(\lambda) : L_1(\mathbb{R}) \cap L_2(\mathbb{R}) \to \mathbb{C}^2, \tag{2.9}$$

and define the operator $\mathcal{F}_0 : L_1(\mathbb{R}) \cap L_2(\mathbb{R}) \to L_2(\mathbb{R}_+; \mathbb{C}^2)$ by the formula $(\mathcal{F}_0 f)(\lambda) = \Gamma_0(\lambda)f$. This operator extends by continuity to a unitary mapping $\mathcal{F}_0 : \mathcal{H} \to L_2(\mathbb{R}_+; \mathbb{C}^2)$. Then $\mathcal{F}_0 H_0 = A\mathcal{F}_0$, where A is the operator of multiplication by λ in the space $L_2(\mathbb{R}_+; \mathbb{C}^2)$.

Now we discuss generalizations of these objects for the operator H. Let us set

$$\Gamma_\pm(\lambda)f = \Gamma_0(\lambda)(I - VR(\lambda \pm i0))f, \quad f \in L_1(\mathbb{R}) \cap L_2(\mathbb{R}), \quad \lambda \in \Lambda. \tag{2.10}$$

According to Theorem 2.4, $\Gamma_\pm(\lambda)f$ is a continuous function of $\lambda \in \Lambda$. A proof of the following result can be found in [**6**] or [**16**].

THEOREM 2.6. *Let assumption* (1.2) *hold. Define the mapping* $\mathcal{F}_\pm$ *on the set* $L_1(\mathbb{R}) \cap L_2(\mathbb{R})$ *by the equalities* (2.10) *and*

$$(\mathcal{F}_\pm f)(\lambda) = \Gamma_\pm(\lambda)f. \tag{2.11}$$

This mapping extends by continuity to a bounded operator $\mathcal{F}_\pm : \mathcal{H} \to L_2(\mathbb{R}_+; \mathbb{C}^2)$, *satisfies the relations*

$$\mathcal{F}_\pm \mathcal{F}_\pm^* = I, \quad \mathcal{F}_\pm^* \mathcal{F}_\pm = P^{(c)} \tag{2.12}$$

and diagonalizes H, *that is,*

$$\mathcal{F}_\pm H = A\mathcal{F}_\pm. \tag{2.13}$$

Time-dependent wave operators for the pair H_0, H are defined as strong limits

$$W_\pm = W_\pm(H, H_0) = \text{s-lim}_{t\to\pm\infty} e^{iHt}e^{-iH_0t}. \tag{2.14}$$

Recall that, by the spectral theorem,

$$R(\lambda \pm i\varepsilon) = \pm i \int_0^\infty e^{-\varepsilon t \pm i\lambda t} e^{\mp iHt} dt$$

so that, by the Parseval identity,

$$2\varepsilon \int_0^\infty e^{-2\varepsilon t}(e^{\mp itH_0}f_0, e^{\mp itH}f)dt = \pi^{-1}\varepsilon \int_{-\infty}^\infty (R_0(\lambda \pm i\varepsilon)f_0, R(\lambda \pm i\varepsilon)f)d\lambda \tag{2.15}$$

for all $f_0, f \in \mathcal{H}$.

Let us show that this expression has a limit as $\varepsilon \to 0$. This entails the existence of the weak wave operators understood, moreover, in the abelian sense. Such wave operators are defined by the limit of the left-hand side of (2.15). It suffices to verify the existence of the limit for $f_0, f \in L_1(\mathbb{R}) \cap L_2(\mathbb{R})$.

Let us consider the right-hand side of (2.15). Note that, by the resolvent identity (2.4),

$$\pi^{-1}\varepsilon(R_0(\lambda \pm i\varepsilon)f_0, R(\lambda \pm i\varepsilon)f) = (\delta_\varepsilon(H_0 - \lambda)f_0, (I - VR(\lambda \pm i\varepsilon))f),$$

where

$$\delta_\varepsilon(H_0 - \lambda) = (2\pi i)^{-1}(R_0(\lambda + i\varepsilon) - R_0(\lambda - i\varepsilon))$$

is an "approximate" operator-valued delta-function. It follows from the spectral theorem, standard properties of the Cauchy-type singular integrals and formulas (2.8), (2.9) that

$$\lim_{\varepsilon \to 0}(\delta_\varepsilon(H_0 - \lambda)f_0, f) = d(E_0(\lambda)f_0, f)/d\lambda = \langle \Gamma_0(\lambda)f_0, \Gamma_0(\lambda)f \rangle \tag{2.16}$$

where $\langle \cdot, \cdot \rangle$ is the scalar product in $\mathbb{C}^2$. Therefore, according to Proposition 2.3, we have

$$\lim_{\varepsilon \to 0} \mathbf{G}_0 \delta_\varepsilon(H_0 - \lambda)f_0 = \mathbf{G}_0 \Gamma_0^*(\lambda)\Gamma_0(\lambda)f_0. \tag{2.17}$$

The convergence in (2.16) and (2.17) is uniform on compact intervals of $\mathbb{R}_+$. Furthermore, Theorem 2.4 ensures that there exists

$$\lim_{\varepsilon \to 0} \mathbf{G}R(\lambda \pm i\varepsilon)f = \mathbf{G}R(\lambda \pm i0)f. \tag{2.18}$$

Combining relations (2.16), (2.17) and (2.18), we obtain that

$$\begin{aligned} &\lim_{\varepsilon \to 0}(\delta_\varepsilon(H_0 - \lambda)f_0, (I - VR(\lambda \pm i\varepsilon))f) = \langle \Gamma_0(\lambda)f_0, \Gamma_0(\lambda)f \rangle \\ &\quad - (\mathbf{G}_0\Gamma_0^*(\lambda)\Gamma_0(\lambda)f_0, \mathbf{G}R(\lambda \pm i0)f) = \langle \Gamma_0(\lambda)f_0, \Gamma_\pm(\lambda)f \rangle. \end{aligned} \tag{2.19}$$

The convergence in (2.18) and (2.19) is uniform with respect to λ from compact intervals of the set Λ.

It remains to justify a passage to the limit $\varepsilon \to 0$ in the integral on the right-hand side of (2.15). Note that, by the Schwarz inequality, for any Borel set $X \subset \mathbb{R}$,

$$\begin{aligned} &\Big| \int_X \varepsilon(R_0(\lambda \pm i\varepsilon)f_0, R(\lambda \pm i\varepsilon)f)d\lambda \Big|^2 \\ &\leq \int_X \varepsilon \|R_0(\lambda \pm i\varepsilon)f_0\|^2 d\lambda \int_{-\infty}^{\infty} \varepsilon \|R(\lambda \pm i\varepsilon)f\|^2 d\lambda \\ &= \pi^2 \int_X (\delta_\varepsilon(H_0 - \lambda)f_0, f_0)d\lambda \, \|f\|^2. \end{aligned} \tag{2.20}$$

Since the function $(\delta_\varepsilon(H_0 - \lambda)f_0, f)$ is the Poisson integral of the function (2.16) belonging to the space $L_1(\mathbb{R})$, we see (see, e.g., [**9**]) that the convergence in (2.16) holds true in the sense of $L_1(\mathbb{R})$. Therefore the right-hand side of (2.20) tends to zero as $|X| \to 0$ or $X = (N, \infty)$ and $N \to \infty$ uniformly with respect to $\varepsilon \in (0,1)$. Moreover, it tends to zero as $\varepsilon \to 0$ if $X = (-\infty, 0)$.

Thus, we have shown that the integral on the right-hand side of (2.15) has a limit which equals the integral of function (2.19) over $\lambda \in \mathbb{R}_+$. It follows that there exists a limit

(2.21)

$$\lim_{\varepsilon \to 0} 2\varepsilon \int_0^\infty e^{-2\varepsilon t}(e^{\mp itH_0}f_0, e^{\mp itH}f)dt = \int_0^\infty \langle \Gamma_0(\lambda)f_0, \Gamma_\pm(\lambda)f \rangle d\lambda = (\mathcal{F}_0 f_0, \mathcal{F}_\pm f)$$

and hence the abelian weak wave operators for the pair H_0, H exist and are equal to $\mathcal{F}_\pm^* \mathcal{F}_0$.

The strong wave operators (2.14) also exist. This fact can be deduced from general results of [**6**] or [**16**]. Alternatively, Theorem 2.4 entails that the operators $\mathbf{G}_0$ and $\mathbf{G}$ are H-smooth (as well as H_0-smooth) in the sense of Kato (see, e.g., [**13**] or [**16**]) on all compact intervals $X \subset \Lambda$ which also implies the existence of strong limits (2.14). Finally, we note that under assumption (1.2) the difference $R(z) - R_0(z)$ belongs to the trace class. Therefore the existence and completeness of wave operators (2.14) is a consequence of the Birman-Kreĭn theorem obtained in [**5**].

The results discussed above can be summarized in the following assertion.

THEOREM 2.7. *Let assumption* (1.2) *hold. Then the wave operators* (2.14) *exist and satisfy the relation* $W_\pm = \mathcal{F}_\pm^* \mathcal{F}_0$. *The operators* $W_\pm$ *are isometric and complete, that is, their ranges* $\operatorname{Ran} W_\pm = \mathcal{H}^{(c)}$. *The intertwining property* $HW_\pm = W_\pm H_0$ *holds.*

2.3. Since the scattering operator $\mathcal{S} = W_+^* W_-$ commutes with the operator H_0, the operator

$$\mathcal{F}_0 \mathcal{S} \mathcal{F}_0^* = \mathcal{F}_+ \mathcal{F}_-^* \tag{2.22}$$

acts in the space $L_2(\mathbb{R}_+; \mathbb{C}^2)$ as multiplication by a 2×2 matrix-valued function

$$S(\lambda) = \begin{pmatrix} s_{11}(\lambda) & s_{12}(\lambda) \\ s_{21}(\lambda) & s_{22}(\lambda) \end{pmatrix} \tag{2.23}$$

known as the scattering matrix. According to (2.11) equality (2.22) means that

$$S(\lambda)\Gamma_-(\lambda) f = \Gamma_+(\lambda) f, \quad \lambda \in \Lambda, \quad \forall f \in L_1(\mathbb{R}) \cap L_2(\mathbb{R}). \tag{2.24}$$

This equation determines the scattering matrix uniquely. Let us show that its solution is given by the formula

$$S(\lambda) = I - 2\pi i \Gamma_0(\lambda)(V - VR(\lambda + i0)V)\Gamma_0^*(\lambda), \quad \lambda \in \Lambda. \tag{2.25}$$

Let us set $\Gamma_0 = \Gamma_0(\lambda)$, $R_0 = R_0(\lambda + i0)$ and $R = R(\lambda + i0)$. By virtue of definition (2.10), we have to check that

$$(I - 2\pi i \Gamma_0 (V - VRV)\Gamma_0^*)\Gamma_0(I - VR^*) = \Gamma_0(I - VR).$$

In view of identity (2.16), it suffices to verify that

$$(I - RV)(R_0 - R_0^*)(I - VR^*) = R - R^*.$$

This equality is a direct consequence of the resolvent identity (2.4). Thus, we have proven the following result.

THEOREM 2.8. *Let assumption* (1.2) *hold. Then the scattering matrix* $S(\lambda)$ *for the pair* H_0, H *admits representation* (2.25).

According to (2.25) the scattering matrix $S(\lambda)$ is a continuous function of $\lambda \in \Lambda$.

Since $v_j(x) = \overline{v_j(x)}$ for $j = 0, 1$, the resolvent $R(z)$ commutes with the complex conjugation which can formally be written in terms of its kernel as $\overline{R(x, y; z)} = R(x, y; \bar{z})$. Also taking into account that $R(\bar{z}) = R^*(z)$, we see that the Green function $R(x, y; z)$ is symmetric, that is,

$$R(x, y; z) = R(y, x; z). \tag{2.26}$$

In view of representation (2.25), it follows from this relation that

$$s_{11}(\lambda) = s_{22}(\lambda). \tag{2.27}$$

3. A homogeneous equation

3.1. Let us study the structure of the exceptional set $\mathcal{N}$. Suppose that for one of the signs, equation (2.7) is satisfied. Set

$$\psi = R_0(\lambda \pm i0)\mathbf{G}^*\mathbf{f}. \tag{3.1}$$

Then it follows from (2.7) that

$$\mathbf{f} + \mathbf{G}_0\psi = 0 \tag{3.2}$$

and hence

$$\psi + R_0(\lambda \pm i0)V\psi = 0. \tag{3.3}$$

Taking into account definition (2.5) of the operator $\mathbf{G}$, we see that

$$V\psi = -\mathbf{G}^*\mathbf{f} = \varphi_0 + D\varphi_1, \quad \text{where} \quad \varphi_j = -\sqrt{|v_j|}f_j \in L_1(\mathbb{R}), \quad j = 0, 1. \tag{3.4}$$

The functions $R_0(\lambda \pm i0)\varphi_0$ and $R_0(\lambda \pm i0)D\varphi_1$ are well defined by Proposition 2.2.

PROPOSITION 3.1. *Suppose that assumption* (1.2) *holds. Let* $\mathbf{f} \in \mathcal{G}$, $\mathbf{f} \neq 0$, *satisfy equation* (2.7), *and let* ψ *be defined by formula* (3.1). *Then* ψ *is not identically zero,* $\psi \in C^2(\mathbb{R})$, *the functions* $\psi''(x)$ *and* $\psi'''(x) - v_1(x)\psi'(x)$ *are absolutely continuous and the differential equation* (1.3) *is satisfied.*

PROOF. If $\psi = 0$, then $\mathbf{f} = 0$ according to equation (3.2). By Proposition 2.2, the inclusion $\psi \in C^2(\mathbb{R})$ and the absolute continuity of $\psi''(x)$ follow from equation (3.3) and representation (3.4). For $\theta \in C_0^\infty(\mathbb{R})$, we have that

$$(R_0(\lambda \pm i0)V\psi, (D^4 - \lambda)\theta) = (v_1\psi', R_0(\lambda \mp i0)(D^4 - \lambda)\theta') + (v_0\psi, R_0(\lambda \mp i0)(D^4 - \lambda)\theta).$$

Using equation (3.3) and the fact that $R_0(\lambda \mp i0)(D^4 - \lambda)\theta = \theta$ for all $\theta \in C_0^\infty(\mathbb{R})$, we can rewrite this equality as

$$-(\psi, (D^4 - \lambda)\theta) = (v_1\psi', \theta') + (v_0\psi, \theta).$$

Thus,

$$(-\psi''' + v_1\psi', \theta') = (\lambda\psi - v_0\psi, \theta),$$

where $-\psi''' + v_1\psi' \in L_1^{(loc)}(\mathbb{R})$ and $\lambda\psi - v_0\psi \in L_1^{(loc)}(\mathbb{R})$. It follows that the derivative in the sense of distributions of the function $\psi''' - v_1\psi'$ equals $\lambda\psi - v_0\psi$. This implies that the function $\psi''' - v_1\psi'$ is absolutely continuous and the differential equation (1.3) is satisfied. □

Below the differential equation (1.3) is always understood in the sense specified in Proposition 3.1.

Next we find the asymptotic behavior of $\psi(x)$ as $|x| \to \infty$. To that end, we need the following standard assertion.

LEMMA 3.2. *Suppose that, for one of the signs, equation* (2.7) *is satisfied. Then*

$$\Gamma_0(\lambda)\mathbf{G}^*\mathbf{f} = 0. \tag{3.5}$$

PROOF. It follows from equation (2.7) that

$$\lim_{\varepsilon\to 0}(\mathbf{f} + \mathbf{G}_0 R_0(\lambda \pm i\varepsilon)\mathbf{G}^*\mathbf{f}, \mathbf{G}R_0(\lambda \pm i\varepsilon)\mathbf{G}^*\mathbf{f}) = 0.$$

Taking here the imaginary part and using that the operator $V = \mathbf{G}^*\mathbf{G}_0$ is symmetric, we obtain the equality

$$\lim_{\varepsilon\to 0}((R_0(\lambda + i\varepsilon) - R_0(\lambda - i\varepsilon))\mathbf{G}^*\mathbf{f}, \mathbf{G}^*\mathbf{f}) = 0.$$

Now (3.5) is a consequence of relation (2.16). □

Below integrals containing derivatives of L_1-functions (for example, $\varphi_1'(x)$) are understood in the sense of distributions, that is, integration by parts is tacitly assumed. Using formula (2.8) and representation (3.4), we can reformulate Lemma 3.2 in the following way.

COROLLARY 3.3. *Suppose that, for one of the signs, equation* (2.7) *is satisfied. Define the function* ψ *by formula* (3.1). *Then for both signs "* $\pm$ *",*

$$\int_{-\infty}^{\infty} e^{\pm ikx}(V\psi)(x)dx = 0, \quad k = \sqrt[4]{\lambda}. \tag{3.6}$$

We use also the following simple result.

LEMMA 3.4. *Let* $\varphi \in L_1(\mathbb{R})$. *Then*

$$(R_0(\lambda + i0)\varphi)(x) = \frac{i}{4k^3}e^{\pm ikx}\int_{-\infty}^{\infty} e^{\mp iky}\varphi(y)dy + o(1) \tag{3.7}$$

as $x \to \pm\infty$.

PROOF. Suppose for definiteness that $x \to +\infty$. According to (2.3) the function $-4ik^3(R_0(\lambda + i0)\varphi)(x)$ consists of two terms. The first of them equals

$$\begin{aligned}\int_{-\infty}^{\infty} e^{ik|x-y|}\varphi(y)dy &= e^{ikx}\int_{-\infty}^{\infty} e^{-iky}\varphi(y)dy \\ &- e^{ikx}\int_x^{\infty} e^{-iky}\varphi(y)dy + e^{-ikx}\int_x^{\infty} e^{iky}\varphi(y)dy.\end{aligned} \tag{3.8}$$

Since $\varphi \in L_1(\mathbb{R})$, both integrals over (x,∞) on the right-hand side tend to zero as $x \to +\infty$. The second term equals

$$\int_{-\infty}^{\infty} e^{-k|x-y|}\varphi(y)dy = e^{-kx}\int_{-\infty}^{x} e^{ky}\varphi(y)dy + e^{kx}\int_x^{\infty} e^{-ky}\varphi(y)dy. \tag{3.9}$$

Since

$$\Big|\int_{-\infty}^{x} e^{ky}\varphi(y)dy\Big| \le e^{kx/2}\int_{-\infty}^{x/2}|\varphi(y)|dy + e^{kx}\int_{x/2}^{x}|\varphi(y)|dy$$

and

$$\Big|\int_x^{\infty} e^{-ky}\varphi(y)dy\Big| \le e^{-kx}\int_x^{\infty}|\varphi(y)|dy,$$

both terms on the right-hand side of (3.9) tend to zero as $x \to +\infty$. □

Of course the asymptotics of $(R_0(\lambda - i0)\varphi)(x)$ is obtained from (3.7) by complex conjugation.

In view of Lemma 3.4, equation (3.3) and condition (3.6) imply that

$$\lim_{|x|\to\infty}\psi(x) = 0. \tag{3.10}$$

Let us formulate the results obtained in the following intermediary assertion.

PROPOSITION 3.5. *Under the assumptions of Proposition* 3.1 *condition* (3.10) *is satisfied.*

3.2. In this subsection we use specific methods of ordinary differential equations. Let us first of all rewrite equation (1.3) as a system of four equations of the first order. We set

$$\mathbf{u} = (u_1 u_2, u_3, u_4), \quad \text{where} \quad u_1 = u,\ u_2 = u',\ u_3 = u'',\ u_4 = u''' - v_1 u', \tag{3.11}$$

and

$$A = \begin{pmatrix} 0 & 1 & 0 & 0 \\ 0 & 0 & 1 & 0 \\ 0 & 0 & 0 & 1 \\ \lambda & 0 & 0 & 0 \end{pmatrix}, \quad K(x) = \begin{pmatrix} 0 & 0 & 0 & 0 \\ 0 & 0 & 0 & 0 \\ 0 & v_1(x) & 0 & 0 \\ -v_0(x) & 0 & 0 & 0 \end{pmatrix}. \tag{3.12}$$

Then equation (1.3) is equivalent to the system

$$\mathbf{u}'(x) = A\mathbf{u}(x) + K(x)\mathbf{u}(x). \tag{3.13}$$

Clearly, the matrix A has eigenvalues $\gamma_1 = ik$, $\gamma_2 = -ik, \gamma_3 = -k, \gamma_4 = k$. We denote by $\mathbf{p}_j = (1, \gamma_j, \gamma_j^2, \gamma_j^3)$, $j = 1, 2, 3, 4$, the corresponding eigenvectors. Let the (nonorthogonal) projectors P_j be defined by the relation

$$P_j \mathbf{f} = c_j \mathbf{p}_j \quad \text{if} \quad \mathbf{f} = \sum_{l=1}^{4} c_l \mathbf{p}_l.$$

Then $P_j^2 = P_j$, $P_j P_l = 0$ if $j \neq l$, $AP_j = \gamma_j P_j$ and

$$I = \sum_{j=1}^{4} P_j, \quad e^{Ax} = \sum_{j=1}^{4} e^{\gamma_j x} P_j.$$

Although the following result is a particular case of Problem 29, Chapter 3, of [**8**], we give its proof for completeness of our presentation.

PROPOSITION 3.6. *Let assumption* (1.2) *hold. Then, for each of the signs "$\pm$", system* (3.13) *has four solutions* $\mathbf{u}_j^{(\pm)}(x, \lambda)$, $j = 1, 2, 3, 4$, *such that*

$$\mathbf{u}_j^{(\pm)}(x, \lambda) = e^{\pm\gamma_j x}(\mathbf{p}_j + o(1)) \tag{3.14}$$

as $x \to \pm\infty$. *Estimates of the remainders in* (3.14) *are uniform with respect to* λ *from compact subintervals of* $\mathbb{R}_+$.

PROOF. We suppose that $x \to +\infty$ and omit the upper index "$\pm$". Pick some $j = 1, 2, 3, 4$. Let us set

$$Y_1(x) = Y_2(x) = e^{-kx} P_3, \quad Y_3(x) = 0, \quad Y_4(x) = \sum_{l=1}^{3} e^{\gamma_l x} P_l$$

and

$$Z_j(x) = e^{Ax} - Y_j(x).$$

Notice that

$$|Y_j(x)| \leq C e^{(\sigma_j - k)x} \quad \text{for} \quad x \geq 0 \quad \text{and} \quad |Z_j(x)| \leq C e^{\sigma_j x} \quad \text{for} \quad x \leq 0, \tag{3.15}$$

where $\sigma_j = \operatorname{Re}\gamma_j$. Let us choose a number a such that

$$2C\int_a^\infty |K(y)|dy \leq 1 \tag{3.16}$$

and consider an integral equation

$$\mathbf{u}_j(x) = e^{\gamma_j x}\mathbf{p}_j + \int_a^x Y_j(x-y)K(y)\mathbf{u}_j(y)dy - \int_x^\infty Z_j(x-y)K(y)\mathbf{u}_j(y)dy. \tag{3.17}$$

Below we also omit the index j. Let us show that equation (3.17) has a solution $\mathbf{u}(x)$ satisfying an estimate

$$|\mathbf{u}(x)| \leq 2pe^{\sigma x}, \quad p = \mathbf{p}, \quad x \geq a. \tag{3.18}$$

We use the method of successive approximations, setting $\mathbf{u}^{(0)}(x) = \mathbf{p}e^{\gamma x}$ and

$$\mathbf{u}^{(l+1)}(x) = \mathbf{p}e^{\gamma x} + \int_a^x Y(x-y)K(y)\mathbf{u}^{(l)}(y)dy - \int_x^\infty Z(x-y)K(y)\mathbf{u}^{(l)}(y)dy. \tag{3.19}$$

Let us check that, for all l,

$$|\mathbf{u}^{(l)}(x) - \mathbf{u}^{(l-1)}(x)| \leq 2^{-l}pe^{\sigma x}. \tag{3.20}$$

Supposing (3.20) for some l and using definition (3.19), we obtain an estimate

$$|\mathbf{u}^{(l+1)}(x) - \mathbf{u}^{(l)}(x)| \leq 2^{-l}p\Big(\int_a^x |Y(x-y)||K(y)|e^{\sigma y}dy + \int_x^\infty |Z(x-y)||K(y)|e^{\sigma y}dy\Big).$$

By inequalities (3.15) and condition (3.16) this expression does not exceed

$$2^{-l}p\,Ce^{\sigma x}\int_a^\infty |K(y)|dy \leq 2^{-l-1}pe^{\sigma x}.$$

This proves estimate (3.20) for $l+1$ in place of l and hence for all l. Thus, the sequence $\mathbf{u}^{(l)}(x)$ converges as $l \to \infty$ to a function $\mathbf{u}(x)$ satisfying estimate (3.18). Passing in (3.19) to the limit $l \to \infty$, we get equation (3.17).

To prove the asymptotics (3.14) for the function $\mathbf{u}(x)$, we combine inequalities (3.15) and (3.18). Obviously, the last integral on the right-hand side of (3.17) is $o(e^{\sigma x})$ as $x \to \infty$. The first integral is estimated by

$$2C\,pe^{\sigma x}\Big(e^{-kx/2}\int_a^{x/2} |K(y)|dy + \int_{x/2}^x |K(y)|dy\Big),$$

which is also $o(e^{\sigma x})$.

Finally, a direct differentiation shows that a solution of the integral equation (3.17) satisfies also system (3.13). □

Let $u_j^{(\pm)}(x,\lambda)$ be the first component of the vector $\mathbf{u}_j^{(\pm)}(x,\lambda)$. Proposition 3.6 can be reformulated in terms of solutions of equation (1.3).

PROPOSITION 3.7. *Let assumption* (1.2) *hold. Then, for each of the signs "±", the differential equation* (1.3) *has four solutions* $u_j^{(\pm)}(x,\lambda)$, $j = 1,2,3,4$, *such that*

$$\begin{aligned} u_1^{(\pm)}(x,\lambda) &= e^{\pm ikx}(1+o(1)), \quad u_2^{(\pm)}(x,\lambda) = e^{\mp ikx}(1+o(1)), \\ u_3^{(\pm)}(x,\lambda) &= e^{\mp kx}(1+o(1)), \quad u_4^{(\pm)}(x,\lambda) = e^{\pm kx}(1+o(1)) \end{aligned} \tag{3.21}$$

as $x \to \pm\infty$. *Estimates of the remainders in* (3.21) *are uniform with respect to* λ *from compact subintervals of* $\mathbb{R}_+$.

REMARK 3.8. It follows from formulas (3.11) that the asymptotic relations (3.21) are two times differentiable with respect to x. Moreover,

$$d^3u_j^{(\pm)}(x,\lambda)/dx^3 - v_1(x)du_j^{(\pm)}(x,\lambda)/dx = \pm\gamma_j^3 e^{\pm\gamma_j x}(\mathbf{p}_j + o(1))$$

as $x \to \pm\infty$.

Every solution $u(x)$ of equation (1.3) is a linear combination of the solutions $u_j^{(\pm)}(x,\lambda)$, $j = 1,2,3,4$. Therefore if $u(x) \to 0$ as $x \to \pm\infty$, then $u(x) = c^{(\pm)}u_3^{(\pm)}(x,\lambda)$ for some constant $c^{(\pm)}$ and hence belongs to $L_2(\mathbb{R}_\pm)$. In particular, we have

PROPOSITION 3.9. *Suppose that a function $u(x)$ satisfies equation* (1.3) *and $u(x) = o(1)$ as $x \to \pm\infty$. Then $u(x) = 0$ if λ is not an eigenvalue of the operator H.*

Combining Propositions 3.1, 3.5 and 3.9, we obtain that every $\lambda \in \mathcal{N}$ is necessarily an eigenvalue of the operator H. Taking also into account Theorem 2.7, we see that the singular continuous spectrum of the operator H is empty. Conversely, if ψ is an eigenfunction of the operator H, then it satisfies equation (3.3) and hence $\mathbf{f}$ defined by (3.2) satisfies equation (2.7). The positive eigenvalues of H are of course simple.

Let us finally show that the eigenvalues of the operator H might accumulate at the point zero only. Suppose on the contrary that the eigenvalues $\lambda_n = k_n^4 \to \lambda_0 = k_0^4 > 0$. Let ψ_n be the corresponding normalized eigenfunctions. By Proposition 3.7, we have that

$$\psi_n(x) = a_n^{(\pm)} e^{-k_n|x|}(1 + o(1))$$

as $x \to \pm\infty$. The estimate of the remainder here is uniform with respect to n. Since $\|\psi_n\| = 1$, we have that $|a_n^{(\pm)}| \leq C < \infty$. Therefore, for all $\varepsilon > 0$, and sufficiently large $R = R(\varepsilon)$,

$$\int_{|x|\geq R} |\psi_n(x)|^2 dx < \varepsilon \tag{3.22}$$

uniformly with respect to n. Moreover, we have that $h[\psi_n, \psi_n] = \lambda_n$ and hence $\|\psi_n\|_{\mathsf{H}^2(\mathbb{R})} \leq C < \infty$ according to estimate (2.1). Together with (3.22), this ensures compactness of the set of functions ψ_n in the space $L_2(\mathbb{R})$, which contradicts their orthogonality.

Thus, we have obtained

THEOREM 3.10. *Let assumption* (1.2) *hold. Then $\mathcal{N}$ coincides with the set of positive eigenvalues of the operator H, the singular continuous spectrum of the operator H is empty and the eigenvalues of the operator H might accumulate at the point zero only.*

According to Theorem 3.10 the absolutely continuous subspace $\mathcal{H}^{(c)}$ of the operator H equals $\mathcal{H}^{(c)} = \mathcal{H} \ominus \mathcal{H}^{(p)}$, where $\mathcal{H}^{(p)}$ is the subspace spanned by eigenfunctions (of the point spectrum) of the operator H.

4. Eigenfunctions of the continuous spectrum

4.1. The eigenfunctions $\psi_j^{(\pm)}(x,\lambda)$, $j = 1,2$, of the continuous spectrum of the operator H are defined by the formula

$$\psi_j^{(\pm)}(\lambda) = \psi_j^{(0)}(\lambda) - R(\lambda \mp i0)V\psi_j^{(0)}(\lambda), \quad j = 1,2, \tag{4.1}$$

where $\psi_1^{(0)}(x,\lambda) = e^{ikx}$, $\psi_2^{(0)}(x,\lambda) = e^{-ikx}$ and $k = \sqrt[4]{\lambda} > 0$. Set $\psi_j(x,\lambda) = \psi_j^{(-)}(x,\lambda)$, $j = 1, 2$. By virtue of property (2.26) we have that $\psi_1^{(+)}(x,\lambda) = \overline{\psi_2(x,\lambda)}$ and $\psi_2^{(+)}(x,\lambda) = \overline{\psi_1(x,\lambda)}$. The functions $\psi_j(x,\lambda)$ are also known as wave functions. The resolvent identity (2.4) implies that the functions $\psi_j^{(\pm)}(x,\lambda)$ satisfy the Lippmann–Schwinger equation

$$\psi_j^{(\pm)}(\lambda) = \psi_j^{(0)}(\lambda) - R_0(\lambda \mp i0)V\psi_j^{(\pm)}(\lambda), \quad j = 1, 2. \tag{4.2}$$

Observe that the right-hand side here is correctly defined. Indeed, it follows from Theorem 2.4 that $g\psi_j(\lambda), g\psi_j'(\lambda) \in L_2(\mathbb{R})$ for an arbitrary function $g \in L_2(\mathbb{R})$ and hence the functions $V\psi_j(\lambda)$ admit representation (3.4).

Similarly to the proof of Proposition 3.1, it is easy to deduce from (4.2) that the wave functions also satisfy differential equation (1.3). Their asymptotics as $|x| \to \infty$ can be found with the help of Lemma 3.4. Thus, we obtain the following result.

PROPOSITION 4.1. *Let assumption* (1.2) *hold. Suppose that* $\lambda = k^4$ *is not an eigenvalue of the operator* H. *Let matrix* (2.23) *be defined by equation* (2.25), *and let the solutions* $\psi_1(x,\lambda)$ *and* $\psi_2(x,\lambda)$ *of equation* (1.3) *be defined by formula* (4.1). *Then the asymptotic relations*

$$\begin{cases} \psi_1(x,\lambda) = e^{ikx} + s_{21}(\lambda)e^{-ikx} + o(1), & x \to -\infty, \\ \psi_1(x,\lambda) = s_{11}(\lambda)e^{ikx} + o(1), & x \to \infty \end{cases} \tag{4.3}$$

and

$$\begin{cases} \psi_2(x,\lambda) = e^{-ikx} + s_{12}(\lambda)e^{ikx} + o(1), & x \to \infty, \\ \psi_2(x,\lambda) = s_{22}(\lambda)e^{-ikx} + o(1), & x \to -\infty \end{cases} \tag{4.4}$$

hold.

REMARK 4.2. Here and below all asymptotic relations are differentiable in the sense of Remark 3.8. Actually, for example, the first relation (4.3) entails, by virtue of Proposition 3.7, that

$$\psi_1(x,\lambda) = s_{21}(\lambda)u_1^{(-)}(x,\lambda) + u_2^{(-)}(x,\lambda),$$

where the functions $u_1^{(-)}$ and $u_2^{(-)}$ are differentiable according to Remark 3.8.

Theorem 2.6 can be reformulated as an expansion of an arbitrary function in a generalized Fourier integral over the eigenfunctions $\psi_1^{(\pm)}(x,\lambda)$ and $\psi_2^{(\pm)}(x,\lambda)$ of the operator H. Indeed, according to definition (4.1) for an arbitrary $f \in L_1(\mathbb{R}) \cap L_2(\mathbb{R})$,

$$((I - VR(\lambda \pm i0))f, \psi_j^{(0)}(\lambda)) = (f, \psi_j^{(\pm)}(\lambda)), \quad j = 1, 2,$$

and according to definitions (2.9) and (2.10),

$$\Gamma_\pm(\lambda)f = 2^{-1}(2\pi)^{-1/2}\lambda^{-3/8}((f, \psi_1^{(\pm)}(\lambda)), (f, \psi_2^{(\pm)}(\lambda))). \tag{4.5}$$

Therefore, formula (2.11) reads as $(\mathcal{F}_\pm f)(\lambda) = (\tilde{f}_1(\lambda), \tilde{f}_2(\lambda))$, where

$$\tilde{f}_j(\lambda) = 2^{-1}(2\pi)^{-1/2}\lambda^{-3/8}\int_{-\infty}^{\infty} \overline{\psi_j^{(\pm)}(x,\lambda)} f(x)dx.$$

It follows that the relation $P^{(c)}f = \mathcal{F}_\pm^*\mathcal{F}_\pm f$ can (formally) be written as

$$(P^{(c)}f)(x) = 2^{-1}(2\pi)^{-1/2}\sum_{j=1}^{2}\int_0^\infty \psi_j^{(\pm)}(x,\lambda)\tilde{f}_j(\lambda)\lambda^{-3/8}d\lambda.$$

By representation (4.5), equality (2.24) is equivalent to the relations

$$\left\{\begin{array}{lcl} s_{11}(\lambda)\overline{\psi_1(x,\lambda)} + s_{12}(\lambda)\overline{\psi_2(x,\lambda)} & = & \psi_2(x,\lambda), \\ s_{21}(\lambda)\overline{\psi_1(x,\lambda)} + s_{22}(\lambda)\overline{\psi_2(x,\lambda)} & = & \psi_1(x,\lambda). \end{array}\right. \tag{4.6}$$

4.2. Let us show that the asymptotics (4.3) or (4.4) determine uniquely solutions of equation (1.3). We start with an auxiliary assertion which is true without assumption (1.2).

LEMMA 4.3. *Suppose that the functions v_0 and v_1 are real. Set*

$$F_u(r) = (u'''(r) - v_1(r)u'(r))\overline{u(r)} - u''(r)\overline{u'(r)}. \tag{4.7}$$

Then, for an arbitrary solution $u(x)$ of equation (1.3) *and all r, we have*

$$\operatorname{Im} F_u(r) = \operatorname{Im} F_u(-r). \tag{4.8}$$

PROOF. It follows from equation (1.3) that

$$\operatorname{Im} \int_{-r}^{r} (u''' - v_1 u')' \bar{u} dx = 0.$$

Integrating here by parts, we get equality (4.8). □

Now we can formulate the uniqueness result.

PROPOSITION 4.4. *Let assumption* (1.2) *hold. Suppose that $\lambda = k^4$ is not an eigenvalue of the operator H. If a solution $u(x)$ of equation* (1.3) *satisfies the conditions*

$$\left\{\begin{array}{ll} u(x) = \sigma_+ e^{\pm ikx} + o(1), & x \to \infty, \\ u(x) = \sigma_- e^{\mp ikx} + o(1), & x \to -\infty \end{array}\right. \tag{4.9}$$

for one of the signs and some numbers σ_+ and σ_-, then $\sigma_+ = \sigma_- = 0$ and $u(x) = 0$.

PROOF. It follows (see Remark 4.2) from relations (4.9) that function (4.7) has the asymptotics

$$F_u(r) = 2(\pm ik)^3 |\sigma_+|^2 + o(1), \quad r \to \infty,$$

and

$$F_u(-r) = 2(\mp ik)^3 |\sigma_-|^2 + o(1), \quad r \to -\infty.$$

Using Lemma 4.3, we find that $|\sigma_+|^2 + |\sigma_-|^2 = 0$ and hence $u(x) \to 0$ as $|x| \to \infty$. Thus, $u(x) = 0$ by Proposition 3.9. □

COROLLARY 4.5. *If a solution $\tilde{\psi}_1(x)$ of equation* (1.3) *has the asymptotics* (4.3) *with some coefficients s_{11} and s_{21}, then necessarily s_{11} and s_{21} are the entries of the scattering matrix and $\tilde{\psi}_1 = \psi_1$. Similarly, if a solution $\tilde{\psi}_2(x)$ of equation* (1.3) *has the asymptotics* (4.4) *with some coefficients s_{12} and s_{22}, then necessarily s_{12} and s_{22} are the entries of the scattering matrix and $\tilde{\psi}_2 = \psi_2$.*

Formulas (4.3) and (4.4) give us the definition of the scattering matrix in terms of solutions of the differential equation (1.3). Similarly to the Schrödinger equation, the numbers $s_{11}(\lambda)$, $s_{22}(\lambda)$ can be interpreted as transmission coefficients and $s_{21}(\lambda)$, $s_{12}(\lambda)$ can be interpreted as reflection coefficients for a plane wave coming from minus or plus infinity and interacting with the potentials $v_0(x)$ and $v_1(x)$.

4.3. Here we find asymptotics of the wave functions up to terms decaying super-exponentially at infinity. We say that a function $\varphi(x)$ decays super-exponentially if

$$\varphi(x) = O(e^{-\varkappa|x|}), \quad \forall \varkappa > 0, \quad |x| \to \infty. \tag{4.10}$$

The following result supplements Lemma 3.4.

LEMMA 4.6. *Let a function φ satisfy condition (4.10). Then*

$$(R_0(\lambda + i0)\varphi)(x) = \frac{i}{4k^3} e^{\pm ikx} \int_{-\infty}^{\infty} e^{\mp iky} \varphi(y) dy$$
$$- \frac{1}{4k^3} e^{-k|x|} \int_{-\infty}^{\infty} e^{\pm ky} \varphi(y) dy + O(e^{-\varkappa|x|}), \quad \forall \varkappa > 0,$$

as $x \to \pm\infty$.

PROOF. Suppose again for definiteness that $x \to +\infty$. According to (2.3) the function $-4ik^3(R_0(\lambda + i0)\varphi)(x)$ consists of the terms (3.8) and

$$\int_{-\infty}^{\infty} e^{-k|x-y|} \varphi(y) dy = e^{-kx} \int_{-\infty}^{\infty} e^{ky} \varphi(y) dy - e^{-kx} \int_{x}^{\infty} e^{ky} \varphi(y) dy + e^{kx} \int_{x}^{\infty} e^{-ky} \varphi(y) dy. \tag{4.11}$$

The integrals over (x, ∞) on the right-hand sides of (3.8) and (4.11) decay super-exponentially by virtue of condition (4.10). □

Let us return to the Lippmann–Schwinger equation (4.2) and take into account that the functions $\psi_j(x, \lambda)$ and $\psi_j'(x, \lambda)$, $j = 1, 2$, are bounded. Lemma 4.6 now yields a more precise form of Proposition 4.1.

PROPOSITION 4.7. *Let the functions v_0 and v_1 satisfy condition (4.10). Suppose that $\lambda = k^4$ is not an eigenvalue of the operator H. Let the matrix (2.23) be defined by equation (2.25), and let the solutions $\psi_1(x, \lambda)$ and $\psi_2(x, \lambda)$ of equation (1.3) be defined by formula (4.1). Then we have the asymptotic relations*

$$\begin{cases} \psi_1(x, \lambda) = s_{11}(\lambda) e^{ikx} + b_{11}(\lambda) e^{-kx} + O(e^{-\varkappa x}), & x \to \infty, \\ \psi_1(x, \lambda) = e^{ikx} + s_{21}(\lambda) e^{-ikx} + b_{21}(\lambda) e^{kx} + O(e^{\varkappa x}), & x \to -\infty, \end{cases} \tag{4.12}$$

and

$$\begin{cases} \psi_2(x, \lambda) = e^{-ikx} + s_{12}(\lambda) e^{ikx} + b_{12}(\lambda) e^{-kx} + O(e^{-\varkappa x}), & x \to \infty, \\ \psi_2(x, \lambda) = s_{22}(\lambda) e^{-ikx} + b_{22}(\lambda) e^{kx} + O(e^{\varkappa x}), & x \to -\infty, \end{cases} \tag{4.13}$$

where $\varkappa$ is arbitrary and

$$b_{1l}(\lambda) = \frac{1}{4k^3} \int_{-\infty}^{\infty} e^{ky} (V\psi_l(\lambda))(y) dy, \quad b_{2l}(\lambda) = \frac{1}{4k^3} \int_{-\infty}^{\infty} e^{-ky} (V\psi_l(\lambda))(y) dy,$$

$l = 1, 2$. The asymptotic coefficients $b_{jl}(\lambda)$, $j, l = 1, 2$, are continuous functions of $\lambda \in \Lambda$.

Let us find a relation between the sets of coefficients $s_{jl}(\lambda)$ and $b_{jl}(\lambda)$. To that end, we plug the asymptotics (4.12) and (4.13) into system (4.6). Neglecting super-exponentially decaying terms, we have that, as $x \to \infty$,

$$s_{11}(\overline{s_{11}} e^{-ikx} + \overline{b_{11}} e^{-kx}) + s_{12}(\overline{s_{12}} e^{-ikx} + \overline{b_{12}} e^{-kx}) = e^{-ikx} + b_{12} e^{-kx} \tag{4.14}$$

and

$$s_{21}(\overline{s_{11}}e^{-ikx} + \overline{b_{11}}e^{-kx}) + s_{22}(e^{ikx} + \overline{s_{12}}e^{-ikx} + \overline{b_{12}}e^{-kx}) \\ = s_{11}e^{ikx} + b_{11}e^{-kx}. \tag{4.15}$$

Similarly, if $x \to -\infty$, we have that

$$s_{11}(e^{-ikx} + \overline{s_{21}}e^{ikx} + \overline{b_{21}}e^{kx}) + s_{12}(\overline{s_{22}}e^{ikx} + \overline{b_{22}}e^{kx}) \\ = s_{22}e^{-ikx} + b_{22}e^{kx} \tag{4.16}$$

and

$$s_{21}(\overline{s_{21}}e^{ikx} + \overline{b_{21}}e^{kx}) + s_{22}(\overline{s_{22}}e^{ikx} + \overline{b_{22}}e^{kx}) = e^{ikx} + b_{21}e^{kx}. \tag{4.17}$$

Comparing the coefficients at e^{-ikx} on the left- and right-hand sides of equations (4.14) and (4.15), we find that

$$|s_{11}|^2 + |s_{12}|^2 = 1 \tag{4.18}$$

and

$$s_{21}\overline{s_{11}} + s_{22}\overline{s_{12}} = 0. \tag{4.19}$$

Comparing the coefficients at e^{ikx} on the left- and right-hand sides of equations (4.16) and (4.17), we obtain again relations (4.19) and

$$|s_{21}|^2 + |s_{22}|^2 = 1. \tag{4.20}$$

Identities (4.18), (4.19) and (4.20) show that the scattering matrix is a unitary operator in $\mathbb{C}^2$. This result has already been obtained in Section 2 as a consequence of the completeness of the wave operators.

Comparing the coefficients at e^{ikx} (or at e^{-ikx}) on the left- and right-hand sides of equation (4.15) (or of equation (4.16)), we recover relation (2.27), which has already been obtained in Section 2 as a consequence of the stationary representation (2.25) and of relation (2.26).

Finally, comparing the coefficients at exponentially decreasing terms in equations (4.14)–(4.17), we obtain four relations between the coefficients s_{jl} and b_{jl}. We write them in the matrix form as

$$\begin{pmatrix} s_{11}(\lambda) & s_{12}(\lambda) \\ s_{21}(\lambda) & s_{22}(\lambda) \end{pmatrix} \begin{pmatrix} \overline{b_{11}}(\lambda) & \overline{b_{21}}(\lambda) \\ \overline{b_{12}}(\lambda) & \overline{b_{22}}(\lambda) \end{pmatrix} = \begin{pmatrix} b_{12}(\lambda) & b_{22}(\lambda) \\ b_{11}(\lambda) & b_{21}(\lambda) \end{pmatrix}. \tag{4.21}$$

We formulate this result in the following assertion.

THEOREM 4.8. *Let the functions v_0 and v_1 satisfy condition* (4.10). *Suppose that $\lambda = k^4$ is not an eigenvalue of the operator H. Then the asymptotic coefficients s_{jl} and b_{jl} in* (4.12) *and* (4.13) *are linked by the relation* (4.21).

We emphasize that relations (4.21) as well as (2.27) are consequences of the invariance of the problem with respect to complex conjugation.

4.4. The approach developed for the problem on the whole line works of course for a similar problem in the space $L_2(\mathbb{R}_+)$. In this case we have to add a boundary condition at the point $x = 0$, which we choose as

$$u(0) = u'(0) = 0. \tag{4.22}$$

The operator H is defined by the differential expression (1.1) on functions satisfying (4.22). As a "free" operator, we take $H_0 = D^4$ with the same boundary condition. Since the operator H has a simple spectrum, there is now only one wave function

$\psi(x,\lambda)$ and the scattering operator acts in the space $L_2(\mathbb{R}_+)$ as multiplication by the function (scattering matrix) $s(\lambda)$.

Let us construct the function $\psi(x,\lambda)$. Remark first that the function

$$\psi_0(x,\lambda) = \cos(kx+\pi/4) - 2^{-1/2}e^{-kx} \tag{4.23}$$

satisfies the equation $u^{(4)}(x) = \lambda u(x)$ and the boundary condition (4.22). Assume as usual that λ is not an eigenvalue of H. A solution $\psi(x,\lambda)$ of equation (1.3) is then defined (cf. (4.1)) by the formula $\psi(\lambda) = \psi_0(\lambda) - R(\lambda + i0)V\psi_0(\lambda)$. Similarly to Proposition 4.1, it can be shown under the short-range assumption (1.2) that

$$\psi(x,\lambda) = 2^{-1}(s(\lambda)e^{ikx+\pi i/4} + e^{-ikx-\pi i/4}) + o(1) \tag{4.24}$$

as $x \to \infty$. Then we use (cf. Lemma 4.3) that $\operatorname{Im} F_u(r) = 0$ for all solutions $u(x)$ of equation (1.3) satisfying (4.22). Applying this result to the difference of two functions obeying (4.24), we see that actually condition (4.24) distinguishes a unique solution of (1.3). In particular,

$$\overline{\psi(x,\lambda)} = s(\lambda)\psi(x,\lambda). \tag{4.25}$$

Moreover, the identity $\operatorname{Im} F_\psi(r) = 0$ implies that $|s(\lambda)| = 1$.

If the functions $v_1(x)$ and $v_0(x)$ decay super-exponentially, then the remainder $o(1)$ in (4.24) can be replaced by a more precise term $-2^{-1/2}b(\lambda)e^{-kx} + O(e^{-\varkappa x})$, where $\varkappa$ is an arbitrary number. Relation (4.25) entails that

$$s(\lambda)\overline{b(\lambda)} = b(\lambda), \tag{4.26}$$

which plays the role of identity (4.21).

5. Perturbation by a boundary condition

Here we discuss the Hamiltonian $H = D^4$ in the space $L_2(\mathbb{R}_+)$ with some selfadjoint boundary conditions at the point $x = 0$ and calculate explicitly its resolvent. In contrast to the previous sections, here we avoid references to general results of scattering theory and give direct proofs of all assertions.

5.1. Selfadjoint extensions of a symmetric operator D^4 with domain $C_0^\infty(\mathbb{R}_+)$ in the space $L_2(\mathbb{R}_+)$ are defined by the formula $(Hu)(x) = u^{(4)}(x)$ on functions $u(x)$ from the Sobolev space $\mathsf{H}^4(\mathbb{R}_+)$ satisfying some boundary conditions at the point $x = 0$. Let us describe all of them. "Generic" selfadjoint boundary conditions have the form

$$\begin{cases} u''(0) &= \alpha u(0) + \alpha_1 u'(0), \\ u'''(0) &= -\alpha_2 u(0) - \bar\alpha u'(0), \end{cases} \tag{5.1}$$

where α_1 and α_2 are arbitrary real numbers and α is an arbitrary complex number. This family depends on four real constants. We introduce also an "exceptional" three-parameter family

$$u'(0) = \alpha u(0), \quad -u'''(0) + \bar\alpha u''(0) = \alpha_2 u(0), \quad \alpha \in \mathbb{C}, \quad \alpha_2 \in \mathbb{R}, \tag{5.2}$$

and a one-parameter family

$$u(0) = 0, \quad u''(0) = \alpha_1 u'(0), \quad \alpha_1 \in \mathbb{R}, \tag{5.3}$$

of boundary conditions. To exhaust all selfadjoint extensions, we have to add the boundary condition (4.22). The operator D^4 with this boundary condition will be denoted H_0. We also use a special notation H_{00} for the operator D^4 with boundary condition (5.3) where $\alpha_1 = 0$, that is,

$$u(0) = 0, \quad u''(0) = 0. \tag{5.4}$$

The quadratic form of the operator $H = H(\alpha, \alpha_1, \alpha_2)$ with boundary conditions (5.1) is given by the expression

$$h[u, u] = \int_0^\infty |u''(x)|^2 dx + \alpha_2 |u(0)|^2 + 2\,\mathrm{Re}(\alpha u(0)\bar{u}'(0)) + \alpha_1 |u'(0)|^2. \tag{5.5}$$

It is closed on the set $\mathsf{H}^2(\mathbb{R}_+)$. In the case (5.2), the form is defined by the expression

$$h[u, u] = \int_0^\infty |u''(x)|^2 dx + \alpha_2 |u(0)|^2 \tag{5.6}$$

on functions from $\mathsf{H}^2(\mathbb{R}_+)$ satisfying the condition $u'(0) = \alpha u(0)$. In the case (5.3), expression (5.5) where $\alpha = \alpha_2 = 0$ remains true if the form is restricted on the functions $u \in \mathsf{H}^2(\mathbb{R}_+)$ satisfying the condition $u(0) = 0$. Finally, the quadratic form of the operator H_0 is given by expression (5.6), where $\alpha_2 = 0$ on functions $u \in \mathsf{H}^2(\mathbb{R}_+)$ such that $u(0) = u'(0) = 0$.

Note that $H \leq H_0$ for all boundary conditions because the quadratic form of the operator H_0 is defined on the smallest possible set. Similarly, $H \leq H_{00}$ for the boundary conditions (5.1) and all $\alpha_1, \alpha_2, \alpha$ because the quadratic form of the operator H_{00} is defined on a smaller (the boundary condition $u(0) = 0$ is added) set than that of H.

As a "free" operator, it is natural to take the operator H_0 (the same operator as in subsection 4.4). However, technically, it is more convenient to work with the "intermediary free" operator H_{00}. The reason for this choice of the free operator is that H_{00} is the square of the operator D^2 with the boundary condition $u(0) = 0$.

We start with a construction of the resolvent $R(z)$ of the operator H. As a preliminary step, we shall find an explicit expression (cf. Lemma 2.1) for the resolvent $R_{00}(z)$ of the operator H_{00}.

LEMMA 5.1. *Let $z \in \mathbb{C} \setminus [0, \infty)$ and $\zeta^4 = z$, $\arg \zeta \in (0, \pi/2)$. Then*

$$(R_{00}(z)f)(x) = \frac{1}{4\zeta^3} \int_0^\infty (ie^{i\zeta|x-y|} - e^{-\zeta|x-y|} - ie^{i\zeta(x+y)} + e^{-\zeta(x+y)}) f(y) dy. \tag{5.7}$$

PROOF. Using the Fourier sine transform, we see that

$$(R_{00}(z)f)(x) = 2\pi^{-1} \int_0^\infty dk \sin(kx)(k^4 - z)^{-1} \int_0^\infty dy \sin(ky) f(y).$$

Therefore the resolvent kernel equals

$$R_{00}(x, y; z) = -(2\pi)^{-1} \int_{-\infty}^\infty e^{ik(x+y)}(k^4 - z)^{-1} dk + (2\pi)^{-1} \int_{-\infty}^\infty e^{ik|x-y|}(k^4 - z)^{-1} dk.$$

Both integrals can be complemented in the upper half-plane by a big half-circle and then calculated by residues at the points ζ and $i\zeta$. □

The solutions from $L_2(\mathbb{R}_+)$ of the equation

$$u^{(4)}(x) = zu(x) + f(x)$$

satisfying different boundary conditions at the point $x = 0$ differ by a linear combination of the solutions $e^{i\zeta x}$ and $e^{-\zeta x}$ of the homogeneous equation $u^{(4)}(x) = zu(x)$. Therefore we seek the resolvent $R(z)$ of H in the form (known as the Kreĭn formula)

$$R(z) = R_{00}(z) + P(z), \tag{5.8}$$

where $P(z)$ is a two-dimensional operator defined by the equalities

$$\begin{aligned}(P(z)f)(x) = (p_{11}(\zeta)e^{i\zeta x} + p_{12}(\zeta)e^{-\zeta x})Q_+(z)f \\ + (p_{21}(\zeta)e^{i\zeta x} + p_{22}(\zeta)e^{-\zeta x})Q_-(z)f\end{aligned} \tag{5.9}$$

and

$$Q_+(z)f = \int_0^\infty e^{i\zeta y} f(y)dy, \quad Q_-(z)f = \int_0^\infty e^{-\zeta y} f(y)dy. \tag{5.10}$$

Calculating derivatives of expressions (5.7) and (5.9) at $x = 0$, we find that
(5.11)

$$\begin{cases} (R(z)f)(0) &= (p_{11}(\zeta) + p_{12}(\zeta))Q_+(z)f + (p_{21}(\zeta) + p_{22}(\zeta))Q_-(z)f, \\ (R(z)f)'(0) &= 2^{-1}\zeta^{-2}(Q_+(z)f - Q_-(z)f) \\ & \quad +\zeta((ip_{11}(\zeta) - p_{12}(\zeta))Q_+(z)f + (ip_{21}(\zeta) - p_{22}(z))Q_-(\zeta)f), \\ (R(z)f)''(0) &= \zeta^2((-p_{11}(\zeta) + p_{12}(\zeta))Q_+(z)f \\ & \quad +(-p_{21}(\zeta) + p_{22}(\zeta))Q_-(z)f), \\ (R(z)f)'''(0) &= -2^{-1}(Q_+(z)f + Q_-(z)f) \\ & \quad -\zeta^3((ip_{11}(\zeta) + p_{12}(\zeta))Q_+(z)f + (ip_{21}(\zeta) + p_{22}(\zeta))Q_-(z)f). \end{cases}$$

The coefficients $p_{jl}(\zeta)$ in (5.9) are determined by boundary conditions on the functions $u(x) = (R(z)f)(x)$ at the point $x = 0$. We consider only generic boundary conditions (5.1) although the formulas obtained above allow one to also treat easily the cases (5.2) and (5.3). Plugging expressions (5.11) into (5.1) and equating coefficients at Q_+f and Q_-f, we obtain two systems of equations for p_{11}, p_{12} and for p_{21}, p_{22}:

$$\begin{cases} a_{11}(\zeta)p_{j1}(\zeta) + a_{12}(\zeta)p_{j2}(\zeta) = q_{j1}(\zeta), \\ a_{21}(\zeta)p_{j1}(\zeta) + a_{22}(\zeta)p_{j2}(\zeta) = q_{j2}(\zeta), \end{cases} \tag{5.12}$$

where $j = 1, 2$. The coefficients of these systems are given by the explicit formulas

$$\begin{cases} a_{11}(\zeta) = \alpha + i\alpha_1\zeta + \zeta^2, \quad a_{12}(\zeta) = \alpha - \alpha_1\zeta - \zeta^2, \\ a_{21}(\zeta) = -\alpha_2 - i\bar{\alpha}\zeta + i\zeta^3, \quad a_{22}(\zeta) = -\alpha_2 + \bar{\alpha}\zeta + \zeta^3, \end{cases} \tag{5.13}$$

and

$$\begin{cases} q_{11}(\zeta) = -(2\zeta^2)^{-1}\alpha_1, \quad q_{12}(\zeta) = (2\zeta^2)^{-1}\bar{\alpha} - 2^{-1}, \\ q_{21}(\zeta) = (2\zeta^2)^{-1}\alpha_1, \quad q_{22}(\zeta) = -(2\zeta^2)^{-1}\bar{\alpha} - 2^{-1}. \end{cases}$$

Of course, system (5.12) can easily be solved. Let us set $\alpha_0 = \alpha_1\alpha_2 - |\alpha|^2$ and

$$\begin{aligned}\Omega(\zeta) &= -2^{-1/2}e^{-\pi i/4}\zeta^{-1}(a_{11}(\zeta)a_{22}(\zeta) - a_{12}(\zeta)a_{21}(\zeta)) \\ &= \alpha_0 + (1-i)\alpha_2\zeta + 2i\operatorname{Re}\alpha\,\zeta^2 - (1+i)\alpha_1\zeta^3 - \zeta^4.\end{aligned} \tag{5.14}$$

Then

$$p_{jl}(\zeta) = 2^{-3/2}e^{-\pi i/4}\zeta^{-3}\Omega(\zeta)^{-1}\mathbf{p}_{jl}(\zeta), \quad j, l = 1, 2, \tag{5.15}$$

where

$$(5.16)\qquad \begin{cases} \mathbf{p}_{11}(\zeta) = -\alpha_0 - 2\operatorname{Re}\alpha\,\zeta^2 + 2\alpha_1\zeta^3 + \zeta^4, \\ \mathbf{p}_{12}(\zeta) = \alpha_0 + 2i\operatorname{Im}\alpha\,\zeta^2 + \zeta^4, \\ \mathbf{p}_{21}(\zeta) = \alpha_0 - 2i\operatorname{Im}\alpha\,\zeta^2 + \zeta^4, \\ \mathbf{p}_{22}(\zeta) = -\alpha_0 + 2\operatorname{Re}\alpha\,\zeta^2 + 2i\alpha_1\zeta^3 + \zeta^4. \end{cases}$$

Note that

$$\overline{\Omega(\zeta)} = \Omega(i\bar{\zeta}), \quad \zeta \in \mathbb{C},$$

so that the zeros of the function $\Omega(\zeta)$ are symmetric with respect to the line $\operatorname{Re}\zeta = \operatorname{Im}\zeta$.

Let us summarize the results obtained.

THEOREM 5.2. *Let $z \in \mathbb{C}\setminus[0,\infty)$ and $\zeta^4 = z$, $\arg\zeta \in (0,\pi/2)$. Then the resolvent $R(z)$ of the operator $H = d^4/dx^4$ with the boundary conditions (5.1), (5.2) or (5.3) is given by formula (5.8) where the resolvent $R_{00}(z)$ of the operator H_{00} is determined by formula (5.7) and the operator $P(z)$ is determined by formulas (5.9) and (5.10). In the case (5.1), the coefficients $p_{jl}(\zeta)$ are defined by the formulas (5.14), (5.15) and (5.16).*

Note a particular case of this result. If $\alpha = \alpha_1 = \alpha_2 = 0$, then $\Omega(\zeta) = -\zeta^4$ and

$$(5.17)\qquad P(x,y;z) = -2^{-3/2}e^{-\pi i/4}\zeta^{-3}(e^{i\zeta x} + e^{-\zeta x})(e^{i\zeta y} + e^{-\zeta y}).$$

Of course explicit formulas for the coefficients $p_{jl}(\zeta)$ in (5.9) can easily be written down also for other boundary conditions. For example, let us consider the operators H_0 and H_1 determined by the boundary conditions (4.22) and

$$(5.18)\qquad u'(0) = u'''(0) = 0,$$

respectively. Using formulas (5.11), we see that their resolvents are given by the equality

$$(5.19)\qquad R_j(z) = R_{00}(z) + P_j(z), \quad j = 0,1,$$

where $P_j(z)$ are integral operators with kernels

$$(5.20)\qquad P_0(x,y;z) = (4\zeta^3)^{-1}(i-1)(e^{i\zeta x} - e^{-\zeta x})(e^{i\zeta y} - e^{-\zeta y})$$

and

$$(5.21)\qquad P_1(x,y;z) = (2\zeta^3)^{-1}(ie^{i\zeta(x+y)} - e^{-\zeta(x+y)}).$$

5.2. Let us discuss the eigenvalues of the operator H corresponding to the boundary conditions (5.1). Recall that the function Ω was defined by formula (5.14).

PROPOSITION 5.3. *The zeros of the function $\Omega(\zeta)$ from the sector $\arg\zeta \in (0,\pi/2)$ lie on the ray $\arg\zeta = \pi/4$. A point $\lambda = -k^4$, $k > 0$, is an eigenvalue of the operator H if and only if $\Omega(e^{i\pi/4}k) = 0$, that is,*

$$k^4 + \sqrt{2}\alpha_1 k^3 - 2\operatorname{Re}\alpha\,k^2 + \sqrt{2}\alpha_2\,k + \alpha_0 = 0.$$

The multiplicities of an eigenvalue λ of the operator H and of the zero $e^{i\pi/4}k$ of the function Ω coincide. A point $\lambda_0 = -k_0^4$ is a degenerate (of multiplicity 2) eigenvalue of the operator H if and only if

$$(5.22)\qquad \alpha_1 = -\sqrt{2}k_0, \quad \alpha = -k_0^2, \quad \alpha_2 = -\sqrt{2}k_0^3.$$

In this case the eigenfunctions are defined by the equations

$$\psi_1(x) = \exp((-1+i)k_0x/\sqrt{2}), \quad \psi_2(x) = \exp((-1-i)k_0x/\sqrt{2})$$

and necessarily

$$\Omega(\zeta) = -(\zeta - e^{i\pi/4}k_0)^2(\zeta^2 + ik_0^2). \tag{5.23}$$

PROOF. Clearly, a solution

$$u(x) = p_1e^{i\zeta x} + p_2e^{-\zeta x} \tag{5.24}$$

of the equation

$$u^{(4)}(x) = \zeta^4u(x) \tag{5.25}$$

satisfies the boundary conditions (5.1) if and only if (cf. (5.12))

$$\begin{cases} a_{11}(\zeta)p_1 + a_{12}(\zeta)p_2 = 0, \\ a_{21}(\zeta)p_1 + a_{22}(\zeta)p_2 = 0, \end{cases} \tag{5.26}$$

where the coefficients $a_{jl}(\zeta)$ are defined by the formulas (5.13). Since both functions $e^{i\zeta x}$ and $e^{-\zeta x}$ belong to $L_2(\mathbb{R}_+)$, the point $\lambda = \zeta^4$ is an eigenvalue of the operator H if and only if system (5.26) has a nontrivial solution, that is, $\Omega(\zeta) = 0$. It follows that $\arg\zeta = \pi/4$ because necessarily $\lambda < 0$.

A point $\lambda_0 = -k_0^4$ is an eigenvalue of multiplicity 2 if and only if system (5.26), where $\zeta_0 = e^{\pi i/4}k_0$, is satisfied for all numbers p_1 and p_2, that is,

$$a_{11}(\zeta_0) = a_{12}(\zeta_0) = 0, \quad a_{21}(\zeta_0) = a_{22}(\zeta_0) = 0. \tag{5.27}$$

Using (5.13) and solving the first two equations (5.27), we obtain expressions (5.22) for α_1 and α. Similarly, considering the last two equations (5.27), we obtain expression (5.22) for α_2. Plugging expressions (5.22) into (5.14), we get representation (5.23) so that the point ζ_0 is a double zero of the function $\Omega(\zeta)$.

Conversely, if $\Omega(\zeta_0) = \Omega'(\zeta_0) = 0$, then $\mathbf{p}_{jl}(\zeta_0) = 0$ for all $j, l = 1, 2$, because the functions $p_{jl}(\zeta)$ can have only simple poles. This is a consequence of representations (5.8), (5.9) for the resolvent $R(z)$ whose poles are simple. By virtue of (5.16) the equations $\mathbf{p}_{12}(\zeta_0) = \mathbf{p}_{21}(\zeta_0) = 0$ yield $\operatorname{Im}\alpha = 0$, $\alpha_0 + \zeta_0^4 = 0$. Therefore the equations $\mathbf{p}_{11}(\zeta_0) = \mathbf{p}_{22}(\zeta_0) = 0$ can be written as

$$\begin{cases} -\alpha + \alpha_1\zeta_0 + \zeta_0^2 = 0, \\ \alpha + i\alpha_1\zeta_0 + \zeta_0^2 = 0, \end{cases}$$

whence $(1+i)\alpha_1 + 2\zeta_0 = 0$ and $(1-i)\alpha_1\zeta_0 = 2\alpha$. This implies equations (5.22) so that equations (5.27) for the numbers (5.13) are also satisfied. Thus, function (5.24) where $\zeta = \zeta_0$ satisfies the boundary conditions (5.1) for all p_1, p_2 and hence $-k_0^4$ is an eigenvalue of multiplicity 2. □

Since the rank of the operator $R(z) - R_{00}(z)$ equals 2 and $H_{00} \geq 0$, the operator H might have at most 2 negative eigenvalues with multiplicity taken into account. The following result (cf. [**17**]) makes this assertion more precise.

PROPOSITION 5.4. *The total numbers of negative eigenvalues* (*counted with their multiplicity*) *of the operators* H *and*

$$A = \begin{pmatrix} \alpha_2 & \bar{\alpha} \\ \alpha & \alpha_1 \end{pmatrix} : \mathbb{C}^2 \to \mathbb{C}^2$$

coincide.

PROOF. Let $E(\mathbb{R}_-)$ and $\mathcal{E}(\mathbb{R}_-)$ be spectral projectors of the operators H and A, respectively, corresponding to the set $\mathbb{R}_- = (-\infty, 0)$. We have to check that

$$N := \dim E(\mathbb{R}_-)\mathcal{H} = \dim \mathcal{E}(\mathbb{R}_-)\mathbb{C}^2 =: n.$$

Let us define the mapping $J : \mathsf{H}^2(\mathbb{R}_+) \to \mathbb{C}^2$ by the relation $Ju = (u(0), u'(0))$. Then formula (5.5) can be written as

$$h[u,u] = \int_0^\infty |u''(x)|^2 dx + \langle AJu, Ju \rangle. \tag{5.28}$$

Therefore, if $u \in E(\mathbb{R}_-)\mathcal{H}$, $u \neq 0$, then $\langle AJu, Ju \rangle < 0$. Thus, the quadratic form of the operator A is negative on the subspace $JE(\mathbb{R}_-)\mathcal{H}$. This subspace has dimension N because $Ju \neq 0$ for $u \neq 0$ and hence $n \geq N$.

Conversely, pick a function $\varphi \in C^\infty(\mathbb{R}_+)$ such that $\varphi(x) = 1$ in a neighborhood of the point $x = 0$ and $\varphi(x) = 0$ for sufficiently large x and set

$$(\tilde{J}_\varepsilon a)(x) = (a_1 + a_2 x)\varphi(\varepsilon x), \quad a = (a_1, a_2), \quad \tilde{J}_\varepsilon : \mathbb{C}^2 \to \mathsf{H}^2(\mathbb{R}_+).$$

Note that there exists a constant $\gamma > 0$ such that $\langle Aa, a \rangle \leq -\gamma |a|^2$ for all $a \in \mathcal{E}(\mathbb{R}_-)\mathbb{C}^2$. According to (5.28) we have that

$$h[\tilde{J}_\varepsilon a, \tilde{J}_\varepsilon a] = \int_0^\infty |(\tilde{J}_\varepsilon a)''(x)|^2 dx + \langle Aa, a \rangle, \tag{5.29}$$

where

$$\int_0^\infty |(\tilde{J}_\varepsilon a)''(x)|^2 dx \leq \varepsilon |a|^2.$$

Thus, expression (5.29) is negative if $\varepsilon < \gamma$ so that $\tilde{J}_\varepsilon \mathcal{E}(\mathbb{R}_-)\mathbb{C}^2$ is a subspace of dimension n on which the quadratic form h is negative. This implies that $n \leq N$. □

COROLLARY 5.5. *If $\alpha_0 = \alpha_1\alpha_2 - |\alpha|^2 < 0$, then the operator H has precisely one negative eigenvalue. If $\alpha_0 > 0$, then the operator H has two negative eigenvalues for $\alpha_1 < 0$ (or equivalently $\alpha_2 < 0$) and $H \geq 0$ for $\alpha_1 > 0$ (or equivalently $\alpha_2 > 0$).*

PROOF. If $\alpha_0 = \operatorname{Det} A < 0$, then the eigenvalues μ_1 and μ_2 of the operator A have different signs. If $\alpha_0 = \operatorname{Det} A > 0$, then $\mu_1\mu_2 > 0$, $\alpha_1\alpha_2 > 0$ and $\mu_1 + \mu_2 = \alpha_1 + \alpha_2$. Thus, all four numbers μ_1, μ_2 and α_1, α_2 have the same sign. □

Let us further consider zeros of the function $\Omega(\zeta)$ on the half-axis $\zeta = k$ (or equivalently $\zeta = ik$) where $k > 0$.

PROPOSITION 5.6. *The function $\Omega(k) = 0$ for $k > 0$ if and only if $\lambda = k^4$ is an eigenvalue of the operator H.*

PROOF. Put $\zeta = k$. Recall that solution (5.24) of equation (5.25) satisfies the boundary conditions (5.1) if and only if the coefficients p_1 and p_2 satisfy system (5.26). If λ is an eigenvalue of H, then $\psi(x) = \exp(-kx)$ is the eigenfunction of H and hence system (5.26) is satisfied with $p_1 = 0$ and $p_2 = 1$. It follows that $a_{12}(k) = a_{22}(k) = 0$ and hence $\Omega(k) = 0$.

Conversely, if $\Omega(k) = 0$, then system (5.26) has a nontrivial solution p_1, p_2 so that solution (5.24) of equation (5.25) satisfies the boundary conditions (5.1). It remains only to show that $p_1 = 0$. Let us multiply (cf. the proof of Lemma 4.3) equation (5.25) by $\overline{u(x)}$ and integrate it over an interval $(0, r)$. Then we integrate by parts and take the imaginary part. Since the nonintegral terms disappear at $x = 0$, we obtain the identity

$$\operatorname{Im}(u'''(r)\overline{u(r)} - u''(r)\overline{u'(r)}) = 0.$$

Applying this identity to function (5.24) and neglecting terms exponentially decaying as $r \to \infty$, we see that $p_1 = 0$ and hence $u(x) = e^{-kx}$ is an eigenfunction of the operator H. □

Positive eigenvalues are not exceptional for the operator H. It is easy to give simple necessary and sufficient conditions for their existence.

PROPOSITION 5.7. *A point $\lambda = k^4$, $k > 0$, is an eigenvalue of the operator H if and only if*

$$\alpha = \overline{\alpha} \quad \text{and} \quad \alpha_1 = (\alpha - k^2)k^{-1}, \quad \alpha_2 = (\alpha + k^2)k. \tag{5.30}$$

In this case,

$$\Omega'(k) = -2k(\alpha - ik^2) \neq 0 \tag{5.31}$$

so that k is a simple zero of the function $\Omega(\zeta)$.

PROOF. The eigenfunction of H is necessarily e^{-kx}, and the boundary conditions (5.1) for this function are equivalent to equations (5.30). Differentiating (5.14) and using expressions (5.30) for α_1 and α_2, we arrive at (5.31). □

COROLLARY 5.8. *The operator H cannot have more than one positive eigenvalue.*

PROOF. If $\lambda = k^4$, $k > 0$, is an eigenvalue of H, then it follows from equations (5.30) that $k^4 + \alpha_0 = 0$, which determines $k > 0$ uniquely. □

Thus, for each given $\lambda > 0$, there is a one-dimensional manifold in the four-dimensional space of parameters $\alpha, \alpha_1, \alpha_2$ parametrized by $\alpha = \overline{\alpha}$ such that the corresponding operators $H = H(\alpha, \alpha_1, \alpha_2)$ have an eigenvalue at the point λ. If k varies over $\mathbb{R}_+$, then equations (5.30) determine (parametrically) a surface in the space of parameters such that the operators H have a positive eigenvalue. Note that the condition $\alpha_0 < 0$ is necessary for the existence of a positive eigenvalue.

5.3. Next we calculate the spectral measure $E(\lambda)$ of the operator H corresponding to the boundary conditions (5.1).

PROPOSITION 5.9. *Suppose that $\lambda = k^4 > 0$ is not an eigenvalue of the operator H. Set*

$$s(\lambda) = \overline{\Omega(k)}/\Omega(k) \tag{5.32}$$

and

$$b(\lambda) = (\alpha_0 + 2i \operatorname{Im} \alpha\, k^2 + k^4)/\Omega(k), \tag{5.33}$$

where the function $\Omega(k)$ is defined by formula (5.14). Then $dE(\lambda)/d\lambda$ is the integral operator with kernel

$$dE(x, y; \lambda)/d\lambda = (2\pi)^{-1}k^{-3}\psi(x, \lambda)\overline{\psi(y, \lambda)}, \tag{5.34}$$

where

$$\psi(x, \lambda) = 2^{-1}(s(\lambda)e^{ikx+\pi i/4} + e^{-ikx-\pi i/4}) - 2^{-1/2}b(\lambda)e^{-kx}. \tag{5.35}$$

PROOF. Passing to the limit $z \to \lambda \pm i0$ in equality (5.8) and using the relation between the boundary values of the resolvent and the spectral measure (see formula (2.16)), we find that

(5.36)
$$dE(x,y;\lambda)/d\lambda = dE_{00}(x,y;\lambda)/d\lambda + (2\pi i)^{-1}(P(x,y;\lambda+i0) - \overline{P(y,x;\lambda+i0)}).$$

The kernel $P(x,y;z)$ of the operator $P(z)$ is defined by formulas (5.9) and (5.10). Since $E_{00}(\lambda) = E_{\sqrt{H_{00}}}(\sqrt{\lambda})$ where $\sqrt{H_{00}} = D^2$ with the boundary condition $u(0) = 0$, we have that

$$dE_{00}(x,y;\lambda)/d\lambda = (2\pi)^{-1}k^{-3}\sin kx \sin ky.$$

Therefore, relation (5.36) implies that

$$\begin{aligned} dE(x,y;\lambda)/d\lambda = &-(8\pi)^{-1}k^{-3}(e^{ikx}-e^{-ikx})(e^{iky}-e^{-iky}) \\ &+ (2\pi i)^{-1}\Big(p_{11}(k)e^{ik(x+y)} - \overline{p_{11}(k)}e^{-ik(x+y)} \\ &+ p_{12}(k)e^{-k(x-iy)} - \overline{p_{21}(k)}e^{-k(x+iy)} \\ &+ p_{21}(k)e^{-k(-ix+y)} - \overline{p_{12}(k)}e^{-k(ix+y)} \\ &+ (p_{22}(k) - \overline{p_{22}(k)})e^{-k(x+y)}\Big), \end{aligned} \tag{5.37}$$

where the coefficients p_{jl} are defined by formulas (5.14)–(5.16).

Let us plug (5.35) into the right-hand side of (5.34). Taking into account definitions (5.32) and (5.33), we see that the coefficients at all terms $e^{ik(x+y)}$, $e^{-ik(x+y)}$, $e^{ik(-x+y)}$, $e^{ik(x-y)}$, $e^{-k(x-iy)}$, $e^{-k(x+iy)}$, $e^{-k(-ix+y)}$, $e^{-k(ix+y)}$ and $e^{-k(x+y)}$ are the same as in (5.37). This proves relation (5.34). □

The functions $\psi(x,\lambda)$ defined by formula (5.35) satisfy of course the boundary conditions (5.1). They are known as eigenfunctions of the continuous spectrum of the operator H. These functions describe an interaction of a plane wave $e^{-ikx-\pi i/4}$ coming from $+\infty$ with a "point" potential at $x = 0$. The coefficient $s(\lambda)$ at the reflected wave $e^{ikx+\pi i/4}$ is known as the scattering matrix. Note that if $\operatorname{Im}\alpha = 0$ (in this case the problem is not only selfadjoint but is also real), then according to equalities (5.32) and (5.33) the functions $s(\lambda)$ and $b(\lambda)$ are related by formula (4.26). Thus, the scattering matrix $s(\lambda)$ can be recovered from the coefficient $b(\lambda)$ at the exponentially decaying mode of eigenfunction (5.35).

Proposition 5.9 implies that the positive spectrum of the operator H is absolutely continuous except, possibly, at a single eigenvalue λ_0. Moreover, we have the following result.

PROPOSITION 5.10. *The functions $s(\lambda)$ and $b(\lambda)$ are infinitely differentiable for all $\lambda > 0$. If λ_0 is a positive eigenvalue of the operator H, then*

$$s(\lambda_0) = \frac{\alpha + i\sqrt{\lambda_0}}{\alpha - i\sqrt{\lambda_0}}, \quad b(\lambda_0) = -2\frac{\sqrt{\lambda_0}}{\alpha - i\sqrt{\lambda_0}}. \tag{5.38}$$

PROOF. Let us proceed from formulas (5.32) and (5.33). Clearly, the functions $s(\lambda)$ and $b(\lambda)$ are infinitely differentiable away from the point $\lambda_0 = k_0^4$ (the eigenvalue of H) where $\Omega(k_0) = 0$. As far as their behavior at λ_0 is concerned, the result for $s(\lambda)$ follows from formulas (5.31) and

$$s(\lambda_0) = \overline{\Omega'(k_0)}/\Omega'(k_0).$$

To consider $b(\lambda)$, we take into account additionally that according to Proposition 5.7 a point $\lambda_0 = k_0^4$ is a positive eigenvalue of the operator H if and only if conditions (5.30) are satisfied for $k = k_0$. Therefore the numerator in (5.33) equals zero at $k = k_0$ and its derivative equals $4k_0^3$. This yields relation (5.38) for $b(\lambda_0)$. $\square$

We emphasize that both functions $is(\lambda_0)e^{ik_0x} + e^{-ik_0x}$ and e^{-k_0x} satisfy the boundary condition (5.1) if λ_0 is an eigenvalue.

REMARK 5.11. The function $b(\lambda)$ might have a zero only at one point $\lambda_0 = -\alpha_0$. According to Proposition 5.7, this condition is satisfied if λ_0 is an eigenvalue of H. However, according to Proposition 5.10, in this case $b(\lambda_0) \neq 0$. On the contrary, $b(\lambda_0) = 0$ if $\lambda_0 = -\alpha_0$, $\alpha = \bar{\alpha}$, but the last two conditions (5.30) are violated at $k = k_0$. For example, if $\alpha = k_0^2$, $\alpha_1 = 0$ but $\alpha_2 \neq 2k_0$, then $b(\lambda_0) = 0$ for $\lambda_0 = k_0^4$.

5.4. Now we are in a position to establish an expansion in eigenfunctions of the operator H.

THEOREM 5.12. *Let $\psi_-(x,\lambda) = \psi(x,\lambda)$ be defined by the formulas (5.32), (5.33), (5.35), and let*

$$\psi_+(x,\lambda) = \overline{s(\lambda)}\psi_-(x,\lambda). \tag{5.39}$$

Define on the set $L_1(\mathbb{R}_+) \cap L_2(\mathbb{R}_+)$ the mappings $\mathcal{F}_\pm$ by the formula

$$(\mathcal{F}_\pm f)(\lambda) = 2^{-1}(2\pi)^{-1/2}\lambda^{-3/8}\int_0^\infty \overline{\psi_\pm(x,\lambda)}f(x)dx. \tag{5.40}$$

These mappings extend to bounded operators on the space $L_2(\mathbb{R}_+)$ and satisfy the relations (2.12) and (2.13).

PROOF. The intertwining property (2.13) holds because the functions $\psi_\pm(x,\lambda)$ satisfy the equation $\psi_\pm^{(4)} = \lambda\psi_\pm$ and the boundary conditions (5.1). The second relation (2.12) is obtained by integration of representation (5.34) over $\lambda \in \mathbb{R}_+$. Now for the proof of the first relation (2.12), we have to check that the kernel of the operator $\mathcal{F}_\pm^*$ is trivial. Supposing that $\mathcal{F}_\pm^* g = 0$ and hence $\mathcal{F}_\pm^* E_A(\alpha,\beta)g = 0$ for all $\alpha, \beta \in \mathbb{R}_+$, we have

$$\int_\alpha^\beta \psi(x,\lambda)g(\lambda)d\lambda = 0, \quad \forall x \geq 0.$$

Let us differentiate this relation twice, use (5.35) and set $x = 0$. Since α and β are arbitrary, we obtain that

$$(s(\lambda) - i - 2^{1/2}e^{-\pi i/4}b(\lambda))g(\lambda) = 0,$$
$$(s(\lambda) + i - 2^{1/2}e^{\pi i/4}b(\lambda))g(\lambda) = 0,$$
$$(s(\lambda) - i + 2^{1/2}e^{-\pi i/4}b(\lambda))g(\lambda) = 0$$

for a.e. $\lambda > 0$. If $g(\lambda) \neq 0$ for some λ, then the first and third equations imply that $s(\lambda) = i$ and $b(\lambda) = 0$, which contradicts the second equation. Thus, $g(\lambda) = 0$ for a.e. $\lambda > 0$. $\square$

REMARK 5.13. If $\alpha = \bar{\alpha}$, then $\psi_+(x,\lambda) = \overline{\psi_-(x,\lambda)}$.

Consider now the operator $H_0 = D^4$ corresponding to the boundary conditions (4.22). Its eigenfunctions are defined by formula (4.23). It follows from formulas (5.19) and (5.20) that the spectral measure $E_0(\lambda)$ of the operator H_0 satisfies relation (5.34), where the role of ψ is played by ψ_0. Theorem 5.12 applies of course to the operator $\mathcal{F}_0$ defined by the formula

$$(\mathcal{F}_0 f)(\lambda) = 2^{-1}(2\pi)^{-1/2}\lambda^{-3/8}\int_0^\infty \psi_0(x,\lambda)f(x)dx. \tag{5.41}$$

Moreover, the operator $\mathcal{F}_0$ is unitary because the operator H_0 does not have eigenvalues.

Theorem 5.12 allows us to construct directly the time-dependent scattering theory in the same way as for the second order differential operators. We also need the following auxiliary assertion.

LEMMA 5.14. *If $u \in C_0^\infty(\mathbb{R}_+)$, then*

$$\lim_{t\to\pm\infty}\int_0^\infty dx\left|\int_0^\infty \exp(\mp ikx - ik^4 t)u(k)dk\right|^2 = 0 \tag{5.42}$$

and

$$\lim_{|t|\to\infty}\int_0^\infty dx\left|\int_0^\infty \exp(-kx - ik^4 t)u(k)dk\right|^2 = 0. \tag{5.43}$$

PROOF. Both relations (5.42) and (5.43) are obtained by a direct integration by parts, which shows that the integral over k is bounded by $C(x+|t|)^{-1}$. □

THEOREM 5.15. *The wave operators $W_\pm = W_\pm(H, H_0)$ exist, are complete and satisfy the equality $W_\pm = \mathcal{F}_\pm^*\mathcal{F}_0$. The scattering operator $\mathcal{S}$ for the pair H_0, H acts in the space $L_2(\mathbb{R}_+)$ as multiplication by the function $s(\lambda)$ defined by the formulas (5.14), (5.32).*

PROOF. According to Theorem 5.12 all results about the wave operators $W_\pm$ follow from the relation

$$\lim_{t\to\pm\infty}\|(\mathcal{F}_\pm^* - \mathcal{F}_0^*)\exp(-iAt)g\| = 0, \tag{5.44}$$

where $g(k)$ is an arbitrary function from $L_2(\mathbb{R}_+)$. It suffices to check (5.44) on the set $C_0^\infty(\mathbb{R}_+)$. Using (5.40), (5.41), we obtain that

$$\begin{aligned}&((\mathcal{F}_\pm^* - \mathcal{F}_0^*)\exp(-iAt)g)(x)\\ &= 2^{-1}(2\pi)^{-1/2}\int_0^\infty (\psi_\pm(x,\lambda) - \psi_0(x,\lambda))\exp(-i\lambda t)g(\lambda)\lambda^{-3/8}d\lambda.\end{aligned} \tag{5.45}$$

Let us check (5.44), for example, for the sign "$-$". It follows from (4.23) and (5.35) that

$$\psi_-(x,\lambda) - \psi_0(x,\lambda) = 2^{-1}(s(\lambda)-1)e^{ikx+\pi i/4} - 2^{-1/2}(b(\lambda)-1)\exp(-kx).$$

The contributions to (5.45) of the first and second terms on the right-hand side tend in $L_2(\mathbb{R}_+)$ to zero as $t \to -\infty$ according to relations (5.42) and (5.43), respectively.

The equality $W_\pm = \mathcal{F}_\pm^*\mathcal{F}_0$ implies that the scattering operator $\mathcal{S}$ satisfies the relation (2.22). It remains to remark that $(\mathcal{F}_+ f)(\lambda) = s(\lambda)(\mathcal{F}_- f)(\lambda)$ according to (5.39). □

REMARK 5.16. Since the operator $\mathcal{F}_0$ as well as the Fourier transform are bounded operators, it follows from formula (4.23) that the integral operator T with kernel $\exp(-kx)$ is bounded in $L_2(\mathbb{R}_+)$. This result is a by-product of our considerations, but it is not of course new. Indeed, we have that

$$\|Tu\|^2 = (\mathcal{C}u, u),$$

where $\mathcal{C}$ is the integral operator with kernel $(k + k')^{-1}$ known as Carleman's operator. A proof of its boundedness can be found, e.g., in [**12**]. We note also that boundedness of the Fourier transform and of the operator T imply that relations (5.42) and (5.43) remain true for all $u \in L_2(\mathbb{R}_+)$.

6. The perturbation determinant and the spectral shift function

6.1. The mathematical theory of the spectral shift function was constructed by M. Kreĭn. Let us recall here briefly its basic notions (see [**7**] and [**16**], for details). Suppose that the selfadjoint operators H_0 and H are semi-bounded from below and that the difference of their resolvents belongs to the trace class.

The (generalized) perturbation determinant $D(z)$ for the pair H_0, H is defined by the equation

$$\operatorname{Tr}(R(z) - R_0(z)) = -D'(z)D(z)^{-1}, \quad z \notin \sigma(H_0) \cup \sigma(H), \tag{6.1}$$

which fixes $D(z)$ up to a constant factor. We set

$$\ln D(z) = \int_{z_0}^{z} \operatorname{Tr}(R_0(z') - R(z'))dz', \tag{6.2}$$

where z_0 is some real point lying below

$$\nu = \min\{\inf \sigma(H_0), \inf \sigma(H)\}$$

and the integral is taken over some contour in the upper (lower) half-plane if $\operatorname{Im} z > 0$ ($\operatorname{Im} z < 0$). Then the function $\ln D(z)$ is determined up to a real constant (and hence the function $D(z)$ is determined up to a constant positive factor) which is inessential. Clearly, $\arg D(z) = 0$ for $z = \bar{z} < \nu$ and $D(\bar{z}) = \overline{D(z)}$.

The limit of $\arg D(\lambda + i\varepsilon)$ exists for a.e. $\lambda \in \mathbb{R}$, and the spectral shift function $\xi(\lambda)$ for the pair H_0, H is defined by the formula

$$\xi(\lambda) = \pi^{-1} \arg D(\lambda + i0). \tag{6.3}$$

The function $\xi(\lambda)$ assumes constant integral values on component intervals of the set of common regular points of the operators H_0 and H, $\xi(\lambda) = 0$ if $\lambda < \nu$ and

$$\xi(\lambda + 0) - \xi(\lambda - 0) = n_0 - n \tag{6.4}$$

if λ is an isolated eigenvalue of multiplicity n_0 of the operator H_0 and of multiplicity n of the operator H. The spectral shift function satisfies the condition

$$\int_{-\infty}^{\infty} |\xi(\lambda)|(1 + \lambda^2)^{-1} d\lambda < \infty.$$

The trace formula

$$\operatorname{Tr}(\varphi(H) - \varphi(H_0)) = \int_{-\infty}^{\infty} \xi(\lambda)\varphi'(\lambda)d\lambda$$

holds for functions $\varphi \in C^2(\mathbb{R})$ such that

$$(\lambda^2\varphi'(\lambda))' = O(\lambda^{-1-\varepsilon}), \quad \lambda \to +\infty,$$

for some $\varepsilon > 0$. In particular, the trace formula is true for the function $\varphi(\lambda) = (\lambda - z)^{-1}$ when $\varphi(H) = R(z)$.

For the pair H_0, H, the wave operators (2.14) exist, are complete and the relation between the scattering matrix $S(\lambda)$ and the spectral shift function is given by the Birman-Kreĭn formula [**5**]

$$\operatorname{Det} S(\lambda) = e^{-2\pi i \xi(\lambda)}. \tag{6.5}$$

6.2. Now we are in a position to construct explicitly the perturbation determinant and the spectral shift function for the pair H_0, H considered in the previous section. According to equalities (5.8) and (5.19) we have

$$\operatorname{Tr}(R(z) - R_0(z)) = \operatorname{Tr} P(z) - \operatorname{Tr} P_0(z). \tag{6.6}$$

The kernel $P(x, y; z)$ of the operator $P(z)$ is determined by formulas (5.9) and (5.10) so that

$$P(x, x; z) = p_{11}(\zeta)e^{2i\zeta x} + (p_{12}(\zeta) + p_{21}(\zeta))e^{(-1+i)\zeta x} + p_{22}(\zeta)e^{-2\zeta x}.$$

It follows that

$$\operatorname{Tr} P(z) = \int_0^\infty P(x, x; z)dx = (2\zeta)^{-1}(ip_{11}(\zeta) + (1+i)(p_{12}(\zeta) + p_{21}(\zeta)) + p_{22}(\zeta)).$$

Using formulas (5.15) and (5.16) for the coefficients p_{jl}, we find that

$$\operatorname{Tr} P(z) = (4\zeta^4\Omega(\zeta))^{-1}\big(\alpha_0 + 2i \operatorname{Re} \alpha\zeta^2 + 2(1+i)\alpha_1\zeta^3 + 3\zeta^4\big).$$

Similarly, it follows from (5.20) that

$$\operatorname{Tr} P_0(z) = (4z)^{-1}. \tag{6.7}$$

Therefore equality (6.6) yields the following result.

PROPOSITION 6.1. *Suppose that the boundary conditions* (5.1) *are satisfied. Then*

$$\operatorname{Tr}(R(z) - R_0(z)) = -4^{-1}\zeta^{-3}\Omega(\zeta)^{-1}\Omega'(\zeta), \tag{6.8}$$

where the function Ω *is defined by formula* (5.14).

Let us now consider the boundary conditions (5.2) and (5.3). Remark that formally conditions (5.2) can be obtained from (5.1) if we set $\alpha_1 = N$, replace α by $-\alpha N$, replace α_2 by $|\alpha|^2 N + \alpha_2$ and pass to the limit $N \to \infty$. We plug these expressions into (5.14) and observe that the limit of the right-hand side of (6.8) is determined only by the terms containing the factor N. Thus, equation (6.8) is true if we set

$$\Omega(\zeta) = \alpha_2 + (1-i)|\alpha|^2\zeta - 2i \operatorname{Re} \alpha\, \zeta^2 - (1+i)\zeta^3. \tag{6.9}$$

Similarly, in case (5.3), we set $\alpha = 0$, $\alpha_2 = N$ and take the limit $N \to \infty$. This yields

$$\Omega(\zeta) = \alpha_1 + (1-i)\zeta. \tag{6.10}$$

All results obtained in Section 5 for the boundary conditions (5.1) can easily be carried over to the cases (5.2) and (5.3). In particular, formula (5.32) remains true.

Comparing equations (6.1) and (6.8), we obtain

PROPOSITION 6.2. *The perturbation determinant for the pair H_0, H is given by the equality*

$$D(z) = \Omega(\zeta), \quad z = \zeta^4. \tag{6.11}$$

Since all the functions (5.14), (6.9) and (6.10) satisfy the condition $\Omega(e^{\pi i/4}k) > 0$ for large $k > 0$, we can set $\arg \Omega(e^{\pi i/4}k) = 0$ for such a k. Thus, $\Omega(\zeta)$ determines by formulas (6.3), (6.11) the spectral shift function $\xi(\lambda)$ for the pair H_0, H. Clearly, $\xi(\lambda) = 0$ below the lowest eigenvalue of H. Proposition 5.3 implies that the spectral shift function has a jump -1 at a simple eigenvalue of H and a jump -2 at an eigenvalue of multiplicity 2. This is of course consistent with the general formula (6.4). Similarly, relation (5.32) reduces to formula (6.5).

Let us write formulas (5.14), (6.9) and (6.10) in a unified way as

$$D(z) = \sum_{j=0}^{4} \omega_j \zeta^j, \quad z = \zeta^4. \tag{6.12}$$

We put

$$\gamma_0 = \min_j \{j : \omega_j \neq 0\}, \quad \gamma_1 = \max_j \{j : \omega_j \neq 0\}. \tag{6.13}$$

Results pertaining to the behavior of $\xi(\lambda)$ for $\lambda > 0$ are collected in the following assertion.

PROPOSITION 6.3. *For $\lambda > 0$, the spectral shift function for the pair H_0, H is infinitely differentiable away from a positive eigenvalue λ_0 of the operator H (if it exists). The limits of $\xi(\lambda)$ as $\lambda \to \lambda_0 \pm 0$ exist and*

$$\xi(\lambda_0 + 0) - \xi(\lambda_0 - 0) = -1. \tag{6.14}$$

If an interval $(\lambda_1, \lambda_2) \subset \mathbb{R}_+$ does not contain an eigenvalue of H, then

$$\xi(\lambda_2) - \xi(\lambda_1) = \pi^{-1} \int_{k_1}^{k_2} \operatorname{Im}(\Omega(k)^{-1}\Omega'(k))dk, \quad \lambda_j = k_j^4. \tag{6.15}$$

Moreover, there exists a limit $\xi(+0)$,

$$\delta := \xi(+0) - \xi(-0) = -\gamma_0/4 \tag{6.16}$$

and

$$\lim_{\lambda \to +\infty} \xi(\lambda) = -\gamma_1/4. \tag{6.17}$$

PROOF. According to Proposition 5.6 the perturbation determinant $D(\lambda + i0) = 0$ if and only if $\lambda = \lambda_0$ is an eigenvalue of the operator H. Since $D(\lambda + i0)$ is a C^∞-function for $\lambda > 0$, its argument is also a C^∞-function away from the point λ_0. According to Proposition 5.7, λ_0 is a simple zero of the function $D(z)$ so that

$$D(z) = d(z - \lambda_0) + O(|z - \lambda_0|^2), \quad z \to \lambda_0, \quad \operatorname{Im} z \geq 0,$$

for some $d \neq 0$. Therefore the function $\arg D(\lambda + i0)$ has finite limits as $\lambda \to \lambda_0 \pm 0$ and the variation of $\arg D(z)$ as z passes from $\lambda_0 - \varepsilon$ to $\lambda_0 + \varepsilon$ in the clockwise direction over a semi-circle $C_\varepsilon^+(\lambda_0) = \{|z - \lambda_0| = \varepsilon, \operatorname{Im} z \geq 0\}$ equals

$$\operatorname{var}_{C_\varepsilon^+(\lambda_0)} \arg D(z) = -\pi + o(1) \tag{6.18}$$

as $\varepsilon \to 0$. This proves formula (6.14). Relation (6.15) is a direct consequence of formulas (6.3) and (6.11).

The existence of $\xi(+0)$ and relation (6.16) follow from formula (6.12) and the equality

$$\mathrm{var}_{C_\varepsilon^+(0)} \arg D(z) = -\pi\gamma_0/4 + o(1) \tag{6.19}$$

as $\varepsilon \to 0$. Similarly, relation (6.17) follows from formula (6.12) and the equality

$$\mathrm{var}_{C_R^+(0)} \arg D(z) = -\pi\gamma_1/4 + o(1) \tag{6.20}$$

as $R \to \infty$. □

If the boundary conditions (5.1) are satisfied, then $\gamma_1 = 4$, but γ_0 might equal (see (5.14)) any number between 0 and 4.

7. Zero-energy resonances

Here we analyse resolvent singularities and the behavior of the spectral shift function at the bottom of the continuous spectrum. The notion of a zero-energy resonance is introduced and discussed.

7.1. The resolvents of all operators considered in Section 5 admit asymptotic expansions (for fixed x and y) in powers of ζ as $|\zeta| \to 0$. Let us find their singular parts. Below the symbol " $\simeq$ " means an equality valid up to regular terms.

According to formula (5.7),

$$R_{00}(x, y; z) = 2^{-1/2} xy(-z)^{-1/4} + 12^{-1}(|x-y|^3 - (x+y)^3) + O(|z|^{1/4}), \tag{7.1}$$

according to formula (5.20),

$$P_0(x, y; z) = -2^{-1/2} xy(-z)^{-1/4} + 2^{-1} xy(x+y) + O(|z|^{1/4}) \tag{7.2}$$

and according to formula (5.21),

$$P_1(x, y; z) \simeq 2^{-1/2}(-z)^{-3/4} - 2^{-3/2}(x+y)^2(-z)^{-1/4}. \tag{7.3}$$

It is slightly more difficult to find the singular part of the kernel $P(x, y; z)$. It follows from equations (5.9), (5.10) and (5.15) that

$$P(x, y; z) = 2^{-3/2} e^{-\pi i/4} \zeta^{-3} \Omega(\zeta)^{-1} \sum_{n=0}^{\infty} (n!)^{-1} L_n(x, y; \zeta)\zeta^n, \tag{7.4}$$

where

$$\begin{aligned} L_n(x, y; \zeta) = \mathbf{p}_{11}(\zeta) i^n (x+y)^n + \mathbf{p}_{12}(\zeta)(-x+iy)^n \\ + \mathbf{p}_{21}(\zeta)(ix-y)^n + \mathbf{p}_{22}(\zeta)(-1)^n(x+y)^n \end{aligned}$$

and the coefficients $\mathbf{p}_{jl}(\zeta)$, $j, l = 1, 2$, are defined by formulas (5.16). In particular, we have

$$L_0(x, y; \zeta) = 2(1+i)\alpha_1\zeta^3 + 2\zeta^4, \tag{7.5}$$

$$L_1(x, y; \zeta) = -2(1+i)(\alpha x + \bar{\alpha} y)\zeta^2 + 2(-1+i)(x+y)\zeta^4 \tag{7.6}$$

and

$$L_2(x, y; \zeta) = -4i\alpha_0 xy + 2(\alpha x^2 + 4\,\mathrm{Re}\,\alpha xy + \bar{\alpha} y^2)\zeta^2 + 2(i-1)\alpha_1(x^2+y^2)\zeta^3 - 4ixy\zeta^4. \tag{7.7}$$

First let $\alpha_0 \neq 0$. Then $\Omega(0) = \alpha_0 \neq 0$ and the only singular term in (7.4) comes from the first term on the right-hand side of (7.7), which yields

$$P(x, y; z) \simeq -2^{-1/2} xy(-z)^{-1/4}, \quad \alpha_0 \neq 0. \tag{7.8}$$

If $\alpha_0 = 0$ but $\alpha_2 \neq 0$, then $\mathbf{p}_{jl}(\zeta) = O(|\zeta|^2)$, $\Omega(\zeta) = (1-i)\alpha_2\zeta(1+O(|\zeta|)$ as $|\zeta| \to 0$ and the singular terms in (7.4) come from the first terms on the right-hand sides of (7.5) and (7.6). Thus, we have

$$(7.9) \qquad P(x, y; z) \simeq 2^{-3/2}\alpha_2^{-1}(2\alpha_1 - \alpha x - \bar{\alpha} y)(-z)^{-1/4}, \quad \alpha_0 = 0, \quad \alpha_2 \neq 0.$$

If $\alpha_0 = \alpha_2 = 0$ (and hence $\alpha = 0$) but $\alpha_1 \neq 0$, then $\mathbf{p}_{jl}(\zeta) = O(|\zeta|^2)$, $\Omega(\zeta) = -(1+i)\alpha_1\zeta^3(1+O(|\zeta|)$ as $|\zeta| \to 0$ and the singular terms in (7.4) come from (7.5) and the second term on the right-hand side of (7.6). Thus, in this case we have

$$(7.10) \; P(x, y; z) \simeq 2^{-1/2}(-z)^{-3/4} - (\alpha_1^{-1}2^{-1/2}(x+y) + 2^{-3/2}(x^2+y^2))(-z)^{-1/4}.$$

Finally let $\alpha = \alpha_1 = \alpha_2 = 0$. Then it follows from formula (5.17) that

$$(7.11) \qquad P(x, y; z) \simeq 2^{-1/2}(-z)^{-3/4} - 2^{-1}(x+y)(-z)^{-1/2} + 2^{-1/2}xy(-z)^{-1/4}.$$

7.2. Here we discuss different types of zero-energy resonances. Let us distinguish several cases.

1°. The operator H_0 does not have zero-energy resonances. Since the singular terms in (7.1) and (7.2) are compensated in (5.19), the resolvent kernel $R_0(x, y; z)$ is a continuous function as $z \to 0$ and

$$R_0(x, y; 0) = 12^{-1}(|x-y|^3 - (x+y)^3) + 2^{-1}xy(x+y).$$

Linear functions (except zero) do not satisfy the boundary conditions (4.22). All boundary conditions (5.1), (5.2) and (5.3) change the domain of the quadratic form of the operator H_0, and hence the corresponding operators H cannot be considered as small perturbations of H_0.

2°. The operator H_{00} has a zero-energy resonance. Formula (7.1) shows that its resolvent kernel has the singularity $2^{-1/2}xy(-z)^{-1/4}$ as $z \to 0$ while it is of order z^{-1} at an eigenvalue of H. The function $u(x) = x$ satisfies the boundary conditions (5.4). According to equality (6.10) where $\alpha_1 = 0$ the perturbation determinant for the pair H_0, H_{00} equals $D_{00}(z) = (-z)^{1/4}$, and the spectral shift function equals 0 for $\lambda < 0$ and $-1/4$ for $\lambda > 0$. Note that by virtue of (6.4) its jump at an eigenvalue of H equals -1. Thus, it is natural to say that the operator H_{00} has a quarter-bound state at zero energy. The operator H_{00} belongs to the family (5.3) for $\alpha_1 = 0$. One negative eigenvalue appears for an arbitrary $\alpha_1 < 0$.

Next we consider the operator H with the boundary conditions (5.1). According to formula (5.8) the singular part of its resolvent equals the sum of the singular parts of (7.1) and of the kernel $P(x, y; z)$. The jump δ of the spectral shift function $\xi(\lambda)$ at the point $\lambda = 0$ is determined by formula (6.16).

3°. If $\alpha_0 \neq 0$, then the operator H does not have zero-energy resonances. Since the singularities in (7.1) and (7.8) are compensated, the kernel $R(x, y; z)$ is a continuous function as $z \to 0$. Linear functions $u(x) = Ax + B$ satisfy the boundary conditions (5.1) only for $A = B = 0$. Since $\Omega(0) = \alpha_0 \neq 0$, the spectral shift function is continuous at the point $\lambda = 0$. According to Proposition 5.4 new negative eigenvalues of the operator H cannot appear under small perturbations of the coefficients α, α_1 and α_2.

4°. Let $\alpha_0 = 0$ but $\alpha_2 \neq 0$. Then the formulas (7.1), (7.9) show that the resolvent kernel has a singularity at $z = 0$ of the same order $-1/4$ as $R_{00}(z)$. The linear function $u(x) = \alpha_2 x - \bar{\alpha}$ satisfies the boundary conditions (5.1). Now $\omega_0 = 0$ but $\omega_1 = (1-i)\alpha_2 \neq 0$ in (6.12) and hence $\gamma_0 = 1$. It follows from (6.16) that $\delta = -1/4$. Thus, the operator H has a zero-energy resonance of the same "strength" (that is, 1/4-bound state) as the operator H_{00}.

5°. Let $\alpha = \alpha_2 = 0$ but $\alpha_1 \neq 0$. Then formula (7.10) shows that its resolvent kernel has the singularity $2^{-1/2}(-z)^{-3/4}$ as $|z| \to 0$. The equation $u^{(4)}(x) = 0$ has a solution $u(x) = 1$ satisfying the boundary condition (5.1). Now $\omega_0 = \omega_1 = \omega_2 = 0$ but $\omega_3 = -(1+i)\alpha_1 \neq 0$ and hence $\gamma_0 = 3$. It follows from (6.16) that $\delta = -3/4$. To comply with these results, we say that the operator H has a 3/4-bound state at energy zero.

In the cases 4^0 and 5^0, we have that $\operatorname{Det} A = 0$ but $A \neq 0$ so that the matrix A has exactly one zero eigenvalue. Therefore the operator H has an additional negative eigenvalue for an arbitrary negative perturbation of A.

6°. Let $\alpha = \alpha_1 = \alpha_2 = 0$. Then both functions $u(x) = 1$ and $u(x) = x$ satisfy the boundary conditions (5.1). It follows from formula (5.14) that $D(z) = -z$ and $\xi(\lambda) = 0$ for $\lambda < 0$, $\xi(\lambda) = -1$ for $\lambda > 0$. The operator H has both 3/4- and 1/4-bound states at energy zero. This is consistent with formula $\delta = -1$ as well as with the fact that the operator H has two negative eigenvalues for an arbitrary matrix $A < 0$. According to (7.11) the resolvent kernel contains more singularities than in the previous cases.

The families (5.2) and (5.3) can be considered in a similar but simpler way. We discuss only the operator H_1.

7°. Let the boundary conditions (5.18) be satisfied. Formula (7.3) shows that the singularity $2^{-1/2}(-z)^{-3/4}$ of the resolvent kernel is the same as in case 5^0. The function $u(x) = 1$ satisfies conditions (5.18). According to equality (6.9) where $\alpha = \alpha_2 = 0$ the perturbation determinant for the pair H_0, H_1 equals $D_1(z) = (-z)^{3/4}$, and the spectral shift function equals 0 for $\lambda < 0$ and $-3/4$ for $\lambda > 0$. Thus, the operator H has a 3/4-bound state at energy zero. Its quadratic form is given by expression (5.6) where $\alpha_2 = 0$ on functions $u \in \mathsf{H}^2(\mathbb{R}_+)$ such that $u'(0) = 0$. It follows from this expression that the operator H corresponding to the boundary conditions (5.2) where $\alpha = 0$ has a negative eigenvalue for all $\alpha_2 < 0$.

7.3. We finish with an analogue of the Levinson theorem. Consider the closed contour which consists of a small circle $C_\varepsilon = \{|z| = \varepsilon\}$, a big circle $C_R = \{|z| = R\}$ and two intervals (ε, R) lying on the upper and lower edges of the cut along $[0, \infty)$. Moreover, if the operator H has a positive eigenvalue λ_0 we go over it by small semi-circles $C_\varepsilon^\pm(\lambda_0)$, where $|z - \lambda_0| = \varepsilon$ and $\pm \operatorname{Im} z \geq 0$ (lying in the upper and lower half-planes). Let us pass this contour in the positive direction and apply the argument principle to the function $D(z)$. Taking into account the direction of motion and the identity $\overline{D(z)} = D(\bar{z})$, we obtain that

$$\operatorname{var}_{C_\varepsilon^+} \arg D(z) - \operatorname{var}_{C_R^+} \arg D(z) + \arg D(R + i0) - \arg D(\varepsilon + i0) = \pi N_-, \tag{7.12}$$

where N_- is the number of negative eigenvalues of the operator H. Moreover, if H has a positive eigenvalue, then the term

$$\arg D(\lambda_0 - \varepsilon + i0) - \arg D(\lambda_0 + \varepsilon + i0) + \operatorname{var}_{C_\varepsilon^+(\lambda_0)} \arg D(z) \tag{7.13}$$

should be added to the left-hand side.

Let us choose $\arg s(\lambda)$ as a continuous function of $\lambda > 0$. According to formulas (5.32), (6.3) we have that

$$\arg D(R+i0) - \arg D(\lambda_0+\varepsilon+i0) = -2^{-1}(\arg s(R) - \arg s(\lambda_0+\varepsilon)),$$
$$\arg D(\lambda_0-\varepsilon+i0) - \arg D(\varepsilon+i0) = -2^{-1}(\arg s(\lambda_0-\varepsilon) - \arg s(\varepsilon)).$$

Taking the limits $\varepsilon \to 0$, $R \to \infty$ in (7.13) and using relations (6.18)–(6.20), we finally obtain that

$$\arg s(\infty) - \arg s(+0) = -2\pi N + \pi(\gamma_1 - \gamma_0)/2, \tag{7.14}$$

where N is the total number of eigenvalues (including eventually a positive eigenvalue) of the operator H and the numbers γ_0, γ_1 are defined by formulas (6.12), (6.13). We emphasize that an isolated eigenvalue and an eigenvalue embedded in the continuous spectrum give the same contributions to the Levinson formula (7.14).

References

[1] S. Agmon, *Spectral properties of Schrödinger operators and scattering theory*, Ann. Scuola Norm. Sup. Pisa **2** no. 4 (1975), 151–218.

[2] R. Beals, P. Deift, and C. Tomei, *Direct and inverse scattering on the line*, Mathematical Surveys and Monographs, vol. 28, Amer. Math. Soc., Providence, RI, 1988.

[3] M. Sh. Birman, *On the spectrum of singular boundary-value problems*, Matem. Sb. **55**, no. 2 (1961), 125–174; English transl., Eleven Papers on Analysis, Amer. Math. Soc. Transl. (2), vol. 53, Amer. Math. Soc., Providence, RI, 1966, 23–60.

[4] ———, *Scattering problems for differential operators with constant coefficients*, Funct. Anal. Appl. **3** no. 3 (1969), 167–180.

[5] M. Sh. Birman and M. G. Kreĭn, *On the theory of wave operators and scattering operators*, Soviet Math. Dokl. **3** (1962), 740–744.

[6] M. Sh. Birman and D. R. Yafaev, *A general scheme in the stationary theory of scattering*, Problemy Math. Phys. **12** (1987), 89-117. English transl., Amer. Math. Soc. Transl. (Ser.2) vol. 157, Amer. Math. Soc., Providence, RI, 1993, 87–112.

[7] ———, *The spectral shift function. The papers of M. G. Kreĭn and their further development*, St. Petesburg Math. J. **4** no. 5 (1993), 833–870.

[8] E. A. Coddington and N. Levinson, *Theory of ordinary differential equations*, McGraw-Hill, New York, 1955.

[9] K. Hoffman, *Banach spaces of analytic functions*, Prentice-Hall, Inc., Englewood Cliffs, NJ, 1962.

[10] S. T Kuroda, *Scattering theory for differential operators*, J. Math. Soc. Japan **25** no. 1,2 (1973), 75–104, 222-234.

[11] ———, *An introduction to scattering theory*, Lect. Notes Series No. 51, Aarhus University, 1978.

[12] S. G. Power, *Hankel operators on Hilbert space*, Research Notes in Math., Vol. 64, Pitman, Boston, 1982.

[13] M. Reed and B. Simon, *Methods of Modern Mathematical Physics* IV, Academic Press, 1978.

[14] L. D. Faddeev, *Properties of the S-matrix of the one-dimensional Schrödinger equation*, Amer. Math. Soc. Transl. (Ser.2), vol 65, Amer. Math. Soc., Providence, RI, 1967, 139–166.

[15] D. R. Yafaev, *The virtual level of the Schrödinger equation*, Zap. Naučhn. Sem. LOMI **51** (1975), 203-216. English transl., J. Sov. Math. **11** (1979), 501-510.

[16] ———, *Mathematical scattering theory*, General theory, Translations of Mathematical Monographs, vol. 105, Amer. Math. Soc., Providence, Rhode Island, 1992.

[17] ———, *On a zero-range interaction of a quantum particle with the vacuum*, J. Phys. A, **25** (1992), 963-978.

[18] ———, *Scattering theory: Some old and new problems*, Lecture Notes Math., vol. 1735, Springer-Verlag, Heidelberg, 2000.

IRMAR, Université de Rennes I, Campus de Beaulieu, 35042 Rennes Cedex, France
E-mail address: yafaev@univ-rennes1.fr

Titles in This Series

225 **T. Suslina and D. Yafaev, Editors,** Spectral theory of differential operators (M. Sh. Birman 80th anniversary collection)

224 **V. M. Buchstaber and I. M. Krichever, Editors,** Geometry, topology, and mathematical physics (S. P. Novikov's seminar: 2006–2007)

223 Selected Papers on Analysis and Related Topics

222 **N. N. Uraltseva, Editor,** Proceedings of the St. Petersburg Mathematical Society, XIII

221 **Yu. Neretin, M. Olshanetsky, and A. Rosly, Editors,** Seminar on Mathematical Physics, II

220 **M. S. Birman and N. N. Uraltseva, Editors,** Nonlinear Equations and Spectral Theory

219 **N. N. Uraltseva, Editor,** Proceedings of the St. Petersburg Mathematical Society, Volume XII

218 **N. N. Uraltseva, Editor,** Proceedings of the St. Petersburg Mathematical Society, Volume XI

217 **V. Kaimanovich and A. Lodkin, Editors,** Representation theory, dynamical systems, and asymptotic combinatorics

216 **M. V. Karasev, Editor,** Quantum Algebras and Poisson Geometry in Mathematical Physics

214 **N. N. Uraltseva, Editor,** Proceedings of the St. Petersburg Mathematical Society, Volume X

213 **Ernest Vinberg, Editor,** Lie Groups and Invariant Theory

212 **V. M. Buchstaber and I. M. Krichever, Editors,** Geometry, Topology, and Mathematical Physics. S. P. Novikov's Seminar: 2002–2003

211 **K. Nomizu, Editor,** Selected Papers on Analysis and Differential Equations

210 **S. G. Gindikin, Editor,** Lie Groups and Symmetric Spaces. In Memory of F. I. Karpelevich

209 **N. N. Uraltseva, Editor,** Proceedings of the St. Petersburg Mathematical Society, Volume IX

208 **M. V. Karasev, Editor,** Asymptotic Methods for Wave and Quantum Problems

207 **Yu. M. Suhov, Editor,** Analytic Methods in Applied Probability

206 **M. A. Shubin and M. S. Agranovich, Editors,** Partial Differential Equations (Mark Viskik's Seminar)

205 **N. N. Uraltseva, Editor,** Proceedings of the St. Petersburg Mathematical Society, Volume VIII

204 **K. Nomizu, Editor,** Selected Papers on Classical Analysis

203 **N. N. Uraltseva, Editor,** Proceedings of the St. Petersburg Mathematical Society, Volume VII

202 **V. Turaev and A. Vershik, Editors,** Topology, Ergodic Theory, Real Algebraic Geometry. Rokhlin's Memorial

201 **Michael Semenov-Tian-Shansky, Editor,** L. D. Faddeev's Seminar on Mathematical Physics

200 **L. Lerman, G. Polotovskiĭ, and L. Shilnikov, Editors,** Methods of Qualitative Theory of Differential Equations and Related Topics

199 **N. N. Uraltseva, Editor,** Proceedings of the St. Petersburg Mathematical Society, Volume VI

198 **R. A. Minlos, Senya Shlosman, and Yu. M. Suhov, Editors,** On Dobrushin's Way. From Probability Theory to Statistical Physics

197 **Vladimir Arnold, Maxim Kontsevich, and Anton Zorich, Editors,** Pseudoperiodic Topology

For a complete list of titles in this series, visit the AMS Bookstore at **www.ams.org/bookstore/**.